DATE			

ANALYTICAL, NUMERICAL, AND COMPUTATIONAL METHODS FOR SCIENCE AND ENGINEERING

ANALYTICAL, NUMERICAL, AND COMPUTATIONAL METHODS FOR SCIENCE AND ENGINEERING

Gene H. Hostetter

Mohammed S. Santina
Rockwell International Corporation
University of California, Irvine

Paul D'Carpio-Montalvo
California State University, Long Beach

PRENTICE HALL, Englewood Cliffs, New Jersey 07632

Library of Congress Cataloging-in-Publication Data

Hostetter, G. H.
   Analytical, numerical, and computational methods for science and
engineering / Gene H. Hostetter, Mohammed S. Santina, Paul D'Carpio
-Montalvo.
      p.   cm.
   Includes index.
   ISBN 0-13-026055-X
   1. Science—Data processing.   2. Engineering—Data processing.
3. Numerical analysis.   I. Santina, Mohammed S.   II. D'Carpio
-Montalvo, Paul.   III. Title.
Q183.9.H67   1991
502.85—dc20   90-40453
                                                                  CIP

Editorial/production supervision and
      interior design: Michael R. Steinberg
Cover design: Butler/Udell Design
Manufacturing buyers: Linda Behrens/Patrice Fraccio

© 1991 by Prentice-Hall, Inc.
A Division of Simon & Schuster
Englewood Cliffs, New Jersey 07632

Printed in the United States of America

10 9 8 7 6 5 4 3 2 1

ISBN 0-13-026055-X

PRENTICE-HALL INTERNATIONAL (UK) LIMITED, London
PRENTICE-HALL OF AUSTRALIA PTY. LIMITED, Sydney
PRENTICE-HALL CANADA INC., Toronto
PRENTICE-HALL HISPANOAMERICANA, S.A., Mexico
PRENTICE-HALL OF INDIA PRIVATE LIMITED, New Delhi
PRENTICE-HALL OF JAPAN, INC., Tokyo
SIMON & SCHUSTER ASIA PTE. LTD., Singapore
EDITORA PRENTICE-HALL DO BRASIL, LTDA., Rio de Janeiro

To our families:

Donna, Colleen, Kristen,
my father John, and my brother Lane

Dalia, Elham, Fatima,
Abed, and Chafic

My mother Georgina,
and my grandfather Angel

CONTENTS

FOREWORD

While writing this textbook, Dr. Gene Huber Hostetter, Professor of Electrical Engineering, University of California, Irvine, passed away on July 30, 1988. Gene was born on September 14, 1939 in Spokane, WA. He received the B.S. and M.S. degrees in electrical engineering in 1962 and 1963 from University of Washington, Seattle, and the Ph.D. degree in engineering from the University of California, Irvine, in 1973.

Before coming to UCI, Gene worked successively for the Boeing Airplane Company and as Director of Engineering for the Seattle Broadcasting Company. Then, he was a faculty member for 14 years at the California State University, Long Beach, serving two terms as Chairman of the Department of Electrical Engineering. Gene spent his last seven years as a faculty member at UC Irvine, during which time he also served as the first Chairman of Electrical Engineering, and subsequently served as Acting Dean of Engineering. As a teacher, he was consistently judged to be excellent. To his students he was not only a technical advisor but also a confidant and close personal friend who was exemplary in his generous contribution of time and effort.

In 1986 Gene was elected as Fellow of the IEEE for his "Contributions to the theory and algorithms for observers of complex control systems, and to engineering education." He was very active in the IEEE Control Systems Society, and was a member of the Board of Governors in 1986, Technical Associate Editor of the Control Systems Magazine until 1987, Chairman of the Microprocessor-Based Control Technical Committee in 1986, and Publications Chairman of the Conference on Decision and Control in 1983. He served as Chairman of the IEEE Orange County Section, and as Member at Large for the Los Angeles Council.

Gene wrote four textbooks: *Fundamentals of Network Analysis*; *Design of Feedback Control Systems* (with C. J. Savant and R. T. Stefani); *Engineering Network Analysis*; and *Digital Control System Design*. He also wrote over a hundred scholarly technical papers, and numerous reviews and review articles.

Jose B. Cruz, Jr.
James H. Mulligan, Jr.
Allen R. Stubberud

PREFACE

Engineering and scientific design involves a combination of analytical and numerical methods. It is today appropriate to join these topics in an integrated course of study. This book builds on a standard mathematics background, joining to analysis the numerical methods necessary for performing the fundamental operations needed in modern practice. As most people best learn how to do things by *doing* them, computation is important to attaining real capability with the basic tools of design.

We believe strongly in the need for today's textbooks to have a scope that substantially exceeds the minimum. It is of the utmost importance that the horizons and perspective of a text be beyond those of the student, although reachable in time and with effort. We hope the reader will find this book to be stimulating and challenging, yet clear with solid emphasis on the essentials.

INTENDED AUDIENCE

This book is intended for use by these groups:

1. *Undergraduate engineering and science students* who are completing the usual engineering–scientific background in algebra and calculus with analytic geometry. Some modest experience in computer programming is desirable. Two undergraduate courses at the sophomore and junior levels are suggested, after which the student can be expected to apply these tools in courses to follow without the need for additional time spent on analytical, numerial, and computational fundamentals. At the same time, students obtain enough computer algorithm and programming experience to be well prepared for a follow-on programming course that can begin at a well-defined level.

2. *Graduate students* at an early stage, to establish a critical base level of requisite understanding, capability, and experience. The graduate viewpoint is, of course, more sophisticated, so a single course covering the nine chapters is appropriate.

3. *Graduate scientists and engineers in industry* who would like easy access to the basics of analysis and computation, together with a fully explained set of short source-language programs for performing key numerical operations.

At the University of California, Irvine, we offer a sophomore two-course sequence covering the material of this text. These bring the analytical tools of the university calculus sequence together with associated numerical concepts to give comprehensive, coherent experience in the development and application of fundamental problem-solving methods. Students are thereafter routinely expected to be able to do such things as factor high-degree polynomials, perform matrix computations, find extrema, and solve simultaneous nonlinear equations. Because of this experience, they are expected to be competent in programming and obtaining solutions to related but not specifically covered problems, such as frequency response, optimal layout, and system simulation.

For entering graduate students, the material is given again, this time with a beginning graduate viewpoint and sophistication. Because so many of our graduate students come from other institutions, a significant number from other countries where objectives and preparation differ markedly from that of our own B.S. graduates, this general requirement (among others) is quite important to ensuring a minimal breadth of knowledge and experience. Students who have taken the undergraduate courses, then the graduate course covering the same topics, appreciate the opportunity to reexamine important concepts and methods in greater detail and from a graduate perspective.

The situation at California State University, Long Beach, is similar, except that the undergraduate courses are offered at the sophomore and junior levels.

OUTLINE

The book is organized into nine chapters, as follows:

Chapter 1: Introduction. Outline of relevant aspects of the BASIC and FORTRAN77 languages. For those readers who are familiar with BASIC and FORTRAN77 languages, the material of this chapter may serve as a review and can be skipped without loss of continuity.

Chapter 2: Simultaneous Linear Algebraic Equations. Review of complex algebra, including roots and logarithms of complex numbers. Properties and solutions of linear algebraic equations with example applications to finding forces in a static structure and solving for the currents in a resistive electrical network.

Chapter 3: Single-Variable Searches for Zeros, Maxima, and Minima. Solution of nonlinear equations involving a single variable. Dependable numerical methods for finding the zeros and the extrema of single-variable functions. Examples of finding the radiation pattern of an antenna array and determining optimal automobile suspension damping.

Chapter 4: Multivariable Functions and Searches. Solution of simultaneous non-linear equations. Good numerical methods for locating and tracking zero-crossing curves, finding simultaneous zero crossings, and determining multidimensional extrema. Applications to robot arm positioning and optimal manufacturing.

Chapter 5: Polynomials and Factoring. Polynomial properties, manipulation, and factoring. Development of two compact, general factoring routines, suitable for small computing systems. Examples of solving for control system modes and numerical root loci construction.

Chapter 6: Matrix Algebra. Fundamental matrix methods and properties. Computational methods for finding the determinant, inverse, and adjugate. The matrix viewpoint of vector analysis, vector spaces, and coordinate transformations.

Chapter 7: The Characteristic Value Problem. Eigenvalues, eigenvectors, and similarity transformations. The Cayley–Hamilton theorem and finding functions of square matrices. Efficient algorithms and compact computer programs for eigenvalue and eigenvector calculation. Example application to finding the vibrational modes of lossless systems.

Chapter 8: Quadratic Forms and Least Squares. Investigation of the special properties of symmetric matrices and the quadratic functions they define. Transformation to a pure sum of squares and finding the principal axes of quadratic and linear-quadratic functions. Least squares estimation, curve fitting, and recursive least squares. Example applications include identifying polarization axes, matrix rank determination using singular values, economic modeling, and navigation.

Chapter 9: Differential Equations. Analytical, numerical, and computational aspects of differential equation solution. Classical and Laplace transform solution of linear, constant-coefficient equations. Numerical integration and solution of general differential equations, including boundary-value problems. Example applications to finding the response of translational and rotational mechanical systems, Bessel's equation, and Van der Pol oscillations.

COMPUTATION

A great deal of thought and experience has gone into designing sound and effective ways of incorporating the computational aspect into this course of study. In teaching the fundamentals of numerical methods and computation, it is not so much the language as the *accessibility* of computation that is key to efficient learning. Hence we use the world's most available scientific programming languages, BASIC and FORTRAN77. In the early development of the text, the computer programs were designed and written in BASIC and then were translated to FORTRAN77. So we present both versions here; however, our discussions and treatment of the subject matter are centered around the BASIC version. This is not to say that we expect BASIC to dominate science-related industry in the future, as FORTRAN77 still does at present, and it is not an endorsement of BASIC as a "best" language, although it is very good for our purposes. We feel that BASIC and FORTRAN77 are two of a number of languages that should be familiar to science and engineering graduates.

Our needs are for little in the way of equipment investment and for no low-productivity, nonlearning activity (maintenance of computer accounts, passwords, operating systems, etc.) by instructors, staff, or students. Several modest desktop personal computers, readily available, are quite adequate for even relatively large classes.

The emphasis here is on writing and using scientific calculation programs that work and are easily understood. We do not encourage artificial programming complexities and difficult constraints in the name of "readability" or one-to-one correspondence with some way of describing an algorithm. On the other hand, we strongly discourage obscure "tricks" and rat's nest coding. A reasonable and practical balance between programming extremes is sought and is reinforced often through examples. Our concerns are quite different from those connected with teaching large-scale software development teamwork.

At no cost, a floppy disk containing BASIC and FORTRAN77 language source code files for the more than 70 programs developed and discussed here is available from the publisher with the instructor's manual. Although copyrighted, there are no restrictions on disk copying and the use of these programs for educational purposes. The disk is available for the IBM PC/XT/AT computer systems, and compatibles.

We have given some thoughts to users who might want to translate these programs to another language such as Pascal or even convert them to another computer system. We have written the programs in a way that such translation or conversion is as simple as possible.

The computer programs associated with this book are not a substitute for good, integrated numerical methods packages and libraries. Rather, they provide complementary capabilities for relatively simple tasks. They do not require significant learning (and relearning) time by the user. Not needing extensive memory or blinding computational speed, each is easily run on the simplest personal computer. They illustrate and give alternatives to the methods of more extensive programs and can be linked by the user via data files to do successions of tasks, if desired.

PROBLEM SOLUTIONS

The instructor's Manual is an extension of the text, for discretionary use by course instructors. The same care has gone into verification of accuracy of the many text examples extends to the statements and solutions of the chapter problems. Several additional programs, related to chapter problem solutions, are included on the program disk.

ACKNOWLEDGMENTS

We are greatly indebted to many students and colleagues who have contributed in many ways to the preparation of this book. We are especially grateful to professors Jose B. Cruz, Jr., James M. Mulligan, Jr. and Allen R. Stubberud for their inspiration and valuable suggestions during the development of this book. Also, our personal thanks go to Professors Benham Bavarian, Neil Bershad, Jeffrey Burl, Roland Schinzinger, and to James Sampson, Jr., Dr. Ted Nishimoto, Dominic Camillone, Jerry Fowler, Alexander Cormack III, David Wilson, Paul Andrews, Hari Hablani, Milton Allione, Ja Sung Lee, Mike Cross, and Leanne Chun. We also like to extend our thanks to Professors Sandy Cynar, John Lane, Edward Evans, Tim Jordanides, Morton Schwartz, Dennis Volper, Joel Caraissimo, Ken James and Mike Mahoney. Special thanks are due to Fayad M. Khatib for his assistance in preparing the instructor's manual, to Jeff Finley who developed the FORTRAN programs and compiled the problem solutions, and to Michael Steinberg in production for his efforts in making this book a reality.

Finally, we would like to thank Mr. Tim Bozik, the executive editor at Prentice Hall for key suggestions and support during the development and preparation of our book.

G. H. Hostetter
M. S. Santina
P. D'Carpio-Montalvo

ANALYTICAL, NUMERICAL, AND COMPUTATIONAL METHODS FOR SCIENCE AND ENGINEERING

1

INTRODUCTION

1.1 PREVIEW

This is a book about analysis and numerical computation and their complementary application to fundamental classes of engineering and scientific problems. In the first four chapters, we investigate and develop sound methods for the solution of linear algebraic equations, nonlinear algebraic equations, and simultaneous nonlinear algebraic equations.

This introductory chapter begins with a review of the BASIC and FORTRAN77 computer languages that emphasizes those aspects relevant to the programs that are developed later. We have chosen these two languages because they are readily available on a wide variety of inexpensive processors and they are adequate for our needs. For those readers who are familiar with BASIC and FORTRAN77 languages, the material of this chapter may be skipped without loss of continuity. Those without prior computer experience should begin a collateral study of these programming languages right away. The programs in this book are written in the BASIC and FORTRAN77 languages. They are available with the instructor's manual, at no cost from the publisher, on a floppy disk.

Why the BASIC and FORTRAN77 instead of, say, Pascal, Ada, or APL or some other language? Because BASIC and FORTRAN77 are almost universally available for computers ranging from the early, now very inexpensive microprocessor systems, to the more modern personal computers (such as the Macintosh, the IBMs, and their clones), through expensive computer workstations and large mainframe computers. Even the most modest of these will be quite adequate for the programs in this book, many of which are fairly sophisticated.

For most systems, BASIC is also very easy to use. It takes little time before a person is able to write programs that work, and there is little overhead in the form of complicated procedures for running a program that need to be memorized. Our philosophy is that one should use the most convenient (for the user) tools available to get the job done. Although it might be neat to work with a language that allows subroutines to call themselves or one that forces a certain kind of organization to a program, no single language is likely to emerge as universally acclaimed in the near future. In the meantime, a working knowledge of several diverse languages is the best goal for most readers, and BASIC and FORTRAN77 are very good choices for two of those languages. (In this book, unless otherwise specified, the term FORTRAN stands for FORTRAN77).

Our philosophy, too, is that it is wasteful at this stage to take a long, long time to write the world's greatest, best organized program for, say, matrix inversion. Better to write an adequate program that can easily be understood by a colleague or, a few days later, by the original writer. Then the writer can move on to new ideas and new horizons. This is not to encourage sloppiness or rat's nests of instructions in programming; it is simply a recognition of a trade-off between perfection and expediency. Perhaps, too, the perfectionist in all of us needs to be reminded of the diminishing return (and satisfaction) of overly emphasizing and investing in perfection for its own sake.

1.2 THE BASIC LANGUAGE

Compilation is the process of converting the human-oriented instructions comprising a program to the codes for the operations to be performed by the machine. It is done by another program, the *compiler*, that accepts the BASIC instructions, stores them, and produces the corresponding string of machine language code. (Yes, computer scientists sometimes call the BASIC compilation process *translation*, but the distinction is quite artificial and seldom very useful.)

Normally, BASIC is compiled line by line, as it is run, instead of all at once before it is run. This means that incomplete programs and parts of programs can easily be tested. When an error in an instruction has been made, program execution can simply be stopped (with a message stating the type of error) at the instruction in error. The additional time needed to compile a BASIC program as it is running need be no more than the compile and run time if the program were first entirely compiled, then run. In fact, it can be shorter because any portions of a program that are not used in a given run are not compiled.

When it is desired to speed the execution of a BASIC program somewhat, a compiler that compiled the entire program before it is run, similar to the compilers used for FORTRAN, Pascal, and other languages, can be employed. The resulting *compiled BASIC* program in the computer's machine language is simply loaded into memory and executed instead of being regenerated each time the program is run. Depending on the particular instructions that comprise the program, one might expect a two- or threefold increase in execution speed with compiled BASIC. There is, however, the need to recompile the program any time a change is made, and there is the added complication of dealing explicitly with both the source code and the machine-language versions of the program. None of the source code programs in this book run terribly slowly even on the most primitive microprocessors.

The remainder of this section is an outline and review of some of the instructions available in BASIC. The language has many other capabilities—for instance, character string manipulation, data list, and operations with logical variables—but we will concentrate here on those features of immediate use. We have given some thought, too, to readers who, especially because of the large existing base of available programs, might want to translate these programs to Pascal. Wherever it does not complicate the resulting BASIC program, we have written the programs in a way that such translation is as painless as possible.

1.2.1 Composing a Program

Table 1-1 lists important commands used to compose a BASIC program. To begin, one types

NEW

which causes erasure of any program previously stored in memory. Each instruction begins with a line number, and the lines are arranged in line-number order, regardless of the order in which they are entered. To insert an instruction between two instructions entered previously, one simply gives the new instruction a line number that is between the line numbers of the other two instructions. To change an instruction, the changed instruction is entered with the same line number as the instruction to be changed and the new entry replaces the old one. To delete an instruction, its line number is entered. This causes the replacement of the original instruction by no instruction. To delete several instructions in a row, the DEL command is used. For example,

DEL 150, 200

causes all instructions with line numbers from 150 to 200 to be erased.

To list on the display all the entered instructions so far, the

LIST

command is used. The LIST command followed by a single line number lists only the instruction with that line number, if one exists. The command

TABLE 1-1   Some BASIC Program Composition Commands

DEL [line number]	Deletes the line having this line number. Actually, this replaces the original line with an incomplete line having no statement.
DEL $\begin{bmatrix} \text{starting} \\ \text{line number} \end{bmatrix}$, $\begin{bmatrix} \text{ending} \\ \text{line number} \end{bmatrix}$	Deletes all lines beginning with the starting line through the ending line.
LIST	Lists the program on the computer display or terminal.
LIST $\begin{bmatrix} \text{starting} \\ \text{line number} \end{bmatrix}$,	Lists the part of the program from the line number to the end.
LIST $\begin{bmatrix} \text{starting} \\ \text{line number} \end{bmatrix}$, $\begin{bmatrix} \text{ending} \\ \text{line number} \end{bmatrix}$	Lists all the lines from the starting line number to the ending line number.
NEW	Deletes any program previously in computer memory and awaits the entry of new numbered instructions.
RUN	Executes the program stored in memory.

```
LIST 10, 300
```

for example, lists all the instructions with line numbers from 10 to 300.

Once a complete program is stored in the computer memory, the command RUN begins the program execution. To return control smoothly to the operating system after execution, the program's last instruction should be

```
[line number] END
```

Once a program is stored in memory, most computer systems have a provision for transferring that program to more permanent storage, such as on floppy disk. Often this is done with a SAVE command that also includes a name the user has selected to distinguish the program. A command, typically LOAD and the program's name, will then place the program back into computer memory for execution.

1.2.2 Numbers and Variables

Numbers are entered and printed in BASIC either in decimal form, such as

```
-342.078
```

or in exponent form, for example

```
1.2345678E-12
```

which means $1.2345678 \times 10^{-12}$, or

```
-47.39E+8
```

which means -4.739×10^9.

A variable is a symbol or group of symbols used to represent a number. Some BASIC compilers allow names like PHI, EXPONENT, and ITEM24 to be used as variables. Names such as NEXT and IF that conflict with the symbols used for instructions are not allowed. Other BASIC compilers restrict variables to single letters, or letter–digit combinations. We will use single letters for all variables so that there will be no incompatibility with any reasonably complete BASIC compiler. In addition, we will, wherever practical, use the variables I, J, K, L, M, and N to represent integers, as is commonly done in the FORTRAN language.

An arithmetic expression in BASIC is a formula involving variables, parentheses, numbers, and the arithmetic operations listed in Table 1-2. For example, the arithmetic expression

TABLE 1-2   Arithmetic Operations
                   in BASIC

+	Addition
−	Subtraction and negation
*	Multiplication
/	Division
** (or ↑ or ∧)	Exponentiation by an integer

$$W + (3 * X + 7 * Y) / (((X + Z) * * 2) + 4)$$

is the formula

$$w + \frac{3x + 7y}{(x + z)^2 + 4}$$

Parentheses are used to group terms as needed, and within parentheses the operations are evaluated in the reverse order of Table 1-2. Thus

$$3 * X + 7 * Y$$

is the formula

$$3x + 7y$$

not

$$3(x + 7)y$$

Two arithmetic operations should never appear next to one another, except for the minus sign. A minus sign preceding a variable always means the negative of that variable, so

$$Y * -Z$$

which means that

$$y(-z)$$

is acceptable.

An *assignment instruction* in BASIC is of the form

$$[line number] [variable] = [arithmetic expression].$$

It causes the variable to the left to be given the value of the arithmetic expression on the right. For example for

$$200 \ A = X * X - Y$$

if $X = 3$ and $Y = 2.4$, then A is given the value 6.6. Of course, all the variables in the arithmetic expression must have been assigned values before reaching this instruction; otherwise, the assignment cannot be made and an error message will probably be printed. In some BASIC compilers, variables with values not previously assigned are given the value of zero, but it is not a good idea to rely on this assumption. It is also dangerous if an error message is not given, because an obvious check on a program's correctness is bypassed.

When the same variable appears on both sides of an assignment statement, as in the instruction

$$A = A + 1$$

the original value of the variable is used in the arithmetic expression on the right. After the instruction is executed, the variable is given the new value.

1.2.3 Control Instructions

Table 1-3 lists some BASIC control instructions in alphabetical order. When arrays of variables are to be used, storage space in memory must be reserved for them. The dimension instruction, DIM, performs this function. For example,

```
DIM A(3,2), B(8)
```

sets aside memory space for the 12 two-dimensional variables A(0,0), A(0,1), A(0,2), A(1,0), A(1,1), A(1,2), A(2,0), A(2,1), A(2,2), A(3,0), A(3,1), A(3,2), and the nine one-dimensional variables B(0), . . ., B(8). It is, of course, necessary to give this instruction before any variables are used; in fact, many BASIC compilers require that dimension instructions occur before any variables at all are used. The dimension of an array can always be set larger than needed, but reserving too much memory space for array variables will risk not leaving enough memory for other needed functions.

A set of FOR . . . NEXT instructions defines a looping operation where the variable in the FOR instruction is assigned the starting value, the instructions between the FOR and NEXT are executed, the variable is incremented, those instructions executed again, and so on, until incrementing the variable causes it to exceed the ending value. For example, in the sequence

```
        ⋮
300 FOR I=1 TO 3
310 E=E+(1/I)
320 PRINT E
330 NEXT I
        ⋮
```

instructions 310 and 320 are executed with $I = 1$, then they are executed with $I = 2$, then with $I = 3$, after which the next instruction after 330 is executed.

The starting and ending values of the incremented variable can be other variables or arithmetic expressions, as in the instruction

```
440 FOR K=A TO C*SQR(B+1)
        ⋮
500 NEXT K
```

In this form, each increment of the variable is by unity, but if the instruction is written in a form such as

```
110 FOR P=-3.5 TO 7.1 STEP 0.4
```

P will begin with the value -3.5 and be incremented by steps of 0.4 each time the loop is traversed until another increment results in P exceeding 7.1.

The IF . . . THEN instruction is a conditional branching operation. For example,

```
390 IF X>Y THEN PRINT X
```

causes the value of the variable X to be printed only if $X > Y$. For the IF . . . THEN instruction in

TABLE 1-3 Some BASIC Control Instructions

DIM

Array variable definition.

FOR $\begin{bmatrix} \text{index} \\ \text{variable} \end{bmatrix}$ = $\begin{bmatrix} \text{starting variable} \\ \text{or arithmetic} \\ \text{expression or number} \end{bmatrix}$ TO $\begin{bmatrix} \text{ending variable} \\ \text{or arithmetic} \\ \text{expression or number} \end{bmatrix}$

...

NEXT $\begin{bmatrix} \text{same index} \\ \text{variable} \end{bmatrix}$

Begins with the index variable equal to the value of the starting quantity, executes the instructions between the FOR and NEXT lines, increments the index variable by 1, executes the instructions between FOR and NEXT, and so on. The execution continues until incrementing the index variable causes it to exceed the ending quantity. Program execution then resumes with the next instruction after the NEXT line.

FOR $\begin{bmatrix} \text{index} \\ \text{variable} \end{bmatrix}$ = $\begin{bmatrix} \text{starting variable} \\ \text{or arithmetic} \\ \text{expression or number} \end{bmatrix}$ TO $\begin{bmatrix} \text{ending variable} \\ \text{or arithmetic} \\ \text{expression or number} \end{bmatrix}$ STEP $\begin{bmatrix} \text{step variable} \\ \text{or arithmetic} \\ \text{expression or number} \end{bmatrix}$

...

NEXT $\begin{bmatrix} \text{same index} \\ \text{variable} \end{bmatrix}$

Same as the commands above except that the index variable is incremented by the step value (instead of 1) each time the loop is traversed.

GOSUB [line number]

Jump to the subroutine that begins at the given line.

GOTO [line number]

Causes the program execution to jump to the given line.

IF $\begin{bmatrix} \text{tested} \\ \text{variable} \end{bmatrix}$ $\begin{bmatrix} \text{equality or} \\ \text{inequality} \end{bmatrix}$ $\begin{bmatrix} \text{variable} \\ \text{or number} \end{bmatrix}$ THEN [instruction]

If the equality or inequality holds, the instruction after THEN is executed. If the equality or inequality does not hold, the instruction after THEN is not executed. The symbols for equalities and inequalities are $=$, $>$, $<$, $<>$ (meaning $\neq$), $>=$ (meaning $\geq$), and $<=$ (meaning $\leq$).

(continued)

TABLE 1-3 Some BASIC Control Instructions (continued)

INPUT [variable]

Assigns the number entered on the computer keyboard to the variable.

PRINT $\begin{bmatrix} \text{alphanumeric string} \\ \text{in quotes or variable} \\ \text{or arithmetic expression} \end{bmatrix} \begin{bmatrix} \text{control} \\ \text{character} \end{bmatrix} \begin{bmatrix} \text{alphanumeric string} \\ \text{in quotes or variable} \\ \text{or arithmetic expression} \end{bmatrix} \begin{bmatrix} \text{control} \\ \text{character} \end{bmatrix} \cdots$

Prints alphanumeric strings and/or the values of variables or expressions on the computer display or printer. Any alphanumeric string to be printed is placed within quotation marks. The control characters determine how the components are joined to one another, as follows:

Carriage return: End the line (and the PRINT instruction).

;   Place one component right after another.

,   Tab before printing the next component.

REM [alphanumeric string]

Such a line is listed when the program is listed but is not executed. A convenient way of including comments in the program listing.

RETURN

Causes the program execution to jump from the end of the subroutine back to the next instruction after the subroutine call.

END

Ends program execution and returns control to the BASIC compiler or the operating system.

```
         ⋮
900 IF A*A <= 1E−10 THEN GOTO 990
910 A=0
         ⋮
990 PRINT "VALUE OF A IS ";A
         ⋮
```

program execution jumps to instruction 990 if $A^2 < 10^{-10}$; otherwise, execution continues with instruction 910. Some BASIC compilers allow execution of multiple instructions when the IF condition is satisfied by separating them by colons after the THEN, as, for example, in

```
700 IF A<B THEN PRINT A:C=C−B:D=3
```

We will not assume that this feature is available, so this combination of conditional instructions would be performed here by

```
700 IF A<B THEN PRINT A
701 IF A<B THEN C=C−B
702 IF A<B THEN D=3
```

instead, or by some other method.

The GOTO instruction causes program execution to jump to the instruction having the line number specified. For example, in the sequence of instructions

```
         ⋮
210 C=A+B
220 GOTO 250
230 PRINT "RESULT IS AS FOLLOWS:"
240 PRINT A,SQR(C)
250 D=D*C
         ⋮
```

the GOTO instruction of line 220 causes the instructions to be processed in the order 210, 220, 250, skipping lines 230 and 240. The GOTO instruction is often used to form a loop of instructions that is exited when some condition is satisfied. For example, the instruction sequence

```
         ⋮
300 D=D+0.1
310 I=I+1
320 IF D<E THEN GOTO 340
330 GOTO 300
340 PRINT X
         ⋮
```

causes instructions 300 and 310 to be executed repeatedly until the variable D is less than some number E, after which the variable X is printed.

A subroutine is a sequence of instructions ending with the RETURN instruction. A GOSUB instruction such as

```
286 GOSUB 1000
```

causes program execution to jump to the instruction on line 1000. The instructions after line 1000 are executed in the normal order until a RETURN instruction is encountered.

Then program execution jumps to the next instruction after the one on line 286.

Subroutines are used whenever the same string of instructions would otherwise occur several places in a program. They are also used to divide a program into "blocks" of instructions, so the main program can then consist primarily of a series of subroutine calls. Furthermore, when the same set of instructions for a certain function is to be used in more than one program, it is helpful to write them as a subroutine common to each program.

Some advanced BASIC compilers allow subroutines to use *local variables*, variables with the same symbols as those in the main program and in other subroutines, but without the subroutine affecting the values of the same-named variables outside the subroutine. In other words, a variable X inside a subroutine is treated as a different variable from an X in the main program. This is not the case for most BASIC compilers. For them, all variables are *global variables*. One must then take care that only variables that should be changed by a subroutine are affected by that subroutine.

The PRINT instruction causes a message to be produced on the computer display or printer. A message enclosed in quotes is printed, as for example in the instruction

```
260 PRINT "PROGRAM TO DIVIDE TWO POLYNOMIALS"
```

when the symbol for a variable or when an arithmetic expression appears in a PRINT instruction, the value of the variable or expression is printed. For example,

```
710 PRINT X
```

causes the value of the variable X to be printed and

```
490 PRINT 3+COS(X*Y)
```

prints the value of $3 + cos XY$.

Items separated by commas in a PRINT instruction are printed side by side, with present tabulations between them. For example,

```
620 PRINT "SOLUTION IS",X,Y
```

causes an output that might look like the following:

SOLUTION IS	43.924	2.732E+7
←*left margin*	←*first tab*	←*second tab*

When more items are listed than can be printed on a single line, they are continued on the next line. If the items are separated by a semicolon, one item is printed immediately after the other. The instruction

```
25 PRINT "SOLUTION IS X=";X;" AND Y=";Y
```

causes an output

```
SOLUTION IS X=43.924 AND Y=2.732E+7
```

A carriage return at the end of a PRINT instruction causes the display to move to the next line. When a PRINT instruction ends with a semicolon, then a carriage return, the display or printer just moves to the next space.

The INPUT instruction is used to assign numbers to variables from the computer keyboard. The instruction

```
180 INPUT X
```

for example, causes a question mark to be printed on the display device, indicating that program execution is awaiting an input from the keyboard. When a number is entered by the user, that number is assigned to the variable X. REM is used to include comments in the program listing without affecting the program's operation. Every time a REM instruction is encountered, the compiler simply ignores it.

Functions such as exponentiation are built into the BASIC language. Table 1-4 lists the most common built-in functions. In BASIC, the argument of a built-in function can be any arithmetic expression, including those involving built-in functions themselves. For example, the instruction

```
405 Z= SQR(X*X+Y*Y)
```

assigns the value of the square root of the sum of the squares of X and Y to the variable Z. The instruction

```
613 Z=ATN(X/SQR(1-X*X))
```

assigns the value of

$$\arctan \frac{X}{\sqrt{1 - X^2}}$$

to the variable Z.

TABLE 1-4   Some Built-In BASIC Functions

ABS $(\cdot)$	Returns the absolute value of the expression in parentheses.
ATN $(\cdot)$	Returns the principal (first and fourth quadrant) arctangent of the expression in parentheses. The returned angle is in radians, between $-\pi/2$ and $\pi/2$.
COS $(\cdot)$	Returns the cosine of the expression in parentheses. The expression is taken to be in radians.
EXP $(\cdot)$	Returns the exponential of the expression in parentheses.
INT $(\cdot)$	Returns the largest integer less than the expression in parentheses.
LOG $(\cdot)$	Returns the natural logarithm of the expression in parentheses. The expression should be positive.
RND $(\cdot)$	Returns a random real number between 0 and 1. If the expression in parentheses is positive, RND returns a new random number each time it is called.
SGN $(\cdot)$	Returns -1 if the expression in parentheses is negative, 0 if the expression is zero, and 1 if the expression is positive.
SIN $(\cdot)$	Returns the sine of the expression in parentheses. The expression is taken to be in radians.
SQR $(\cdot)$	Returns the positive square root of the expression in parentheses. The expression should not be negative.

1.2.4 Some Illustrative Examples

A simple BASIC program involving a FOR . . . NEXT loop is the following:

```
100 PRINT "FACTORIAL CALCULATION"
110 X=1
120 FOR I=1 TO 20
130 X=X*I
140 PRINT I;"! = ";X
150 NEXT I
160 END
```

It computes and prints $X = I!$ for values of I from 1 through 20. The simple BASIC program

```
100 PRINT "GUESS AN INTEGER FROM 1 TO 20"
110 X=17
120 PRINT "WHAT IS YOUR GUESS?"
130 INPUT Y
140 IF Y>X THEN PRINT "TOO HIGH"
150 IF Y<X THEN PRINT "TOO LOW"
160 IF Y=X THEN GOTO 180
170 GOTO 120
180 PRINT "CORRECT! THE NUMBER IS ";X
190 END
```

has an IF . . . THEN instruction. It asks the user to guess the integer X. The following program uses a subroutine to compute an approximate value of $Y = \exp(x)$ by summing the first 21 terms in the power-series expansion

$$\exp(x) \simeq 1 + \frac{x}{1!} + \frac{x^2}{2!} + \frac{x^3}{3!} + \cdots + \frac{x^{20}}{20!}$$

where x is a user-entered number:

```
100  PRINT "EXPONENTIAL CALCULATION"
110  PRINT "PLEASE ENTER THE ARGUMENT"
120  INPUT X
130  GOSUB 1000
140  PRINT "EXP(";X;") = ";Y
150  GOTO 110
160  END
1000 REM SUBROUTINE TO EXPONENTIATE
1010 Z=1
1020 Y=1
1030 FOR I=1 TO 20
1040 Z=Z*X/I
1050 Y=Y+Z
1060 NEXT I
1070 RETURN
```

Finally, a BASIC program that uses arrays is the following:

```
100 PRINT "ELEMENTARY SPREADSHEET"
110 DIM A(5,6), B(6)
120 REM ENTER VALUES IN THE ARRAY
```

```
130 PRINT "ENTER ITEM AMOUNTS"
140 FOR J = 1 TO 6
150 FOR I = 1 TO 5
160 PRINT "ROW ";I;"COLUMN ";J;
170 INPUT A(I,J)
180 NEXT I
190 NEXT J
200 REM ADD COLUMNS
210 FOR J = 1 TO 6
220 B(J) = 0
230 FOR I = 1 TO 5
240 B(J) = B(J) + A(I,J)
250 NEXT I
260 PRINT COLUMN ";J;" TOTAL IS ";B(J)
270 NEXT J
280 END
```

This simple program asks the user to enter numbers into the five rows and six columns of a two-dimensional array A, then adds each of the array columns, prints the result, and stores them in a single-dimensional array B. These operations are elementary ones that might be used in data processing, for example in a company's monthly profit and loss statement.

1.3 THE FORTRAN LANGUAGE

FORTRAN is generally used for those applications that involve mathematical computations and other manipulation of numerical data. The name FORTRAN comes from Formula Translator, which indicates the type of application for which the language was designed.

The *source program* (also called *source code*) is the set of language elements the programmer writes in order to accomplish the computations. In a process called *compilation*, the FORTRAN compiler analyzes the source program statements and translates them into a machine language program called the *object program* or *object code*. The FORTRAN compiler is itself a large program that treats the source code as data.

A source statement performs one of three functions:

1. Performs operations
2. Specifies the nature of the data
3. Specifies the characters of the source program

Traditionally, the FORTRAN program is written on a FORTRAN coding sheet to be typed later on the terminal. The format is as follows:

1. Columns 1 to 5 are to be used for line numbers if the program requires them.
2. Column 6 is to be left blank except to show continuation of the preceding line. Any character may be used for this purpose except zero (0) or blank.
3. Columns 7 to 72 are allocated for the FORTRAN statements.
4. Columns 73 to 80 may be used, if desired, for sequencing numbers for ease of identification or sequencing. These columns are ignored by the compiler.

5. Comment lines should start with a letter "C" in column 1. The comment itself may be written on columns 2 to 72 or 80, as desired.

FORTRAN variables consist of six characters. The first variable must be a letter, and all other variables must be alphanumeric. No special characters or blanks are allowed. The first letter also denominates what type of variable it is: integer or real/alphanumeric. Variables starting with letters I through N are integers. All other variables may be real or alphanumeric. (This, however, may be overridden by the Integer or Real statements when dealing with numeric data). The use of meaningful variable names serves as an aid in documenting a program.

The data the program manipulates is generally written in a separate storage area. It is called the *dataset*.

1.3.1 Numbers and Variables

There are two major types of numbers in FORTRAN: REAL and INTEGER. REAL numbers have a decimal point in them such as

$$12.04, \text{ and } 296.3316$$

INTEGER numbers, however, are written with no decimal points in them;

$$13,614 \text{ and } 1639$$

are examples. In computer languages, numbers are called *numerical constants*, or simply *constants*. The reason for this is that a constant's value does not change.

Corresponding to REAL and INTEGER constants, there are REAL and INTEGER variables. A REAL variable is the name (more accurately, the address) of a memory location that contains a REAL constant. Similarly, an INTEGER variable is the name of a memory location that contains an INTEGER constant. In most FORTRAN compilers, a variable may consist of one to six characters. The first character must be a letter of the alphabet, while each of the succeeding characters may be either a digit or a letter. A REAL variable is one that starts with any letter of the alphabet except I, J, L, K, M, or N. Thus

```
DATA6, COLL31, and A19561
```

are all REAL variables. On the other hand, an INTEGER variable is one that begins with I, J, K, L, M, or N. Thus

```
INDEX, M, and J29364
```

are all INTEGER variables.

In general, a mathematical expression occupies more than one line of print. For example, the expression appearing in the following equation:

$$Y = \frac{(X + 2)^2 \cdot (Z + 6)}{W + 2.17}$$

requires one line for the

```
exponent 2,
```

one line for

$$(X + 2)**2 \cdot (Z + 6),$$

add one line for the divisor,

$$W + 2.17.$$

Similar to BASIC, the operations available in FORTRAN are tabulated in Table 1-5.

TABLE 1-5   Arithmetic Operations
              in FORTRAN

+	Addition
−	Subtraction and negation
*	Multiplication
/	Division
**	Exponentiation

1.3.2 Control Instructions

The **C** instruction allows one to write comments anywhere in the program provided that a letter **C** is typed in the first column of the comment line. This line will not be executed.

The **END** instruction in a program indicates to the compiler that the program is completed and that no more statements will follow. The END instruction should appear as the last instruction in any FORTRAN program.

The **STOP** instruction instructs the computer to terminate the program.

The **OUTPUT** instruction is used to write data to records in a file. It has two forms:

```
WRITE (cilist) output list
PRINT f,output list
```

where

- cilist is a control information list which, in its simplest form, is u, f.
  - u is a unit identifier.
  - f is a format identifier.
- output list is, in its simplest form, a list of variable names or expressions separated by commas. The output list may be omitted.

A **WRITE** instruction with an asterisk as a unit identifier and the **PRINT** instruction both refer to the designated output unit. To make the output more easily understood, it is often helpful to include text along with the numeric information. The simplest way to output character information is to use the nonrepeatable apostrophe edit descriptor. It has the form of characters of the constant to be outputed. (The apostrophe edit descriptor cannot be used on input.) The instruction

```
WRITE (22,'("YOU MUST BE STUDYING")')
```

will output the characters

YOU MUST BE STUDYING

to the file connected to unit 22. The instruction

PRINT'("COMPUTER PROGRAM IN FORTRAN77")'

will print the characters

COMPUTER PROGRAM IN FORTRAN77

at the start of a new line on the designated output unit. Notice that the **PRINT** instruction does not imply that printing will occur, for example, if the designated output device is not specified for printing. The **PRINT** instruction will print the numbers assigned to CAR1 and CAR2. To do this in FORTRAN, we write

PRINT*,CAR1,CAR2

or

```
C PROGRAM USES THE PRINT INSTRUCTION
  CAR1=19900.0
  CAR2=21999.0
  PRINT*,CAR1,CAR2
  STOP
  END
```

This instruction prints the value of CAR1(19900), followed by the value of CAR2(21999).

The **READ** instruction enables the computer to assign values to variables by reading constants from a terminal during the execution of the program. To show this, consider a portion of a program that evaluates the polynomial

$$X = a_0 + a_1 T + a_2 T^2$$

for any values of a_0, a_1, a_2, and T. If we assign values to the variables a_0, a_1, a_2, and T, by reading the information from a terminal, we would write

READ*,a_0,a_1,a_2,T

The variables in the READ instruction must be separated by commas; and an asterisk followed by a comma must follow the word READ. As an example,

```
C EVALUATES A POLYNOMIAL
  READ*,a_0,a_1,a_2,T,
  PRINT*,'a_0=',a_0,'a_1=',a_1,'a_2=',a_2,'T=',T

C . . .THE POLYNOMIAL IS
  X=a_0+a_1*T+a_2*T**2
  PRINT*,'X=',X,'FOR T=',T
  STOP
  END
```

When using an "actual" array in a FORTRAN program unit, one must assign:

1. The name of the array (i.e., any combination of one to six letters and digits beginning with a letter

2. The type (INTEGER, REAL, DOUBLE PRECISION, COMPLEX, LOGICAL, or CHARACTER) of the array (all the elements of an array in FORTRAN must be of the same type)

3. The number of dimensions of the array

4. The lower and upper bounds on each dimension (in FORTRAN no lower bound may be greater in value than the corresponding upper bound)

Essentially, this information is conveyed in a program unit by means of an array declarator in a type instruction, a DIMENSION instruction, or a third which will not be considered here. Type statements may contain array declarators as in the following example:

```
REAL L,M,X(10),Y(-4:4)
INTEGER R,FLAG,A(10,10)
```

specifying that

1. L and M are the names of two variables of type REAL.

2. X1, X2, . . ., X10 is a one-dimensional array, the elements of which are of type REAL.

3. $Y-4$, $Y-3$, . . ., Y4 is a one-dimensional array, the elements of which are also of type REAL.

4. R,FLAG are the names of two variables of type INTEGER.

5. The matrix **A**

$$\begin{bmatrix} A(1,1) & A(1,2) & \cdots & A(1,10) \\ A(2,1) & A(2,2) & \cdots & A(2,10) \\ \vdots & \vdots & \vdots & \vdots \\ A(10,1) & A(10,2) & \cdots & A(10,10) \end{bmatrix}$$

has elements of INTEGER type.

Alternatively, DIMENSION statements (possibly in conjunction with type statements) may be used to convey the same information. Each DIMENSION instruction has the form

```
DIMENSION array declarator 1, array declarator 2, ...
```

where the first letter of each array name indicates the type of the array via the implicit naming convention unless the array name also occurs in a type instruction in the same program unit. The dimension instruction

```
DIMENSION A(15),B(0,3),K(5,6)
```

specifies three arrays—thus

- A(1), . . ., A(15) of type REAL
- B(0), . . ., B(3) of type REAL

$$\bullet \begin{bmatrix} K(1,1) & K(1,2) & \cdots & K(1,6) \\ K(2,1) & K(2,2) & \cdots & K(2,6) \\ \vdots & \vdots & \vdots & \vdots \\ K(5,1) & K(5,2) & \cdots & K(5,6) \end{bmatrix} \text{ of INTEGER type}$$

unless the type, indicated by the implicit naming convention in each case, has been altered by the presence of one or more of the names A, B, and K in type statements in the same program unit. For example,

```
INTEGER A
REAL K
DIMENSION A(15), B(0:3),K(5,6)
```

would specify the type of the elements of the array A to be INTEGER and the type of the elements of the array K to be REAL.

Three basic structures are sufficient to describe an algorithm:

1. A *processing* instruction, which causes the stated action to be taken

2. A *decision* structure, which causes different actions to be taken according to the values of the specified data (there are two types of decision structures: the **if-then** instruction and the **if-then-else** instruction)

3. A *looping* structure, which causes a group of statements to be processed a number of times (there are two types of looping structure: the **loop-while** instruction and the **loop-for** instruction).

In moving from the final refinement of an algorithm to the program code it is necessary to use suitable statements to implement these basic algorithmic structures. In a given language it is usually possible to adopt a reasonably standard approach to this, so the coding state should be straightforward. In FORTRAN, for example, a processing instruction is usually implemented using one or more assignment, input, or output statements. Decision and looping structures, on the other hand, require the use of *control statements*. Control statements are used to control the sequence in which the other program statements are executed.

Logical expressions are used in both the decision and looping structures. Here only *relational expressions*, which are the most commonly used form of logical expressions, will be considered. A relational expression is used to compare the values of two arithmetic expressions using one of the *relational operators* shown in Table 1-6. Note that each operator starts and ends with a decimal point. Table 1-7 lists examples of

TABLE 1-6 Relational Operations in FORTRAN

Operator	Mathematical Description
.LT.	$<$ (less than)
.LE.	$\leq$ (less than or equal to)
.EQ.	$=$ (equal to)
.NE.	$\neq$ (not equal to)
.GT.	$>$ (greater than)
.GE.	$\geq$ (greater than or equal to)

TABLE 1-7 Some Examples of Relational Expressions
in FORTRAN

Expression	Mathematical Equivalent
B**2.LT.4.0* A*C	$b^2 < 4ac$
COST.GE.50.0	Cost. ≥ 50
EXP (X+Y) .LE. (X+1.0)**7	$e^{x+y} \leq (x + 1)^7$
I.EQ.J	$i = j$

relational expressions in FORTRAN. Evaluation of a relational expression produces a result of type LOGICAL with a value of **true** or **false**.

Control instructions for decisions are the *block IF instruction* along with the *END IF instruction* and, if necessary, the *ELSE* and *ELSE if statements*. The block **IF** instruction has the form

$$IF \ (\cdot) \ THEN$$

where the quantity within parentheses ia a *logical expression*. A block **IF** instruction always precedes an *IF-block*, which is defined to be all the executable statements after the block **IF** instruction up to, but not including, the corresponding **END IF, ELSE,** or **ELSE IF** instruction. The effect of a block **IF** is as follows:

1. If the value of the logical expression is **true**, the **IF**-block is executed, after which control is transferred to the corresponding **END IF** instruction.

2. If the value of the logical expression is **false**, control is transferred to the corresponding **END IF, ELSE,** or **ELSE IF** instruction.

The **END IF** instruction has the form

$$END \ IF$$

For each block **IF** instruction, there must be a corresponding **END IF** instruction. The algorithmic decision instruction

if the bank balance is overdrawn then print it

could be expressed in FORTRAN as

```
IF (BLANCE.LT.0) THEN
        PRINT*,BLANCE
END IF
```

In this case, the **IF**-block consists of *one* statement only. Notice that if BLANCE $\geq$ 0, then the **IF** -block is not executed and control passes to the **END IF** instruction.

The **ELSE** instruction is used when coding an if-then-else decision instruction. The **ELSE** instruction

$$ELSE$$

precedes an **ELSE**-block, defined to be all the executable statements after the **ELSE** instruction up to, but not including, the corresponding **END IF** instruction. The **ELSE** instruction is used only with the block **IF** (or **ELSE IF**) statements. If the value of the logical expression of the block **IF** (or **ELSE IF**) instruction is **false**, the **ELSE**-block is executed. When finding the roots of a quadratic equation $ax^2 + bx + c = 0$, a negative discriminant $b^2 - 4ac$ indicates imaginary roots. A code for solving quadratic equations might contain the statements

```
DISC = B**2 - 4.0*A*C
IF(DISC.LT.0) THEN
        PRINT*,'IMAG ROOTS'
        STOP
                ELSE
        TEMP = SQRT(DISC)
        X1 = ( -B + TEMP)/(2.0*A)
        X2 = ( -B - TEMP)/(2.0*A)
END IF
```

An instruction **LABEL** is a means of referring to individual statements and is written as sequence of up to five digits, at least one of which must be nonzero. The label precedes the instruction and is written anywhere in columns 1 to 5 of the initial line of the instruction. Any instruction may be labeled, but the same label must not be given to more than one instruction in a program unit. Although statements labels are numbers, they have no arithmetic significance; that is, they are only a means of identifying statements.

The unconditional **GO TO** instruction has the form

```
GO TO r
```

where r is the instruction label of an executable instruction that appears in the same program unit as the unconditional **GO TO** instruction. Execution of an unconditional **GO TO** instruction transfers control to the instruction identified by the instruction label. The instruction

```
GO TO 32
```

means "transfer control to the instruction with instruction label 32, that is, execute next the instruction label 32."

The **DO** instruction specifies a loop, known as a **DO**-loop. The **DO** instruction has the form

```
DO  r  i  =  exp1,exp2,exp3
```

where

- r is an instruction label (optionally followed by a comma).
- i is the name of the integer, real, or double-precision variable called the **DO**-variable.
- exp1, exp2, and exp3 are each integer, real, or double-precision expressions; exp3 is optional and may be omitted together with the preceeding comma, in which case the default will be a 1.

For example, the instruction

```
DO 45 I=2,10,2
```

means "execute the **DO**-loop for I taking the values 2, 4, 6, 8, and 10."
The instruction

```
DO 51 J=N,1,-1
```

means "execute the **DO**-loop for J taking the values N, $N-1$, $N-2$, . . ., 3, 2, 1."
The values of exp1, exp2, and exp3 are used as the values of the initial parameter, the
final parameter, and the incrementation parameter, respectively. The value of exp3 must
not be zero, and if exp3 and the preceding comma are omitted, a value of 1 is assumed
for the incrementation parameter.

The **CONTINUE** instruction

```
CONTINUE
```

is always used as the terminal instruction of a **DO**-loop. For example, the program

```
        DO 30 N=1,50
            I=N*(N+1)*(2*N+1)/6
            PRINT*,N,I
    30  CONTINUE
        END
```

computes and prints

$$1^2 + 2^2 + 3^2 + \cdots + n^2 = \tfrac{1}{6}n(n + 1)(2n + 1)$$

for values of n between 1 and 50.

A **subroutine** is a subprogram which may be called periodically from the main
program—or calling program. Information is transferred to and from the main program
only through the arguments (and COMMON statements) of the subroutine.

The **CALL** instruction has the form

```
CALL subname(arg)
```

where

- subname is the name of a subroutine (or a dummy procedure).
- arg is a list of actual arguments separated by commas.

The **RETURN** instruction

```
RETURN
```

terminates the execution of the subprogram and returns control to the referencing program
unit.

The **Arguments** of a subprogram are the primary means of passing information
between a subprogram and another program unit. As an example, the program

```
C SUBROUTINE IS USED TO CALCULATE THE FACTORIAL
    INTEGER M,N
```

(continued)

```
      READ*,N
      PRINT*,'VALUE OF N IS',N
      CALL FACT(M,N)
      PRINT*,N,'!=',REAL(M)
      STOP
      END
      SUBROUTINE FACT(L,N)

C     . . .CALCULATES N!
      INTEGER J,L,N
      PRINT*,'VALUE OF N PASSED IS:',N
      IF(N.LT.0)THEN
          PRINT*,N,'IS NEGATIVE. STOP PROGRAM.'
                STOP
      END IF
      L=1
      DO 10 J=1,N
          L=L*J
10    CONTINUE
      RETURN
      END
```

uses a subroutine called FACT to calculate N!. We transfer control from the main program to the subroutine by using a **CALL** instruction:

```
CALL FACT(M,N)
```

The value of N is the number whose factorial we want to find. After the subroutine has been executed, the value of M will be the required factorial. In the first line of the subroutine FACT,

```
SUBROUTINE FACT(L,N)
```

L and N are dummy arguments. The variable L corresponds to M in the main program, and we have used N in both the main program and the subroutine. Since subroutine names are not used in expressions, we do not have to concern ourselves with whether to make them REAL or INTEGER. The rule for forming a subroutine name is the same as the rule for forming a variable name. Note that we cannot use the same name for a subroutine and for a variable, and cannot list the name in a declaration instruction.

1.3.3 More on Instructions

The following is a simple FORTRAN program involving some additional instructions:

```
C
C THIS FORTRAN PROGRAM CALCULATES COST AND PRINTS ITEMS OVER
C $100.00. ALSO CALCULATES AN ITEM COUNT.
C
C INITIALIZING COUNTER
      NP=0
      WRITE(6,100)

C MAIN PROCESSING
10    READ (5,200,END=30) PART, PRICE, QTY
```

```
              COST = PRICE * QTY
              IF (COST.GT.100.0) GO TO 20
              GO TO 10
      20      NP = NP + 1
              WRITE (6,300) PART, PRICE, QTY, COST
              GO TO 10

C OUTPUT PROCESSING
      30      WRITE (6,400) NP

C
     100 FORMAT ('LIST OF PART NUMBERS COSTING > $100.00',//)
     200 FORMAT (5A4, F8.2, F4.0)
     300 FORMAT (10X,5A4,3X,F10.2,T45,F4.0,3X,F10.2)
     400 FORMAT (//,'TOTAL ITEMS IS: ', I5)
              END
```

FORTRAN has many input/output statements. There are three types: nonexecutable (FORMAT and NAMELIST), sequential (READ, WRITE, PUNCH, PRINT, END FILE, REWIND, BACKSPACE), and direct-access (DEFINE FILE, READ, WRITE, and FIND). There are various requirements for input/output statements.

1. The command itself, which tells what is to be done.

2. The dataset reference number, which defines the dataset. It may be any number from 1 to 99. As a general rule, 5 denominates standard input and 6 denominates printed output.

3. The list of variables, which has the variable names in the order they are to be processed and the quantity of data involved.

The READ statement causes the transfer of information from the dataset into main storage. The form is

```
READ (5,200,END=30) PART, PRICE, QTY
```

where 5 is the dataset reference number, 200 is the line number, END = 30 indicates what statement processing moves to after reaching end of dataset (30 is a line number), and PART, PRICE, and QTY are input variables. It must be noted that the line number points to the pertaining FORMAT statement.

The FORMAT instruction is nonexecutable. It merely specifies the type and quantity of the data named in the main I/O instruction. It is advisable to place them together at the bottom of the program as shown for readability and to be used more than once, as is the case with several I/O statements. The form is

```
200 FORMAT (5A4, F8.2, F4.0)
```

where 200 is the line number of the pertaining READ instruction. The parentheses describe one by one the quantity, type, and format of the input variables being read (in this case). Let's examine each. The first block says that there are five alphanumeric (A) sets of four digits for the variable PART. This means that it contains 20 alphanumeric characters ($5 \times 4 = 20$). The second block says that PRICE is one real integer (F) made up of eight digits, two of which are to the right of the decimal point. Note that the "1" is implied. The third block says that QTY is one real integer (F) made up of four digits, of which none is to the right of the decimal point. Here is the list of format specifications:

```
FORMAT  DESCRIPTION
     A  Literal data.
     D  Real*8 data.
     E  Real*4 data.
     F  Real*4, Real*8, and Real*16 data without decimal exponent.
     G  Real, integer, logical, or complex data.
     H  Used with literal data such as carriage control.
     I  Integer data.
     L  Logical data.
     Q  Real*16 data.
     T  Data position pointer.
     X  Skips characters.
     Z  Hexadecimal data
     /  Skips lines
```

FORTRAN encompasses various functions such as SQRT, which calculates the square root of the variable or expression in the parentheses next to it. Table 1-8 lists the most common built-in functions.

TABLE 1-8 Some Built-in FORTRAN Functions

ABS $(\cdot)$	Returns the absolute value of the expression in parentheses.
TAN $(\cdot)$	Returns the principal (first and fourth quadrant) arctangent of the expression in parentheses. The returned angle is in radians, between $-\pi/2$ and $\pi/2$.
COS $(\cdot)$	Returns the cosine of the expression in parentheses. The expression is taken to be in radians.
EXP $(\cdot)$	Returns the exponential of the expression in parentheses.
INT $(\cdot)$	Returns the largest integer less than the expression in parentheses.
LOG $(\cdot)$	Returns the natural logarithm of the expression in parentheses. The expression should be positive.
LOG10 $(\cdot)$	Returns the logarithm to the base 10 of the expression in parentheses. The expression should be positive.
SIN $(\cdot)$	Returns the sine of the expression in parentheses. The expression is taken to be in radians.
SQRT $(\cdot)$	Returns the positive square root of the expression in parentheses. The expression should not be negative.
MIN $(x_1, x_2, \ldots, x_n)$	Returns the smallest value of $x_1, x_2, \ldots, x_n$; $n \geq 2$.
MAX $(x_1, x_2, \ldots, x_n)$	Returns the largest value of $x_1, x_2, \ldots, x_n$; $n \geq 2$.

1.3.4 Illustrative Example

The following FORTRAN program computes and prints the mean, standard deviation, and range for a given number of light bulbs and their lifetime. The relevant equations used are

$$\text{Mean} = \sum_{i=1}^{N} \frac{x_i}{N} = \text{sum of} \frac{\text{lifetime}}{\text{number of light bulbs}}$$

$$\text{Standard deviation} = \sqrt{\frac{\sum\limits_{i=1}^{N} (x_i - \text{Mean})^2}{N}}$$

Range: $x \geq x_i$ gives the highest limit; $x \leq x_i$ gives the smallest limit.

The following data are used in the program:

Light bulb number:	1	2	3	4	5	6	7	8	9	10
Lifetime (hours):	120	138	265	111	196	310	482	173	201	160

PROGRAM LISTING

```
C    PURPOSE: TO CALCULATE STATISTICAL INFORMATION LIKE
C    THE MEAN, THE STANDARD DEVIATION, AND THE RANGE
C    OF LIFETIMES FOR A GIVEN NUMBER OF LIGHT BULBS
C
C    *******************VARIABLE LIST*********
C    X(I):LIFETIME OF EACH LIGHT BULB
C    I:LIGHT BULB NUMBER
C    N:NUMBER OF LIGHT BULBS
C    SUMX:SUM OF LIFETIMES
C    DIFF:DIFFERENCE BETWEEN EACH LIFETIME AND THE MEAN
C    SUMDIFF:SUM OF THE SQUARES OF DIFF
C    HIGH:HIGHEST LIFETIME
C    LOW:LOWEST LIFETIME
C
C    ****************PROGRAM*********************
     DIMENSION X(20),DIFF(20)
     REAL X,SUMX,MEAN,DIFF,SUMDIFF
     REAL STANDARD DEVIATION
     INTEGER HIGH,LOW
C
C    SET UP INITIAL VALUES AND READ STATEMENTS
C
     I=1
     SUMX=0.0
     PRINT*,'N=?'
     READ*,N
C
     DO 5 I=1,N,1
     PRINT*,'LIFETIME OF LIGHT BULB',I,'='
     READ*,X(I)
   5 CONTINUE
C
C    COMPUTE SUM OF LIFETIMES
C
     DO 10 I=1,N,1
     SUMX=SUMX+X(I)
  10 CONTINUE
C
C    FIND THE VALUE OF THE MEAN
```

(continued)

```
C
      MEAN = SUMX/N
      PRINT*, 'MEAN = ', MEAN
C
C     COMPUTE DIFFERENCE BETWEEN X(I) AND MEAN
C
      DO 15 I = 1, N, 1
      DIFF(I) = X(I) - Mean
  15 CONTINUE
C
C     CALCULATE SUM OF DIFF SQUARED OF ALL LIFETIMES
C
      SUMDIFF = 0.0
      DO 20 I = 1, N, 1
      SUMDIFF = SUMDIFF + DIFF(I)**2
  20 CONTINUE
C
C     COMPUTE STANDARD DEVIATION
C
      STANDARD DEVIATION = (SUMDIFF/N)**.5
C
      PRINT*'STANDARD DEVIATION = ', STANDARD DEVIATION
C
C     FIND HIGHEST VALUE OF LIFETIME
C
      HIGH = 0.0
      DO 50 I = 1, N, 1
      IF(X(I).GT.HIGH) HIGH = X(I)
  50 CONTINUE
      PRINT*, 'HIGH = ', HIGH, 'HOURS'
C
C     FIND LOWEST VALUE OF LIFETIME
C
      LOW = HIGH
      DO 55 I = 1, N, 1
      IF(X(I).LT.LOW) LOW = X(I)
  55 CONTINUE
      PRINT*, 'LOW = ', LOW, 'HOURS'
C
      STOP
      END
```

The program computes and prints

```
      MEAN  =  +215.6
      STANDARD DEVIATION  =  106.73631
      HIGH  =  482 HOURS
      LOW  =  111 HOURS
```

2

SIMULTANEOUS LINEAR ALGEBRAIC EQUATIONS

2.1 PREVIEW

This chapter begins with a review of complex algebra, including Euler's and DeMoivre's theorems. The material is important to those parts of our study that involve complex numbers, beginning with most of the solutions of quadratic equations, given by the quadratic formula.

Sections 2.3 and 2.4 consist of a detailed discussion of the properties and solutions of linear and simultaneous linear algebraic equations. Whereas most of the reader's prior experience has probably been with equal numbers of equations and variables, having a unique solution, that happy circumstance is not necessarily met in many practical applications. We consider carefully the important cases of equation inconsistency, linear dependence, and multiple solutions.

In Section 2.5, numerical solution of simultaneous linear algebraic equations is done using Gauss–Jordan pivoting in such a way that inconsistent and linearly dependent equations are readily identified, as are those variables that can be assigned arbitrary values. When a unique solution to the equations exists, it is found easily. An interactive computer program for performing this solution process is developed and applied to the solution of a structural static force problem.

In the final section of the chapter we discuss other matters relating to the solution of simultaneous linear algebraic equations. When adequate solution accuracy cannot be obtained with the amount of numerical round-off or truncation error of a computer, improved precision results from dealing additionally with the simultaneous linear algebraic equations describing the previous solution's errors. A common iterative approximate solution method that is the basis for ''relaxation'' solutions of partial differential equations is described and illustrated with an electrical circuit solution example.

2.2 COMPLEX ALGEBRA

The variables describing systems are sometimes complex numbers. In this section we review complex algebra, including Euler's relation and logarithms and roots of complex numbers.

2.2.1 Complex Numbers and Operations

A complex number $s = x + iy$ consists of a real part plus $i = \sqrt{-1}$ times an imaginary part:

$$\operatorname{Re}[s] = x \qquad \operatorname{Im}[s] = y$$

The algebra of complex numbers is the same as real number algebra, where i, the imaginary unit, is treated as the constant it is. Powers of i are listed in Table 2-1.

TABLE 2-1   Complex Algebra

POWERS OF THE IMAGINARY UNIT

$$i = \sqrt{-1}$$
$$i^2 = -1$$
$$i^3 = -i$$
$$i^4 = 1$$
$$i^5 = i$$
$$\vdots$$

ALGEBRAIC OPERATIONS

For complex numbers

$$s_1 = x_1 + iy_1 \qquad \text{and} \qquad s_2 = x_2 + iy_2$$

and a real number k, then

(1) $ks_1 = (kx_1) + i(ky_1)$

(2) $s_1^* = x_1 - iy_1$

(3) $s_1 \pm s_2 = (x_1 + x_2) \pm i(y_1 + y_2)$

(4) $s_1 + s_1^* = 2x_1$

(5) $s_1 s_2 = (x_1 + iy_1)(x_2 + iy_2) = (x_1 x_2 - y_1 y_2) + i(x_1 y_2 + x_2 y_1)$

(6) $\dfrac{s_1}{s_2} = \dfrac{x_1 + iy_1}{x_2 + iy_2} = \dfrac{(x_1 + iy_1)(x_2 - iy_2)}{(x_2 + iy_2)(x_2 - iy_2)} = \dfrac{x_1 x_2 + y_1 y_2}{x_2^2 + y_2^2} + i\,\dfrac{x_2 y_1 - x_1 y_2}{x_2^2 + y_2^2}$

A real number k times a complex number gives

$$ks = (kx) + i(ky)$$

The *complex conjugate* of a complex number s is denoted by s^* and is defined to be

$$s^* = (x + iy) = x - iy = x + i(-y)$$

It has the same real part as the original number, but the negative of its imaginary part. The solutions of a second-degree polynomial equation with real coefficients,

$$f(s) = as^2 + bs + c = 0$$

given by the quadratic formula,

$$s_1, s_2 = \frac{-b \pm \sqrt{b^2 - 4ac}}{2a}$$

are complex conjugates if $b^2 - 4ac < 0$.

The sum of two complex numbers is

$$s_1 + s_2 = x_1 + \underline{i}y_1 + x_2 + \underline{i}y_2 = (x_1 + x_2) + \underline{i}(y_1 + y_2)$$

and their difference is

$$s_1 - s_2 = x_1 + \underline{i}y_1 - (x_2 + \underline{i}y_2) = (x_1 - x_2) + \underline{i}(y_1 - y_2)$$

The sum of a complex number and its complex conjugate is twice the real part of either number:

$$s_1 + s_1^* = x_1 + \underline{i}y_1 + x_1 - \underline{i}y_1 = 2x_1 = 2\,\mathrm{Re}[s_1] = 2\,\mathrm{Re}[s_1^*]$$

The product and quotient of two complex numbers have real and imaginary parts that are found using the ordinary rules of algebra. For the product,

$$s_1 s_2 = (x_1 + \underline{i}y_1)(x_2 + \underline{i}y_2) = x_1 x_2 + \underline{i}x_1 y_2 + \underline{i}y_1 x_2 + \underline{i}^2 y_1 y_2$$

$$= (x_1 x_2 - y_1 y_2) + \underline{i}(x_1 y_2 + y_1 x_2)$$

The product of a complex number with its complex conjugate is a real number:

$$s_1 s_1 = (x_1 + \underline{i}y_1)(x_1 - \underline{i}y_1) = x_1^2 - \underline{i}x_1 y_1 + \underline{i}y_1 x_1 - \underline{i}^2 y_1 y_1$$

$$= x_1^2 + y_1^2$$

The quotient of two complex numbers is then

$$\frac{s_1}{s_2} = \frac{x_1 + \underline{i}y_1}{x_2 + \underline{i}y_2} = \frac{(x_1 + \underline{i}y_1)\,(x_2 - \underline{i}y_2)}{(x_2 + \underline{i}y_2)\,(x_2 - \underline{i}y_2)} \tag{2-1}$$

$$= \frac{x_1 x_2 - \underline{i}x_1 y_2 + \underline{i}y_1 x_2 - \underline{i}^2 y_1 y_2}{x_2^2 - \underline{i}x_2 y_2 + \underline{i}y_2 x_2 - \underline{i}^2 y_2^2}$$

$$= \frac{x_1 x_2 + y_1 y_2}{x_2^2 + y_2^2} + \underline{i}\,\frac{x_2 y_1 - x_1 y_2}{x_2^2 + y_2^2}$$

In equation (2-1), the numerator and denominator of the quotient are each multiplied by the complex conjugate of the denominator, resulting in a real denominator.

Complex numbers are represented on the complex plane by coordinates of the number's real and imaginary parts, as in Figure 2-1(a). Often, an arrow is drawn from the origin to the coordinates. The addition of two complex numbers is easy to perform graphically. It is the arrow along the diagonal of the parallelogram having sides s_1 and s_2 to emphasize their vector nature, as in Figure 2-1(b). Subtraction of one complex number from another,

$$s = s_1 - s_2$$

is done graphically by first constructing $-s_2$, then adding s_1 and $(-s_2)$ with a parallelogram, as in Figure 2-1(c).

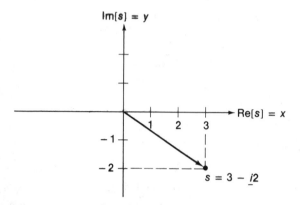

(a) Representation

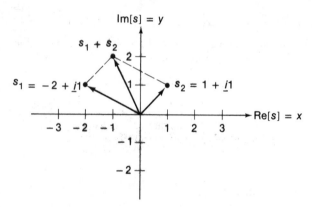

(b) Graphical addition of two complex numbers

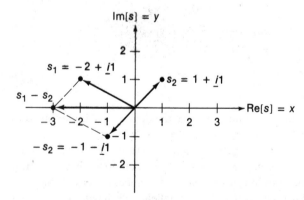

(c) Graphical subtraction of two complex numbers

Figure 2-1   Representing complex numbers on the complex plane.

2.2.2 Euler's Relation

The *magnitude* of a complex number $s = x + \underline{i}y$ is

$$r = |s| = \sqrt{x^2 + y^2}$$

It is the length of the arrow representing the number on the complex plane. The *angle* of a complex number is the counterclockwise angle the arrow makes with the positive real axis:

$$\phi = \angle s = \arctan \frac{y}{x}$$

As the arctangent function is multiple valued, one must know the individual algebraic signs of y and x to fix the quadrant of ϕ. The conversion of a complex number's rectangular coordinates (x, y) to its polar coordinates (r, ϕ) is shown in Figure 2-2(a).

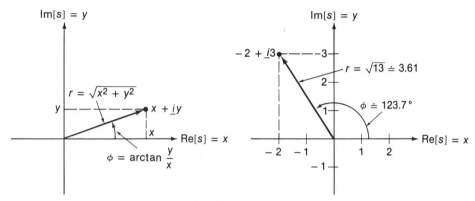

(a) Rectangular-to-polar conversion

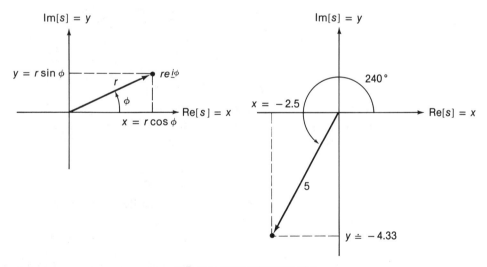

(b) Polar-to-rectangular conversion

Figure 2-2   Polar form of a complex number.

Any complex number can also be expressed in *polar form*,

$$s = x + \underline{i}y = re^{\underline{i}\phi} = r \exp(\underline{i}\phi)$$

Consider the power series for the following functions:

$$e^{\alpha} = \exp(\alpha) = 1 + \frac{\alpha}{1!} + \frac{\alpha^2}{2!} + \frac{\alpha^3}{3!} + \cdots$$

$$\cos \alpha = 1 - \frac{\alpha^2}{2!} + \frac{\alpha^4}{4!} - \frac{\alpha^6}{6!} + \cdots$$

$$\sin \alpha = \frac{\alpha}{1!} - \frac{\alpha^3}{3!} + \frac{\alpha^5}{5!} - \frac{\alpha^7}{7!} + \cdots$$

The angle α must be expressed in *radians* for the series to be in these simple forms. If, in the series $\exp(\alpha)$, α is replaced by $\underline{i}\phi$, there results

$$e^{\underline{i}\phi} = \exp(\underline{i}\phi) = 1 + \frac{\underline{i}\phi}{1!} + \frac{\underline{i}^2\phi^2}{2!} + \frac{\underline{i}^3\phi^3}{3!} + \frac{\underline{i}^4\phi^4}{4!} + \frac{\underline{i}^5\phi^5}{5!} + \cdots$$

$$= \left(1 - \frac{\phi^2}{2!} + \frac{\phi^4}{4!} - \frac{\phi^6}{6!} + \cdots\right)$$

$$+ \underline{i}\left(\frac{\phi}{1!} - \frac{\phi^3}{3!} + \frac{\phi^5}{5!} - \frac{\phi^7}{7!} + \cdots\right)$$

$$= \cos \phi + \underline{i} \sin \phi$$

This result is known as *Euler's relation*.

A complex number can thus be expressed in terms of its magnitude and angle as

$$s = x + \underline{i}y = \sqrt{x^2 + y^2}\left(\frac{x}{\sqrt{x^2 + y^2}} + \underline{i}\,\frac{y}{\sqrt{x^2 + y^2}}\right)$$

$$= r(\cos \phi + \underline{i} \sin \phi) = re^{\underline{i}\phi}$$

For example,

$$s = -2.5 - \underline{i}4.33 = 5e^{\underline{i}240°}$$

as shown in Figure 2-2(b). In practice, one expresses the angle of a complex number in degrees when it is expedient to do so, indicating this with the degree sign.

Using exponential properties,

$$e^a e^b = e^{a+b} \qquad \text{and} \qquad \frac{e^a}{e^b} = e^{a-b}$$

multiplication and division of complex numbers is easily performed in the polar form. The product of two complex numbers is

$$(r_1 e^{\underline{i}\phi_1})(r_2 e^{\underline{i}\phi_2}) = r_1 r_2 e^{\underline{i}(\phi_1 + \phi_2)}$$

It has a magnitude that is the product of the individual magnitudes and angle that is the sum of the individual angles. The quotient of two complex numbers,

$$\frac{r_1 e^{\underline{i}\phi_1}}{r_2 e^{\underline{i}\phi_2}} = \frac{r_1}{r_2} e^{\underline{i}(\phi_1 - \phi_2)}$$

has magnitude that is the quotient of the individual magnitudes and angle that is the numerator angle minus the denominator angle.

2.2.3 Applications of Euler's Relation

Using Euler's relation,

$$e^{ix} = \cos x + i \sin x$$

$$e^{-(ix)} = \cos (-x) + i \sin (-x) = \cos x - i \sin x$$

the following identities result:

$$\cos x = \frac{e^{ix} + e^{-ix}}{2}$$

$$\sin x = \frac{e^{ix} - e^{-ix}}{2i}$$

With these, it is easy to prove various trigonometric identities. For example,

$$\cos (x + y) = \tfrac{1}{2} e^{i(x+y)} + \tfrac{1}{2} e^{-i(x+y)} = \tfrac{1}{2} e^{ix} e^{iy} + \tfrac{1}{2} e^{-ix} e^{-iy}$$

$$= \tfrac{1}{2} (\cos x + i \sin x)(\cos y + i \sin y)$$

$$+ \tfrac{1}{2} (\cos x - i \sin x)(\cos y - i \sin y)$$

$$= \cos x \cos y - \sin x \sin y$$

The natural logarithm ln s of a complex number s is defined by

$$e^{\ln s} = s$$

Expressing s in polar form and ln s in rectangular form,

$$s = re^{i\phi} \qquad \ln s = u + iv$$

then

$$e^{u + iv} = e^u e^{iv} = re^{i\phi}$$

so that

$$u = \ln r$$
$$v = \phi + 2\pi m \qquad m = 0, \pm 1, \pm 2, \ldots$$

and

$$\ln (s = re^{i\phi}) = \ln r + i(\phi + 2\pi m) \qquad m = 0, \pm 1, \pm 2, \ldots$$

the complex logarithm (even of a real number) is multiple valued since any multiple of 2π radians can be added or subtracted from the angle of a complex number.

The square root $\sqrt{s}$ of a complex number s is defined by

$$(\sqrt{s})^2 = s$$

Expressing s and $\sqrt{s}$ in polar form, explicitly noting that the angle of s can include any multiple of 2π radians,

$$s = re^{i(\phi + 2\pi m)} \qquad m = 0, \pm 1, \pm 2, \ldots$$

$$\sqrt{s} = \rho e^{i\sigma}$$

then

$$(\sqrt{s})^2 = \rho^2 e^{i2\sigma} = re^{i(\phi + 2\pi m)} \qquad m = 0, \pm 1, \pm 2, \ldots$$

so that

$$\sqrt{s} = \sqrt{r}\, e^{i[(\phi/2) + \pi m]} \qquad m = 0, \pm 1, \pm 2, \ldots$$

There are two different square roots of s,

$$\sqrt{r}\, e^{i(\phi/2)} \qquad \text{and} \qquad \sqrt{r}\, e^{i[(\phi/2) + \pi]} = -\sqrt{r}\, e^{i(\phi/2)}$$

For example, the two square roots of

$$s = -3 + \underline{i}2 = 3.60 e^{\underline{i}2.55}$$

are

$$\sqrt{s} = 1.90 e^{\underline{i}1.275} \qquad \text{and} \qquad 1.90 e^{\underline{i}4.417}$$

as illustrated in Figure 2-3(a). When a numerical value is given for a quantity, it obviously

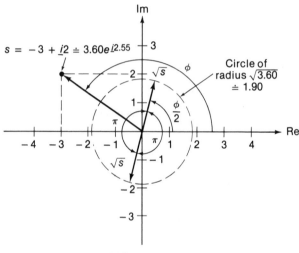

(a) Square root

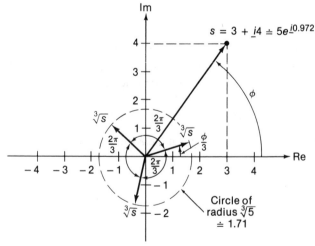

(b) Cube root

Figure 2-3   Roots of complex numbers.

might not be as precise, because of the finite number of digits used to represent it. Under these circumstances and when the numerical representation has been rounded further, the dot over the equal sign will be used.

Similarly, the three cube roots of a complex number s are defined by

$$(\sqrt[3]{s})^3 = \rho^3 e^{i3\sigma} = s = re^{i(\phi + m\,2\pi)} \qquad m = 0, \pm 1, \pm 2, \ldots$$

so

$$\sqrt[3]{s} = \sqrt[3]{r}\, e^{i(\phi/3)} \qquad \sqrt[3]{r}\, e^{i[(\phi/3)+(2\pi/3)]} \qquad \sqrt[3]{r}\, e^{i[(\phi/3)+(4\pi/3)]}$$

The three cube roots are spaced by angles $2\pi/3 = 120°$ on the complex plane. For example, the three cube roots of

$$s = 3 + \underline{i}4 = 5e^{\underline{i}0.927}$$

are

$$\sqrt[3]{s} = 1.71e^{\underline{i}0.309} \qquad 1.71e^{\underline{i}2.403} \qquad 1.71e^{\underline{i}4.498}$$

as shown in Figure 2-3(b).

In general, there are n roots of a complex number. They are separated by angles $2\pi/n$ on the complex plane:

$$\sqrt[n]{s} = \sqrt[n]{re^{i(\phi+m2\pi)}} = \sqrt[n]{r}\, e^{i[(\phi+m2\pi)/n]}$$

$$= \sqrt[n]{r}\, e^{i\frac{\phi}{n}}, \sqrt[n]{r}\, e^{i[(\phi+2\pi)/n]}, \sqrt[n]{r}\, e^{i[(\phi+4\pi)/n]}, \ldots, \sqrt[n]{r}\, e^{i\{[\phi+2(n-1)\pi]/n\}}$$

This result is known as *DeMoivre's theorem*.

2.3 LINEAR ALGEBRAIC EQUATIONS

Most people who are pursuing a scientific/engineering career learn a great deal about linear algebraic equations early in their studies. Often, however, that knowledge is, at first, a scattered collection of results. The purpose of this section is to solidify the reader's understanding of linear algebra and perhaps to extend it a little.

2.3.1 Equations and Properties

A set of m simultaneous linear algebraic equations in the n variables (or unknowns) x_1, $x_2, \ldots, x_n$ has the form

$$
\begin{aligned}
a_{11}x_1 + a_{12}x_2 + \cdots + a_{1n}x_n &= b_1 \\
a_{21}x_1 + a_{22}x_2 + \cdots + a_{2n}x_n &= b_2 \\
&\vdots \\
a_{m1}x_1 + a_{m2}x_2 + \cdots + a_{mn}x_n &= b_m
\end{aligned}
$$

The a's are the coefficients of the equations, and the b's are the equation's knowns. Any combination of the variables that satisfies all m equations is termed a *solution* to the equations.

A set of simultaneous equations can have a unique solution, as does the set

$$
\begin{aligned}
2x_1 - x_2 &= 7 \\
3x_1 + 4x_2 &= -6
\end{aligned}
$$

the solution to which is $x_1 = 2$, $x_2 = -3$. A set of simultaneous equations can have no solutions, as is the case with

$$2x_1 - x_2 = 7$$
$$4x_1 - 2x_2 = 5$$

Or, a set of simultaneous linear algebraic equations can have a whole family of solutions. For example, for the set

$$2x_1 - x_2 + x_3 = 7$$
$$3x_1 + 4x_2 - x_3 = -6$$

$$x_1 = 2 - \tfrac{3}{11} x_3 \qquad x_2 = -3 + \tfrac{5}{11} x_3$$

is a solution for any value of the variable x_3.

An equation

$$a_{11}x_1 + a_{12}x_2 + \cdots + a_{1n}x_n = b_1$$

is *linearly dependent* on some other equations,

$$a_{21}x_1 + a_{22}x_2 + \cdots + a_{2n}x_n = b_2$$
$$a_{31}x_1 + a_{32}x_2 + \cdots + a_{3n}x_n = b_3$$
$$\vdots$$
$$a_{m1}x_1 + a_{m2}x_2 + \cdots + a_{mn}x_n = b_m$$

if it can be expressed as a linear combination of the other equations:

$$a_{11}x_1 + a_{12}x_2 + \cdots + a_{1n}x_n = k_1(a_{21}x_1 + a_{22}x_2 + \cdots + a_{2n}x_n)$$
$$+ k_2(a_{31}x_1 + a_{32}x_2 + \cdots + a_{3n}x_n) + \cdots$$
$$+ k_{m-1}(a_{m1}x_1 + a_{m2}x_2 + \cdots + a_{mn}x_n)$$

and

$$b_1 = k_1 b_2 + k_2 b_3 + \cdots + k_{m-1} b_m$$

for some numbers $k_1, k_2, \ldots, k_{m-1}$. Any linearly dependent equation is redundant and can be deleted from a set of equations without affecting the solution. Indeed, it is highly desirable to make such deletions to simplify the set of equations and to place the set in a nonredundant form, for which further results apply most directly.

An equation

$$a_{11}x_1 + a_{12}x_2 + \cdots + a_{1n}x_n = b_1$$

is *inconsistent* with some other equations,

$$a_{21}x_1 + a_{22}x_2 + \cdots + a_{2n}x_n = b_2$$
$$a_{31}x_1 + a_{32}x_2 + \cdots + a_{3n}x_n = b_3$$
$$\vdots$$
$$a_{m1}x_1 + a_{m2}x_2 + \cdots + a_{mn}x_n = b_m$$

if the unknowns side of that equation can be expressed as a linear combination of the unknown sides of the other equations,

$$a_{11}x_1 + a_{12}x_2 + \cdots + a_{1n}x_n = k_1(a_{21}x_1 + a_{22}x_2 + \cdots + a_{2n}x_n)$$
$$+ k_2(a_{31}x_1 + a_{32}x_2 + \cdots + a_{3n}x_n) + \cdots$$
$$+ k_{m-1}(a_{m1}x_1 + a_{m2}x_2 + \cdots + a_{mn}x_n)$$

but the same linear combination on the knowns side is *not* equal:

$$b_1 \neq k_1 b_2 + k_2 b_3 + \cdots + k_{m-1} b_m$$

The presence of one or more inconsistent equations in a set implies that the set *has no solution*.

A set of homogeneous equations, a set with all the knowns zero,

$$\begin{aligned}
a_{11}x_1 + a_{12}x_2 + \cdots + a_{1n}x_n &= 0 \\
a_{21}x_1 + a_{22}x_2 + \cdots + a_{2n}x_n &= 0 \\
&\vdots \\
a_{m1}x_1 + a_{m2}x_2 + \cdots + a_{mn}x_n &= 0
\end{aligned}$$

cannot be inconsistent since any linear combination of elements on the knowns side always gives zero. Another way of viewing this is that for homogeneous equations,

$$\begin{aligned}
x_1 &= 0 \\
x_2 &= 0 \\
&\vdots \\
x_n &= 0
\end{aligned}$$

is always a solution.

2.3.2 Cramer's Rule

The solution of a set of n independent linear algebraic equations in n variables,

$$\begin{aligned}
a_{11}x_1 + a_{12}x_2 + \cdots + a_{1n}x_n &= b_1 \\
a_{21}x_1 + a_{22}x_2 + \cdots + a_{2n}x_n &= b_2 \\
&\vdots \\
a_{n1}x_1 + a_{n2}x_2 + \cdots + a_{nn}x_n &= b_n
\end{aligned} \tag{2-2}$$

is given by *Cramer's rule*,

$$x_i = \frac{|\mathbf{A}_i|}{|\mathbf{A}|}$$

where the *determinant* of the equations is

$$|\mathbf{A}| = \begin{vmatrix}
a_{11} & a_{12} & \cdots & a_{1n} \\
a_{21} & a_{22} & \cdots & a_{2n} \\
& & \vdots & \\
a_{n1} & a_{n2} & \cdots & a_{nn}
\end{vmatrix}$$

and $|\mathbf{A}_i|$ is the same as $|\mathbf{A}|$ except that the ith column is replaced by the column of knowns:

$$|\mathbf{A}_i| = \begin{vmatrix}
a_{11} & \cdots & b_1 & \cdots & a_{1n} \\
a_{21} & \cdots & b_2 & \cdots & a_{2n} \\
& & \vdots & & \\
a_{n1} & \cdots & b_2 & \cdots & a_{nn}
\end{vmatrix}$$

The *cofactor* of any element of a square array is the determinant of the array formed by deleting the row and column of that element and multiplying by 1 or -1, as given by the "checkerboard" of signs.

$$\begin{vmatrix}
+ & - & + & - & \cdots \\
- & + & - & + & \cdots \\
+ & - & + & - & \cdots \\
& & \vdots & &
\end{vmatrix}$$

For example, the cofactor of the element in row 3, column 2, in

$$\begin{vmatrix} -2 & 3 & 1 \\ 4 & 7 & 5 \\ 9 & -8 & 7 \end{vmatrix}$$

is

$$|C_{32}| = -\begin{vmatrix} -2 & 1 \\ 4 & 5 \end{vmatrix}$$

The value of a 2 × 2 determinant is

$$\begin{vmatrix} a_{11} & a_{12} \\ a_{21} & a_{22} \end{vmatrix} = a_{11}a_{22} - a_{21}a_{12}$$

Values of determinants of larger square arrays can be found by Laplace expansions. A determinant can be Laplace expanded along any row or any column. The value of the determinant is the sum of the products of each element in the row or column times their cofactors. The following is an example of Laplace expansion:

$$\begin{vmatrix} 5 & 3 & -2 & 2 \\ 0 & 6 & 9 & 1 \\ 4 & -3 & 0 & 0 \\ -2 & 2 & 0 & 6 \end{vmatrix} = -2\begin{vmatrix} 0 & 6 & 1 \\ 4 & -3 & 0 \\ -2 & 2 & 6 \end{vmatrix} - 9\begin{vmatrix} 5 & 3 & 2 \\ 4 & -3 & 0 \\ -2 & 2 & 6 \end{vmatrix}$$

$$= -2\left(-6\begin{vmatrix} 4 & 0 \\ -2 & 6 \end{vmatrix} - 3\begin{vmatrix} 0 & 1 \\ -2 & 6 \end{vmatrix} - 2\begin{vmatrix} 0 & 1 \\ 4 & 0 \end{vmatrix} \right)$$

$$- 9\left(-4\begin{vmatrix} 3 & 2 \\ 2 & 6 \end{vmatrix} - 3\begin{vmatrix} 5 & 2 \\ -2 & 6 \end{vmatrix} \right)$$

$$= -2[-6(24) - 3(2) - 2(-4)] - 9[-4(14) - 3(34)]$$
$$= 284 + 1422 = 1706$$

When a unique solution to a set of n linear algebraic equations in n unknowns [equation (2-2)] exists, it is given by Cramer's rule. If a unique solution to the equations does not exist, or if the equations are inconsistent,

$$|\mathbf{A}| = 0$$

and

$$\mathbf{x}_i = \frac{|\mathbf{A}_i|}{|\mathbf{A}|}$$

cannot be found.

For the equations

$$\begin{array}{rcr} 2x_1 + 4x_2 - x_3 &=& 4 \\ x_1 + 0 + x_3 &=& 0 \\ -3x_1 + 2x_2 &=& -5 \end{array}$$

the solution using Cramer's rule is

$$x_1 = \frac{\begin{vmatrix} 4 & 4 & -1 \\ 0 & 0 & 1 \\ -5 & 2 & 0 \end{vmatrix}}{\begin{vmatrix} 2 & 4 & -1 \\ 1 & 0 & 1 \\ -3 & 2 & 0 \end{vmatrix}} = \frac{-28}{-18} \qquad x_2 = \frac{\begin{vmatrix} 2 & 4 & -1 \\ 1 & 0 & 1 \\ -3 & -5 & 0 \end{vmatrix}}{\begin{vmatrix} 2 & 4 & -1 \\ 1 & 0 & 1 \\ -3 & 2 & 0 \end{vmatrix}} = \frac{3}{-18}$$

$$x_3 = \frac{\begin{vmatrix} 2 & 4 & 4 \\ 1 & 0 & 0 \\ -3 & 2 & -5 \end{vmatrix}}{\begin{vmatrix} 2 & 4 & -1 \\ 1 & 0 & 1 \\ -3 & 2 & 0 \end{vmatrix}} = \frac{28}{-18}$$

For the equations,

$$3x_1 - 2x_2 = 1$$
$$-6x_1 + 4x_2 = 0$$

Cramer's rule gives

$$x_1 = \frac{\begin{vmatrix} 1 & -2 \\ 0 & 4 \end{vmatrix}}{\begin{vmatrix} 3 & -2 \\ -6 & 4 \end{vmatrix}} = \frac{4}{0}$$

indicating that a unique solution for x_1 and x_2 does not exist.

Because of its complexity, Cramer's rule is not well suited to computer implementation; Gauss–Jordan pivoting is much more efficient. Nonetheless, Cramer's rule is simple and straightforward to apply in the case of a small number of equations, say two or three. Furthermore, Cramer's rule is a good analytical tool for showing equation properties. For example, Laplace expansion of $|\mathbf{A}_i|$ along the ith column

$$x_i = \frac{|\mathbf{A}_i|}{|\mathbf{A}|} = \frac{|\mathbf{A}_{1i}|}{|\mathbf{A}|} b_1 + \frac{|\mathbf{A}_{2i}|}{|\mathbf{A}|} b_2 + \cdots + \frac{|\mathbf{A}_{ni}|}{|\mathbf{A}|} b_n$$

demonstrates that any solution of n equations in n unknowns is a linear combination of the knowns, $b_1, b_2, \ldots, b_n$.

2.3.3 Derivation of Cramer's Rule

We now derive Cramer's rule. Consider the n equations and n unknowns [equation (2-2)]. Multiplying the first equation by the cofactor $|\mathbf{A}_{11}|$, the second equation by $|\mathbf{A}_{21}|$, the third by $|\mathbf{A}_{31}|$, and so on, gives the set of equations

$$a_{11}\,|\mathbf{A}_{11}|x_1 + a_{12}\,|\mathbf{A}_{11}|x_2 + \cdots + a_{1n}\,|\mathbf{A}_{11}|x_n = |\mathbf{A}_{11}|b_1$$
$$a_{21}\,|\mathbf{A}_{21}|x_1 + a_{22}\,|\mathbf{A}_{21}|x_2 + \cdots + a_{2n}\,|\mathbf{A}_{21}|x_n = |\mathbf{A}_{21}|b_2$$
$$\vdots$$
$$a_{n1}\,|\mathbf{A}_{n1}|x_1 + a_{n2}\,|\mathbf{A}_{n1}|x_2 + \cdots + a_{nn}\,|\mathbf{A}_{n1}|x_n = |\mathbf{A}_{n1}|b_n$$

Adding all these equations gives

$$(a_{11}\,|\mathbf{A}_{11}| + a_{21}\,|\mathbf{A}_{21}| + \cdots + a_{n1}\,|\mathbf{A}_{n1}|)x_1$$
$$+ (a_{12}\,|\mathbf{A}_{11}| + a_{22}\,|\mathbf{A}_{21}| + \cdots + a_{n2}\,|\mathbf{A}_{n1}|)x_2$$
$$+ \cdots + (a_{1n}\,|\mathbf{A}_{11}| + a_{2n}\,|\mathbf{A}_{21}| + \cdots + a_{nn}\,|\mathbf{A}_{n1}|)x_n \tag{2-3}$$
$$= |\mathbf{A}_{11}|b_1 + |\mathbf{A}_{21}|b_2 + \cdots + |\mathbf{A}_{n1}|b_n$$

The first expression in parentheses is the determinant of the array:

$$a_{11}\,|\mathbf{A}_{11}| + a_{21}\,|\mathbf{A}_{21}| + \cdots + a_{n1}\,|\mathbf{A}_{n1}|$$

Each of the other expressions in parentheses except the last is zero because each is the Laplace expansion of a determinant with two identical columns. Any determinant having two or more proportional rows or two or more proportional columns is zero. The coefficient of x_2 is, for instance, the Laplace expansion

$$\begin{vmatrix} a_{12} & a_{12} & a_{13} & \cdots & a_{1n} \\ a_{22} & a_{22} & a_{23} & \cdots & a_{2n} \\ \vdots & \vdots & \vdots & \vdots & \vdots \\ a_{n2} & a_{n2} & a_{n3} & \cdots & a_{nn} \end{vmatrix} = a_{12}\,|\mathbf{A}_{11}| + a_{22}\,|\mathbf{A}_{21}| + \cdots + a_{n2}\,|\mathbf{A}_{n1}| = 0$$

There then results

$$|\mathbf{A}|\,x_1 = |\mathbf{A}_{11}|b_1 + |\mathbf{A}_{21}|b_2 + \cdots + |\mathbf{A}_{n1}|b_n = \begin{vmatrix} b_1 & a_{12} & a_{13} & \cdots & a_{1n} \\ b_2 & a_{22} & a_{23} & \cdots & a_{2n} \\ \vdots & \vdots & \vdots & \vdots & \vdots \\ b_n & a_{n2} & a_{n3} & \cdots & a_{nn} \end{vmatrix}$$

or

$$x_1 = \frac{|\mathbf{A}_1|}{|\mathbf{A}|}$$

provided that

$$|\mathbf{A}| \neq 0$$

In a similar way, the general result

$$x_i = \frac{|\mathbf{A}_i|}{|\mathbf{A}|}$$

holds provided that the determinant of the coefficient array is nonzero.

2.3.4 Complex Equations

When the coefficients and knowns of a set of m linear algebraic equations,

$$a_{11}s_1 + a_{12}s_2 + \cdots + a_{1n}s_n = b_1$$
$$a_{21}s_1 + a_{22}s_2 + \cdots + a_{2n}s_n = b_2$$
$$\vdots$$
$$a_{m1}s_1 + a_{m2}s_2 + \cdots + a_{mn}s_n = b_m$$

are complex numbers, solutions for the variables $s_1, s_2, \ldots, s_n$ are generally complex numbers. Separating the real and imaginary parts of each coefficient,

$$a_{ij} = \alpha_{ij} + i\beta_{ij} \qquad i = 1, 2, \ldots, m; \quad j = 1, 2, \ldots, n$$

each known,

$$b_i = \chi_i + i\delta_i \qquad i = 1, 2, \ldots, m$$

and each unknown,

$$s_j = x_j + iy_j \qquad j = 1, 2, \ldots, n$$

these become

$$(\alpha_{11} + i\beta_{11})(x_1 + iy_1) + (\alpha_{12} + i\beta_{12})(x_2 + iy_2) + \cdots + (\alpha_{1n} + i\beta_{1n})(x_n + iy_n) = \chi_1 + i\delta_1$$
$$(\alpha_{21} + i\beta_{21})(x_1 + iy_1) + (\alpha_{22} + i\beta_{22})(x_2 + iy_2) + \cdots + (\alpha_{2n} + i\beta_{2n})(x_n + iy_n) = \chi_2 + i\delta_2$$
$$\vdots$$
$$(\alpha_{m1} + i\beta_{m1})(x_1 + iy_1) + (\alpha_{m2} + i\beta_{m2})(x_2 + iy_2) + \cdots + (\alpha_{mn} + i\beta_{mn})(x_n + iy_n) = \chi_m + i\delta_m$$

Equating real and imaginary parts, there results the equivalent set of $2m$ equations with real coefficients and knowns in the $2n$ real unknowns, x and y:

$$\alpha_{11}x_1 - \beta_{11}y_1 + \alpha_{12}x_2 - \beta_{12}y_2 + \cdots + \alpha_{1n}x_n - \beta_{1n}y_n = \chi_1$$
$$\beta_{11}x_1 + \alpha_{11}y_1 + \beta_{12}x_2 + \alpha_{12}y_2 + \cdots + \beta_{1n}x_n + \alpha_{1n}y_n = \delta_1$$
$$\alpha_{21}x_1 - \beta_{21}y_1 + \alpha_{22}x_2 - \beta_{22}y_2 + \cdots + \alpha_{2n}x_n - \beta_{2n}y_n = \chi_2$$
$$\beta_{21}x_1 + \alpha_{21}y_1 + \beta_{22}x_2 + \alpha_{22}y_2 + \cdots + \beta_{2n}x_n + \alpha_{2n}y_n = \delta_2$$
$$\vdots$$
$$\alpha_{m1}x_1 - \beta_{m1}y_1 + \alpha_{m2}x_2 - \beta_{m2}y_2 + \cdots + \alpha_{mn}x_n - \beta_{mn}y_n = \chi_m$$
$$\beta_{m1}x_1 + \alpha_{m1}y_1 + \beta_{m2}x_2 + \alpha_{m2}y_2 + \cdots + \beta_{mn}x_n + \alpha_{mn}y_n = \delta_m$$

2.4 GAUSS–JORDAN PIVOTING

An especially important method of finding solutions of simultaneous linear algebraic equations, if they exist, involves replacing the original set of equations with a new set that has the same solution but is simpler than the original set. One simplification method, called *back substitution*, is to solve one of the equations for one of the variables, then substitute for that variable in the other equations. For example, for the equations

$$2x_1 + x_2 - x_3 = 5$$
$$x_1 - x_2 + x_3 = 0$$
$$3x_1 + 2x_2 + 2x_3 = -3$$

solving the first equation for x_1,

$$x_1 = \tfrac{5}{2} - \tfrac{1}{2}x_2 + \tfrac{1}{2}x_3$$

Substituting for x_1 in the second and third of these equations gives

$$\frac{5}{2} - \frac{1}{2}x_2 + \frac{1}{2}x_3 - x_2 + x_3 = 0 \qquad -\frac{3}{2}x_2 + \frac{3}{2}x_3 = -\frac{5}{2}$$

$$\frac{15}{2} - \frac{3}{2}x_2 + \frac{3}{2}x_3 + 2x_2 + 2x_3 = -3 \qquad \frac{1}{2}x_2 + \frac{7}{2}x_3 = -\frac{21}{2}$$

An equivalent set of equations, one that has the same solution(s) as the original set, consists of the original first equation and the new second and third equations:

$$2x_1 + x_2 - x_3 = 5$$

$$-\frac{3}{2}x_2 + \frac{3}{2}x_3 = -\frac{5}{2}$$

$$\frac{1}{2}x_2 + \frac{7}{2}x_3 = -\frac{21}{2}$$

These are simpler than the original equations because they have more coefficients that are zero. In a similar way, the new second equation can be solved for x_2 and the variable x_2 eliminated from the first and third equations, and so on.

Gauss–Jordan pivoting is a more efficient way of obtaining results similar to those with back substitution by combining equations.

2.4.1 Equivalent Equations

Any equation in a set can be replaced by the sum of that equation and a constant times any other equation in the set. The new set of equations is equivalent to the original set; that is, it has the same solution(s). For example, the sets

$$x_1 + 2x_2 - x_3 = 5$$
$$3x_1 - x_2 + 4x_3 = -2$$
$$2x_1 + x_2 = 6$$

and

$$7x_1 + 7x_3 = 1$$
$$3x_1 - x_2 + 4x_3 = -2$$
$$2x_1 + x_2 = 6$$

are equivalent. Twice the second equation was added to the first equation to obtain the new first equation. The second set of equations is simpler than the first because it has more coefficients that are zero.

Gauss–Jordan pivoting is a systematic procedure for obtaining a simplest set of equivalent algebraic equations. Multiples of one equation in a set are added to every other equation in the set to make coefficients of the first variable in the other equations zero. The equation used to eliminate terms in the other equations is called the *pivot equation* for this set of operations, a *pivot cycle*.

Then another equation (the next pivot equation) is selected and multiples of that equation are added to every other equation to make the coefficients of the second variable in all other equations zero. The second pivot cycle is then complete. The process is continued until there are no equations left that have not been used as pivot equations or until each variable has been eliminated from all but one equation, or both. For example, with the equations

$$x_1 + 2x_2 = -5$$
$$-3x_1 + 4x_2 = 6$$

adding three times the first equation to the second equation gives

$$x_1 + 2x_2 = -5$$
$$10x_2 = -9$$

The variable x_1 now appears in only one equation in the set. The first pivot cycle is now complete. Normalizing the second equation by dividing it by 10, the coefficient of the variable x_2 gives the equivalent set

$$x_1 + 2x_2 = -5$$
$$x_2 = -\frac{9}{10}$$

Adding -2 times the second equation from the first eliminates the x_2 variable from all but the second equation and gives the *fully pivoted equations*

$$x_1 = -\frac{16}{5}$$
$$x_2 = -\frac{9}{10}$$

For the equations

$$2x_1 + 2x_2 \qquad\quad + 4x_4 = 2$$
$$-3x_1 \qquad\quad + 24x_3 + 6x_4 = 30$$
$$x_1 - x_2 - 20x_3 - 8x_4 = -25$$
$$-2x_1 - 5x_2 - 23x_3 - 14x_4 = -34$$

a step-by-step set of pivoting operations are as follows. Dividing the first equation by 2, the coefficient of x_1, gives

$$x_1 + x_2 \qquad\quad + 2x_4 = 1$$
$$-3x_1 \qquad\quad + 24x_3 + 6x_4 = 30$$
$$x_1 - x_2 - 20x_3 - 8x_4 = -25$$
$$-2x_1 - 5x_2 - 23x_3 - 14x_4 = -34$$

Adding three times the first equation to the second equation, -1 times the first equation to the third equation, and 2 times the first equation to the fourth equation gives

$$x_1 + x_2 \qquad\quad + 2x_4 = 1$$
$$3x_2 + 24x_3 + 12x_4 = 33$$
$$-2x_2 - 20x_3 - 10x_4 = -26$$
$$-3x_2 - 23x_3 - 10x_4 = -32$$

The first pivot cycle is complete. One variable, x_1, has been eliminated from all but one equation, the first equation.

Dividing the second equation by 3, the coefficient of x_2, gives

$$x_1 + x_2 \qquad\quad + 2x_4 = 1$$
$$x_2 + 8x_3 + 4x_4 = 11$$
$$-2x_2 - 20x_3 - 10x_4 = -26$$
$$-3x_2 - 23x_3 - 10x_4 = -32$$

Adding -1 times the second equation to the first equation, 2 times the second equation to the third equation, and 3 times the second equation to the fourth equation completes the second pivot cycle:

$$
\begin{aligned}
x_1 \quad\quad - 8x_3 - 2x_4 &= -10 \\
x_2 + \quad 8x_3 + 4x_4 &= 11 \\
-4x_3 - 2x_4 &= -4 \\
x_3 + 2x_4 &= 1
\end{aligned}
$$

Dividing the third equation by -4, the coefficient of x_3, gives

$$
\begin{aligned}
x_1 - 8x_3 - 2x_4 &= -10 \\
x_2 + 8x_3 + 4x_4 &= 11 \\
x_3 + \tfrac{1}{2}x_4 &= 1 \\
x_3 + 2x_4 &= 1
\end{aligned}
$$

Adding 8 times the third equation to the first equation, -8 times the third equation to the second equation, and -1 times the third equation to the fourth equation eliminates x_3 from all but the third equation and completes the third pivot cycle:

$$
\begin{aligned}
x_1 + \quad\quad 2x_4 &= -2 \\
x_2 \quad\quad\quad &= 3 \\
x_3 + \tfrac{1}{2}x_4 &= 1 \\
\tfrac{3}{2}x_4 &= 0
\end{aligned}
$$

Finally, dividing the fourth equation by 3/2, the coefficient of x_4, gives

$$
\begin{aligned}
x_1 + \quad\quad 2x_4 &= -2 \\
x_2 \quad\quad\quad &= 3 \\
+ x_3 + \tfrac{1}{2}x_4 &= 1 \\
x_4 &= 0
\end{aligned}
$$

Adding -2 times the fourth equation to the first equation, and $-1/2$ times the fourth equation to the third equation, results in the fully pivoted equivalent set of equations:

$$
\begin{aligned}
x_1 \quad\quad\quad\quad &= -2 \\
x_2 \quad\quad\quad &= 3 \\
x_3 \quad &= 1 \\
x_4 &= 0
\end{aligned}
$$

which is the solution to the original equations.

2.4.2 Linear Dependence

A linearly dependent equation is indicated by an equation with an entire row of zero coefficients and a zero known after Gauss–Jordan pivoting. The corresponding equation in the original set (or any equivalent set) is redundant and can be deleted. For example, the equations

$$2x_1 + x_2 - 2x_3 = 4$$
$$-x_1 + 3x_2 + x_3 = 3$$
$$3x_1 + 5x_2 - 3x_3 = 11$$

are linearly dependent since the third equation equals twice the first equation plus the second equation. Results of each Gauss–Jordan pivoting cycle are as follows:

$$x_1 + \tfrac{1}{2}x_2 - x_3 = 2$$
$$\tfrac{7}{2}x_2 = 5$$
$$\tfrac{7}{2}x_2 = 5$$

$$x_1 - x_3 = \tfrac{9}{7}$$
$$x_2 = \tfrac{10}{7}$$
$$0 = 0$$

The last equation above indicates that the third equation in the original set is linearly dependent on the other equations in the set.

In this example, any one of the three equations is linearly dependent on the other two equations, so any one of the three original equations could be considered to be the redundant one. Which equation is indicated as the linearly dependent one after Gauss–Jordan pivoting depends on the choice of pivot elements and order of pivoting.

If there is a linearly dependent equation in a set, it is not safe, however, to delete just any equation in the set. For example, in the set

$$4x_1 - 2x_2 + x_3 = 5$$
$$8x_1 - 4x_2 + 2x_3 = 10$$
$$x_1 + x_2 = -3$$

the first two equations are linearly dependent and either can be deleted from the set, but the third equation must be retained.

2.4.3 Inconsistency

Inconsistent equations are indicated by an equation with an entire row of zero coefficients and a nonzero known in the Gauss–Jordan pivoted equations. As in the case of linearly dependent equations, which equation is indicated as being inconsistent with the others is dependent on the ordering of the pivoting. It is possible, however, that some of the equations in a set are perfectly consistent, while others are inconsistent, as in the set

$$x_1 = 2$$
$$2x_2 + x_3 = 5$$
$$-4x_2 - 2x_3 = 6$$

where the first equation is consistent with the others but the second and third equations are inconsistent with one another.

As another example, the equations

$$2x_1 + x_2 - 2x_3 = 4$$
$$-x_1 + 3x_2 + x_3 = 3$$
$$3x_1 + 5x_2 - 3x_3 = 8$$

are inconsistent. Results of each Gauss–Jordan pivot cycle are as follows:

$$x_1 + \tfrac{1}{2}x_2 - x_3 = 2$$
$$\tfrac{7}{2}x_2 \qquad = 5$$
$$\tfrac{7}{2}x_2 \qquad = 2$$

$$x_1 \qquad - x_3 = \tfrac{9}{7}$$
$$x_2 \qquad = \tfrac{10}{7}$$
$$0 = 2$$

The last equation in the fully pivoted set indicates that the last equation in the original set is inconsistent with the other equations. That is, if the last equation were deleted from the set, the remaining equation would be consistent. As consistency is a property of a set of equations, a different pivoting order would likely identify one of the other equations as being inconsistent with the rest.

2.4.4 Underdetermined Solutions

Suppose that the equations comprising a set are consistent and that any linearly dependent equations have been deleted so that the set consists entirely of linearly independent equations. If there are n variables, the set cannot consist of more than n linearly independent equations. If there were more than n equations, those in excess of n equations could be shown, using Gauss–Jordan pivoting, to be linearly dependent on the others. If there are exactly n consistent, linearly independent equations in the set, there is a *unique* solution to the set, which can be found by Gauss–Jordan pivoting.

If there are less than n consistent, linearly independent equations in the set, the set has a whole spectrum of different solutions. If there are $(n - p)$ linearly independent equations in the set, the solutions may be found by assigning arbitrary values to p of the variables and solving for the resulting $(n - p)$ variables. For example, the set of equations

$$x_1 + 2x_2 + 3x_3 = -4$$
$$-2x_1 + x_2 - x_3 = 5$$

can be rewritten as

$$x_1 + 3x_3 = -4 - 2x_2$$
$$-2x_1 - x_3 = 5 - x_2$$

Assigning arbitrary values to x_2 (in this case x_1 or x_3 could just as well have been chosen) gives the following as some of the solutions:

$$x_1 = -\frac{11}{5}$$

$$x_2 = 0$$

$$x_3 = -\frac{3}{5}$$

$$x_1 = -\frac{6}{5}$$

$$x_2 = 1$$

$$x_3 = -\frac{8}{5}$$

$$x_1 = -\frac{21}{5}$$

$$x_2 = -2$$

$$x_3 = \frac{7}{5}$$

and so on.

For the equations

$$
\begin{aligned}
x_1 + x_2 &= 2 \\
x_1 - x_2 &= -3 \\
x_2 - x_3 + x_4 &= 0
\end{aligned}
$$

the variables x_1 and x_2 are determined by the first two equations, so only variables x_3 or x_4 can be assigned arbitrary values in this case. Equivalent equations after Gauss–Jordan pivoting are as follows:

$$
\begin{aligned}
x_1 &= -\frac{1}{2} \\
x_2 &= \frac{5}{2} \\
-x_3 + x_4 &= -\frac{5}{2}
\end{aligned}
$$

2.5 NUMERICAL EQUATION SOLUTION

We now develop a digital computer program in BASIC/FORTRAN for fully pivoting any set of m linear algebraic equations in n variables. When the equations are consistent and there is an equal number of linearly independent equations and variables, the unique solution to the equations is computed, and any linearly dependent equations are identified by a

$$0 = 0$$

equivalent equation in the fully pivoted set. When there are inconsistent equations, these are identified by equivalent

$$0 = \text{(some nonzero number)}$$

in the fully pivoted set. When the set does not have a unique solution, the equivalent fully pivoted equations are in a form in which it is easy to solve for some of the variables in terms of the others.

To make the program design as simple as possible at this early stage, we first develop a simple program for entering and printing the coefficients of a set of linear algebraic equations. After testing, this initial program is augmented with the additional instructions needed for full Gauss–Jordan pivoting. Unlike other commonly used programs, the one we develop does not change the order of the equations in obtaining the equivalent fully pivoted set. The user can then, by inspection, determine which equations are linearly dependent on and/or inconsistent with the others. This Gauss–Jordan pivoting program easily and automatically solves the equations when they have a unique solution, but its capabilities go beyond that special case.

2.5.1 Equation Entry and Printing

Table 2-2 lists a program in BASIC/FORTRAN called ENTER that prompts the user to enter the number of equations M, the number of variables N, the equation coefficients, and the knowns. The coefficients and knowns are arranged in the two-dimensional array $A(\cdot, \cdot)$ so that the equations are

$$
\begin{aligned}
A(1,1)x_1 + A(1,2)x_2 + \cdots + A(1,N)x_N &= A(1,N+1) \\
A(2,1)x_1 + A(2,2)x_2 + \cdots + A(2,N)x_N &= A(2,N+1) \\
&\vdots \\
A(M,1)x_1 + A(M,2)x_2 + \cdots + A(M,N)x_N &= A(M,N+1)
\end{aligned}
$$

TABLE 2-2

Computer Program in BASIC for Linear Algebraic Equation Entry and Printing (ENTER)

VARIABLES USED

M = number of equations

N = number of variables

$A(20,21)$ = equation coefficients (including the knowns as variable number $N+1$) as a function of the equation number and the variable number

I = equation number index

J = variable number index

LISTING

```
100     PRINT "EQUATION ENTRY AND PRINTING"
110     DIM A(20,21)
120     PRINT "PLEASE ENTER NO. OF EQUATIONS";
130     INPUT M
140     PRINT "PLEASE ENTER NO. OF VARIABLES";
150     INPUT N
160   REM ENTER EQUATIONS
170     FOR I = 1 TO M
180     PRINT "ENTER COEFFICIENTS OF EQUATION NO.";I
190     FOR J = 1 TO N
200     PRINT "COEFFICIENT OF VARIABLE NO.";J;
210     INPUT A(I,J)
```

```
220      NEXT J
230      PRINT "KNOWN FOR EQUATION";I;
240      INPUT A(I,N + 1)
250      NEXT I
260   REM PRINT EQUATIONS
270      PRINT "EQUATIONS ENTERED ARE"
280      GOSUB 1000
290      END
1000  REM SUBROUTINE TO PRINT ARRAY A(M,N)
1010  REM ROW BY ROW
1020     FOR I = 1 TO M
1030     PRINT "COEFFICIENTS OF EQUATION ";I
1040     FOR J = 1 TO N
1050     PRINT A(I,J)
1060     NEXT J
1070     PRINT "KNOWN IS ";A(I,N + 1)
1080     NEXT I
1090     RETURN
1100     PRINT A(I,N + 1)
1110     NEXT I
1120     RETURN
```

Computer Program in FORTRAN for Linear Algebraic Equation Entry and Printing (ENTER)

```
C ********************************************************************
C PROGRAM ENTER -- EQUATION ENTRY AND PRINTING
C ********************************************************************
      REAL A(20,21)

      WRITE(*,'(1X,''EQUATION ENTRY AND PRINTING'')')
      WRITE(*,'(1X,''PLEASE ENTER NO. OF EQUATIONS '',\)')
      READ(*,*) M
      WRITE(*,'(1X,''PLEASE ENTER NO. OF VARIABLES '',\)')
      READ(*,*) N

C *** ENTER EQUATIONS
      DO 10 I=1, M
         WRITE(*,'(1X,''ENTER COEFFICIENTS OF EQUATION NO.'',I3)') I
         DO 20 J=1, N
            WRITE(*,'(1X,''COEFFICIENT OF VARIABLE NO.'',
     *                I3,'' = '',\)') J
            READ(*,*) A(I,J)
 20      CONTINUE
         WRITE(*,'(1X,''KNOWN FOR EQUATION'',I3,'' = '',\)') I
         READ(*,*) A(I,N+1)
 10   CONTINUE

C *** PRINT EQUATIONS
      WRITE(*,'(1X,''EQUATIONS ENTERED ARE'')')
      CALL PRTARR(A,M,N)
      STOP
      END

C ********************************************************************
      SUBROUTINE PRTARR(A,M,N)
C ********************************************************************
      REAL A(20,21)

C *** SUBROUTINE TO PRINT ARRAY A(M,N)
C *** ROW BY ROW
      DO 10 I=1, M
         WRITE(*,'(1X,''COEFFICIENTS OF EQUATION'',I3,)') I
         DO 20 J=1, N
```

(continued)

```
         WRITE(*,'(1X,E15.8)') A(I,J)
20       CONTINUE
         WRITE(*,'(1X,''KNOWN IS '',E15.8)') A(I,N+1)
10   CONTINUE
     RETURN
     END
```

This program then prints the equations entered by listing the array **A** row by row. For arrays having more rows than can be printed side by side by the user (sometimes as few as two), the row-by-row printing is much easier to read than other simple formats. The printing is arranged as a subroutine because we will later reuse these steps to print more than a single array.

The dimension statement used,

$$110 \text{ DIM}(20,21)$$

limits the number of equations and variables to 20 or less, but this is easily increased in the event that the user wishes to enter more than this number of equations and/or variables. Many of the programs listed in this book will be listed with relatively large but arbitrary dimensions which are, similarly, simply increased if need be.

The equations

$$-3x_1 + x_2 = 0$$
$$-2x_1 + 4x_2 = 5$$
$$6x_1 - 7x_2 = 0$$

when entered in the ENTER program give the following printing:

```
EQUATION ENTRY AND PRINTING          KNOWN FOR EQUATION3?0
PLEASE ENTER NO. OF EQUATIONS?3      EQUATIONS ENTERED ARE
PLEASE ENTER NO. OF VARIABLES?2      COEFFICIENTS OF EQUATION 1
ENTER COEFFICIENTS OF EQUATION NO.    -3
1                                     1
COEFFICIENT OF VARIABLE NO. 1?-3     KNOWN IS 0
COEFFICIENT OF VARIABLE NO. 2?1      COEFFICIENTS OF EQUATION 2
KNOWN FOR EQUATION1?0                 -2
ENTER COEFFICIENTS OF EQUATION NO.    4
2                                    KNOWN IS 5
COEFFICIENT OF VARIABLE NO. 1?-2     COEFFICIENTS OF EQUATION 3
COEFFICIENT OF VARIABLE NO. 2?4      6
KNOWN FOR EQUATION2?5                 -7
ENTER COEFFICIENTS OF EQUATION NO.   KNOWN IS 0
3
COEFFICIENT OF VARIABLE NO. 1?6
COEFFICIENT OF VARIABLE NO. 2?-7
```

2.5.2 Gauss–Jordan Pivoting Program

An algorithm for performing Gauss–Jordan pivoting is given in Table 2-3. Algorithms are computational recipes for performing desired functions. There are many ways of stating algorithms, and we will use several of the best ones during this course of study. We will try to use the mode of expression that is best suited to understanding each case. This algorithm is expressed as a series of steps.

TABLE 2-3 Algorithm for Gauss–Jordan Pivoting

DEFINITION

The original equations are

$$
\begin{aligned}
a_{11}x_1 + a_{12}x_2 + \cdots + a_{1n}x_n &= a_{1(n+1)} \\
a_{21}x_1 + a_{22}x_2 + \cdots + a_{2n}x_n &= a_{2(n+1)} \\
&\vdots \\
a_{m1}x_1 + a_{m2}x_2 + \cdots + a_{mn}x_n &= a_{m(n+1)}
\end{aligned}
$$

ALGORITHM

For the ith equation, beginning with $i = 1$ and ending with $i = m$ (lines 290–300 and 520):

1. Find the coefficient in equation i with the largest square. Set p to the value of that coefficient and set q to the variable number of that coefficient. If all coefficients are zero, go to the next equation (lines 310–390).

2. Normalize equation i by dividing each coefficient and the known in that equation by the coefficient a_{iq} of the pivot variable:

$$
a_{ij} \leftarrow \frac{a_{ij}}{p} \qquad j = 1, 2, \ldots, n + 1
$$

   The number $a_{i,n+1}$ is the known (lines 400–430).

3. Pivot on coefficient $a_{iq} = 1$. For the kth equation, beginning with $k = 1$, skipping $k = i$, and ending with $k = m$:

   a. Set p to the coefficient of the qth variable in equation k:

$$
p \leftarrow a_{kq}
$$

   b. Subtract p times each coefficient and the known from the corresponding coefficient or known of the kth equation:

$$
a_{kj} \leftarrow a_{kj} - p*a_{ij} \qquad j = 1, 2, \ldots, n + 1
$$

   Since $a_{iq} = 1$, the coefficient of the qth variable in equation k is $a_{kq} - a_{kq} = 0$ (lines 440–510).

Notation of the form

$$
b \leftarrow a
$$

means "replace b by a." The variable on the left is to be given the value of the variable or expression on the right. Because many keyboards lack the arrow symbol, this assignment of a value to a variable is performed by a statement such as

$$
b = a
$$

in the BASIC language.

From a numerical standpoint, one does not want to pivot on excessively small coefficients, so the largest coefficient in each equation is chosen for pivoting at the beginning of each pivot cycle. The equations are used from first to last, and a variable previously used for pivoting will not be used again later because, after the first pivoting, all coefficients of that variable in later equations are zero. In this table, the line numbers of the relevant steps in the corresponding computer program are given in parentheses, for convenience.

A computer program in BASIC/FORTRAN, called PIVOT, that uses this algorithm is listed in Table 2-4. The equation entry and array printing steps are identical to those of the ENTRY program, so that it is easier for the reader to concentrate on the new steps involved for the pivoting, lines 290–560. The line

```
390   IF P*P < 1E-10 THEN GOTO 520
```

is especially important because it prevents pivoting on an extremely small coefficient, which may actually be zero, although small computational inaccuracies have caused its numerical value to be small but nonzero. The choice of 1E-10 as "too small" is, of course, arbitrary. This choice may need to be adjusted for the pivoting of equations with very small or very large coefficients. If the limit chosen is not small enough, the program simply does not pivot on an element it should pivot on, an event that is easily noted by the user.

TABLE 2-4

Computer Program in BASIC for Gauss–Jordan Pivoting (PIVOT)

VARIABLES USED

M = number of equations

N = number of variables

$A(20,21)$ = equation coefficients (including the knowns as variable number $N+1$) as a function of the equation number and the variable number

P = Value of potential pivot coefficient in pivot equation when finding pivot and normalizing the pivot equation; value of pivot coefficient in other equations when pivoting

Q = variable number of pivot coefficient

I, K = equation number indices

J = variable number index

LISTING

```
100     PRINT "GAUSS-JORDAN PIVOTING"
110     DIM A(20,21)
120     PRINT "PLEASE ENTER NO. OF EQUATIONS";
130     INPUT M
140     PRINT "PLEASE ENTER NO. OF VARIABLES";
150     INPUT N
160   REM ENTER EQUATIONS
170     FOR I = 1 TO M
180     PRINT "ENTER COEFFICIENTS OF EQUATION NO."; I
190     FOR J = 1 TO N
200     PRINT "COEFFICIENT OF VARIABLE NO.";J;
210     INPUT A(I,J)
220     NEXT J
230     PRINT "KNOWN FOR EQUATION";I;
240     INPUT A(I,N+1)
250     NEXT I
260   REM PRINT EQUATIONS
270     PRINT "EQUATIONS ENTERED ARE"
280     GOSUB 1000
290   REM PIVOT EQUATIONS FROM FIRST TO LAST
300     FOR I = 1 TO M
310   REM FIND COEFFICIENT IN EQUATION I
320   REM WITH LARGEST SQUARE
```

```
330      P = 0
340      Q = 0
350      FOR J = 1 TO N
360      IF A(I,J) * A(I,J) > P * P THEN Q = J
370      IF A(I,J) * A(I,J) > P * P THEN P = A(I,J)
380      NEXT J
390      IF P * P < 1E-10 THEN GOTO 520
400    REM NORMALIZE EQUATION I
410      FOR J = 1 TO N + 1
420      A(I,J) = A(I,J) / P
430      NEXT J
440    REM PIVOT ON COEFFICIENT NO. Q IN EQUATION I
450      FOR K = 1 TO M
460      IF K = I THEN GOTO 510
470      P = A(K,Q)
480      FOR J = 1 TO N + 1
490      A(K,J) = A(K,J)- P * A(I,J)
500      NEXT J
510      NEXT K
520      NEXT I
530    REM    PRINT PIVOTED EQUATIONS
540      PRINT "FULLY PIVOTED EQUATIONS ARE"
550      GOSUB 1000
560      END
1000   REM    SUBROUTINE TO PRINT ARRAY A(M,N)
1010   REM    ROW BY ROW
1020     FOR I = 1 TO M
1030     PRINT "COEFFICIENTS OF EQUATION";I
1040     FOR J = 1 TO N
1050     PRINT A(I,J)
1060     NEXT J
1070     PRINT "KNOWN IS ";A(I,N + 1)
1080     NEXT I
1090     RETURN
1100     PRINT A(I,N + 1)
1110     NEXT I
1120     RETURN
```

Computer Program in FORTRAN for Gauss–Jordan Pivoting (PIVOT)

```
C ********************************************************************
C PROGRAM PIVOT -- GAUSS-JORDAN PIVOTING
C ********************************************************************
      REAL A(20,21)

      WRITE(*,'(1X,''GAUSS-JORDAN PIVOTING'')')
      WRITE(*,'(1X,''PLEASE ENTER NO. OF EQUATIONS '',\)')
      READ(*,*) M
      WRITE(*,'(1X,''PLEASE ENTER NO. OF VARIABLES '',\)')
      READ(*,*) N

C *** ENTER EQUATIONS
      DO 10 I=1, M
         WRITE(*,'(1X,''ENTER COEFFICIENTS OF EQUATION NO. '',I2)') I
         DO 20 J=1, N
            WRITE(*,'(1X,''COEFFICIENT OF VARIABLE NO. '',
     *         I2,'' = '',\)') J
            READ(*,*) A(I,J)
 20      CONTINUE
         WRITE(*,'(1X,''KNOWN FOR EQUATION '',I2,'' = '',\)') I
         READ(*,*) A(I,N+1)
 10   CONTINUE
```

(continued)

```
C *** PRINT EQUATIONS
      WRITE(*,'(1X,''EQUATIONS ENTERED ARE'')')
      CALL PRNT(N,M,A)

C *** PIVOT EQUATIONS FROM FIRST TO LAST
      DO 30 I=1, M

C *** FIND COEFFICIENT IN EQUATION I
C *** WITH LARGEST SQUARE
          P = 0.0
          Q = 0.0
          DO 40 J=1, N
              IF (A(I,J)*A(I,J).GT.(P*P)) THEN
                  Q = J
                  P = A(I,J)
              ENDIF
 40       CONTINUE
          IF ((P*P).LT.1.0E-10) GOTO 30

C *** NORMALIZE EQUATION I
          DO 50 J=1, N+1
              A(I,J) = A(I,J)/P
 50       CONTINUE

C *** PIVOT ON COEFFICIENT NO. Q IN EQUATION I
          DO 60 K=1, M
              IF (K.EQ.I) GOTO 60
              P = A(K,Q)
              DO 70 J=1, N+1
                  A(K,J) = A(K,J) - P*A(I,J)
 70           CONTINUE
 60       CONTINUE
 30   CONTINUE

C *** PRINT PIVOTED EQUATIONS
      WRITE(*,'(1X,''FULLY PIVOTED EQUATIONS ARE'')')
      CALL PRNT(N,M,A)
      STOP
      END

C ********************************************************************
      SUBROUTINE PRNT(N,M,A)
C ********************************************************************
      REAL A(20,21)

C *** SUBROUTINE TO PRINT ARRAY A(M,N)
C *** ROW BY ROW
      DO 10 I=1, M
          WRITE(*,'(1X,''COEFFICIENTS OF EQUATION '',I2)') I
          DO 20 J=1, N
              WRITE(*,'(1X,E15.8)') A(I,J)
 20       CONTINUE
          WRITE(*,'(1X,''KNOWN IS '',E15.8)') A(I,N+1)
 10   CONTINUE
      RETURN
      END
```

For the equations

$$
\begin{aligned}
x_1 + 3x_2 - x_3 &= 1 \\
-3x_1 - 2x_2 + 2x_3 &= 4 \\
x_1 \qquad - x_3 &= -2
\end{aligned}
$$

for example, the PIVOT program gives their solution,

```
FULLY PIVOTED EQUATIONS ARE
COEFFICIENTS OF EQUATION 1
0
1
0
KNOWN IS 1
COEFFICIENTS OF EQUATION 2
1
0
0
KNOWN IS -2
COEFFICIENTS OF EQUATION 3
0
0
1
KNOWN IS 0
```

or

$$x_2 = 1$$
$$x_1 = -2$$
$$x_3 = 0$$

The equations

$$x_1 + x_2 + 2x_3 = 0$$
$$3x_1 + 2x_3 = -4$$
$$2x_1 - x_2 = -4$$
$$2x_2 - x_3 = -1$$

are found by PIVOT to have the third equation linearly dependent on the others,

```
FULLY PIVOTED EQUATIONS ARE
COEFFICIENTS OF EQUATION 1
0
0
1
KNOWN IS 1
COEFFICIENTS OF EQUATION 2
1
0
0
KNOWN IS -2
COEFFICIENTS OF EQUATION 3
0
0
0
KNOWN IS 0
COEFFICIENTS OF EQUATION 4
0
1
0
KNOWN IS 0
```

or

$$x_3 = 1$$
$$x_1 = -2$$
$$0 = 0$$
$$x_2 = 0$$

For the equations

$$x_1 - x_2 - 2x_3 = 3$$
$$-x_1 + 2x_2 - x_3 = 4$$
$$-2x_1 + 3x_2 + x_3 = -2$$

PIVOT finds the third equation inconsistent with the other two,

```
FULLY PIVOTED EQUATIONS ARE
COEFFICIENTS OF EQUATION 1
-.2
0
1
KNOWN IS -2
COEFFICIENTS OF EQUATION 2
-.6
1
0
KNOWN IS 1
COEFFICIENTS OF EQUATION 3
0
0
0
KNOWN IS -3
```

or

$$-0.2x_1 \qquad + x_3 = -2$$
$$-0.6x_1 + x_2 \qquad = 1$$
$$0 = -3$$

The importance of line 390 to the PIVOT program is illustrated by the homogeneous equations

$$4x_1 + 2x_2 - 12x_3 = 0$$
$$-6x_1 - x_2 + x_3 = 0$$
$$22x_1 + 7x_2 - 32x_3 = 0$$

for which the result is

```
FULLY PIVOTED EQUATIONS ARE
COEFFICIENTS OF EQUATION 1
0
-.117647059
1
KNOWN IS 0
COEFFICIENTS OF EQUATION 2
1
.147058824
```

```
                    0
                    KNOWN IS 0
                    COEFFICIENTS OF EQUATION 3
                    0
                    −7.09405868E−10
                    0
                    KNOWN IS 0
```

or

$$-0.117647059x_2 + x_3 = 0$$
$$x_1 + \quad 0.147058824x_2 \quad = 0 \quad\quad (2\text{-}4)$$
$$-(7.09405868\text{E-}10)x_2 \quad = 0$$

The small size of the coefficient involving E-10 indicates that the third equation of (2-4) is

$$0 = 0$$

or very nearly so, as it actually is. This means that the third equation of (2-4) is linearly dependent on the other two equations. If line 390 were, instead,

```
            390 IF P=0 THEN GOTO 520
```

the result would have been

```
                    FULLY PIVOTED EQUATIONS ARE
                    COEFFICIENTS OF EQUATION 1
                    0
                    0
                    1
                    KNOWN IS 0
                    COEFFICIENTS OF EQUATION 2
                    1
                    0
                    0
                    KNOWN IS 0
                    COEFFICIENTS OF EQUATION 3
                    0
                    1
                    0
                    KNOWN IS 0
```

which identifies the trivial solution but gives no indication of the likely linear dependence of the third equation. The subject of when, numerically, an equation is practically linearly dependent or inconsistent is examined further in Section 8.5.

Because problems involving n linearly independent and consistent equations in n unknowns occur frequently in practice, it is useful to have at hand a program to handle the solution of this special case. The program named SOLVE performs this function. It is developed and explained in Section 6.4.2 after considerably more background on linear equations and their properties. The main difficulty in using a routine such as PIVOT for the function of just printing the solution when a unique one exists is in having the program decide when the equations are linearly dependent or inconsistent. A poor decision process can easily lead to misleading results for the user.

2.5.3 Solution for Static Forces in a Structure

A truss is a lightweight structure with large load-carrying capacity. Although its elements are usually fastened together with rivets or welds, their slenderness is such that they transmit significant forces only if they were fastened with pin joints (or ball joints, in three dimensions). Figure 2-4(a) shows a plane truss that is part of a railroad drawbridge. The length of each member in meters is indicated. The applied vertical force F_1 represents the weight of the platform. The horizontal force F_2 represents wind loading, and the vertical force F_3 is due to the weight of a train crossing the bridge. The forces F_4, F_5, and F_6 are as yet unknown horizontal and vertical reaction forces at the bridge supports. Each force is expressed in newtons.

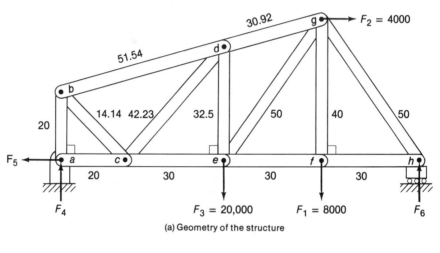

(a) Geometry of the structure

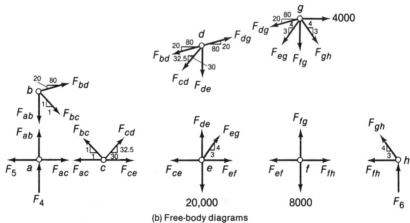

(b) Free-body diagrams

Figure 2-4   Analyzing a plane truss.

Free-body diagrams for each of the joints are shown in Figure 2-4(b). Equating horizontal forces and equating vertical forces gives two linear algebraic equations for each joint. For example, for joint c, the equations are

$$F_{ac} + 0.707F_{bc} - 0.678F_{cd} - F_{ce} = 0$$
$$0.707F_{bc} + 0.735F_{cd} = \quad\quad\quad 0$$

The remaining seven pins give these 14 additional simultaneous linear algebraic equations:

$$F_5 - F_{ac} = \quad\quad\quad 0$$
$$F_4 + F_{ab} = \quad\quad\quad 0$$
$$F_{ab} + 0.707F_{bc} - 0.242F_{bd} = \quad\quad\quad 0$$
$$0.707F_{bc} + 0.970F_{bd} = \quad\quad\quad 0$$
$$0.970F_{bd} + 0.678F_{cd} - 0.970F_{dg} = \quad\quad\quad 0$$
$$0.242F_{bd} + 0.735F_{cd} + F_{de} - 0.242F_{dg} = \quad\quad\quad 0$$
$$F_{ce} - F_{ef} - 0.6F_{eg} = \quad\quad\quad 0$$
$$F_{de} + 0.8F_{eg} = \quad\quad 20000$$
$$F_{ef} - F_{fh} = \quad\quad\quad 0$$
$$F_{fg} = \quad\quad 8000$$

$$0.970F_{dg} + 0.6F_{eg} - 0.6F_{gh} = \quad\quad 4000$$
$$0.242F_{dg} + 0.8F_{eg} + F_{fg} + 0.8F_{gh} = \quad\quad 0$$
$$F_{fh} + 0.6F_{gh} = \quad\quad\quad 0$$
$$F_6 + 0.8F_{gh} = \quad\quad\quad 0$$

Using the PIVOT program, the solution of these equations is

$F_4 \approx 11{,}633$	$F_5 = 4{,}000$	$F_6 \approx 16{,}367$
$F_{ab} \approx -11{,}633$	$F_{ac} = 4{,}000$	$F_{bc} \approx 13{,}169$
$F_{bd} \approx -9{,}599$	$F_{cd} \approx -12{,}667$	$F_{ce} \approx 21{,}899$
$F_{de} \approx 7{,}168$	$F_{dg} \approx -18{,}453$	$F_{ef} \approx 12{,}275$
$F_{eg} \approx 16{,}040$	$F_{fg} = 8{,}000$	$F_{fh} \approx 12{,}275$
$F_{gh} \approx -20{,}458$		

Negative solutions represent forces in the opposite sense from that given by the arrow on the free-body diagram.

2.6 OTHER EQUATION SOLUTION METHODS

2.6.1 Triangularization with Back Substitution

If a set of linear algebraic equations has a solution, it might be easier to reduce the equations to a *triangular form* than to pivot them fully. A triangular form has one equation containing just one variable, another containing just that variable and one other, another containing just the previous two variables and one other, and so on. The following is an example:

$$x_1 - 2x_2 + x_3 = 4$$
$$x_2 - x_3 = 2$$
$$x_3 = -7$$

From this form, the solution for one variable is given by the equation, which contains only one variable. Substituting this variable into the equation that contains two variables gives the solution for a second variable, and so forth. This method of solution is known as *Gaussian elimination with back substitution*. Solving for the variables in the example above gives

$$
\begin{aligned}
x_3 &= -7 \\
x_2 &= \ \ 2 + x_3 = -5 \\
x_1 &= \ \ 4 + 2x_2 - x_3 = 1
\end{aligned}
$$

2.6.2 Error Equations

Pivoting solution methods with the usual numerical accuracy of a compiler are usually adequate for solution of up to about 20 equations in 20 unknowns. There are, of course, smaller, pathological sets of equations for which great inaccuracy results. And there are much larger equation sets for which no numerical difficulty will be encountered. When floating-point calculation is not adequate, error equations can be used to improve solution precision considerably.

Denoting the inaccurate solution to a set of n linear algebraic equations in n unknowns,

$$
\begin{aligned}
a_{11}x_1 + a_{12}x_2 + \cdots + a_{1n}x_n &= y_1 \\
a_{21}x_1 + a_{22}x_2 + \cdots + a_{2n}x_n &= y_2 \\
&\vdots \\
a_{n1}x_1 + a_{n2}x_2 + \cdots + a_{nn}x_n &= y_n
\end{aligned}
\tag{2-5}
$$

by $x_1', x_2', \ldots, x_n'$, then substituting these solutions back into the equations will give

$$
\begin{aligned}
a_{11}x_1' + a_{12}x_2' + \cdots + a_{1n}x_n' &= y_1' \\
a_{21}x_1' + a_{22}x_2' + \cdots + a_{2n}x_n' &= y_2' \\
&\vdots \\
a_{n1}x_1' + a_{n2}x_2' + \cdots + a_{nn}x_n' &= y_n'
\end{aligned}
$$

Letting

$$
\begin{aligned}
\Delta x_1 &= x_1 - x_1' & x_1 &= x_1' + \Delta x_1 \\
\Delta x_2 &= x_2 - x_2' & x_2 &= x_2' + \Delta x_2 \\
&\vdots & &\vdots \\
\Delta x_n &= x_n - x_n' & x_n &= x_n' + \Delta x_n
\end{aligned}
$$

and substituting into (2-5) gives

$$
\begin{aligned}
a_{11}\,\Delta x_1 + a_{12}\,\Delta x_2 + \cdots + a_{1n}\,\Delta x_n &= y_1 - y_1' = \Delta y_1 \\
a_{21}\,\Delta x_1 + a_{22}\,\Delta x_2 + \cdots + a_{2n}\,\Delta x_n &= y_2 - y_2' = \Delta y_2 \\
&\vdots \\
a_{n1}\,\Delta x_1 + a_{n2}\,\Delta x_2 + \cdots + a_{nn}\,\Delta x_n &= y_n - y_n' = \Delta y_n
\end{aligned}
\tag{2-6}
$$

The solution of the error equations (2-6) gives a correction to the previous solution which tends to increase the computational precision. For example, for the equations

$$2x_1 - x_2 - 3x_3 = \frac{7}{3}$$

$$\frac{9}{8}x_2 + x_3 = 0$$

$$5x_1 + 6x_2 - 4x_3 = -\frac{31}{7}$$

the solution with normal arithmetic precision is found to be

$$x_1 = 3.83757339$$
$$x_2 = -2.24918461$$
$$x_3 = 2.53033268$$

Suppose that our computations only resulted in two significant decimal digits of accuracy and that the numerical solution obtained is, instead,

$$x_1' = 3.8$$
$$x_2' = -2.2$$
$$x_3' = 2.5$$

Substituting these values into the original equations gives

$$2x_1' - x_2' - 3x_3' = y_1' = 2.3$$

$$\frac{9}{8}x_2' + x_3' = y_2' = 0.025$$

$$5x_1' + 6x_2' - 4x_3' = y_3' = -4.2$$

and

$$\Delta y_1 = y_1 - y_1' = 0.033333333$$
$$\Delta y_2 = y_2 - y_2' = -0.025$$
$$\Delta y_3 = y_3 - y_3' = -0.228571429$$

If these computations, too, only had two decimal digits of accuracy, the values of the Δy's obtained might be

$$\Delta y_1 = 0.033$$
$$\Delta y_2 = -0.025$$
$$\Delta y_3 = -0.23$$

Solving the error equations,

$$2\Delta x_1' - \Delta x_2' - 3\Delta x_3' = \Delta y_1' = 0.033$$

$$\frac{9}{8}\Delta x_2' + \Delta x_3' = \Delta y_2' = -0.025$$

$$5\Delta x_1' + 6\Delta x_2' - 4\Delta x_3' = \Delta y_3' = -0.23$$

gives

$$\Delta x_1' = 0.0375616435$$
$$\Delta x_2' = -0.0493150685$$
$$\Delta x_3' = 0.030479452$$

or with two-decimal-digit accuracy,

$$\Delta x_1 = \quad 0.038$$
$$\Delta x_2 = -0.049$$
$$\Delta x_3 = \quad 0.030$$

The improved equation solutions are then

$$x_1 \approx x_1' + \Delta x_1' = \quad 3.8 + 0.038 = \quad 3.838$$
$$x_2 \approx x_2' + \Delta x_2' = -2.2 - 0.049 = -2.249$$
$$x_3 \approx x_3' + \Delta x_3' = \quad 2.5 + 0.030 = \quad 2.530$$

which are seen to be improved considerably.

2.6.3 Gauss–Seidel Method

Gauss–Jordan pivoting, possibly with error equations, becomes impractical in many situations where hundreds or even thousands of simultaneous linear algebraic equations must be solved. Such large numbers of equations commonly occur in the description of electrical power systems and in the numerical solution of partial differential equations such as the one describing heat flow. One of the most useful techniques for solving large numbers of simultaneous linear equations is the Gauss–Seidel method. The method does not always converge to a solution, and when it does converge, it might converge slowly. However, for a large class of problems, solutions are obtained in this way when they cannot be obtained with other methods.

The Gauss–Seidel method is described as an algorithm in step-by-step form in Table 2-5. As a numerical example, consider the three-loop electrical network of Figure 2-5. Equating the voltages around each mesh to zero,

$$10 - V_2 + V_1 = \quad 0$$
$$V_1 + V_3 = \quad 0$$
$$12 - V_3 - V_2 + V_4 = 0$$

in terms of the mesh currents and resistor values, these are

$$7I_1 - 3I_2 - 4I_3 = \quad 10$$
$$-3I_1 + 8I_2 - 5I_3 = \quad 0 \qquad (2\text{-}7)$$
$$-4I_1 - 5I_2 + 11I_3 = -12$$

Ordinary solution methods such as Gauss–Jordan pivoting will work nicely on this simple problem, but we will use the Gauss–Seidel solution method for illustration. One can easily imagine a much more complicated network, perhaps involving hundreds of loops, for which other solution methods would be impractical. A problem of this kind is well suited to the Gauss–Seidel method because each equation involves a different variable that has a coefficient that is larger in magnitude than all the other coefficients in the equation. This is a result of the way in which the original equations were formulated.

Arbitrarily choosing $I_1 = 0$, $I_2 = 0$, $I_3 = 0$ as the initial trial solution (actually, the initial I_1 will not be used as the next trial solution for I_1), and using the first equation, gives

$$7I_1 = 3(0) - 4(0) = 10 \qquad I_1 = \tfrac{10}{7}$$

TABLE 2-5   Algorithm for Gauss–Seidel Solution of n Consistent, Independent Linear Algebraic Equations in n Unknowns

DEFINITION

The equations to be solved are

$$a_{11}x_1 + a_{12}x_2 + \cdots + a_{1n}x_n = b_1$$
$$a_{21}x_1 + a_{22}x_2 + \cdots + a_{2n}x_n = b_2$$
$$\vdots$$
$$a_{n1}x_1 + a_{n2}x_2 + \cdots + a_{nn}x_n = b_n$$

ALGORITHM

1. Choose an initial trial solution $x_1, x_2, \ldots, x_n$ for the equations. If an approximate solution is known, it should be used; if not, any values of the n variables can be used initially. The closer this initial choice is to the actual solution, the fewer iterations will generally be needed to obtain a solution with the required accuracy.

2. Solve the first equation for the variable having the coefficient with the largest magnitude, using the initial trial values of the other variables and the known, b_1.

3. Solve the second equation for a different variable, having the coefficient with the largest magnitude of those eligible. Use the calculated value of the variable found in step 2, the initial trial values of the other variables, and the known, b_2.

4. Continue solving for one variable in each equation in this way, always using the values of the calculated variables found from earlier equations. When all equations have been used, resulting in the trial values of all of the variables updated in this way, one updating cycle has been completed.

5. Use the updated trial values of the variables as the new trial solution and repeat steps 2 to 4 to obtain another trial solution, and so on, until the solution is obtained with satisfactory accuracy.

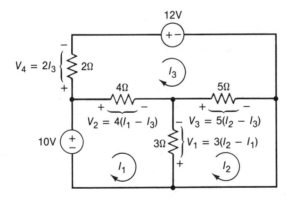

Figure 2-5   A three-loop electrical network with loop currents defined.

Using this value of I_1 and the initial trial value of $I_3 = 0$ in the second equation gives the updated trial value of I_2:

$$-3\left(\tfrac{10}{7}\right) + 8I_2 - 5(0) = 0 \qquad I_2 = \tfrac{15}{28} \approx 0.536$$

and using the new estimates of I_1 and I_2 in the third equation,

$$-4\left(\tfrac{10}{7}\right) - 5\left(\tfrac{15}{28}\right) + 11I_3 = -12 \qquad I_3 = -\frac{101}{(11)(28)} \approx -0.328$$

The first updating cycle is now complete.

Repeating the process for the second updating cycle yields

$$
\begin{aligned}
7I_1 - 3(0.536) - 4(-0.328) &= 10 & I_1 &\approx 1.471 \\
-3(1.471) + 8I_2 - 5(-0.328) &= 0 & I_2 &\approx 0.347 \\
-4(1.471) - 5(0.347) + 11I_3 &= -12 & I_3 &\approx -0.398
\end{aligned}
$$

Continuing in this way will cause the trial values of I_1, I_2, and I_3 to approach the simultaneous solution of the equations, which is $I_1 \approx 0.702$, $I_2 \approx -0.362$, $I_3 \approx -1.00$.

A simple computer program in BASIC for performing these calculations is listed in Table 2-6. In the program, X is I_1, Y is I_2, and Z is I_3. The solution is obtained very easily and does converge, but 58 cycles are required to obtain three-digit accuracy. With a better choice of initial trial solution of $I_1 = \tfrac{10}{7}$, $I_2 = 0$, $I_3 = \tfrac{12}{11}$, only 33 cycles are required for three-digit accuracy.

TABLE 2-6   Computer Program in BASIC to Solve Equations (2-7) by the Gauss–Seidel Method

```
100   X=0
110   Y=0
120   Z=0
130   X=(10+3*Y+4*Z)/7
140   Y=(3*X+5*Z)/8
150   Z=(-12+4*X+5*Y)/11
160   PRINTX,Y,Z
170   GOTO130
180   END
```

PROBLEMS

2-1. Convert the following complex numbers to complex exponential form:

(a) $3 + \underline{i}5$ (b) $7 - \underline{i}8$

(c) $-2.1 + \underline{i}6.2$ (d) $-35 - \underline{i}\tfrac{15}{2}$

2-2. Convert the following complex numbers to rectangular form. All angles are in radians.

(a) $3e^{\underline{i}2}$ (b) $100e^{\underline{i}(1/8)}$

(c) $4.25e^{-\underline{i}3.66}$ (d) $\pi e^{-\underline{i}}$

2-3. Use Euler's relation to prove the following identity:

$$\sin(x \pm y) = \cos x \sin y \pm \sin x \cos y$$

2-4. In an attempt to generalize the concept of a complex number, one might seek a "super"-complex unit $\underline{h}$ for which

$$\underline{h}^3 = -1 \qquad \text{analogous to } \underline{i}^2 = -1$$

But

$$h = \sqrt[3]{-1} = 1e^{i\pi/3}, \quad 1e^{i2\pi/3}, \quad 1e^{i\pi}$$

are other complex numbers that can be manipulated just as any other complex numbers. Taking $\underline{h}$ to be the principal cube root of -1 (just as $\underline{i}$ is its principal square root), express the number

$$m = \frac{3 + \underline{h}}{-4 + 2\underline{h}}$$

in the following forms:

(a) Polar: $m = re^{\underline{h}\theta}$

(b) Rectangular: $m = x + \underline{h}y$

2-5. Write, test, and debug a computer program, from a user-entered complex number magnitude and angle, that produces the real and imaginary parts of that complex number. Let the angle be entered in *degrees*, not radians.

2-6. Write, test, and debug a program in which the user enters the real and imaginary parts of a complex number and the program computes and prints the number's magnitude and angle. The angle found should be in the range from $-\pi$ to π radians. As the ATN($\cdot$) functions in BASIC only produce an angle between 0 and $\pi/2$ for a positive argument and between 0 and $-\pi/2$ for a negative argument, it will be necessary to account for the quadrant of the angle by examining the separate algebraic signs of the real and imaginary parts.

2-7. Write, test, and debug a computer program that when the user enters the real and imaginary parts of a complex number and a positive integer n, computes and prints the n roots of that number.

2-8. Find two equations in three variables with no solution.

2-9. Find three equations in two variables with a unique solution.

2-10. Find three equations in two variables with an under-determined solution.

2-11. Use Cramer's rule to find, if possible, the solutions to the following sets of equations. If the equations have no unique solution, so state.

(a) $2x_1 + 6x_2 = 1$
$3x_1 + 4x_2 = -2$

(b) $x_1 + 2x_2 + 3x_3 = 0$
$4x_1 + 5x_2 + 6x_3 = -2$
$7x_1 + 8x_2 + 9x_3 = -4$

(c) $3x_1 + 2x_2 = 2$
$-x_1 + 3x_2 = 2$
$4x_2 + x_3 = 2$

(d) $2x_1 - 5x_2 + 7x_3 = -1$
$3x_1 + 9x_3 = 0$
$4x_1 + 6x_2 + 8x_3 = -2$

2-12. If there are more variables than equations, some of the variables can be placed on the knowns side of the equation and Cramer's rule used to find solutions in terms of the variables on the knowns side. Use Cramer's rule to find the solution of

$$2x_1 + x_2 - 7x_3 = 4$$
$$5x_1 - 6x_2 + 3x_3 = -2$$

in terms of the variable x_3.

2-13. If there are more equations than unknowns, Cramer's rule can be used to solve simultaneously a number of the equations equal to the number of unknowns. Then this solution can be substituted into the remaining equations to determine whether or not they are consistent. Use this approach to solve

$$x_1 - 2x_2 = 4$$
$$-x_1 + x_2 = -3$$
$$3x_1 - 4x_2 = 10$$

if possible.

2-14. Derive Cramer's rule for the case of three equations and three variables.

2-15. Solve the simultaneous complex equations

$$3s_1 + (2 - i)s_2 = 0$$
$$is_1 + (-4 + 2i)s_2 = 3 + i5$$

for the real and imaginary parts of the unknowns.

2-16. Fully pivot the following sets of equations, showing the results at the end of each pivot cycle. Then interpret the results by identifying linearly dependent and inconsistent equations and giving a unique solution for the variables or a solution for some of the variables in terms of the other variables when solutions exist.

(a) $x_1 + 3x_2 = 4$
      $2x_1 - 2x_2 = 0$

(b) $x_1 + x_2 + 3x_3 = -1$
      $3x_1 + 2x_2 - x_3 = -1$

(c) $3x_1 + x_2 = 2$
     $-2x_1 + 2x_2 = -1$
       $x_1 - 3x_2 = 0$

(d) $x_2 + 2x_3 = 1$
    $-x_1 + 2x_2 - x_3 = -3$
    $-x_1 - 3x_2 - 2x_3 = 3$

(e) $3x_1 - x_2 + 2x_3 = 2$
    $-x_1 \quad\quad - 3x_3 = 0$
      $x_1 - x_2 - 4x_3 = -1$

(f) $-x_1 - x_2 + x_3 + 2x_4 = 3$
     $x_1 + x_2 - 2x_3 - x_4 = 3$
   $-2x_1 - 3x_2 + 3x_3 + 3x_4 = 0$

2-17. Modify the PIVOT program so that it prints the results in order of the variable numbers for those variables pivoted upon.

2-18. Write, test, and debug a computer program that performs Gauss–Jordan pivoting, but instead of pivoting on the variable with largest magnitude coefficient in each *row* of the current coefficient array, pivots on the variable with largest-magnitude coefficient in each *column* of the current array.

2-19. A truss used to support a heavy sign from the side of a building is diagrammed in Figure 2-6. The distances are in meters and the applied force is in newtons. Draw free-body diagrams for each joint, equate vertical and horizontal forces at each joint, and solve the resulting simultaneous linear algebraic equations with PIVOT for all the forces. What happens when the *ad* element is removed? What happens if, in the original structure, an element *bc* is added? The *ad* and *bc* elements cross one another but are not connected together.

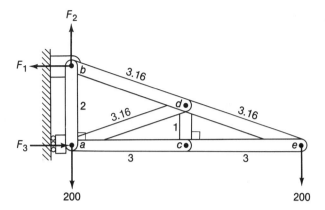

Figure 2-6

2-20. Convert the following to triangular forms by partial Gauss–Jordan pivoting, then solve by back substitution.

(a) $3x_1 + 4x_2 + 2x_3 = -2$
$-x_1 + x_2 - 3x_3 = 0$
$2x_1 - x_2 - x_3 = -1$

(b) $x_2 + 4x_3 = 0$
$-2x_1 + 2x_2 - 3x_3 = 4$
$-2x_1 + x_2 - 7x_3 = 4$

2-21. Use the PIVOT program twice to obtain an accurate solution to the equations

$$2.001 \times 10^3 x_1 - 3.666x_2 + 4.023 \times 10^{-3} x_3 = 1$$
$$0.0306x_1 + 1.2 \times 10^{-2} x_2 - 86.799x_3 = 3.022 \times 10^{-2}$$
$$-1.105 \times 10^{-3} x_1 + 5.333x_2 - 0.732x_3 = 6 \times 10^2$$

In the second use of PIVOT, solve the error equations.

2-22. Compare the solution of the equations

$$6.282 \times 10^3 x_1 - 1.333 \times 10^{-3} x_2 + 4.002 \times 10^3 x_3 = 6.237 \times 10^{-3}$$
$$-3.113 \times 10^6 x_1 - 2.782 \times 10^4 x_2 + 9.080 \times 10^5 x_3 = 3.123 \times 10^3$$
$$7.022 \times 10^{-5} x_1 + 8.555 \times 10^{-3} x_2 - 5.556 \times 10^{-4} x_3 = 7.456 \times 10^{-2}$$

using the PIVOT program to the solution of an equivalent set of equations where each has been multiplied by a constant that makes the coefficient with the largest magnitude in that equation unity.

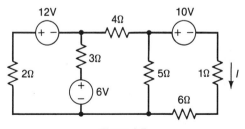

Figure 2-7

2-23. Write, test, and debug a computer program to solve the simultaneous linear algebraic equations

$$-1.2x_1 + 8.4x_2 + 2.2x_3 = 3.0$$
$$7.6x_1 - 0.6x_2 + 0.5x_3 = -4.4$$
$$-0.3x_1 - 1.0x_2 + 10.7x_3 = 1.3$$

by the Gauss–Seidel method.

2-24. Solve the electrical network of Figure 2-7 for the current I.

3

SINGLE-VARIABLE SEARCHES FOR ZEROS, MAXIMA, AND MINIMA

3.1 PREVIEW

In this chapter we develop methods for the numerical solution of a single nonlinear equation involving a single variable. Chapter 4 deals with nonlinear functions of more than one variable and simultaneous nonlinear equations. Although our predecessors, by necessity, applied these methods with hand calculations and graphical constructions, we can now use the powerful tool of digital computation. Consequently, in this and the next chapter, the emphasis on computational methods that began toward the end of Chapter 2 continues.

In Section 3.2 we discuss the nature of nonlinear equations and the ways we can solve them. In Section 3.3 a bisection method for solving a single nonlinear equation in one variable is developed. The algorithm is realized as a computer program in BASIC/FORTRAN which is then applied to a variety of problems, including that of finding the radiation pattern of a broadcast antenna array. Other solution methods are discussed in Section 3.4.

Our attention then turns to locating the maxima and minima of a function of a single variable. A bisection search is, again, one of the best general methods and this is developed and programmed in Section 3.5. In Section 3.6, Newton's method for extrema location is discussed and programs are given. Even though Newton's method has shortcomings, it is important to our work with functions of more than one variable, especially in Chapter 4. The concluding section concerns higher-order extrema location methods and the use of extrema location in finding those solutions to nonlinear equations

$$f(x) = 0$$

that do not cross the x-axis.

3.2 NONLINEAR FUNCTIONS

If a function is not linear, it is *nonlinear*. Having discussed the solution of linear and simultaneous linear equations in Chapter 2, our attention now turns to the solution of nonlinear equations. There are basically two methods of solving nonlinear equations:

   1. Identities and "tricks," when they can be found and when they apply

   2. Approximate numerical solution

As a simple example of method 1, consider the single nonlinear equation involving a single variable,

$$\sin^2 x + 2 \cos^2 x = \tfrac{3}{2} \tag{3-1}$$

Using the trigonometric identity

$$\sin^2 x + \cos^2 x = 1$$

equation (3-1) is equivalent to

$$\cos^2 x = \tfrac{1}{2}$$

which has solution

$$x = \cos^{-1} \sqrt{\tfrac{1}{2}}$$

    Another simple example of method 1 involves the nonlinear equation in two variables,

$$x^2 + 4x - 3 \exp(y) = 2$$

Completing the square in x gives the equivalent equation

$$(x + 2)^2 - 3 \exp(y) = 6$$

Defining the new variables

$$u = (x + 2)^2 \qquad x = \sqrt{u} - 2$$
$$v = \exp(y) \qquad y = \ln(v)$$

gives an equivalent *linear* equation in terms of u and v:

$$u - 3v = 6$$

Once u and v are found, x and y can be determined.

    The application of method 1 is an art and is often based on careful understanding of a physical situation described by the equations. As one gains experience with a variety of such problems, capabilities increase, but still, the vast majority of problems of this nature encountered in practice cannot be solved in this way. This leaves method 2, approximate numerical solution, the *only* method.

    When a problem involves a simultaneous combination of nonlinear and consistent linear equations, the linear equations in the set can be used to eliminate some of the variables, leaving the solution of the nonlinear equations as the essence of the problem.

For example, the simultaneous three equations, one of them nonlinear,

$$\begin{aligned} x_1^2 + 3x_1 x_2 + x_2 \sin(x_3 \sqrt{x_2}) &= 5 \\ x_1 + x_2 - x_3 &= 4 \\ 2x_1 - x_2 &= -3 \end{aligned}$$

are equivalent to the three equations

$$x_1^2 + 3x_1 x_2 + x_2 \sin(x_3 \sqrt{x_2}) = 5$$
$$x_1 = \tfrac{1}{3} x_3 + \tfrac{1}{3} \tag{3-2}$$
$$x_2 = \tfrac{2}{3} x_3 + \tfrac{11}{3}$$

which, in turn, are equivalent to

$$\tfrac{7}{9} x_3^2 + \tfrac{41}{9} x_3 + \tfrac{34}{9} + \tfrac{1}{3}(11 + 2x_3) \sin\left(x_3 \sqrt{\tfrac{2}{3} x_3 + \tfrac{11}{3}}\right) = 5$$

$$x_1 = \tfrac{1}{3} x_3 + \tfrac{1}{3} \tag{3-3}$$

$$x_2 = \tfrac{2}{3} x_3 + \tfrac{11}{3}$$

In equations (3-3), the second and third equations of (3-2) are used to eliminate the variables x_1 and x_2 from the nonlinear equation. Once the nonlinear equation in (3-3) is solved for x_3, x_1 and x_2 are easily found from the other two linear equations.

3.3 REAL ZERO LOCATION BY BISECTION

A nonlinear equation in terms of a single variable x can always be written in the form

$$f(x) = 0$$

The solutions of the equation are termed the *zeros* (or the *roots*) of the function f. This section concerns a very good numerical method for locating the zeros of a function of a single variable.

3.3.1 Bisection Search for Axis Crossings

A bisection search for zeros of a smooth function $f(x)$ is described in Figure 3-1. The search begins at the left endpoint a, where $f(x_1 = a)$ is calculated. Then the variable is incremented by an amount δ and $f(x_2 = a + \delta)$ is found. If $f(x_1)$ and $f(x_2)$ have opposite algebraic signs, the function must have a zero crossing between x_1 and x_2. If not, the variable is incremented again, the algebraic signs of $f(x_2 = a + \delta)$ and $f(x_3 = a + 2\delta)$ compared, and so on. As soon as a zero crossing of the function is detected in this way, the increment δ is repeatedly halved and the search continues from the left until the zero crossing is located with the desired degree of accuracy.

To illustrate the steps involved, consider the function

$$f(x) = x^2 - e^{-x}$$

Let the search begin at $x_1 = 0$ with an initial increment $\delta = 0.2$. As shown in Table

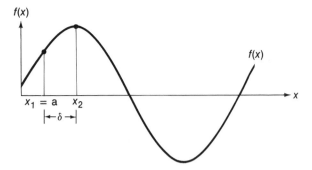

(a) First iteration

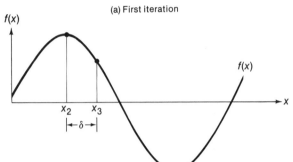

(b) Second iteration

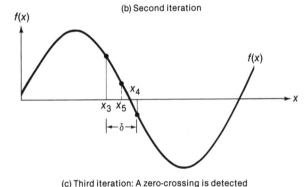

(c) Third iteration: A zero-crossing is detected

Figure 3-1 Bisection search for an axis-crossing.

3-1, a zero of the function is detected between $x = 0.6$ and $x = 0.8$ because the corresponding function values at these points have opposite algebraic signs. The increment δ is halved to 0.1 and the search resumes from $x_1 = 0.6$. Because $f(0.7)$ and $f(0.8)$ have opposite algebraic signs, a zero of the function then lies between 0.7 and 0.8. Again, halving the increment δ to 0.05, starting at $x_1 = 0.7$, and comparing the algebraic signs of $f(0.7)$ and $f(0.75)$ shows that the zero actually lies between 0.7 and 0.75. This process is repeated until the desired degree of accuracy is achieved. In Table 3-1, the value of the function at $x = 0.70347$ is approximately 4.9019×10^{-6} after 21 iterations. If the function can have multiple zeros, the increment δ is then set to its initial value and the search resumes from the right of the most recently located zero crossing until the value of the variable exceeds a specified endpoint, $x = b$.

TABLE 3-1   Zero Location of $f(x) = x^2 - e^{-x}$ Using the Bisection Method

Iteration	x	$f(x)$
1	0	-1
2	0.2	-0.77873
3	0.4	-0.51032
4	0.6	-0.18881
5	0.8	$+0.19067$
6	0.7	-0.00659
7	0.75	$+0.09013$
8	0.725	$+0.04$
.	.	.
.	.	.
.	.	.
19	0.70346	-0.00001
20	0.70348	$+0.00002$
21	0.70347	$+4.9019 \times 10^{-6}$

Bisection is a very good zero-crossing location method for several reasons. First, it is very apparent when it will fail. However, if there are two zeros closer than the initial δ to one another, it is likely that bisection will miss them. If the function has a zero but does not cross the x-axis, just touching it instead, bisection will fail to find the root. Otherwise, the method will work, and it will converge to root location with any desired accuracy (within numerical constraints) in a predictable number of steps. At each step where δ is halved, the precision of the result is doubled, a very nice property.

3.3.2 Bisection Zero-Finding Program

This bisection method for real root location is expressed as a series of steps in the algorithm of Table 3-2. The algorithm is used for the program in BASIC/FORTRAN called BISECT, which is listed in Table 3-3. For use with the other functions and other parameters, the search boundaries (listed from -6 to 12), the initial variable increment (0.00001), and the function f, lines 120–150 and 1010 are to be changed as needed by the user.

The number of function evaluations needed by this program can sometimes be reduced by storing and reusing more than the most recent of the values calculated previously. We have, instead, opted for program simplicity.

An interesting problem involving multiple solutions is the intersection of a sinusoid and a straight line,

$$\cos x = \tfrac{1}{7} x$$

as shown in Figure 3-2(a) on p. 77. The function

$$f(x) = \cos x - \tfrac{1}{7} x$$

TABLE 3-2 Algorithm for Real Root Location by Bisection

DEFINITIONS

For the function $f(x)$, the roots x_0 such that

$$f(x_0) = 0$$

between $x = a$ and $x = b$ are to be located within ε. It is assumed that the function $f(x)$ crosses the x-axis and has intervals larger than c above and below the x-axis on either side of the zero crossing.

ALGORITHM

1. Set the increment δ to c,

$$\delta \leftarrow c$$

   set the variable x to the left endpoint,

$$x \leftarrow a$$

   and set

$$f \leftarrow f(x)$$

   (lines 180–220).

2. Transfer f to g,

$$g \leftarrow f$$

   (lines 230–240).

3. Increment the variable

$$x \leftarrow x + \delta$$

   and check if the right endpoint has been exceeded (i.e., if $x > b$). If so, end. If not, set

$$f \leftarrow f(x)$$

   where x is now incremented (lines 250–290 and 440).

4. If there is no algebraic sign change between g and f, that is, if

$$gf \geq 0$$

   then set

$$g \leftarrow f$$

   and go to step 2 to continue incrementing and testing (lines 300–310).

5. If there is an algebraic sign change between g and f, check to see if the desired precision has been attained (i.e., if $\delta < \varepsilon$). If so, print $x_0 \approx x$ and f, set the right boundary to

$$a \leftarrow x + \delta$$

   and go to step 1 to search for a new zero (lines 320–330 and 380–430).

6. If the desired precision has not been attained, reset x,

$$x \leftarrow x - \delta$$

   halve the search increment,

$$\delta \leftarrow \frac{\delta}{2}$$

   and go to step 2 to refine the search (lines 340–370).

TABLE 3-3

Computer Program in BASIC for Real Root Location by Bisection (BISECT)

VARIABLES USED

X = the search variable

A = left limit of the search variable

B = right limit of the search variable

C = initial increment of the search variable

D = present increment of the search variable

E = desired precision of root location

F = value of the function; value of the function at the right test point

G = value of the function at the left test point

LISTING

```
100     PRINT "BISECTION REAL ROOT SEARCH"
110     REM SET BOUNDARIES, INITIAL INCREMENT, AND PRECISION
120     A =-6
130     B =12
140     C =.1
150     E =.00001
160     REM PRINT HEADING
170     PRINT "X", "F(X)"
180     REM SET INCREMENT TO INITIAL INCREMENT
190     D =C
200     REM EVALUATE FUNCTION AT LEFT BOUNDARY
210     X =A
220     GOSUB 1000
230     REM SET G TO LEFT VALUE OF FUNCTION
240     G =F
250     REM INCREMENT VARIABLE AND EVALUATE FUNCTION
260     REM END IF RIGHT BOUNDARY EXCEEDED
270     X =X +D
280     IF X > B THEN GOTO 440
290     GOSUB 1000
300     REM IF NO FUNCTION SIGN CHANGE, CONTINUE SEARCH
310     IF G*F > 0 THEN GOTO 240
320     REM PRINT ROOT AND RESTART IF INCREMENT IS SMALLER THAN E
330     IF D \< E THEN GOTO 380
340     REM MOVE X BACK, HALVE INCREMENT, AND REEVALUATE
350     X =X -D
360     D =D/2
370     GOTO 250
380     REM PRINT ROOT AND RESTART SEARCH
390     PRINT X,F
400     REM RESET LEFT BOUNDARY TO ONE INITIAL INCREMENT PAST ROOT
410     REM AND CONTINUE SEARCH
420     A =X +C
430     GOTO 180
440     END
1000    REM SUBROUTINE TO EVALUATE FUNCTION
1010    F =COS(X) - (X/7)
1020    RETURN
```

(continued)

Computer Program in FORTRAN for Real Root Location by Bisection (BISECT)

```
C **********************************************************************
C PROGRAM BISECT -- BISECTION REAL ROOT SEARCH
C **********************************************************************
      WRITE(*,'(1X,''BISECTION REAL ROOT SEARCH'')')

C *** SET BOUNDARIES, INITIAL INCREMENT, AND PRECISION
      A =  -6.0
      B =  12.0
      C =   0.1
      E =   0.00001

C *** PRINT HEADING
      WRITE(*,'(1X,''      X              F(X)'')')

C *** LOOP FOR A MAXIMUM OF 100 ROOTS
      DO 30 J=1, 100

C *** SET INCREMENT TO INITIAL INCREMENT
         D = C

C *** EVALUATE FUNCTION AT LEFT BOUNDARY
         X = A
         CALL FUNCT(X,F)

C *** SET G TO LEFT VALUE OF FUNCTION
         G = F

C *** INCREMENT VARIABLE, EVALUATE FUNCTION, AND
C *** END IF RIGHT BOUNDARY EXCEEDED.  REPEAT FOR A
C *** MAXIMUM OF 100 ITERATIONS
         DO 10 I=1, 100
            X = X + D
            IF(X.GT.B) STOP
            CALL FUNCT(X,F)

C *** IF NO FUNCTION SIGN CHANGE, CONTINUE SEARCH
C *** IF SIGN HAS CHANGED, CHECK TO SEE IF ROOT HAS BEEN FOUND
            IF((G*F).LE.0.0) THEN

C *** PRINT ROOT AND RESTART IF INCREMENT IS SMALLER THAN E
               IF(D.LT.E) THEN
                  GO TO 20

C *** MOVE X BACK, HALVE INCREMENT, AND REEVALUATE
               ELSE
                  X = X - D
                  D = D/2
               ENDIF
            ENDIF
 10      CONTINUE

C *** PRINT ROOT AND RESTART SEARCH
 20      WRITE(*,'(1X,2E15.8)') X,F

C *** RESET LEFT BOUNDARY TO ONE INITIAL INCREMENT PAST ROOT
C *** AND CONTINUE SEARCH
         A = X + C

 30   CONTINUE
      STOP
      END
```

```
C *******************************************************************
      SUBROUTINE FUNCT(X,F)
C *******************************************************************

C *** SUBROUTINE TO EVALUALTE FUNCTION
      F = COS(X) - (X/7.0)
      RETURN
      END
```

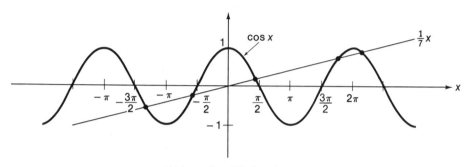

(a) Intersection of the functions

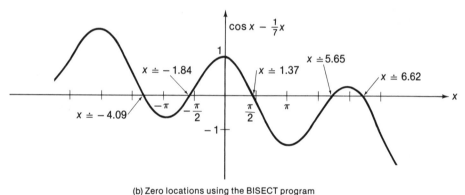

(b) Zero locations using the BISECT program

Figure 3-2   Intersections of a straight line and a sinusoid.

has five zeros, as in Figure 3-2(b). Using this $f(x)$ in the BISECT program gives the following roots:

```
        BISECTION REAL ROOT SEARCH
              X                        F(X)
        -4.08862916              -1.20676123E-06
        -1.83621828               4.32337402E-07
         1.37332762              -1.8045539E-06
         5.65222776               1.89675484E-06
         6.61596677              -7.38538802E-07
```

It is tempting to arrange a root-finding program such as this one to stop bisecting when the function $f(x)$ is within a certain small amount from zero instead of stopping when x is within an acceptable distance of the zero. It is dangerous to do so, however, because it might not be possible to compute $f(x)$ with the needed accuracy, in which event the search will never end.

3.3.3 Bisection with Direct Reversal

A variation on the bisection algorithm is to reverse the search direction each time a zero crossing is passed instead of always searching to the right. This is done easily by reversing the algebraic sign of δ at each step,

$$\delta \leftarrow -\frac{\delta}{2}$$

and also alternating the other value of the function to which the sign of the new value is compared. There is no particular advantage or disadvantage to using this algorithm instead of the one in Table 3-2; it is just another possibility.

Another popular modification is to divide the increment by 10 instead of by 2 at each step. This slows convergence a little, but has the advantage of increasing precision by one decimal digit per step. If the user specifies the number of decimal digits of precision required, that translates easily into the number of division steps to be performed. If the search is made to proceed from the side of the interval for which the function has the smallest magnitude, an average convergence speed can be obtained without the disadvantages of linear interpolation.

3.3.4 Broadcast Antenna Radiation Pattern

Low-frequency broadcast stations employ vertical antenna towers such as the one sketched in Figure 3-3 to radiate their signals. A screen of ground wires beneath the tower acts as a mirror, so that, electrically, the device acts as a dipole antenna. The waves that are

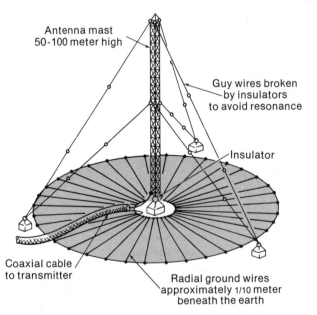

Antenna mast
50-100 meter high

Guy wires broken
by insulators
to avoid resonance

Insulator

Coaxial cable
to transmitter

Radial ground wires
approximately 1/10 meter
beneath the earth

Figure 3-3 A low-frequency broadcast antenna.

launched tend to stay "stuck" to the ground so that they follow the hills, valleys, and the curvature of the earth, unlike higher-frequency waves, which propagate in straight lines, like light.

When several nearby antenna towers are driven with the radio signal, their contributions to the strength of the radiation at each point in the distance is very nearly the sum of the contributions of the individual towers. In some directions, the individual contributions tend to aid one another and in others they tend to cancel. It is thus possible to control the directions in which the radiation is the strongest and weakest by using multiple towers and properly spacing and driving them. This allows two stations using the same or a nearby channel to reduce the area in which they interfere (allowing them to be located closer to one another than would otherwise be desirable) and for a station to "beam" a larger amount of its power to a selected area.

Because radio broadcast channels consist of relatively narrow bands of frequencies, the behavior of the signal can be analyzed as if it were a sinusoid of a single frequency. Figure 3-4 shows the geometry of a certain three-tower antenna array, as viewed from above. At a point sufficiently distant from the towers, the angle θ to each of the towers is virtually the same, as shown. The distances to each tower are not quite the same, and if the tower spacing is a significant part of the wavelength, the phase of the sinusoidal signal contribution of each tower will depend greatly on the angle θ.

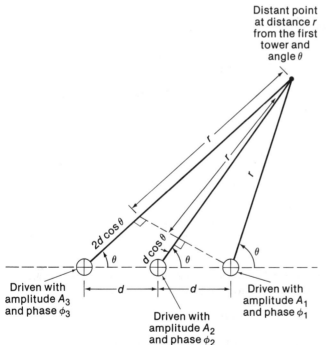

Figure 3-4 A three-tower antenna array viewed from above.

Let the three towers be driven with sinusoidal signals of frequency f, amplitudes $A_1, A_2,$ and A_3, respectively, and corresponding phases $\phi_1, \phi_2,$ and ϕ_3. Then the radiated field at a point at distance r and angle θ will be

$$E(r, \theta, t) = \frac{1}{r} \left\{ \left[A_1 \cos\left(2\pi ft + \phi_1 + \frac{2\pi fr}{c}\right) \right. \right.$$

$$+ A_2 \cos\left(2\pi ft + \phi_2 + \frac{2\pi fd}{c} \cos\theta + \frac{2\pi fr}{c}\right)$$

$$\left. \left. + A_3 \cos\left(2\pi ft + \phi_3 + \frac{4\pi fd}{c} \cos\theta + \frac{2\pi fr}{c}\right) \right] g(r, \phi) \right\}$$

where c is the speed of propagation (very nearly the speed of light) and where $g(r, \theta)$ describes the ground conductivity. Ground conductivity is a measure of how well the waves "stick" to the ground in the area.

The amplitude of the sinusoidal signal E, which consists of the sum of sinusoidal components from each of the three towers, depends on r and θ and is given by

$\varepsilon(r, \theta)$

$$= \frac{g(r, \theta)}{r} \sqrt{ \begin{array}{l} [A_1 \cos\phi_1 + A_2 \cos\left(\phi_2 + \frac{2\pi fd}{c}\cos\theta\right) + A_3 \cos\left(\phi_3 + \frac{4\pi fd}{c}\cos\theta\right)]^2 + \\[2mm] [A_1 \sin\phi_1 + A_2 \sin\left(\phi_2 + \frac{2\pi fd}{c}\cos\theta\right) + A_3 \sin\left(\phi_3 + \frac{4\pi fd}{c}\cos\theta\right)]^2 \end{array} }$$

This is called the *field strength* of the transmitted signal and is measured in microvolts per meter. For a certain station, suppose that

$$g(r, \theta) = 0.7(1 + 0.1 \sin\theta)r^{-1/2}$$

and that $A_1 = 80$, $A_2 = 40$, $A_3 = 120$, $\phi_1 = 0$, $\phi_2 = \pi/4$, $\phi_3 = \pi/8$, and $fd/c = \frac{1}{4}$, so that the field strength is given by

$$\varepsilon(r, \theta) = 56(1 + 0.1 \sin\theta)$$

$$\bullet \sqrt{ \frac{\left[1 + 0.5 \cos\left(\frac{\pi}{4} + \frac{\pi}{2}\cos\theta\right) + 1.5\cos\left(\frac{\pi}{8} + \pi\cos\theta\right)\right]^2 + [0.5\sin(\pi/4 + \pi/2\cos\theta) + 1.5\sin(\pi/8 + \pi\cos\theta)]^2}{r^3} }$$

The area with field strength 100 microvolts per meter and higher is considered an area of very good reception. Using the BISECT program to find the roots r of

$$\varepsilon(r, \theta) - 0.0001 = 0$$

for various angles θ gives the 100-microvolt/meter contour sketched in Figure 3-5. For this solution the angle θ was incremented in a FOR . . . NEXT loop, and the search for each root was from $r = 0.1$ to $r = 10{,}000$. As only a single root is expected, a large initial step size can be used. Other contours are found similarly.

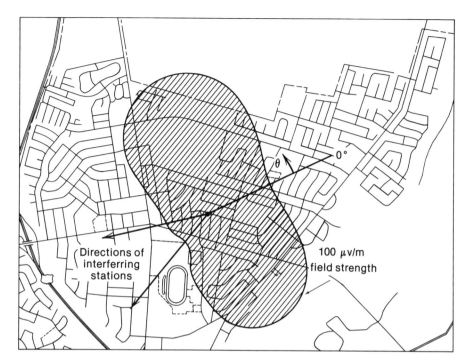

Figure 3-5   The 100-microvolt/meter field strength contour for the example broadcast station.

3.4 OTHER REAL ZERO LOCATION METHODS

There are a number of other ways to find the zeros of a function. Several of these are now considered.

3.4.1 Method of False Position

It seems as though we should be able to improve on bisecting an interval containing a zero crossing, as shown in Figure 3-6(a). Instead of dividing the interval between x_1 and x_2 in half to obtain the next search point x_3, the next search point could be closer to whichever of x_1 and x_2 is closer to zero.

One way of deriving an expression for x_3 is to recognize that the straight line connecting the points $(x_1, f(x_1))$, $(x_2, f(x_2))$ also passes through the point $(x_3, 0)$. If this straight line is of the form

$$y = \alpha x + \beta$$

then

$$f(x_1) = \alpha x_1 + \beta$$
$$f(x_2) = \alpha x_2 + \beta$$

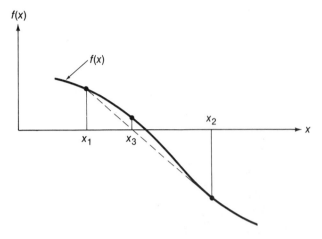

Figure 3-6   Method of false position.

Solving these two equations for α and β gives

$$\alpha = \frac{f(x_2) - f(x_1)}{x_2 - x_1}$$

$$\beta = \frac{f(x_1)x_2 - f(x_2)x_1}{x_2 - x_1}$$

Since $(x_3, 0)$ is also a point on this line,

$$\alpha x_3 + \beta = 0$$

or

$$x_3 = -\frac{\beta}{\alpha} = \frac{f(x_2)x_1 - f(x_1)x_2}{f(x_2) - f(x_1)} \tag{3-4}$$

In general, the zero crossing occurs between either x_1 and x_3 or x_3 and x_2. If $f(x_1)$ and $f(x_3)$ have opposite algebraic signs, the zero crossing is located between x_1 and x_3. However, if $f(x_1)$ and $f(x_3)$ have the same algebraic sign, the zero crossing occurs between x_3 and x_2, in which event x_3 becomes the new x_1, $f(x_3)$ becomes the new $f(x_1)$, and equation (3-4) is applied repeatedly until the zero crossing is located with the desired degree of accuracy. If, on the other hand, $f(x_1)$ and $f(x_3)$ have opposite algebraic signs, x_3 becomes the new x_2, $f(x_3)$ becomes $f(x_2)$, and (3-4) is applied to determine the new value of x_3. This process is repeated over and over until the desired accuracy of the zero crossing is achieved.

The false position method is also called the method of bisection with linear interpolation. It often speeds convergence a little, but per-step speed is slower because it requires a more complicated calculation for the location of x_3.

3.4.2 Newton–Raphson (Slope) Search

Another popular method for determining the zeros of an algebraic or transcendental function is the Newton–Raphson slope search method. Although the search for the zeros of a function using this method has the potential of rapid convergence, there is also the possibility of overlooking zeros and of not converging at all. We start with an initial guess of the zero. If at each step, the function $f(x)$ is approximated with a straight line of the form

$$f_i(x) = a_i x + b_i \qquad (3\text{-}5)$$

having the same value and slope as $f(x)$ at $x = x_i$, then the constants a_i and b_i satisfy

$$f_i(x_i) = a_i x_i + b_i = f(x_i)$$

$$\dot{f}_i(x_i) = a_i = \dot{f}(x_i)$$

or

$$\begin{aligned} a_i &= \dot{f}(x_i) \\ b_i &= f(x_i) - \dot{f}(x_i)x_i \end{aligned} \qquad \text{where } \dot{f} = \frac{df}{dx}$$

The zero crossing using this straight-line approximation occurs at

$$a_i x + b_i = 0$$

or

$$x = -\frac{b_i}{a_i} = x_i - \frac{f(x_i)}{\dot{f}(x_i)}$$

and this is a good choice for the next trial value of x, x_{i+1}. Then

$$x_{i+1} = x_i - \frac{f(x_i)}{\dot{f}(x_i)} \qquad (3\text{-}6)$$

Because this solution is only approximate, x_{i+1} is used to calculate a new trial point,

$$x_{i+2} = x_{i+1} - \frac{f(x_{i+1})}{\dot{f}(x_{i+1})}$$

in an iterative process that it is hoped will converge to the location of a zero. Of course, provision must be made for the possibility that $\dot{f} = 0$.

The results in (3-6) can be arrived at geometrically by examining Figure 3-7, where at each step the slope of the function at the point of evaluation is used to give the next value of the variable to be used. In general, the tangent of a function at point x_1 is given by

$$\dot{f}(x) = \frac{-f(x_1)}{x_2 - x_1}$$

or

$$x_2 = x_1 - \frac{f(x_1)}{\dot{f}(x_1)}$$

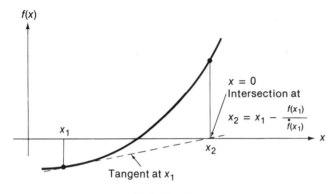

(a) First iteration

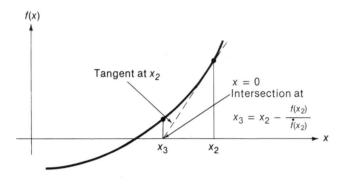

(b) Second iteration

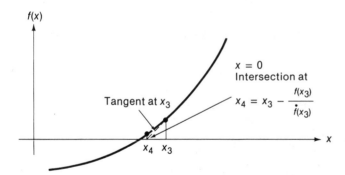

(c) Third iteration

Figure 3-7 Newton–Raphson method of iteratively locating real zeros.

and the tangent of the function at point x_2 is

$$\dot{f}(x_2) = \frac{-f(x_2)}{x_3 - x_2}$$

or

$$x_3 = x_2 - \frac{f(x_2)}{\dot{f}(x_2)}$$

and so on, until

$$x_{i+1} = x_i - \frac{f(x_i)}{\dot{f}(x_i)}$$

As a numerical example of applying the Newton–Raphson method, consider the function examined previously,

$$f(x) = x^2 - e^{-x}$$

Starting with $x = 0$ and using (3-6) repeatedly gives the values in Table 3-4. Examining the results in Tables 3-1 and 3-4 shows that with the Newton–Raphson method, only five iterations are needed compared to the 21 iterations required by the bisection method to achieve the same accuracy.

TABLE 3-4   Zero Location of $f(x) = x^2 - e^{-x}$ Using Newton–Raphson Method

Iteration	x	$f(x)$	$\dot{f}(x)$
1	0	-1	1
2	1	0.63212	2.36788
3	0.73302	0.0569	1.94653
4	0.70381	0.00065	1.90232
5	0.70347	4.9019×10^{-6}	1.90181

To the extent that the function is very nearly a straight line in the interval between x_i and x_{i+1}, this algorithm will converge rapidly to a zero of a function. Even when the function is not nearly a straight line in the interval, it may still converge. However, there are situations, such as the one in Figure 3-8, in which convergence does not occur, and

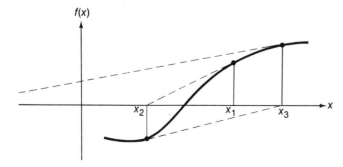

Figure 3-8   Divergence of the Newton–Raphson method.

there are other situations for which convergence is exceedingly slow. There is also a problem of restarting the search to find new zero crossings without converging on the old ones again.

Slope search is not a good general root location method but is sometimes used in practice because, when it does converge, its convergence speed is even faster than that of bisection. Additional speed becomes especially important when it is necessary to find the zeros of a large number of functions. A program using slopes will be developed for a similar problem, that of extrema location, later in this chapter.

3.4.3 Secant Method

When the derivative of a function is not known or when it is inconvenient to calculate, a derivative approximation in Newton–Raphson formula can be used. For a slope search for a zero, where the next trial value of the variable is given by (3-6), replacing $\dot{f}(x_i)$ with

$$\dot{f}(x_i) = \frac{f(x_i) - f(x_{i-1})}{x_i - x_{i-1}}$$

gives an alternative to (3-6):

$$x_{i+1} = x_i - f(x_i) \frac{x_i - x_{i-1}}{f(x_i) - f(x_{i-1})} \tag{3-7}$$

The algorithm defined by (3-7) is called the *secant method*. The geometric interpretation of the secant method is that instead of using a tangent line to the function at each step, as shown in Figure 3-7, a secant line is drawn between the points as shown in Figure 3-9. In the first iteration x_3 is determined as the point of intersection between the secant

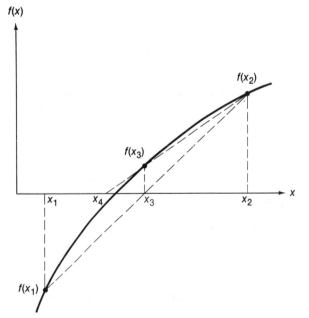

Figure 3-9 Secant method.

line connecting $f(x_1)$ and $f(x_2)$ and the x-axis. In the second iteration, the line connecting $f(x_3)$ and $f(x_2)$ intersects the x-axis at x_4. This process is repeated until the desired degree of accuracy is achieved.

3.4.4 Slope Search Using a Numerical Derivative

When the derivative of a function is not known or when it is inconvenient to calculate, a numerical derivative approximation can be used. For a slope search for a zero, where the next trial value of the variable is given by (3-6), replacing $\dot{f}(x_i)$ with

$$\dot{f}(x_i) \approx \frac{f(x_i + u) - f(x_i)}{u}$$

where u is a small number, gives an alternative to (3-6) that uses a numerical derivative:

$$x_{i+1} = x_i - \frac{uf(x_i)}{f(x_i + u) - f(x_i)} \tag{3-8}$$

The algorithm defined by (3-8), for small u, will work about as well as (3-6) when $|f(x_i + u) - f(x_i)|$ is fairly large. When the magnitude of the change in the function is very small, numerical errors in the evaluations of the function and in computing the numerical derivative can lead to erratic behavior. Also, the accuracy with which the zero is found is likely to be within about u, regardless of the number of iterations.

3.5 ONE-DIMENSIONAL EXTREMA LOCATION BY BISECTION

For a continuous function of a single variable $f(x)$ with derivative $\dot{f}(x)$, a critical point is a point x_c for which

$$\dot{f}(x_c) = 0$$

If the function's second derivative exists at $x = x_c$, then at a critical point, if

$$\ddot{f}(x_c) < 0$$

the point is a relative maximum. If

$$\ddot{f}(x_c) > 0$$

the point is a relative minimum, and if

$$\ddot{f}(x_c) = 0$$

it is a point of inflection.

When the derivatives exist, expanding the function in a Taylor series about a critical point gives

$$f(x) = f(x_c) + \frac{\dot{f}(x_c)}{1!}(x - x_c) + \frac{\ddot{f}(x_c)}{2!}(x - x_c)^2 + \frac{\dddot{f}(x_c)}{3!}(x - x_c)^3 + \cdots$$

but

$$\dot{f}(x_c) = 0$$

For points very near x_c,

$$f(x) \approx f(x_c) + \frac{\ddot{f}(x_c)}{2!}(x - x_c)^2$$

The function $f(x)$ is decreasing on either side of x_c if $\ddot{f}(x_c) < 0$ and increasing on either side of x_c if $\ddot{f}(x_c) > 0$. If

$$\ddot{f}(x_c) = 0$$

then for points very near x_c,

$$f(x) \approx f(x_c) + \frac{\dddot{f}(x_c)}{3!}(x - x_c)^3$$

and $f(x)$ increases on one side of x_c and decreases on the other side of x_c, meaning that x_c is neither a maximum nor a minimum of $f(x)$. It could happen, however, that $\dddot{f}(x_c) = 0$ also, in which event

$$f(x) \approx f(x_c) + \frac{f^{[4]}(x_c)}{4!}(x - x_c)^4$$

for points very near x_c. The critical point x_c is then a relative maximum or a relative minimum according to the algebraic sign of $f^{[4]}(x_c)$; and so on. Unfortunately, some beginning calculus texts and courses can leave one with the idea that $\dddot{f}(x_c) = 0$ means that x_c is not an extremum, a proposition that is easily shown to be false, for example with the function

$$f(x) = -x^4 + 4x^3 - 6x^2 + 4x + 3$$

at $x_c = 1$.

   If the derivative of the function under consideration is known, a bisection or other search for the zeros of the derivative will locate critical points. Evaluation of the function at values of the variable that are a small distance either side of a critical point can be used to determine if the critical point is a relative maximum, a relative minimum, or a point of inflection; or the function's second derivative, if known, can be evaluated.

   When the derivative of a function is not known or when it is inconvenient to find, bisection is a good way of locating relative extrema. In the following, we emphasize finding the relative minima of a function. The relative maxima can be found in similar ways or simply by locating the relative minima of the negative of the function.

   The locations of the global extrema of a function in an interval $[a, b]$ are the values of x between $x = a$ and $x = b$ for which the function is maximum and minimum. These can be found by evaluating the function at each relative extremum. The endpoints of the interval, too, are candidates for global extrema.

3.5.1 Bisection Search for Minima

Figure 3-10 illustrates how to find the relative minima of a function with a bisection search. Beginning at the left endpoint, the values of the function at three successive points are compared. If the value of the function at the middle of the three points is smaller than the values at both sides, a relative minimum must be near. When this occurs, the increment is halved and the search is restarted from the leftmost of the three points.

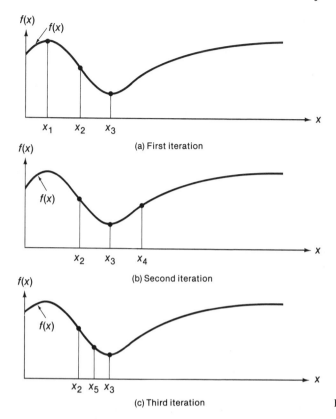

(a) First iteration

(b) Second iteration

(c) Third iteration

Figure 3-10   Bisection search for minima.

After enough successive halvings of the search increment, the relative minimum is found with the needed precision. Then the increment is returned to its original size and the search for another relative minimum continues to the right.

3.5.2 Minima-Finding Bisection Program

This minima search procedure is expressed as an algorithm in Table 3-5 and as a computer program in BASIC/FORTRAN, called MINIMA, in Table 3-6. The program uses some of the same steps as the previous BISECT program. Its listed parameters and function to be tested are the same as in the BISECT listing. The parameters and function are, of course, to be changed by the user as desired.

The function

$$f(x) = \cos x - \tfrac{1}{7} x$$

sketched earlier in Figure 3-2(b), is found by the MINIMA program to have the following relative minima in the range from $x = -6$ to $x = 12$:

```
MINIMA LOCATION BY BISECTION
         X                              F(X)
    -2.9982422                      -.561422592
     3.2849243                      -1.45902049
     9.56809692                     -2.35661839
```

TABLE 3-5   Algorithm for Minima Location by Bisection

DEFINITIONS

For the function $f(x)$, the relative minima x_m between $x = a$ and $x = b$ are to be located within ε. It is assumed that the function $f(x)$ is smooth and that the minima of $f(x)$ are separated by a distance greater than $2c$.

ALGORITHM

1. Set the increment δ to c,

$$\delta \leftarrow c$$

set the variable x to the left endpoint,

$$x \leftarrow a$$

and set

$$g \leftarrow f(x)$$

(lines 180–240).

2. Increment the variable

$$x \leftarrow x + \delta$$

and check if the right endpoint has been exceeded (i.e., if $x > b$). If so, end. If not, set

$$h \leftarrow f(x)$$

where x is now incremented (lines 250–280 and 490).

3. Increment the variable again,

$$x \leftarrow x + \delta$$

and recheck. If the right endpoint has been exceeded, $x > b$. If so, end. If not, set

$$f \leftarrow f(x)$$

(lines 290–310 and 490).

4. If the middle of the three values of $f(x)$ is not less than the other two (i.e., if $h > g$ or if $h > f$), set

$$g \leftarrow h$$
$$h \leftarrow f$$

and go to step 3 to continue incrementing and testing (lines 320–340 and 450–480).

5. If the middle of the three values of $f(x)$ is less than the other two, check to see if the desired precision has been obtained (i.e., if $\delta < \varepsilon$). If so, print $x_m \approx x - \delta$ and $h = f(x - \delta)$, set the right boundary to

$$a \leftarrow x + \delta$$

and go to step 1 to search for a new minimum (lines 350–360 and 410–440).

6. If the desired precision has not been attained, reset x,

$$x \leftarrow x - 2\delta$$

halve the search increment,

$$\delta \leftarrow \frac{\delta}{2}$$

and go to step 2 to refine the search (lines 370–400).

TABLE 3-6

Computer Program in BASIC for Minima Location by Bisection (MINIMA)

ADDITIONAL VARIABLE USED

In addition to the variables used in the BISECT program of Table 3-3, this program uses

$$H = \text{center value of the function for the search interval}$$

LISTING

```
100     PRINT "MINIMA LOCATION BY BISECTION"
110   REM SET BOUNDARIES, INITIAL INCREMENT AND PRECISION
120     A = -6
130     B = 12
140     C = 0.1
150     E = 0.00001
160   REM PRINT HEADING
170     PRINT "     X","F(X)"
180   REM SET INCREMENT TO INITIAL INCREMENT
190     D = C
200   REM EVALUATE FUNCTION AT THREE POINTS. END IF RIGHT
210   REM BOUNDRY WILL BE EXCEEDED.
220     X = A
230     GOSUB 1000
240     G = F
250     X = X + D
260     IF X > B THEN GOTO 490
270     GOSUB 1000
280     H = F
290     X = X + D
300     IF X > B THEN GOTO 490
310     GOSUB 1000
320   REM IF MIDVALUE NOT LOWER THAN OTHER TWO, CONTINUE SEARCH.
330     IF H > G THEN GOTO 450
340     IF H > F THEN GOTO 450
350   REM PRINT RESULT AND RESTART IF INC IS SMALLER THAN E
360     IF D < E THEN GOTO 410
370   REM MOVE X BACK, HALVE THE INCREMENT, AND REEVALUATE
380     X = X -2 * D
390     D =D / 2
400     GOTO 250
410   REM PRINT RESULT AND RESTART SEARCH
420     PRINT X - D, H
430     A = X + D
440     GOTO 180
450   REM SHIFT FUNCTION POINTS AND REEVALUATE
460     G = H
470     H = F
480     GOTO 290
490     END
1000  REM SUBROUTINE TO EVALUATE FUNCTION
1010    F = COS(X) - (X/7)
1020    RETURN
```

Computer Program in FORTRAN for Minima Location by Bisection (MINIMA)

```
C ********************************************************************
C Computer Program in FORTRAN For Minima Location By Bisection (MINIMA)
C ********************************************************************

      WRITE(*,'(1x,''MINIMA LOCATION BY BISECTION'')')
```

(continued)

```
C *** SET BOUNDARIES, INITIAL INCREMENT, AND PRECISION
      A = -6.0
      B = 12.0
      C = 0.1
      E = 0.00001

C *** PRINT HEADING
      WRITE(*,'(1X,''          X              F(X)'')')

180   CONTINUE
C *** SET INCREMENT TO INITIAL INCREMENT
      D = C

C *** EVALUATE FUNCTION AT THREE POINTS. END IF RIGHT
C *** BOUNDARY WILL BE EXCEEDED
      X = A
      CALL FUNCT(X,F)
      G = F
250   X = X + D
      IF (X.GT.B) GOTO 490
      CALL FUNCT(X,F)
      H = F
290   X = X + D
      IF (X.GT.B) GOTO 490
      CALL FUNCT(X,F)
C *** IF MIDVALUE NOT LOWER THAN OTHER TWO, CONTINUE SEARCH
      IF (H.GT.G) GOTO 450
      IF (H.GT.F) GOTO 450
C *** PRINT RESULT AND RESTART IF INC IS SMALLER THAN E
      IF (D.LT.E) GOTO 410
C *** MOVE X BACK, HALVE THE INCREMENT, AND REEVALUATE
      X = X - 2*D
      D = D/2
      GOTO 250
410   CONTINUE
C *** PRINT RESULT AND RESTART SEARCH
      TEMP = X - D
      WRITE(*,'(1X,E15.8,2X,E15.8)') TEMP,H
      A = X + D
      GOTO 180
450   CONTINUE
C *** SHIFT FUNCTION POINTS AND REEVALUATE
      G = H
      H = F
      GOTO 290
490   END

C*******************************************************************
      SUBROUTINE FUNCT(X,F)
C*******************************************************************

C *** SUBROUTINE TO EVALUATE FUNCTION

      F = COS(X) - (X/7.0)
      RETURN
      END
```

As $f(x)$ is easily differentiated,

$$\frac{df}{dx} = -\sin x - \tfrac{1}{7}$$

these relative minima can also be located by finding the roots of the derivative of $f(x)$. The roots of the derivative are found by the BISECT program of Table 3-3 with

```
1010 F = -SIN(X) - (1/7)
```

to be

```
BISECTION REAL ROOT SEARCH
       X                         F(X)
 -2.9982422              2.85753049E-06
 -.14334718             -3.84869054E-07
  3.28494261             2.3652683E-06
  6.13984373            -5.92839206E-06
  9.56812744             1.88167905E-06
```

which also identifies the locations of the relative maxima in that range. Finding the relative minima directly, as with the MINIMA program, has the advantages of not requiring that the derivative be calculated and not needing any interpretation as to whether each point found is a relative minimum, relative maximum, or a point of inflection.

The relative maxima of $f(x)$ between $x = -6$ to $x = 12$ are found with the MINIMA program by reversing the algebraic sign of the function:

```
1010 F = -COS(X) + (X/7)
```

They are found to be

```
MINIMA LOCATION BY BISECTION
       X                         F(X)
 -.143359388            -1.01022154
  6.13982542            -.112623642
```

3.5.3 Optimum Automobile Suspension Damping

The suspension of the rear axle of an automobile is modeled by the spring, mass, and damper shown in Figure 3-11. When the tires encounter a sharp bump in the road at time $t = 0$, the body's position varies as

$$x(t) = A - Ae^{-\sigma t}\left(\cos \omega t + \frac{\sigma}{\omega} \sin \omega t\right)$$

where

$$\sigma = \frac{B}{2M} \qquad \omega = \sqrt{\frac{K}{M} - \frac{B^2}{4M^2}}$$

and where A is the height of the bump. When

$$\frac{B^2}{4M^2} = \frac{K}{M}$$

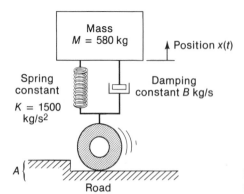

Figure 3-11 Model of the suspension of an automobile's rear axle.

the response is *critically damped*. For

$$\frac{B^2}{4M^2} > \frac{K}{M}$$

$x(t)$ is overdamped and consists of the sum of two real exponentials. With the given values of mass M and spring constant K,

$$\sigma = \frac{B}{1160} \qquad \omega \approx \sqrt{2.586 - 7.43 \times 10^{-7}B^2} \qquad (3\text{-}9)$$

The position reponse of this vehicle for various values of B, which models the action of the shock absorbers, is shown in Figure 3-12(a). When B is small, the oscillations decay slowly. When B is large, the position changes very slowly. After $t = 0$, the error in position is

$$e(t) = A - x(t) = Ae^{-\sigma t}(\cos \omega t + \frac{\sigma}{\omega} \sin \omega t)$$

A popular measure of the quality of an automobile's ''ride'' is the integral of the square of the error under these circumstances:

$$I = \int_0^\infty e^2(t)\, dt = A^2 \int_0^\infty e^{-2\sigma t}(\cos \omega t + \frac{\sigma}{\omega} \sin \omega t)^2\, dt$$

$$= A^2 \int_0^\infty e^{-2\sigma t}[\cos^2\omega t + \frac{2\sigma}{\omega} \cos \omega t \sin \omega t + \frac{\sigma^2}{\omega^2} \sin^2\omega t]\, dt$$

$$= A^2 \int_0^\infty e^{-2\sigma t}\left[\frac{1}{2}(1 + \cos 2\omega t) + \frac{\sigma}{\omega} \sin 2\omega t + \frac{\sigma^2}{2\omega^2}(1 - \cos 2\omega t)\right] dt$$

$$= A^2 \int_0^\infty e^{-2\sigma t}\left[\frac{1}{2}\left(1 + \frac{\sigma^2}{\omega^2}\right) + \frac{1}{2}\left(1 - \frac{\sigma^2}{\omega^2}\right)\cos 2\omega t + \frac{\sigma}{\omega} \sin 2\omega t\right] dt \qquad (3\text{-}10)$$

$$= A^2\left[\frac{1}{2}\left(1 + \frac{\sigma^2}{\omega^2}\right)\frac{e^{-2\sigma t}}{-2\sigma} + \frac{1}{2}\left(1 - \frac{\sigma^2}{\omega^2}\right)\frac{e^{-2\sigma t}(2\omega \sin 2\omega t - 2\sigma \cos 2\omega t)}{4\sigma^2 + 4\omega^2}\right.$$

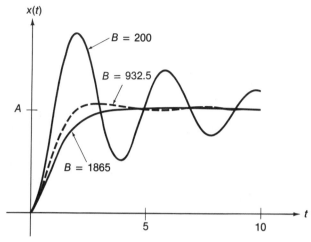

(a) Position response for several damping constants

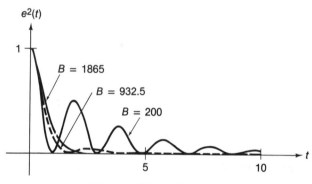

(b) Square of the position response

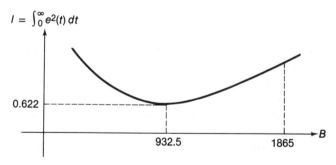

(c) Integral square response as a function of the damping constant

Figure 3-12   Finding an optimum damping constant.

$$+ \frac{\sigma}{\omega} \frac{e^{-2\sigma t}(-2\omega \cos 2\omega t - 2\sigma \sin 2\omega t)}{4\sigma^2 + 4\omega^2}\Bigg]\Bigg|_0^{\infty}$$

$$= A^2\left[\frac{1}{4\sigma}\left(1 + \frac{\sigma^2}{\omega^2}\right) + \frac{3\sigma - \sigma^3/\omega^2}{4(\sigma^2 + \omega^2)}\right]$$

The nature of $e^2(t)$ and of I for various damping constants is illustrated in Figure 3-12(b) and (c). The smaller the value of I, the better. As a function of the damping constant B, the performance measure is given by (3-10), where σ and ω are as in (3-9). This expression holds for all positive B. The location of the minimum of I is not affected by the value of A.

When $I(B)/A^2$ is used as the function in the MINIMA program,

```
1010 S = X / 1160
1020 W = SQR (2.586 - 7.43E-7*X*X)
1030 F=1/(4*S) + S/(4*W*W) + (3*S-S↑3/(W*W)) / (4*(S*S + W*W))
1040 RETURN
```

the minimum is found to be

$$B = X \approx 932.5$$

which is 1/2 times the value of B that results in critical damping. For this calculation, the search extended from $B = 0.1$ to $B = 10000$, with an initial step size of 100.

A more accurate model would include Coulomb as well as viscous damping and perhaps allow for some deliberate nonlinearity of the springs. In this case, a numerical solution for the optimum value of B would probably involve numerical solution of the differential equation governing $x(t)$, numerical integration of $x^2(t)$, then location of the minimum.

3.6 NEWTON'S METHOD FOR EXTREMA LOCATION

Newton's method is a slope method of extrema location. Compared to bisection, it is not a very good general method because it sometimes does not converge. Our primary interest in Newton's method is not so much for ordinary single-variable searches for relative extrema, but for searches of functions of several variables where, because of the large number of possible combinations of the variables, convergence speed is very valuable. Some experience with the method and its problems in the simple single-variable case is quite valuable in understanding the problems and possibilities in more complicated situations.

3.6.1 The Method

In Newton's method for locating relative minima of a function, the slope of the function at each trial point is used to determine the direction and distance of the next trial point.

As illustrated in Figure 3-13, the next trial point x_{i+1} is obtained from the present trial point x_i according to

$$x_{i+1} = x_i - \alpha \dot{f}(x_i) \tag{3-11}$$

where α is a positive constant called the *convergence constant*. If α is too small, convergence to a relative minimum will be exceedingly slow. If α is too large, as in the example of Figure 3-14(a), the process might not converge at all. A good value of α for one function might be a very poor choice for another function. Functions with "sharp" minima such as the absolute value type of function in Figure 3-14(b) are especially difficult to deal with because of the abrupt change in slope.

The best programs of this type monitor the distances $|x_{i+1} - x_i|$ between successive steps and attempt to vary the value of α to obtain rapid convergence to minima. This is complicated to do and even more complicated to analyze in any but a heuristic way.

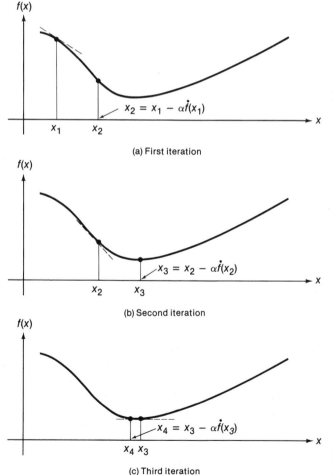

(a) First iteration

(b) Second iteration

(c) Third iteration

Figure 3-13 Newton's method search for a relative minimum.

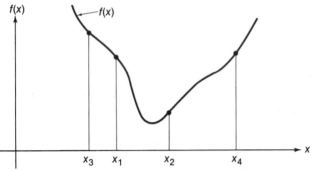

(a) Convergence constant that is too large

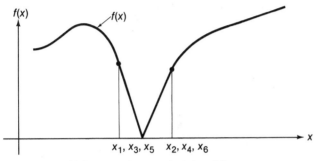

(b) Absolute value type of relative minimum

Figure 3-14   Divergence of a Newton's method search.

3.6.2 Newton's Method Program

A Newton's method search algorithm is given in Table 3-7, and the corresponding computer program in BASIC/FORTRAN is listed in Table 3-8. The function listed is

$$f(x) = \cos x - \frac{x}{7}$$

This function is computed in subroutine 1000, and its derivative,

$$\dot{f}(x) = -\sin x - \tfrac{1}{7}$$

is computed in subroutine 2000. The function itself is not used in the search; its value at the located minima are printed, along with the minima. The convergence constant C, the desired precision E, and the function and its derivative are, of course, to be specified by the user in practice. The program is written so that it loops back and asks the user for a new starting value of x each time a relative minimum is found.

When run with the listed parameters and function, the NEWTON program gives the following results when starting at $x = -6$, $x = 1$, and $x = 7$:

```
RELATIVE MINIMA BY NEWTON'S METHOD
ENTER THE STARTING POINT?-6
MINIMUM                                 FUNCTION
-2.99825503                             -.561422592
```

TABLE 3-7 Algorithm for a Newton's Method Search for Relative Minima

DEFINITIONS

For the function $f(x)$ with known derivative $\dot{f}(x)$, a relative minima, if it exists, is to be located such that, at the point found, $|\dot{f}(x)| < \varepsilon$ has been specified by the user. The search begins at a given point x and proceeds in the x direction having downward slope, in increments of x proportional to the slope. The constant of proportionality used is c, the convergence constant.

ALGORITHM

1. Set x to the starting point of the search (lines 140–160).

2. Compute
$$g \leftarrow \dot{f}(x)$$
(lines 190–200).

3. If $|g| < \varepsilon$, print the values of x and g and go to step 1 for a new starting point (lines 210–230, 280–300).

4. Otherwise, increment x by $-cg$,
$$x \leftarrow x - cg$$
and go to step 2 to continue the search (lines 240–270).

TABLE 3-8

Computer Program in BASIC for a Newton's Method Search for Relative Minima (NEWTON)

VARIABLES USED

C = convergence constant

E = required precision

X = search variable

F = value of the function

G = value of the derivative of the function

LISTING

```
100     PRINT "RELATIVE MINIMA BY NEWTON'S METHOD"
110     REM SET CONVERGENCE CONSTANT AND PRECISION
120     C =.25
130     E=.00001
140     REM ENTER INITIAL VALUE OF THE VARIABLE
150     PRINT "ENTER THE STARTING POINT ";
160     INPUT X
170     REM PRINT HEADING
180     PRINT "MINIMUM","FUNCTION"
190     REM EVALUATE DERIVATIVE
200     GOSUB 2000
210     REM PRINT RESULT AND GO ON IF SLOPE
220     REM MAGNITUDE IS LESS THAN PRECISION
230     IF ABS(G) < E THEN GOTO 270
240     REM INCREMENT VARIABLE AND CONTINUE SEARCH
250     X =X -C*G
260     GOTO 190
270     REM PRINT RESULT AND GO ON
280     GOSUB 1000
```

(continued)

```
290      PRINT X,F
300      GOTO 140
310      END
1000  REM SUBROUTINE TO EVALUATE FUNCTION
1010     F =COS(X) -X/7
1020     RETURN
2000  REM SUBROUTINE TO EVALUATE DERIVATIVE
2010     G =-SIN(X) -1/7
2020     RETURN
```

Computer Program in FORTRAN for a Newton's Method Search for Relative Minima (NEWTON)

```
C ****************************************************************
C PROGRAM NEWTON -- RELATIVE MINIMA BY NEWTON'S METHOD
C ****************************************************************
      WRITE(*,'(1X,''RELATIVE MINIMA BY NEWTONS METHOD'')')

C *** SET CONVERGENCE CONSTANT AND PRECISION
      C =.25
      E=.00001

C *** ENTER INITIAL VALUE OF THE VARIABLE FOR MAXIMUM OF 100 ITERATIONS
      DO 10 J=1, 100
          WRITE(*,'(1X,''ENTER THE STARTING POINT '',\)')
          READ(*,*) X

C *** PRINT HEADING
          WRITE(*,'(1X,''    MINIMUM          FUNCTION'')')

C *** EVALUATE DERIVATIVE FOR MAXIMUM OF 1000 INTERTATIONS
          DO 20 I=1, 1000
              CALL DERIV(X,G)

C *** PRINT RESULT AND GO ON IF SLOPE
C *** MAGNITUDE IS LESS THAN PRECISION
              IF (ABS(G).LT.E) GOTO 30

C *** INCREMENT VARIABLE AND CONTINUE SEARCH
              X = X - C*G
   20         CONTINUE

C *** PRINT RESULT AND GO ON
   30     CALL FUNCT(X,F)
          WRITE(*,'(1X,E15.8,1X,E15.8)') X,F
   10 CONTINUE

      STOP
      END

C ****************************************************************
      SUBROUTINE FUNCT(X,F)
C ****************************************************************

C *** SUBROUTINE TO EVALUATE FUNCTION
      F =COS(X) -X/7.0
      RETURN
      END

C ****************************************************************
      SUBROUTINE DERIV(X,G)
C ****************************************************************

C *** SUBROUTINE TO EVALUATE DERIVATIVE
      G =-SIN(X) -1/7.0
      RETURN
      END
```

```
ENTER THE STARTING POINT?1
MINIMUM                                        FUNCTION
3.2849318                                      -1.45902049
ENTER THE STARTING POINT?7
MINIMUM                                        FUNCTION
9.56811617                                     -2.35661839
```

These compare well with the minima found for the same function with the MINIMA program, and execution time was faster. To find maxima of this function, the function and its derivative are replaced by their negatives in the subroutines,

```
1010 F = -COS(X) + X / 7
2010 G = SIN(X) + 1 / 7
```

resulting in the following:

```
ENTER THE STARTING POINT?1
MINIMUM                                        FUNCTION
-.143339654                                    -1.01022154
ENTER THE STARTING POINT?4
MINIMUM                                        FUNCTION
6.13982964                                     -.112623642
```

These also compare well with the relative maxima found earlier using the MINIMA program.

   To automate the search for minima somewhat, one can move to the right and restart the search after each relative minimum is found. An easy way to do this is to specify a "skip distance" D. After a minimum is found, the program moves D units to the right and searches again. If the search "falls into" the previous minimum again, X is incremented by 2D and the search is restarted, and so on.

3.6.3 Newton's Method Using a Numerical Derivative

When the derivative of a function $f(x)$ is not known or is inconvenient to find or program, a numerical approximation to $\dot{f}$ can be used in place of the derivative itself. Approximating the derivative by

$$\frac{df}{dx} = \dot{f}(x) \approx \frac{f(x + u) - f(x)}{u}$$

where u is small, gives the algorithm

$$x_{i+1} = x_i - \frac{\alpha[f(x_i + u) - f(x_i)]}{u}$$

instead of (3-11).

   Table 3-9 lists a modification of the NEWTON program, called NEWTON2, that uses a numerically calculated slope instead of a known derivative function. To show the modifications clearly, line numbers have not been changed so that the changes and additions are obvious. The original subroutine to calculate the derivative that began at line 2000 has been deleted. A new parameter U, the increment of X used in the numerical slope calculation, is set by the user, then the slope is approximated by

TABLE 3-9

Computer Program in BASIC for a Newton's Method Search for Relative Minima Using a Numerical Derivative (NEWTON2)

This is a modification of the NEWTON program of Table 3-8. The variable G now stores the value of the function and the numerical slope of the function. U is the increment of X used in the numerical slope calculation.

LISTING

```
100     PRINT "NEWTON'S METHOD WITH NUMERICAL DERIVATIVE"
110     REM SET CONVERGENCE CONSTANT AND PRECISION
120       C =.25
130       E =.00001
131     REM SET NUMERICAL DERIVATIVE INTERVAL
132       U =.01
140     REM ENTER INITIAL VALUE OF THE VARIABLE
150       PRINT "ENTER THE STARTING POINT";
160       INPUT X
170     REM PRINT HEADING
180       PRINT "MINIMUM", "FUNCTION"
190     REM COMPUTE NUMERICAL DERIVATIVE
200       GOSUB 1000
201       G =F
202       X =X +U
203       GOSUB 1000
204       X =X -U
205       G =(F -G)/U
210     REM PRINT RESULT AND GO ON IF SLOPE
220     REM MAGNITUDE IS LESS THAN PRECISION
230       IF ABS(G) < E THEN GOTO 270
240     REM INCREMENT VARIABLE AND CONTINUE SEARCH
250       X =X -C*G
260       GOTO 190
270     REM PRINT RESULT AND GO ON
280       GOSUB 1000
290       PRINT X,F
300       GOTO 140
310       END
1000    REM SUBROUTINE TO EVALUATE FUNCTION
1010      F =COS(X) -X/7
1020      RETURN
```

Computer Program in FORTRAN for a Newton's Method Search for Relative Minima Using a Numerical Derivative (NEWTON2)

```
C ******************************************************************
C PROGRAM NEWTON2 -- NEWTON'S METHOD WITH NUMERICAL DERIVATIVE
C ******************************************************************

      WRITE(*,'(1X,''NEWTONS METHOD WITH NUMERICAL DERIVATIVE'')')

C *** SET CONVERGENCE CONSTANT AND PRECISION
      C = 0.25
      E = 0.00001

C *** SET NUMERICAL DERIVATIVE INTERVAL
      U = 0.01

C *** ENTER INITIAL VALUE OF THE VARIABLE, MAXIMUM OF 100 LOOPS
      DO 10 J=1, 100
         WRITE(*,'(1X,''ENTER THE STARTING POINT '',\)')
         READ(*,*) X
```

```
C *** PRINT HEADING
      WRITE(*,'(1X,''    MINIMUM           FUNCTION'')')

C *** COMPUTE NUMERICAL DERIVATIVE, MAXIMUM OF 1000 ITERATIONS
      DO 20 I=1, 1000
          CALL FUNCT(X,F)
          G = F
          X = X + U
          CALL FUNCT(X,F)
          X = X - U
          G = (F - G)/U

C *** PRINT RESULT AND GO ON IF SLOPE
C *** MAGNITUDE IS LESS THAN PRECISION
          IF(ABS(G).LT.E) GO TO 30

C *** INCREMENT VARIABLE AND CONTINUE SEARCH
          X = X - C*G
 20       CONTINUE

C *** PRINT RESULT AND GO ON
 30       CALL FUNCT(X,F)
          WRITE(*,'(1X,E15.8,1X,E15.8)') X,F
 10   CONTINUE

      STOP
      END

C ********************************************************************
      SUBROUTINE FUNCT(X,F)
C ********************************************************************

C *** SUBROUTINE TO EVALUATE FUNCTION
      F =COS(X) - X/7.0
      RETURN
      END
```

$$G = \frac{f(X + U) - f(X)}{U}$$

Results of a minima search for the function listed and used previously, with the same starting points of $x = -6$, $x = 1$, and $x = 7$, are as follows:

```
RELATIVE MINIMA BY NEWTON'S METHOD
ENTER THE STARTING POINT?-6
MINIMUM                              FUNCTION
-3.00325428                          -.561410172
ENTER THE STARTING POINT?1
MINIMUM                              FUNCTION
3.27993244                           -1.45900808
ENTER THE STARTING POINT?7
MINIMUM                              FUNCTION
9.56311687                           -2.35660598
```

These are close to those found using the actual derivative. It is well to bear in mind, however, that the precision E here is the *numerical derivative*. As that calculation involves the interval of U = 0.01, the minima locations found are probably accurate only within about 0.01 unit. There is also a danger in choosing the numerical derivative interval U to be too small. If U is so small that the numerical values of $f(x)$ and $f(x + u)$ are the same, minima might be found where none actually exist.

3.7 MORE ABOUT EXTREMA

When higher derivatives of $f(x)$ are known, one can make improved estimates of a relative minimum location at each iteration by using a higher-degree polynomial approximation to $f(x)$. This subject is now introduced with parabolic approximations. Then we return to the problem of locating the zeros of functions where the function touches, but does not cross, the axis. These "grazing" zeros are relative minima or maxima of the function and so can be located easily by finding and examining the function's extrema.

3.7.1 Higher-Order Methods

If, at each step x, $f(x)$ is approximated by a parabola

$$p_i(x) = a_i x^2 + b_i x + c_i$$

having the same value, slope, and inflection as $f(x)$ at $x = x_i$, then the constants a_i, b_i, and c_i satisfy

$$p_i(x) = a_i x^2 + b_i x + c_i = f(x_i)$$
$$\dot{p}_i(x) = 2a_i x + b_i = \dot{f}(x_i)$$
$$\ddot{p}_i(x) = 2a_i = \ddot{f}(x_i)$$

or

$$a_i = \tfrac{1}{2}\ddot{f}(x_i)$$
$$b_i = \dot{f}(x_i) - 2a_i x_i = \dot{f}(x_i) - \ddot{f}(x_i)x_i$$
$$c_i = f(x_i) - b_i x_i - a_i x_i^2 = f(x_i) - \dot{f}(x_i)x_i + \tfrac{1}{2}\ddot{f}(x_i)x_i^2$$

The parabolic approximation to $f(x)$ is maximum or minimum where

$$\dot{p}(x) = 2a_i x + b_i = 0$$

or

$$x = -\frac{b_i}{2a_i} = \frac{-\dot{f}(x_i) + \ddot{f}(x_i)x_i}{\ddot{f}(x_i)} = x_i - \frac{\dot{f}(x_i)}{\ddot{f}(x_i)}$$

and this is a good choice for the next trial value of x, x_{i+1}. Figure 3-15 shows the idea involved for the starting point of the search x_1 and the next two trial values of x.

This process easily locates both relative maxima and relative minima of $f(x)$, provided that the initial trial value of x is sufficiently close to the extremum that the parabola is a sufficiently good approximation to $f(x)$.

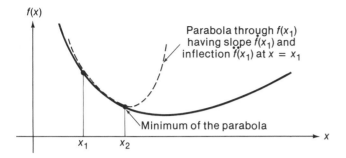

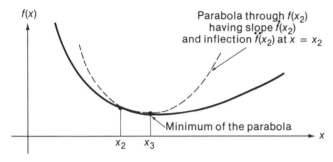

Figure 3-15 Using a second-order method of locating relative minima. The function is approximated by a parabola.

If

$$\ddot{p}(x_i) = 2a_i = \ddot{f}(x_i) > 0$$

the parabola is concave upward, as in Figure 3-15. One would then be searching for a relative minimum. If

$$\ddot{p}(x_i) = 2a_i = \ddot{f}(x_i) < 0$$

the parabola is, instead, concave downward and the search is for a relative maximum.

The parabolic approximation is usually used only after an extremum has been found roughly by Newton's method or bisection. It is used to increase the convergence speed and/or increase the numerical accuracy of the solution obtained. More complicated calculation at each iteration is traded for the relatively simple calculations that generally require more iterations for a given accuracy of the result.

When the derivatives of $f(x)$ are not known, they can be approximated numerically as with Newton's method:

$$\dot{f}(x) \approx \frac{f(x + \Delta x) - f(x)}{\Delta x}$$

A commonly used second derivative approximation is

$$\ddot{f}(x) \approx \frac{\dot{f}(x + \Delta x) - \dot{f}(x)}{\Delta x}$$

$$= \frac{[f(x + 2\Delta x) - f(x + \Delta x)]/\Delta x - [f(x + \Delta x) - f(x)]/\Delta x}{\Delta x}$$

$$= \frac{f(x + 2\Delta x) - 2f(x + \Delta x) + f(x)}{(\Delta x)^2}$$

3.7.2 Finding "Grazing" Zeros

When a function $f(x)$ just touches (or "grazes") the x-axis without crossing it, as in Figure 3-16, a bisection search will not detect the root. Polynomials with repeated real zeros of even multiplicity, such as

$$f(x) = x^4 - 5x^3 - 8x^2 + 48 = (x + 3)(x - 4)^2$$

$f(x) = x^4 + 2x^3 - 3x^2 - 4x + 4$

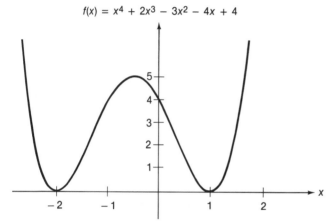

Figure 3-16   Graph of a function with two "grazing" roots.

which has a double zero at $x = 4$ have grazing roots. Other functions, such as

$$f(x) = \sin^2 x$$

also have grazing zeros. An important method of locating grazing zeros is to find the relative maxima and minima of the function and check to see which values of the function at the minima or maxima, if any, are zero or nearly zero.

As an example, the polynomial function

$$f(x) = x^4 + 2x^3 - 3x^2 - 4x + 4 = (x - 1)^2(x + 2)^2$$

is graphed in Figure 3-16. It has grazing zeros at $x = 1$ and $x = -2$. The MINIMA program finds the two relative minima of this function to be as follows:

```
         MINIMA LOCATION BY BISECTION
              X                              F(X)
         -2.00003664                    7.4505806E-09
          .999993883                         0
```

As the value of the function is zero (or very nearly so) at each of these two values of x, they are "grazing" zeros.

PROBLEMS

3-1. Convert the following sets of simultaneous equations to an equivalent set consisting of a nonlinear equation to be satisfied by one of the variables and additional equations, each of which is the solution of one of the other variables in terms of the first variable:

(a) $x^3 + x^2y - 3y^2 + yz - 4xyz = 0$

$\quad x + 3y + 2z \qquad\qquad\qquad = 4$

$\quad 2x - y \qquad\qquad\qquad\qquad = 3$

(b) $r \cos \theta = y$

$\quad r \sin \theta = y^2$

$\quad e^y \qquad = \dfrac{\theta}{2\pi}$

3-2. Use the BISECT program to find the real zeros of the following functions in the intervals indicated.

(a) $f(x) = 0.5 - x \arctan x^3 \quad$ for $-20 \le x \le 0$

(b) $f(x) = \dfrac{\frac{1}{2} + \cos 3x - \frac{1}{4}x}{3 + 2 \cos x - \sin 2x} \quad$ for $-5 \le x \le 5$

(c) $f(x) = \dfrac{1}{x} - \sin x \quad$ for $0 < x \le 50$

3-3. The natural radian frequencies of vibration ω of a cantilever beam of length l are given by

$$\omega = kr^2$$

where r satisfies

$$\cosh rl \cos rl = -1$$

and where the constant k depends on the dimensions of the beam and the properties of the beam material. For $k = 2.7 \times 10^4$ m²/s and $l = 20$ m, find the values of the four smallest radian frequencies.

3-4. Use the BISECT program to find all simultaneous intersections of the pair of functions

$$f(x) = x \arctan x^2$$
$$g(x) = x^2 - 5x - 8$$

3-5. Use the BISECT program to find all solutions of

$$2x - \frac{\cos (x^2 - x)}{x} = 7$$

3-6. Find and sketch the 50-microvolt/meter field strength contour for the broadcast antenna array of Section 3.3.4.

3-7. For the broadcast antenna array of Section 3.3.4, suppose that there is an interruption of the signal of amplitude A_3 that drives the third antenna tower. Assuming that the third tower itself does not affect the radiation pattern under these circumstances, find the distance of the 100-microvolt/meter field strength contour in the $\theta = 170°$ and the $\theta = 210°$ directions.

3-8. Write, test, and debug a program to perform a zero-crossing search using the method of false position. Then use it to find the zeros of the following functions in the intervals given.

(a) $f(x) = \dfrac{1}{x^3} - \sin 4x \quad$ for $1 \le x \le 10$

(b) $f(x) = x \sin x - \arctan 2x \quad$ for $-5 \le x \le 5$

3-9. Modify the BISECT program to use direction reversal, as outlined in Section 3.3.3. Use the modified program to find the zeros of the following functions in the intervals given.

(a) $f(x) = 6 \sin 3x - 10 \cos 2x + 2$ for $-10 \le x \le 10$

(b) $f(x) = \sin(-3x^2 + x + 2)$ for $-2 \le x \le 2$

3-10. Use a combination of analysis and computation to find *all* the zeros of

$$f(x) = \frac{x}{\sqrt{1 + x^2}} \cos \frac{x}{3} + 0.5$$

3-11. Write, test, and debug a computer program to locate the zeros of a function by the Newton–Raphson method. Then use the program to find the zeros of the following functions in the intervals indicated.

(a) $f(x) = x^4 - 3x^3 + 2x^2 - 10$ for $-4 \le x \le 16$

(b) $f(x) = \dfrac{x}{x^2 + 7} - \arctan(0.3x)$ for $-3 \le x \le 3$

3-12. Modify the Newton–Raphson program of Problem 3-11 so that it uses a numerical derivative. Then find the zeros again.

3-13. Use the MINIMA program to help find the relative minima of the following functions in the intervals indicated.

(a) $f(x) = \dfrac{x}{2 + \sin x} + \dfrac{\sin x}{3 - \cos x}$ for $-5 \le x \le 5$

(b) $f(x) = \dfrac{\sin x}{x}$ for $-10 \le x \le 10$

(c) $f(x) = \text{sgn}(\cos \sqrt{x}) + 6x^2 - 10$ for $0 \le x \le 5$

(The signum function is defined as follows: sgn $y = 1$ if $y \ge 0$ and sgn $y = -1$ if $y < 0$.)

3-14. The Newton–Raphson slope search method for finding the zeros of a function $f(x)$ can be extended to achieve better accuracy by approximating the function with a parabola of the form

$$f_i(x) = a_i x^2 + b_i x + c_i$$

having the same value, slope, and inflection as $f(x)$ at each step $x = x_i$. By analogy with equation (3-6), show that the iteration equation for the parabolic approximation is

$$x_{i+1} = x_i - \frac{f(x_i)}{\dot{f}(x_i) - \frac{1}{2}\dfrac{\ddot{f}(x_i)f(x_i)}{\dot{f}(x_i)}}$$

3-15. Use the MINIMA program to help find the extrema (both relative maxima and relative minima) of the following functions in the indicated intervals. Do not consider the interval endpoints to be candidates for extrema.

(a) $f(x) = \dfrac{1}{x} - \dfrac{1}{x^2} + \dfrac{x}{x^2 + 4}$ for $1 \le x \le 20$

(b) $f(x) = \dfrac{1 + \cos 2x}{2 - \sin 3x}$ for $-5 \le x \le 5$

(c) $f(x) = \dfrac{x^2 - 3x + 2}{x^2 - x}$ for $0.5 \le x \le 4.5$

3-16. Use the MINIMA program to determine the zeros of the function

$$f(x) = |x^2 + 3x - 2 + \sin 4x|$$

in the interval $-5 \le x \le 5$.

3-17. For the automobile suspension described in Section 3.5.3, let the mass supported by the rear axle be, instead, $M = 700$ kg. Use the MINIMA program to find the new optimum damping constant B. The average compression of the spring is proportional to the supported mass, so it is possible to use the spring's average compression to adjust B automatically for various mass loads.

3-18. For the automobile suspension of Section 3.5.3 with $M = 450$ kg, $K = 1500$ kg/s^2, and $B = 1700$ kg/s, use the MINIMA program to find the additional mass, if any (maybe in the form of sacks of sand in the car's trunk), that will result in optimum damping.

3-19. Write, test, and debug a bisection program that locates both the relative maxima and the relative minima of a function during a single search. In addition to printing the located values of the variable and the function, the program should print whether it is a relative maximum or relative minimum.

3-20. Use the NEWTON program to find the relative extrema (both relative maxima and relative minima) of the following functions in the indicated intervals. Do not consider the interval endpoints to be candidates for extrema.
(a) $f(x) = x^3 + 14x^2 + 41x - 56$ $0 \le x \le 10$
(b) $f(x) = xe^{-0.2x^2}$ for all x

3-21. Use the NEWTON program to determine if the function

$$f(x) = x^4 - 3.6x^3 - 2.7x^2 - 4.3x - 5.5$$

is ever more negative than -60.

3-22. Study and report on the relative speed and the dependability of the NEWTON2 program in finding relative minimum nearest to $x = 0$ of

$$f(x) = \frac{\cos 3x + x}{3 - \cos x - x}$$

within a precision of 0.0001.

3-23. Use the NEWTON2 program to find the relative minima of the following functions in the intervals indicated.
(a) $f(x) = x \,|\, 3 \sin (x - 1)\,|$ for $-10 \le x \le 10$
(b) $f(x) = \dfrac{x^2 - 2x}{3 + \cos^2(x - 5)}$ for $0 \le x \le 15$

4

MULTIVARIABLE FUNCTIONS AND SEARCHES

4.1 PREVIEW

We now examine numerical methods for the solution of single and simultaneous nonlinear functions of more than a single variable. Many of these methods can be extended in a straightforward way to more nonlinear functions and more variables, but we concentrate on one and two functions of two variables because of their practical importance and to avoid getting bogged down with too many details.

When the variables involved are complex numbers, a function, nonlinear or not, is generally complex. By considering a complex variable

$$s = x + iy$$

to be composed of the two real variables x and y and by separately considering the real and imaginary parts of the function,

$$\Phi(s) = 0$$

such an equation is seen to be equivalent to two simultaneous real equations in two real variables:

$$\text{Re}[\Phi(s)] = f(x, y) = 0$$

$$\text{Im}[\Phi(s)] = g(x, y) = 0$$

In this way, complex functions of complex variables can be accommodated as twice as many simultaneous real functions of twice as many real variables. We will return to this equivalence in Chapter 5 when we apply the present methods to determination of polynomial roots.

In the next three sections, methods of finding solutions to a single nonlinear equation in two variables,

$$f(x, y) = 0$$

are discussed. Generally, these solutions are curves in the $x-y$ plane. In Section 4.2, collections of points on these curves are located by performing series of single-dimensional zero-crossing searches in a grid pattern. In Section 4.3 the solution curves are traced in much the same way as an automobile is steered along a road. In Section 4.4 we discuss successively refining grid searches so that points on the zero-crossing curves in any region are located with increasing precision. Computer programs for each method are designed and tested.

In Section 4.5, a very important method for locating the simultaneous solutions of two nonlinear equations in two variables,

$$f(x, y) = 0$$

$$g(x, y) = 0$$

is developed. The corresponding computer program is both simple and powerful. Positioning of a robot arm provides a nice, practical example of its use, and other applications are explored later. Newton–Raphson methods for zero location are outlined in Section 4.6.

Sections 4.7 and 4.8 concern numerical methods for locating the extrema of a function of more than one variable. When only a single variable is involved, the search for extrema involves only increasing or decreasing that variable. But when there are two or more variables, increments in those variables define a true two-dimentional *direction* of the search.

4.2 TWO-DIMENSIONAL ZERO-CROSSING LOCATION

An equation in two variables,

$$f(x, y) = 0$$

has solutions that are usually curves in the $x-y$ plane. For example, the solutions of a linear equation of the form

$$ax + by + c = 0$$

where a, $b \neq 0$ and c are constants, is a straight line

$$y = -\frac{a}{b}x - \frac{c}{b}$$

The solutions of the nonlinear equation

$$x^2 + \frac{y^2}{4} - 1 = 0$$

is an ellipse, centered on the origin and aligned with the x and y axes, with major axis of length 2 in the y direction and minor axis of length 1.

In the next several sections, numerical methods for locating closely spaced points on the zero-crossing curves of functions are developed. From these points, the user can

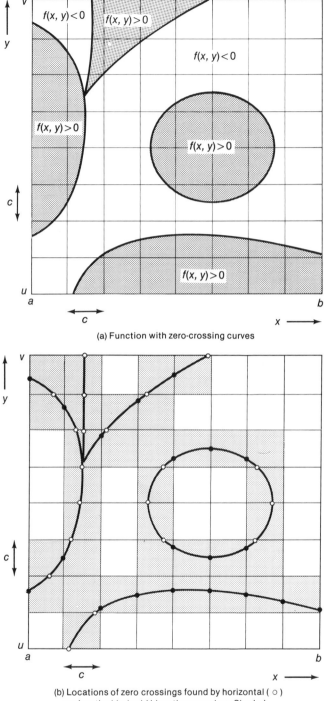

(a) Function with zero-crossing curves

(b) Locations of zero crossings found by horizontal (○) and vertical (●) grid bisection searches. Shaded squares are those where a zero crossing occurs on at least one side. Further search with a finer grid might be done for these squares.

Figure 4-1 Bisection search for zero-crossing curves of a two-variable function along a grid.

visualize the curve shapes. As in the single-variable case, we first look at zero crossings for which the function has different algebraic signs on either side of the crossing; if the function just touches zero (a "grazing" crossing), it will be necessary to find and evaluate the functions minima or use some other method.

4.2.1 Bisection Grid Method

One way of locating the zero-crossing curves of a function of two variables in a region is illustrated in Figure 4-1. A rectangular grid is established in the region, and bisection searches for zero crossings are done along each of the grid lines, both horizontal and vertical. Closely spaced points on nearly horizontal and nearly vertical sections of the curves are found, something that might not happen with a search along only the horizontal or only the vertical grid lines.

The grid lines must be spaced closely enough to achieve an adequate spacing of the zero-crossing points found and, of course, any crossing curve that is entirely within a grid square (or rectangle) will not be detected. Each grid line involves a separate one-dimensional bisection search, so the number of computations can become large. It is usually helpful to search for the zero-crossing curves with a relatively coarse grid first, then use a finer grid spacing in smaller regions of particular interest.

4.2.2 Bisection Program

A computer program in BASIC/FORTRAN for a two-variable bisection search for zero-crossing curves of a function $f(x, y)$ is given in Table 4-1. It is called CURVE. The program locates the values of the x-coordinates of zero crossings to within E for a grid of y values spaced at distance C. Then it locates the values of the y-coordinates of zero

---

TABLE 4-1

Computer Program in BASIC for Two-Variable Grid Bisection Search for Zero-Crossing Curves (CURVE)

VARIABLES USED

X = one search variable
Y = the other search variable
R = lower limit of the X variable
B = upper limit of the X variable
S = lower limit of the Y variable
V = upper limit of the Y variable
C = initial increment of the variables
E = smallest increment of the variables
D = present increment of X or of Y
F = value of the function; value of the function for largest X; value of the function for largest Y
G = value of the function for smallest X; value of the function for smallest Y
A = temporary lower limit of X or of Y

(continued)

LISTING

```
100     PRINT "2-VARIABLE ZERO-CROSSING BISECTION"
110     R = 0
120     B = 1
130     S = 0
140     V = 1
150     C = .1
160     E = .001
170   REM PRINT HEADING
180     PRINT "X","Y","F(X,Y)"
190   REM SEARCH ALONG X FOR EACH Y INCREMENT
200     Y = S
210     GOSUB 1000
220     Y = Y + C
230     IF Y > V THEN GOTO 250
240     GOTO 210
250   REM  SEARCH ALONG Y FOR EACH X INCREMENT
260     X = R
270     GOSUB 3000
280     X = X + C
290     IF X > B THEN GOTO 310
300     GOTO 270
310     END
1000  REM  SUBROUTINE FOR ZERO-CROSSING BISECTION
1010  REM  ALONG X DIRECTION
1020  REM  START AT LEFT BOUNDARY
1030    A = R
1040  REM  SET INCREMENT TO INITIAL INCREMENT
1050    D = C
1060  REM  EVALUATE FUNCTION AT LEFT BOUNDARY
1070    X = A
1080    GOSUB 2000
1090    G = F
1100  REM  INCREMENT VARIABLE AND EVALUATE FUNCTION
1110  REM  END IF RIGHT BOUNDARY EXCEEDED
1120    X = X + D
1130    IF X > B THEN GOTO 1290
1140    GOSUB 2000
1150  REM  IF NO FUNCTION SIGN CHANGE, CONTINUE SEARCH
1160    IF G * F > = 0 THEN GOTO 1090
1170  REM  PRINT ROOT AND RESTART IF INCREMENT IS SMALLER THAN E
1180    IF D < E THEN GOTO 1230
1190  REM  MOVE X BACK, HALVE INCREMENT, AND REEVALUATE
1200    X = X - D
1210    D = D / 2
1220    GOTO 1100
1230  REM  PRINT ROOT AND RESTART SEARCH
1240    PRINT X,Y,F
1250  REM  RESET LEFT BOUNDARY TO ROOT PLUS C
1260  REM  AND CONTINUE SEARCH
1270    A = X + C
1280    GOTO 1040
1290    RETURN
2000  REM  SUBROUTINE TO EVALUATE FUNCTION
2010    F = SIN(6 * X) + 2 * COS(5 * Y)
2020    RETURN
3000  REM  SUBROUTINE FOR ZERO-CROSSING BISECTION
3010  REM  ALONG Y DIRECTION
3020  REM  START AT LOWER BOUNDARY
3030    A = S
3040  REM  SET INCREMENT TO INITIAL INCREMENT
3050    D = C
3060  REM  EVALUATE FUNCTION AT LOWER BOUNDARY
3070    Y = A
```

```
3080      GOSUB 2000
3090      G = F
3100  REM  INCREMENT VARIABLE AND EVALUATE FUNCTION
3110  REM  END IF UPPER BOUNDARY EXCEEDED
3120      Y = Y + D
3130      IF Y > V THEN GOTO 3290
3140      GOSUB 2000
3150  REM  IF NO FUNCTION SIGN CHANGE, CONTINUE SEARCH
3160      IF G * F > = 0 THEN GOTO 3090
3170  REM  PRINT ROOT AND RESTART IF INCREMENT IS SMALLER THAN E
3180      IF D < E THEN GOTO 3230
3190  REM  MOVE Y BACK, HALVE INCREMENT, AND REEVALUATE
3200      Y = Y - D
3210      D = D / 2
3220      GOTO 3100
3230  REM  PRINT ROOT AND RESTART SEARCH
3240      PRINT X,Y,F
3250  REM  RESET LOWER BOUNDARY TO ROOT PLUS C
3260  REM  AND CONTINUE SEARCH
3270      A = Y + C
3280      GOTO 3040
3290      RETURN
```

Computer Program in FORTRAN for Two-Variable Grid Bisection Search for Zero-Crossing Curves (CURVE)

```
C *********************************************************************
C PROGRAM CURVE -- 2-VARIABLE ZERO-CROSSING BISECTION
C *********************************************************************

      WRITE(*,'(1X,''2-VARIABLE ZERO-CROSSING BISECTION'')')

      R = 0.0
      B = 1.0
      S = 0.0
      V = 1.0
      C = 0.1
      E = 0.001

C *** PRINT HEADING
      WRITE(*,'(1X,''         X              Y              F(X,Y)'')')

C *** SEARCH ALONG X FOR EACH Y INCREMENT, MAXIMUM OF 1000 LOOPS
      Y = S
      DO 10 I=1, 1000
         CALL ZCBX(R,C,B,E,X,Y)
         Y = Y + C
         IF(Y.GT.V) GOTO 20
 10   CONTINUE

 20   CONTINUE

C *** SEARCH ALONG Y FOR EACH X INCREMENT, MAXIMUM OF 1000 LOOPS
      X = R
      DO 30 I=1, 1000
         CALL ZCBY(S,C,V,E,X,Y)
         X = X + C
         IF(X.GT.B) STOP
 30   CONTINUE

      STOP
      END
```

(continued)

```
C *********************************************************************
      SUBROUTINE ZCBX(R,C,B,E,X,Y)
C *********************************************************************

C *** SUBROUTINE FOR ZERO-CROSSING BISECTION
C *** ALONG X DIRECTION
C *** START AT LEFT BOUNDARY
      A = R

C *** LOOP MAXIMUM OF 1000 TIMES
C *** SET INCREMENT TO INITIAL INCREMENT
      DO 10 I=1, 1000
         D = C

C *** EVALUATE FUNCTION AT LEFT BOUNDARY
         X = A
         CALL FUNCT(X,Y,F)
         G = F

C *** INCREMENT VARIABLE AND EVALUATE FUNCTION, MAXIMUM OF 1000 LOOPS
C *** END IF RIGHT BOUNDARY EXCEEDED
         DO 20 J=1, 1000
            X = X + D
            IF(X.GT.B) RETURN
            CALL FUNCT(X,Y,F)

C *** IF NO FUNCTION SIGN CHANGE, CONTINUE SEARCH
            IF(G*F.GE.0.0) THEN
               G=F
            ELSE

C *** PRINT ROOT AND RESTART IF INCREMENT IS SMALLER THAN E
               IF(D.LT.E) GOTO 30

C *** MOVE X BACK, HALVE INCREMENT, AND REEVALUATE
               X = X - D
               D = D/2.0
            ENDIF

 20      CONTINUE

C *** PRINT ROOT AND RESTART SEARCH
 30      WRITE(*,'(1X,E15.8,1X,E15.8,1X,E15.8)') X,Y,F

C *** RESET LEFT BOUNDARY TO ROOT PLUS C
C *** AND CONTINUE SEARCH
         A = X + C
 10   CONTINUE
      RETURN
      END

C *********************************************************************
      SUBROUTINE ZCBY(S,C,V,E,X,Y)
C *********************************************************************

C *** SUBROUTINE FOR ZERO-CROSSING BISECTION
C *** ALONG Y DIRECTION
C *** START AT LOWER BOUNDARY
      A = S

C *** LOOP MAXIMUM OF 1000 TIMES
C *** SET INCREMENT TO INITIAL INCREMENT
      DO 10 I=1, 1000
         D = C
```

```
C *** EVALUATE FUNCTION AT LOWER BOUNDARY
      Y = A
      CALL FUNCT(X,Y,F)
      G = F

C *** INCREMENT VARIABLE AND EVALUATE FUNCTION, MAXIMUM OF 1000 LOOPS
C *** END IF UPPER BOUNDARY EXCEEDED
      DO 20 J=1, 1000
         Y = Y + D
         IF(Y.GT.V) RETURN
         CALL FUNCT(X,Y,F)

C *** IF NO FUNCTION SIGN CHANGE, CONTINUE SEARCH
         IF(G*F.GE.0.0) THEN
            G=F
         ELSE

C *** PRINT ROOT AND RESTART IF INCREMENT IS SMALLER THAN E
            IF(D.LT.E) GOTO 30

C *** MOVE Y BACK, HALVE INCREMENT, AND REEVALUATE
            Y = Y - D
            D = D/2.0
         ENDIF

  20     CONTINUE

C *** PRINT ROOT AND RESTART SEARCH
  30     WRITE(*,'(1X,E15.8,1X,E15.8,1X,E15.8)') X,Y,F

C *** RESET LOWER BOUNDARY TO ROOT PLUS C
C *** AND CONTINUE SEARCH
         A = Y + C
  10  CONTINUE
      RETURN
      END

C *****************************************************************
      SUBROUTINE FUNCT(X,Y,F)
C *****************************************************************

C *** SUBROUTINE TO EVALUATE FUNCTION
      F = SIN(6.0*X) + 2.0*COS(5.0*Y)
      RETURN
      END
```

crossings to within E for a grid of x values spaced at distance C. The search extends from $x = $ R to $x = $ B and from $y = $ S to $y = $ V.

The BISECT program of Table 3-3 is renumbered as subroutine 1000, which is simply called repeatedly for the various y values. The limits and parameters of the search are written (in lines 110–160) as part of the program, and these are to be changed as needed by the user. The function itself is computed in subroutine 2000. The function listed is one for an example to follow. For the search in the y direction for various x values, the BISECT is repeated as subroutine 2000, with search limits changed to those for y and the search variable changed from x to y.

This is perhaps a good example with which to point out that programs can be much easier to follow and modify when common sense is used in composing them. It is true that subroutine 3000 could have been omitted and subroutine 2000 used to perform both the x- and y-direction searches. This would require, however, a fairly complicated swapping of variables for at least one of the two search directions and for printing. Although the program is longer (although not greatly so) this way, program memory space is seldom at a premium nowadays for a program of modest requirements such as this one. The program's running time is not appreciably affected by the switching of subroutines.

The CURVE program produces the following results for the function

$$f(x) = \sin 6x + 2 \cos 5y$$

when the search is from $x = 0$ to $x = 1$ and from $y = 0$ to $y = 1$.

TWO-VARIABLE ZERO-CROSSING BISECTION

X	Y	F(X, Y)	X	Y	F(X, Y)
.54765625	.3	$-2.36971785E{-}03$	.3	.84140625	$5.60666212E{-}03$
.1640625	.4	$6.32715679E{-}04$	.4	.38359375	$-5.0174028E{-}03$
.36015625	.4	$-1.43154544E{-}03$	.4	.87421875	$6.04739052E{-}03$
.07265625	.9	$6.8813604E{-}04$	.5	.32890625	$-6.21624647E{-}03$
.4515625	.9	$-2.70604725E{-}03$	.5	.92890625	$5.50868053E{-}03$
0	.31484375	$-6.84483404E{-}03$	.6	.2703125	$-7.55673993E{-}03$
.1	.371875	$-4.53744393E{-}03$	.7	.22421875	$-2.18037294E{-}03$
.1	.8859375	$6.74065553E{-}03$	.8	.21015625	$-2.37816705E{-}03$
.2	.41171875	$-5.3235985E{-}03$	.9	.23515625	$-3.12027614E{-}03$
.2	.84609375	$5.07597336E{-}03$			
.3	.41640625	$-4.66166252E{-}03$			

The zeros found are plotted in Figure 4-2, in which the curves

$$f(x, y) = 0$$

are sketched. Unfortunately, the computer industry still lacks widely accessible, low-cost, machine-independent graphics capability. It is relatively easy for the user to learn a particular system and to add curve plotting to a program such as this one, if desired.

Graphics instructions are normally a part of the BASIC language and involve such commands as CLEAR (which clears the screen), PLOT X,Y (which places a dot at coordinates X,Y), and PLOT X,Y to U,V (which draws a line from coordinates X,Y to coordinates U,V). It is necessary, however, to deal with fixed plotting ranges for the variables, usually related to the number of independent dots (pixels) that can be displayed in the horizontal and vertical directions, or first to define the limits of the variables that the user wants to be the boundaries of the screen. The situation is similar for making plots on paper with a printer or plotter, where the plot boundaries are limited by the paper size. Often, there are also many other instructions available, for drawing and labeling axes, for using color, for automatic scaling, and other functions.

For those readers who wish to use their machine's graphic capabilities, we suggest keeping things simple at first. Begin by drawing your own axes with straight lines, for instance, instead of investing hours in learning the more sophisticated instructions before producing a single graph.

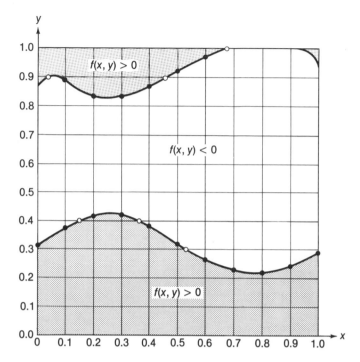

Figure 4-2   Results of the bisection search for zero crossings of the function $f(x, y) = \sin 6x + 2 \cos 5y$.

4.3 ZERO-CURVE TRACKING

Another way of determining closely spaced points on a function's zero-crossing curves is to locate a segment of a curve and then attempt to follow or *track* that curve. Instead of searching for points on the curve all over a large region, only the areas immediately ahead of the previously found points on the curve are examined. This is much like the process of uncovering an overgrown trail in the woods or a jungle. There is, of course, the likelihood of tracing a closed zero-crossing curve over and over again. For a curve with a "figure eight" shape, it is possible that only one loop of the curve will be tracked. If the curve abruptly changes direction, there is danger, too, that it may be "lost." However, when the character of the curve and the shortcomings of the tracking method used are known to the user, tracking can be a good, efficient way of finding closely spaced points on a zero-crossing curve.

4.3.1 Tracking Zero-Crossing Curves

Figure 4-3 shows a simple, effective way of doing zero-crossing curve tracking. First, two nearby points on the curve are located. This can be done by performing a grid search until one point on the curve is located, then moving upward a small distance w and searching between grid boundaries for a second point on the curve, as shown. Then one moves a distance w in the direction of the line through these two points and searches along a path perpendicular to that line. At the next step, the most recent two points found on the curve define the direction of a step of distance w in the approximate direction of

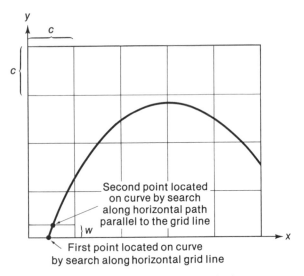

(a) Finding an initial two closely spaced points
on the zero-crossing curve

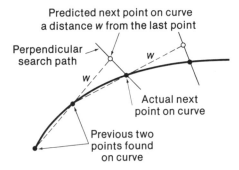

(b) Perpendicular search path

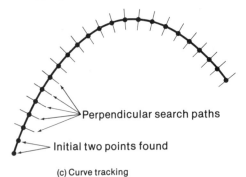

(c) Curve tracking

Figure 4-3 Curve tracking with linear prediction and perpendicular search.

the curve. A point on the curve is found by searching perpendicular to that step, and so on. The points found in this way are spaced distances of approximately w along the curve.

Figure 4-4(a) shows desired geometric relationships between the previous two zero locations found (u, v) and (x, y) and the endpoints of the perpendicular search path for

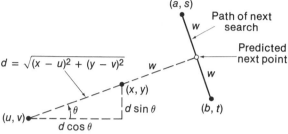

(a) Relationships between the previous two zeros found
and the path of the search for the next zero

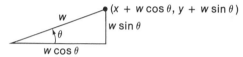

(b) Center point of the search path

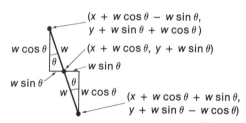

(c) Endpoints of the search path

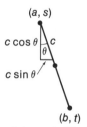

(d) Increments in x and y if the total increment is c

Figure 4-4 Finding the search boundaries and increments.

the next zero-crossing point. The length of the search path has been chosen to be $2w$, centered on the predicted next point. In terms of the coordinates,

$$\cos \theta = \frac{x - u}{d} \qquad \sin \theta = \frac{y - v}{d}$$

where

$$d = \sqrt{(x - u)^2 + (y - v)^2}$$

From Figure 4-4(b) and (c), the starting point of the next search path has coordinates (a, s), where

$$a = x + w \cos \theta - w \sin \theta = x + \frac{w(x - u - y + v)}{d}$$

$$s = y + w \sin \theta + w \cos \theta = y + \frac{w(y - v + x - u)}{d}$$

Similarly, the ending point of this search path has coordinates (b, t), where

$$b = x + w \cos \theta + w \sin \theta = x + \frac{w(x - u + y - v)}{d}$$

$$t = y + w \sin \theta - w \cos \theta = y + \frac{w(y - v - x + u)}{d}$$

From Figure 4-4(d), an increment of c units along this path has x and y components that are

$$p = c \sin \theta = \frac{c(y - v)}{d}$$

and

$$q = -c \cos \theta = \frac{c(u - x)}{d}$$

respectively.

4.3.2 Zero-Crossing Tracking Program

This method of zero-crossing tracking is expressed as step-by-step algorithm in Table 4-2. It is programmed in the TRACK program of Table 4-3. The starting parameters are set in lines 120–160, and the function to be tracked is given in line 2010 of the program. These are to be changed by the user as needed. Subroutine 1000 is a slightly modified version of the bisection subroutines in the previous CURVE program (Table 4-1). In it, X is incremented by P and Y is incremented by Q at each step. Absolute values of P and of Q are used when comparing with the desired precision E, since P and/or Q can be negative.

When the TRACK program is used to trace the zero-crossing curve of the function

$$f(x, y) = x^2 + y^2 - 4$$

with the listed parameters, the result is as follows:

```
ZERO-CROSSING CURVE TRACKING
ZERO-CROSSING COORDINATES
X                   Y
2                   0
1.99755859          .1
1.9900413           .19984627
1.97756727          .299190136
1.96007013          .397778356
```

TABLE 4-2 Algorithm for Tracking Two-Dimensional Zero-Crossing Curves

DEFINITIONS

For a zero-crossing curve $f(x, y) = 0$, points (x, y) spaced a distance of approximately w along the curve are to be found, with the coordinates of every point located with a precision of at least ϵ each. The search begins at coordinates (a, y) and seeks an initial zero crossing with a bisection pattern holding y constant and incrementing x with initial step size c. Other variables used are

p	$=$	present x step size
q	$=$	present y step size
(u, v)	$=$	coordinates of previous zero crossing found
α	$=$	sine of the angle from the present zero crossing to the starting point for the next zero-crossing search
β	$=$	cosine of the angle from the present zero crossing to the starting point for the next zero-crossing search

ALGORITHM

1. Use the bisection process to find a zero of $f(x, y)$ (if there is one) using the starting coordinates (a, y) and incrementing x. Find the x coordinate of the zero crossing within ϵ, using an initial step size c. Store the coordinates of this zero as (u, v):

$$u \leftarrow x$$

$$v \leftarrow y$$

(lines 200–270 and subroutine 1000).

2. Use bisection to find another zero of $f(x, y)$ with the starting coordinates $(a, y + w)$ and incrementing x. Find the x coordinate of the zero crossing within ϵ, using an initial step size c. Let the crossing coordinates found be (x, y) (lines 280–330 and subroutine 1000).

3a. Calculate starting coordinates for a search for the next zero crossing of $f(x, y)$, approximately a distance w from the present crossing, in the sense from the coordinates (u, v) to the coordinates (x, y): Save the last zero-crossing coordinates as (u, v).

$$\alpha = \sin \theta \leftarrow \frac{x - u}{\sqrt{(x - u)^2 + (y - v)^2}}$$

$$\beta = \cos \theta \leftarrow \frac{y - v}{\sqrt{(x - u)^2 + (y - v)^2}}$$

$$u \leftarrow x$$

$$v \leftarrow y$$

$$x \leftarrow w(\alpha - \beta)$$

$$y \leftarrow w(\alpha + \beta)$$

(lines 340–420).

(continued)

3b. Calculate the initial increments in x and in y that are to be taken in the next bisection search:

$$p \leftarrow c\alpha$$

$$q \leftarrow -c\beta$$

(lines 430–440).

3c. Use bisection to find another zero of $f(x, y)$ with the starting coordinates (x, y) incrementing x with an initial step size p and y with an initial step size q. Find the x and y coordinates of the crossing each within ϵ (lines 450–460 and subroutine 1000).

4. Go to step 3 and find another zero crossing (lines 470–480).

TABLE 4-3

Computer Program in BASIC to Track Two-Dimensional Zero-Crossing Curves (TRACK)

VARIABLES USED

A	=	initial x coordinate, cosine of angle θ
X	=	x search variable
Y	=	initial y coordinate, y search variable
C	=	initial step size
E	=	desired precision
W	=	spacing between points to be located on the curve
P	=	present x step size
Q	=	present y step size
U	=	x coordinate of previous zero crossing found
V	=	y coordinate of previous zero crossing found
F	=	present value of the function
G	=	previous value of the function
D	=	distance between previous and present zero-crossing points found
B	=	sine of angle θ

LISTING

```
100     PRINT "ZERO-CROSSING CURVE TRACKING"
110   REM  SET PARAMETERS
120     A = 0
130     Y = 0
140     C = .1
150     E = .0001
160     W = .1
170   REM  PRINT HEADING
180     PRINT "ZERO-CROSSING COORDINATES"
190     PRINT "X","Y"
200   REM  FIND INITIAL ZERO WITH HORIZONTAL SEARCH
210     X = A
220     P = C
230     Q = 0
240     GOSUB 1000
250     PRINT X,Y
260     U = X
270     V = Y
```

```
280    REM  FIND NEXT ZERO WITH HORIZONTAL SEARCH
290       X = A
300       Y = Y + W
310       P = C
320       GOSUB 1000
330       PRINT X,Y
340    REM FIND NEXT ZERO
350       FOR I=1 TO 100
360       D = SQR ((X - U) * (X - U) + (Y - V) * (Y -V))
370       A = (X - U) / D
380       B = (Y - V) / D
390       U = X
400       V = Y
410       X = U + ABS (W) * (A - B)
420       Y = V + ABS (W) * (A + B)
430       P = C * B
440       Q = -C * A
450       GOSUB 1000
460       PRINT X,Y
470    REM  CONTINUE SEARCH
480       NEXT I
490       END
1000   REM  SUBROUTINE TO FIND ROOTS OF A 2-D FUNCTION
1010   REM  EVALUATE FUNCTION
1020      GOSUB 2000
1030      H = F
1040   REM  INCREMENT VARIABLES AND EVALUATE FUNCTION
1050      X = X + P
1060      Y = Y + Q
1070      GOSUB 2000
1080   REM  IF NO SIGN CHANGE, CONTINUE SEARCH
1090      IF H * F > 0 THEN GOTO 1030
1100   REM  REFINE SEARCH IF INCREMENTS EXCEED E
1110      IF ABS (P) > E THEN GOTO 1140
1120      IF ABS (Q) > E THEN GOTO 1140
1130      GOTO 1200
1140   REM  REFINE SEARCH
1150      X = X - P
1160      Y = Y - Q
1170      P = P / 2
1180      Q = Q / 2
1190      GOTO 1040
1200      RETURN
2000   REM  SUBROUTINE TO EVALUATE FUNCTION
2010      F = X * X + Y * Y - 4
2020      RETURN
```

Computer Program in FORTRAN to Track Two-Dimensional Zero-Crossing Curves (TRACK)

```
C ***********************************************************************
C PROGRAM TRACK -- ZERO CROSSING CURVE TRACKING
C ***********************************************************************

      WRITE(*,'(1X,''ZERO-CROSSING CURVE TRACKING'')')

C *** SET PARAMETERS
      A = 0.0
      Y = 0.0
      C = 0.1
      E = 0.0001
      W = 0.1
```

(continued)

```
C *** PRINT HEADING
      WRITE(*,'(1X,''ZERO-CROSSING COORDINATES'')')
      WRITE(*,'(1X,''        X                Y'')')

C *** FIND INITIAL ZERO WITH HORIZONTAL SEARCH
      X = A
      P = C
      Q = 0.0
      CALL ROOT2D(X,Y,P,Q)
      WRITE(*,'(1X,E15.8,1X,E15.8)') X,Y
      U = X
      V = Y

C *** FIND NEXT ZERO WITH HORIZONTAL SEARCH
      X = A
      Y = Y + W
      P = C
      CALL ROOT2D(X,Y,P,Q)
      WRITE(*,'(1X,E15.8,1X,E15.8)') X,Y

C *** FIND NEXT ZERO, MAXIMUM OF 100 ITERATIONS
      DO 10 I=1, 100
         D = SQRT((X - U)*(X - U) + (Y - V)*(Y - V))
         A = (X - U)/D
         B = (Y - V)/D
         U = X
         V = Y
         X = U + ABS(W)*(A - B)
         Y = V + ABS(W)*(A + B)
         P = C*B
         Q = -C*A
         CALL ROOT2D(X,Y,P,Q)
         WRITE(*,'(1X,E15.8,1X,E15.8)') X,Y

C *** CONTINUE SEARCH
   10 CONTINUE

      STOP
      END

C ********************************************************************
      SUBROUTINE ROOT2D(X,Y,P,Q)
C ********************************************************************

C *** SUBROUTINE TO FIND ROOTS OF A 2-D FUNCTION
C *** EVALUATE FUNCTION
      CALL FUNCT(X,Y,F)
      H = F

C *** INCREMENT VARIABLES AND EVALUATE FUNCTION, MAXIMUM OF 1000 LOOPS
      DO 10 II=1, 1000
         X = X + P
         Y = Y + Q
         CALL FUNCT(X,Y,F)

C *** IF NO SIGN CHANGE, CONTINUE SEARCH
         IF (H*F.LE.0.0) THEN

C *** REFINE SEARCH IF INCREMENTS EXCEED E
            IF ((ABS(P).GT.E).OR.(ABS(Q).GT.E)) THEN

C *** REFINE SEARCH
               X = X - P
               Y = Y - Q
               P = P/2.0
```

```
          Q = Q/2.0
      ELSE
          RETURN
      ENDIF

  ELSE
      H=F
  ENDIF

10    CONTINUE

  END

C ********************************************************************
      SUBROUTINE FUNCT(X,Y,F)
C ********************************************************************

C *** SUBROUTINE TO EVALUATE FUNCTION
      F = X*X + Y*Y - 4.0
      RETURN
      END
```

The function $f(x)$ has a zero-crossing curve that is a circle about the origin of radius 2. If left running, the program will trace the circle over and over again. Making W smaller in magnitude results in closer spacing of the points found. For larger spacing, the magnitude of W can be increased to about unity before there is failure to track the curve.

When W is positive, the program searches for zero crossings first in the direction of increasing y. To traverse the curve in the opposite direction, a negative value of the point spacing W should be used, for example

```
160 W = -0.1
```

This causes the second zero crossing found to be at a negative value of Y since the search then begins with a negative increment of Y. The magnitude of W is used, however, in

```
400 X = U + ABS (W) * (A - B)
410 Y = V + ABS (W) * (A + B)
```

in order to proceed in the correct predicted direction.

4.4 REFINED GRID SEARCH

Another way of finding closely spaced points along zero-crossing curves of a function of two variables is to divide a square search region into smaller squares that have a zero crossing of the function along one or more of their sides. Actually, one can just as well deal with a rectangular search region and smaller rectangles, but for simplicity of description, we will deal with squares here.

Having located those smaller squares through which the zero-crossing curves of the function pass, these can in turn be subdivided and tested for border zero crossings. By continuing the subdivision and testing process enough times, a desired degree of accuracy of curve location is achieved. Generally, the number of squares to be tested further at each step is much less than the total number of squares, so the search is much more efficient than simply subdividing into an extremely large number of squares and searching all of them.

4.4.1 Finding Zero-Curve Regions

Table 4-4 lists an algorithm for finding all the squares through which a function's zero-crossing curves pass. A square search region with lower left at coordinates (r, s) and sides of length d is divided into m^2 smaller squares with sides of length $c = d/m$, as shown in Figure 4-5(a) for the case of $m = 4$. The coordinates of each of the smaller squares are those of its lower left corner. These coordinates are expressed in terms of the variables x and y or, alternatively in terms of the integer indices i and j, as in Figure 4-5(b).

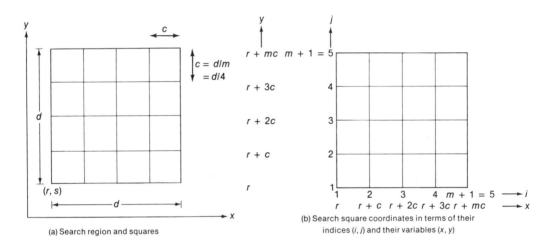

(a) Search region and squares

(b) Search square coordinates in terms of their indices (i, j) and their variables (x, y)

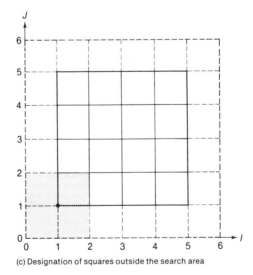

(c) Designation of squares outside the search area

Figure 4-5   Search region and search squares of the refined grid search algorithm for $m = 4$.

TABLE 4-4   Algorithm for Finding Squares Containing a Zero-Crossing Curve

DEFINITIONS

A square region with lower left corner coordinates (r, s) and sides of length d is divided into a grid of smaller squares, m of them to a side of the search region. Each of the smaller squares is searched for zero crossings of $f(x, y)$ on each of its four sides. If there is a zero crossing on any side, the coordinates of the lower left corner of the smaller square are printed.

ALGORITHM

1. Set the values of the array a_{ij} to the function at the coordinates $x = r + (i - 1)c$, $y = s + (j - 1)c$:

$$a_{ij} \leftarrow f[r + (i - 1)c, s + (j - 1)c] \qquad i = 1, 2, \ldots, m + 1; \quad j = 1, 2, \ldots, m + 1$$

and set the array l_{ij} to zero:

$$l_{ij} = 0 \qquad i = 1, 2, \ldots, m + 1; \quad j = 1, 2, \ldots, m + 1$$

(lines 260–360).

2. Identify the squares having zero crossings along their sides (lines 370–520).

   a. If $a_{ij} = 0$, set the four array elements representing the four squares sharing the point with indices (i, j) to unity:

$$l_{ij} = 1$$

$$l_{i-1,j} = 1$$

$$l_{i-1,j-1} = 1$$

$$l_{i,j-1} = 1$$

   (lines 400–440).

   b. If there is a zero crossing between indices (i, j) and $(i + 1, j)$, set the two array elements representing the two squares sharing this side to unity:

$$l_{ij} = 1$$

$$l_{i,j-1} = 1$$

   (lines 450–470).

   c. If there is a zero crossing between indices (i, j) and $(i, j + 1)$, set the two array elements representing the two squares sharing this side to unity:

$$l_{ij} = 1$$

$$l_{i-1,j} = 1$$

   (lines 480–500).

3. Print the coordinates (x, y) of the lower left corners of all the squares having a zero crossing of $f(x, y)$ on their boundary (lines 530–580).

In step 1, the value of the function tested, $f(x, y)$, is found at each of the points on the grid. These are stored as elements of the array a_{ij} in terms of the integer coordinates of the square (i, j). The elements of another array l_{ij} are all set to zero.

In step 2, the values of the function at the grid points, a_{ij}, are compared to determine which of the squares have one or more sides with a zero crossing. If $a_{ij} = 0$, four squares share that point and the four elements of l_{ij} having the coordinates of those squares are then set to unity. If there is an algebraic sign change between a_{ij} and $a_{i+1,j}$, the function has a zero crossing along the side shared by two squares. The two elements of l_{ij} with the coordinates of these squares are then set to unity. If there is an algebraic sign change between a_{ij} and $a_{i,j+1}$, the two elements of l_{ij} with the coordinates of the two squares sharing the side involved are set to unity. In step 3, the (x, y) coordinates of the squares with one or more zero crossings along their sides, for which $a_{ij} = 1$, are printed.

4.4.2 Refined Grid Search Program

A program, called SCAN, that implements the foregoing algorithm is described in Table 4-5. To keep the programming simple, the L array elements for squares bordering the search region are allowed to be set by the program, even though the function is not evaluated at points outside the original grid. For example, a zero-crossing point at the grid intersection along the border of a search region, as in Figure 4-5(c), will result in elements of L corresponding to squares outside the search region being set to zero, even though they will not be considered further.

TABLE 4-5

Computer Program in BASIC to Find Squares Containing a Zero-Crossing Curve (SCAN)

VARIABLES USED

R	=	x coordinate of lower left point of search region
S	=	y coordinate of lower left point of search region
D	=	dimensions of square search region
M	=	number of divisions in search grid; cannot exceed 20 unless dimension instructions are changed
C	=	dimensions of a search grid square
X	=	x coordinate of a point on the search grid
Y	=	y coordinate of a point on the search grid
F	=	value of the function, calculated by subroutine 1000
A (22,22)	=	array of values of the function F at the grid intersections, as a function of grid indices
L (22,22)	=	array of squares containing zero crossings of the function F, as a function of grid indices; if the square at (I,J) has a crossing, L(I,J) = 1; otherwise, L(I,J) = 0
I	=	X grid index, I = 1 corresponds to X = R
J	=	Y grid index, J = 1 corresponds to Y = S

LISTING

```
100      PRINT "SCAN FOR ZERO-CROSSING SQUARES"
110      DIM A(22,22) ,L(22,22)
120   REM   ENTER BOUNDARIES AND GRID
130      PRINT "ENTER SEARCH LOWER LEFT COORDINATES"
140      PRINT "X = ";
150      INPUT R
160      PRINT "Y = ";
170      INPUT S
180      PRINT "ENTER SEARCH DIMENSION";
190      INPUT D
200      PRINT "ENTER NUMBER OF GRID DIVISIONS";
210      INPUT M
220      C = D / M
240      PRINT "LOWER LEFT COORDINATES OF SQUARES ARE"
250      PRINT "X","Y"
260   REM   STORE GRID POINT VALUES OF FUNCTION AND
270   REM   ZERO THE L ARRAY
280      FOR J = 1 TO M + 1
290      FOR I = 1 TO M + 1
300      X = R + (I - 1) * C
310      Y = S + (J - 1) * C
320      GOSUB 1000
330      A(I,J) = F
340      L(I,J) = 0
350      NEXT I
360      NEXT J
370   REM   IDENTIFY SQUARES WITH ZERO CROSSINGS
380      FOR J = 1 TO M + 1
390      FOR I = 1 TO M + 1
400      IF A(I,J) <  > 0 THEN  GOTO 450
410      L(I,J) = 1
420      L(I - 1,J) = 1
430      L(I - 1,J - 1) = 1
440      L(I, J - 1) = 1
450      IF A(I,J) * A(I + 1,J)  > = 0 THEN GOTO 480
460      L(I,J) = 1
470      L(I,J - 1) = 1
480      IF A(I,J) * A(I,J + 1) > = 0 THEN GOTO 510
490      L(I,J) = 1
500      L(I - 1,JJ) = 1
510      NEXT I
520      NEXT J
530   REM   PRINT COORDINATES
540      FOR J = 1 TO M
550      FOR I = 1 TO M
560      IF L(I,J) = 1 THEN  PRINT R +(I - 1) * C,S + (J - 1) * C
570      NEXT I
580      NEXT J
590   REM   PRINT SQUARE SIZE
600      PRINT "SQUARE DIMENSION IS ";C
610      END
1000  REM    SUBROUTINE TO EVALUATE FUNCTION
1010     F = SIN (6 * X) + 2 * COS (5 * Y)
1020     RETURN
```

Computer Program in FORTRAN to Find Squares Containing a Zero-Crossing Curve (SCAN)

```
C ********************************************************************
C PROGRAM SCAN -- SCAN FOR ZERO-CROSSING SQUARES
C ********************************************************************
      REAL A(22,22)
```

(continued)

```
      INTEGER L(22,22)

      WRITE(*,'(1X,''SCAN FOR ZERO-CROSSING SQUARES'')')

C *** ENTER BOUNDARIES AND GRID
      WRITE(*,'(1X,''ENTER SEARCH LOWER LEFT COORDINATES'')')
      WRITE(*,'(1X,''X = '',\)')
      READ(*,*) R
      WRITE(*,'(1X,''Y = '',\)')
      READ(*,*) S
      WRITE(*,'(1X,''ENTER SEARCH DIMENSION '',\)')
      READ(*,*) D
      WRITE(*,'(1X,''ENTER NUMBER OF GRID DIVISIONS '',\)')
      READ(*,*) M
      C = D/M
      WRITE(*,'(1X,''LOWER LEFT COORDINATES OF SQUARES ARE'')')
      WRITE(*,'(1X,''        X                 Y'')')

C *** STORE GRID POINT VALUES OF FUNCTION AND
C *** ZERO THE L ARRAY

      DO 10 J=1, M+1
         DO 20 I=1, M+1
            X = R + (I-1)*C
            Y = S + (J-1)*C
            CALL FUNCT(X,Y,F)
            A(I,J) = F
            L(I,J) = 0
 20      CONTINUE
 10   CONTINUE

C *** IDENTIFY SQUARES WITH ZERO CROSSINGS
      DO 30 J=1, M+1
         DO 40 I=1, M+1
            IF (A(I,J).EQ.0.0) THEN
               L(I+1,J+1) = 1
               L(I,J+1) = 1
               L(I,J) = 1
               L(I+1,J) = 1
            ENDIF
            IF (A(I,J)*A(I+1,J).LT.0.0) THEN
               L(I+1,J+1) = 1
               L(I+1,J) = 1
            ENDIF
            IF (A(I,J)*A(I,J+1).LT.0.0) THEN
               L(I+1,J+1) = 1
               L(I,J+1) = 1
            ENDIF
 40      CONTINUE
 30   CONTINUE

C *** PRINT COORDINATES
      DO 50 J=1, M
         DO 60 I=1, M
            IF (L(I+1,J+1).EQ.1) WRITE(*,'(1X,E15.8,1X,E15.8)')
     *          R+(I-1)*C, S+(J-1)*C
 60      CONTINUE
 50   CONTINUE

C *** PRINT SQUARE SIZE
      WRITE(*,'(1X,''SQUARE DIMENSION IS '',E15.8)') C
      STOP
      END
```

```
C ****************************************************************
      SUBROUTINE FUNCT(X,Y,F)
C ****************************************************************

C *** SUBROUTINE TO EVALUATE FUNCTION
      F = SIN (6 * X) + 2 * COS (5 * Y)
      RETURN
      END
```

When used with the function listed, which was explored in Section 4.2.2, the SCAN program gives these results:

```
SCAN FOR ZERO-CROSSING SQUARES
ENTER SEARCH LOWER LEFT COORDINATES
X  =  ?0
Y  =  ?0
ENTER SEARCH DIMENSION ?1
ENTER NUMBER OF GRID DIVISIONS ?10
LOWER LEFT COORDINATES OF SQUARES ARE
```

X	Y	X	Y
.5	.2	.3	.4
.6	.2	0	.8
.7	.2	.1	.8
.8	.2	.2	.8
.9	.2	.3	.8
0	.3	.4	.8
.1	.3	0	.9
.3	.3	.4	.9
.4	.3	.5	.9
.5	.3	.6	.9
.1	.4	.9	.9
.2	.4		

```
SQUARE DIMENSION IS .1
```

Figure 4-6 shows the squares located, superimposed on the $f(x, y) = 0$ curves. Running the program again, searching the region with lower left coordinates (0.5, 0.2), gives a more detailed view of the zero-crossing curve, as shown in Figure 4-6:

```
SCAN FOR ZERO-CROSSING SQUARES
ENTER SEARCH LOWER LEFT COORDINATES
X  =  ?0.5
Y=  ?0.2
ENTER SEARCH DIMENSION 0.1
ENTER NUMBER OF GRID DIVISIONS 4
LOWER LEFT COORDINATES OF SQUARES ARE
X        Y
```

X	Y
.575	.25
.525	.275
.55	.275
.575	.275

```
SQUARE DIMENSION IS .025
```

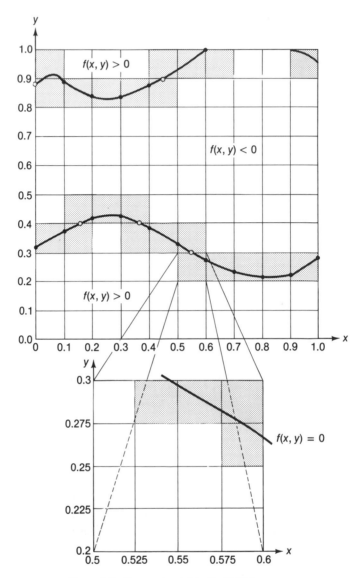

Figure 4-6 Squares located by the SCAN program.

4.4.3 Grid Searches for Relative Minima

Figure 4-7 shows a *contour map* (or elevation map) of a certain function of two variables. It shows the curves

$$f(x, y) = \text{(constant)}$$

for various constants, just as topological maps of the earth's surface show constant elevation curves. The function shown has a minimum in the vicinity of the small oval region outlined by the $f(x, y) = -3$ curve. Contour maps are easily produced by using

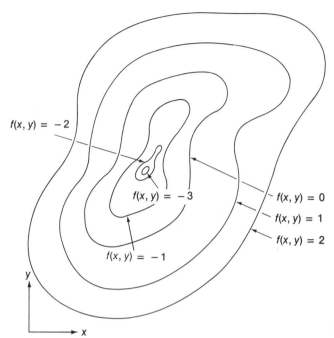

$f(x, y) = -2$

$f(x, y) = -3$

$f(x, y) = 0$

$f(x, y) = 1$

$f(x, y) = 2$

$f(x, y) = -1$

y

x

Figure 4-7   Contour map for a function of two variables.

a program such as CURVE or TRACK to generate closely spaced points on the curves of

$$f(x, y) - \text{constant} = 0$$

for various values of the constant.

The relative minima of a function of two variables can also be located by a refined grid search similar to the previous one for zero crossings. A particularly simple method is to refine the search in those squares for which the value of the function in the center of the square is smaller than any of the values at the neighboring four corners. Of course, minima can be overlooked if the initial grid is too coarse; and one can devise pathological functions that frustrate such a search.

4.5 FINDING SIMULTANEOUS ZERO CROSSINGS

A simultaneous zero crossing of two functions $f(x, y)$ and $g(x, y)$ is a simultaneous solution of the two nonlinear functions,

$$f(x, y) = 0$$

$$g(x, y) = 0$$

One way of locating the vicinity of a simultaneous zero crossing is to track one of the two curves, as with the TRACK program, while looking for an algebraic sign change of the second function. This can be fairly complicated to do in general. For example, it can happen that one zero-crossing curve intersects itself at or near a zero crossing of the other curve. There is, too, the possibility of tracking a nonintersecting closed curve endlessly.

An especially simple and more certain way of locating simultaneous zero crossings is to use the refined grid search method and refine the search in those squares where there are zero crossings of *both* functions on their boundaries. This will be done by modifying the SCAN program, which does not presently refine the search.

For a refined search, it is possible, indeed likely, that some squares having simultaneous zero crossings on their borders will not contain a zero-crossing curve intersection. Their refinement will eventually become a "dead end," making it necessary to decide where and how to restart the refinement process. Fortunately, there is a simple (but powerful) solution to this problem, stacking the tasks to be done.

4.5.1 Simultaneous Grid Search Program

Table 4-6 shows simple modifications to the SCAN program that identify those squares having crossings of both of the two functions F and G. In this, the SCAN2 program, the lines have not been renumbered so that the changes and additions are obvious. Here, the same tests are performed on both functions F and G, with the added tests on G using the arrays B and K. Then the coordinates of those squares having zero crossings of both functions on their boundaries are printed.

TABLE 4-6

Computer Program in BASIC to Find Squares Containing Zero-Crossing Curves of Both of Two Functions (SCAN2)

This is a modification of the SCAN program (Table 4-5) in which two functions, F and G, are each tested to find their zero-crossing squares. The coordinates of those squares having zero crossing on the boundaries of both functions are ultimately stored in the array *l* and printed.

ADDITIONAL VARIABLES USED

$B(22,22)$ = array of values of the function G

$K(22,22)$ = array of squares containing zero crossings of the function G, as a function of grid indices

G = value of the function calculated by subroutine 2000

LISTING

```
100     PRINT "FIND SIMULTANEOUS ZERO-CROSSING SQUARES"
110     DIM A(22,22) ,B(22,22) ,L(22,22) ,K(22,22)
120   REM  ENTER BOUNDARIES AND GRID
130     PRINT "ENTER SEARCH LOWER LEFT COORDINATES"
140     PRINT "X = ";
150     INPUT R
160     PRINT "Y = ";
170     INPUT S
180     PRINT "ENTER SEARCH DIMENSION";
190     INPUT D
200     PRINT "ENTER NUMBER OF GRID DIVISIONS"
210     INPUT M
220     C = D / M
230   REM PRINT HEADING
240     PRINT "LOWER LEFT COORDINATES OF SQUARES ARE"
250     PRINT "X", "Y"
260   REM  STORE GRID POINT VALUES OF FUNCTIONS AND
270   REM ZERO THE L AND K ARRAYS
```

```
280       FOR J = 1 TO M + 1
290       FOR I = 1 TO M + 1
300       X = R + (I - 1) * C
310       Y = S + (J - 1) * C
320       GOSUB 1000
330       A(I,J) = F
331       GOSUB 2000
332       B(I,J) = G
340       L(I,J) = 0
341       K(I,J) = 0
350       NEXT I
360       NEXT J
370   REM  IDENTIFY SQUARES WITH ZERO CROSSINGS
380       FOR J = 1 TO M + 1
390       FOR I = 1 TO M + 1
400       IF A(I,J) < > 0 THEN  GOTO 441
410       L(I,J) = 1
420       L(I - 1,J) = 1
430       L(I - 1,J - 1) = 1
440       L(I,J - 1) = 1
441       IF B(I,J) < > 0 THEN  GOTO 450
442       K(I,J) = 1
443       K(I - 1,J) = 1
444       K(I - 1,J - 1) = 1
445       K(I,J - 1) = 1
450       IF A (I,J) * A(I + 1,J)  > = 0 THEN GOTO 471
460       L(I,J) = 1
470       L(I,J - 1) = 1
471       IF B(I,J) * B(I + 1,J) > = 0 THEN GOTO 480
472       K(I,J) = 1
473       K(I,J - 1) = 1
480       IF A(I,J) * A(I,J + 1) > = 0 THEN GOTO 501
490       L(I,J) = 1
500       L(I - 1,J) = 1
501       IF B(I,J) * B(I,J + 1) > = 0 THEN GOTO 510
502       K(I,J) = 1
503       K(I - 1,J) = 1
510       NEXT I
520       NEXT J
530   REM  PRINT COORDINATES
540       FOR J = 1 TO M
550       FOR I = 1 TO M
560       IF L(I,J) * K(I,J) = 0 THEN GOTO 570
561       PRINT R + (I - 1) * C,S + (J - 1) * C
570       NEXT I
580       NEXT J
590   REM  PRINT SQUARE SIZE
600       PRINT "SQUARE DIMENSION IS ";C
610       END
1000  REM  SUBROUTINE TO EVALUATE FIRST FUNCTION
1010      F = X * X + Y * Y - 4
1020      RETURN
2000  REM  SUBROUTINE TO EVALUATE SECOND FUNCTION
2010      G = X - Y
2020      RETURN
```

Computer Program in FORTRAN to Find Squares Containing Zero-Crossing Curves of Both of Two Functions (SCAN2)

```
C ****************************************************************
C PROGRAM SCAN2 -- SIMULTANEOUS ZERO-CROSSING SQUARES
C ****************************************************************
      REAL    A(22,22), B(22,22)
      INTEGER L(22,22), K(22,22)
```

(continued)

```fortran
      WRITE(*,'(1X,''FIND SIMULTANEOUS ZERO-CROSSING SQUARES'')')

C *** ENTER BOUNDARIES AND GRID
      WRITE(*,'(1X,''ENTER SEARCH LOWER LEFT COORDINATES'')')
      WRITE(*,'(1X,''X = '',\)')
      READ(*,*) R
      WRITE(*,'(1X,''Y = '',\)')
      READ(*,*) S
      WRITE(*,'(1X,''ENTER SEARCH DIMENSION '',\)')
      READ(*,*) D
      WRITE(*,'(1X,''ENTER NUMBER OF GRID DIVISIONS '',\)')
      READ(*,*) M
      C = D/M

C *** PRINT HEADING
      WRITE(*,'(1X,''LOWER LEFT COORDINATES OF SQUARES ARE'')')
      WRITE(*,'(1X,''        X               Y'')')

C *** STORE GRID POINT VALUES OF FUNCTIONS AND
C *** ZERO THE L AND K ARRAYS
      DO 10 J=1, M+1
         DO 20 I=1, M+1
            X = R + (I-1)*C
            Y = S + (J-1)*C
            CALL FUNCT1(X,Y,F)
            A(I,J) = F
            CALL FUNCT2(X,Y,G)
            B(I,J) = G
            L(I,J) = 0
            K(I,J) = 0
 20      CONTINUE
 10   CONTINUE

C *** IDENTIFY SQUARES WITH ZERO CROSSINGS
      DO 30 J=1, M+1
         DO 40 I=1, M+1
            IF (A(I,J).EQ.0.0) THEN
               L(I+1,J+1) = 1
               L(I,J+1) = 1
               L(I,J) = 1
               L(I+1,J) = 1
            ENDIF
            IF (B(I,J).EQ.0.0) THEN
               K(I+1,J+1) = 1
               K(I,J+1) = 1
               K(I,J) = 1
               K(I+1,J) = 1
            ENDIF
            IF (A(I,J)*A(I+1,J).LT.0.0) THEN
               L(I+1,J+1) = 1
               L(I+1,J) = 1
            ENDIF
            IF (B(I,J)*B(I+1,J).LT.0.0) THEN
               K(I+1,J+1) = 1
               K(I+1,J) = 1
            ENDIF
            IF (A(I,J)*A(I,J+1).LT.0.0) THEN
               L(I+1,J+1) = 1
               L(I,J+1) = 1
            ENDIF
            IF (B(I,J)*B(I,J+1).LT.0.0) THEN
               K(I+1,J+1) = 1
               K(I,J+1) = 1
            ENDIF
 40      CONTINUE
 30   CONTINUE
```

```
C *** PRINT COORDINATES
      DO 50 J=1, M
        DO 60 I=1, M
          IF (L(I+1,J+1)*K(I+1,J+1).NE.0) THEN
            WRITE(*,'(1X,E15.8,1X,E15.8)')
     *          R+(I-1)*C, S+(J-1)*C
          ENDIF
60      CONTINUE
50    CONTINUE

C *** PRINT SQUARE SIZE
      WRITE(*,'(1X,''SQUARE DIMENSION IS '',E15.8)') C
      STOP
      END

C ******************************************************************
      SUBROUTINE FUNCT1(X,Y,F)
C ******************************************************************

C *** SUBROUTINE TO EVALUATE FIRST FUNCTION
      F = X * X + Y * Y - 4
      RETURN
      END

C ******************************************************************
      SUBROUTINE FUNCT2(X,Y,G)
C ******************************************************************

C *** SUBROUTINE TO EVALUATE SECOND FUNCTION
      G = X - Y
      RETURN
      END
```

The two functions listed are

$$f(x, y) = x^2 + y^2 - 4$$

which has a zero-crossing curve that is a circle about the origin of radius 2, and

$$g(x, y) = x - y$$

which has a zero-crossing curve that is a straight line through the origin with unity slope. The intersections of these two zero-crossing curves, the simultaneous solution of

$$f(x, y) = 0$$

$$g(x, y) = 0$$

are the points $(\sqrt{2}, \sqrt{2})$ and $(-\sqrt{2}, -\sqrt{2})$. A search in the region of dimension 2 with lower left coordinates (0, 0) gives the following:

```
FIND SIMULTANEOUS ZERO-CROSSING SQUARES
ENTER SEARCH LOWER LEFT COORDINATES
X = ?0
Y = ?0
ENTER SEARCH DIMENSION ?2
ENTER NUMBER OF GRID DIVISIONS ?10
```

(continued)

```
LOWER LEFT COORDINATES OF SQUARES ARE
X       Y
1.4     1.2
1.2     1.4
1.4     1.4
SQUARE DIMENSION IS .2
```

The squares found are pictured in Figure 4-8.

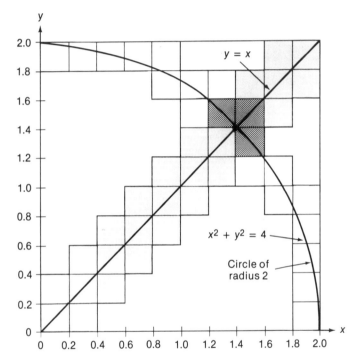

Figure 4-8   Squares containing zero crossings of two functions.

Further searches of these three squares found give

```
FIND SIMULTANEOUS ZERO-CROSSING        FIND SIMULTANEOUS ZERO-CROSSING
SQUARES                                SQUARES
ENTER SEARCH LOWER LEFT COORDINATES    ENTER SEARCH LOWER LEFT COORDINATES
X = ?1.4                               X = ?1.2
Y = ?1.2                               Y = ?1.4
ENTER SEARCH DIMENSION ?0.2            ENTER SEARCH DIMENSION ?0.2
ENTER NUMBER OF GRID DIVISIONS ?10     ENTER NUMBER OF GRID DIVISIONS ?10
LOWER LEFT COORDINATES OF SQUARES      LOWER LEFT COORDINATES OF SQUARES
ARE                                    ARE
X       Y                              X       Y
SQUARE DIMENSION IS .02                SQUARE DIMENSION IS .02
```

which indicate no crossings in the (1.4, 1.2) and (1.2, 1.4) squares, and

```
FIND SIMULTANEOUS ZERO-CROSSING SQUARES
ENTER SEARCH LOWER LEFT COORDINATES
X = ?1.4
Y = ?1.4
ENTER SEARCH DIMENSION ?0.2
ENTER NUMBER OF GRID DIVISIONS ?10
LOWER LEFT COORDINATES OF SQUARES ARE
X          Y
1.4        1.4
1.42       1.4
1.4        1.42
SQUARE DIMENSION IS .02
```

The latter search greatly narrows down the position of the zero crossing, but again lists three possible squares for its location, two of which contain both curves but not their crossing.

4.5.2 Simultaneous Zero-Crossing Program

A program that used grid searches automatically to narrow down the location of a simultaneous zero crossing must be able to handle the possibility that a square containing both curves may not contain their crossing. Figure 4-9 illustrates two such situations. In

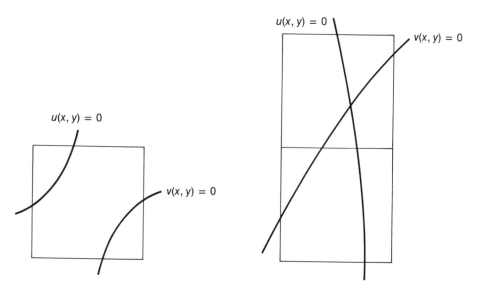

(a) Square containing both functions but no crossing of the two functions

(b) One square containing no crossing and another square with a crossing of the two functions

Figure 4-9   Squares containing both curves do not necessarily contain a crossing of the curves.

the first, the two curves approach one another but do not cross; one can hope that the initial search grid used is sufficiently fine that this does not happen. In the second situation, there are adjacent squares that contain the zero-crossing curves of both functions.

It may seem a simple matter just to keep track of the simultaneous zero-crossing squares found in the original search and the squares found in each subsequent refinement of the search, but this can be fairly involved to do because there can be an entire chain of refinements that terminate in a "dead end." To keep track of the squares of various sizes that are to be searched further for the zero crossing, we will use a *software stack*. The stack is just a list of the coordinates and dimension of each of the squares to be searched further and a *stack pointer*, an index variable, that is the "height" of the stack. When a new square's coordinates and dimension is to be stored, they are added to the list and the stack pointer is incremented so that it "points" to the most recently stacked item. When a previously stored square's coordinates and dimension are retrieved for further search, the stack pointer is decremented. When the stack pointer has value zero, the stack is empty.

An algorithm for locating the simultaneous zero crossings of two functions is given in Table 4-7. Initial squares having simultaneous zero crossings on their borders are found, if there are any. The coordinates and dimensions of these are stacked, then these squares are subdivided and searched for the smaller squares with crossings, and so on, until there is a final set of squares having crossings of dimension less than the user-entered precision ϵ.

TABLE 4-7   Algorithm for Locating a Simultaneous Zero Crossing of Two Functions of Two Variables

DEFINITIONS

The coordinates of each square of dimension ϵ or less that have the zero-crossing curves of two functions within their boundaries are to be found. An original square region selected by the user is divided into m^2 squares, where the integer m is user determined. This first subdivision is to begin with squares sufficiently small that no zero-crossing curve is contained entirely within a square. Thereafter, these squares with zero-crossing curves of both functions on their boundaries are divided into four squares, searched again, and so on.

ALGORITHM

1. Locate the squares having simultaneous zero crossing along their borders with the SCAN2 algorithm. The lower left coordinates of the search area are (r, s). The search consists of m^2 squares, each of dimension c (lines 300–700).

2. Store the coordinates of the squares with simultaneous zero crossing and their dimension in a stack (lines 710–800).

3. If the stack is empty, end (lines 810–820).

4. If the dimension of the square at the top of the stack is less than or equal to the desired precision, take it from the stack, print its coordinates, and go to step 3 (lines 830–870).

5. If the dimension of the square at the top of the stack is greater than the desired precision, take it from the stack, set the coordinates (r, s) to the coordinates of the square, set the search dimension to half the square's dimension, set the number of grid divisions m to 2, and go to step 1 to refine the search of that square (lines 880–940).

The corresponding computer program is listed in Table 4-8. A major part of it is derived from the SCAN2 routine of Table 4-6. The stack pointer is the variable T, and the stack itself is the array Z(3,50). The coordinates of a square to be searched further are stored as Z(1,T) and Z(2,T), and the square's dimension is stored as Z(3,T).

TABLE 4-8

Computer Program in BASIC for Locating a Simultaneous Zero Crossing of Two Functions of Two Variables (CROSS)

VARIABLES USED

The variables used are those of the SCAN2 program and the following:

Z(3,50) = stack of coordinates and dimensions of squares to be searched further
T   = stack pointer

LISTING

```
100      PRINT "SIMULTANEOUS CROSSING OF TWO CURVES"
110      DIM A(22,22) ,B(22,22) ,L(22,22) ,K(22,22) ,Z(3,50)
120   REM  ENTER BOUNDARIES, GRID, AND PRECISION
130      PRINT "ENTER SEARCH LOWER LEFT COORDINATES"
140      PRINT "X = ";
150      INPUT R
160      PRINT "Y = ";
170      INPUT S
180      PRINT "ENTER SEARCH DIMENSION";
190      INPUT D
200      PRINT "ENTER NUMBER OF GRID DIVISIONS";
210      INPUT M
220      C = D / M
230      PRINT "ENTER REQUIRED PRECISION";
240      INPUT E
250   REM  ZERO THE STACK POINTER
260      T = 0
270   REM  PRINT HEADING
280      PRINT " CROSSING COORDINATES ARE"
290      PRINT "X","Y"
300   REM  STORE GRID POINT VALUES OF FUNCTIONS AND
310   REM  ZERO THE L AND K ARRAYS
320      FOR J = 1 TO M + 1
330      FOR I = 1 TO M + 1
340      X = R + (I - 1) * C
350      Y = S + (J - 1) * C
360      GOSUB 1000
370      A(I,J) = F
380      GOSUB 2000
390      B(I,J) = G
400      L(I,J) = 0
410      K(I,J) = 0
420      NEXT I
430      NEXT J
440   REM  IDENTIFY SQUARES WITH ZERO CROSSINGS
450      FOR J = 1 TO M +1
460      FOR I = 1 TO M + 1
470      IF A(I,J) <  > 0 THEN GOTO 520
480      L(I,J) = 1
490      L(I - 1,J) = 1
500      L(I - 1,J - 1) = 1
510      L(I,J - 1) = 1
```

(continued)

```
520     IF B(I,J) < > 0 THEN GOTO 570
530     K(I,J) = 1
540     K(I - 1,J) = 1
550     K(I - 1,J - 1) = 1
560     K(I,J - 1) = 1
570     IF A(I,J) * A(I + 1,J)  > = 0 THEN GOTO 600
580     L(I,J) = 1
590     L(I,J - 1) = 1
600     IF B(I,J) * B(I + 1,J)  > = 0 THEN GOTO 630
610     K(I,J) = 1
620     K(I,J - 1) = 1
630     IF A(I,J) * A(I,J + 1)  > = 0 THEN GOTO 660
640     L(I,J) = 1
650     L(I - 1,J) = 1
660     IF B(I,J) * B(I,J + 1)  > = 0 THEN GOTO 690
670     K(I,J) = 1
680     K(I - 1, J) = 1
690     NEXT I
700     NEXT J
710  REM   STACK COORDINATES AND DIMENSION
720     FOR J = 1 TO M
730     FOR I = 1 TO M
740     IF L(I,J) * K(I,J) = 0 THEN GOTO 790
750     T = T + 1
760     Z(1,T) = R + (I - 1) * C
770     Z(2,T) = S + (J - 1) * C
780     Z(3,T) = C
790     NEXT I
800     NEXT J
810  REM   IF STACK EMPTY, END
820     IF T = 0 THEN GOTO 950
830  REM   IF SUFFICIENT PRECISION, PRINT COORDINATES
840     IF Z(3,T)  > E THEN GOTO 880
850     PRINT Z(1,T),Z(2,T)
860     T = T - 1
870     GOTO 810
880  REM   REFINE SEARCH OF A STACKED SQUARE
890     R = Z(1,T)
900     S = Z(2,T)
910     C = Z(3,T) / 2
920     M = 2
930     T = T - 1
940     GOTO 300
950     END
1000 REM   SUBROUTINE TO EVALUATE FIRST FUNCTION
1010    F = X * X + Y * Y - 4
1020    RETURN
2000 REM   SUBROUTINE TO EVALUATE SECOND FUNCTION
2010    G = X - Y
2020    RETURN
```

Computer Program in FORTRAN for Locating a Simultaneous Zero Crossing of Two Functions of Two Variables (CROSS)

```
C ********************************************************************
C PROGRAM CROSS -- SIMULTANEOUS CROSSING OF TWO CURVES
C ********************************************************************
      REAL A(22,22), B(22,22),  Z(3,50)
      INTEGER T,     L(22,22),  K(22,22)

      WRITE(*,'(1X,''SIMULTANEOUS CROSSING OF TWO CURVES'')')

C ***  ENTER BOUNDARIES, GRID, AND PRECISION
      WRITE(*,'(1X,''ENTER SEARCH LOWER LEFT COORDINATES'')')
      WRITE(*,'(1X,''X = '',\)')
```

```
          READ(*,*) R
          WRITE(*,'(1X,''Y = '',\)')
          READ(*,*) S
          WRITE(*,'(1X,''ENTER SEARCH DIMENSION '',\)')
          READ(*,*) D
          WRITE(*,'(1X,''ENTER NUMBER OF GRID DIVISIONS '',\)')
          READ(*,*) M
          C = D / M
          WRITE(*,'(1X,''ENTER REQUIRED PRECISION '',\)')
          READ(*,*) E

C ***   ZERO THE STACK POINTER
          T = 0

C ***   PRINT HEADING
          WRITE(*,'(1X,''CROSSING COORDINATES ARE'')')
          WRITE(*,'(1X,''         X                 Y'')')

C ***   LOOP FOR A MAXIMUM OF 1000 VALUES
          DO 90 II=1, 1000

C ***   STORE GRID POINT VALUES OF FUNCTIONS AND
C ***   ZERO THE L AND K ARRAYS
             DO 10 J=1, M+1
                DO 20 I=1, M+1
                   X = R + (I - 1) * C
                   Y = S + (J - 1) * C
                   CALL FUNCT1(X,Y,F)
                   A(I,J) = F
                   CALL FUNCT2(X,Y,G)
                   B(I,J) = G
                   L(I,J) = 0
                   K(I,J) = 0
20              CONTINUE
10           CONTINUE

C ***   IDENTIFY SQUARES WITH ZERO CROSSINGS
             DO 30 J=1, M+1
                DO 40 I=1, M+1
                   IF (A(I,J).EQ.0.0) THEN
                      L(I,J) = 1
                      L(I - 1,J) = 1
                      L(I - 1,J - 1) = 1
                      L(I,J - 1) = 1
                   ENDIF
                   IF (B(I,J).EQ.0.0) THEN
                      K(I,J) = 1
                      K(I - 1,J) = 1
                      K(I - 1,J - 1) = 1
                      K(I,J - 1) = 1
                   ENDIF
                   IF ((A(I,J)*A(I+1,J)).LT.0.0) THEN
                      L(I,J) = 1
                      L(I,J - 1) = 1
                   ENDIF
                   IF ((B(I,J)*B(I+1,J)).LT.0.0) THEN
                      K(I,J) = 1
                      K(I,J - 1) = 1
                   ENDIF
                   IF ((A(I,J)*A(I,J+1)).LT.0.0) THEN
                      L(I,J) = 1
                      L(I - 1,J) = 1
                   ENDIF
```

(continued)

```
                  IF ((B(I,J)*B(I,J+1)).LT.0.0) THEN
                     K(I,J) = 1
                     K(I - 1, J) = 1
                  ENDIF
40          CONTINUE
30       CONTINUE

C ***   STACK COORDINATES AND DIMENSION
         DO 50 J=1, M
            DO 60 I=1, M
               IF ((L(I,J)*K(I,J)).NE.0.0) THEN
                  T = T + 1
                  Z(1,T) = R + (I - 1) * C
                  Z(2,T) = S + (J - 1) * C
                  Z(3,T) = C
               ENDIF
60          CONTINUE
50       CONTINUE

C ***   LOOP FOR A MAXIMUM OF 100 VALUES ON THE STACK
         DO 70 I=1, 1000

C ***   IF STACK EMPTY, END
            IF (T.EQ.0) STOP

C ***   IF SUFFICIENT PRECISION, PRINT COORDINATES ELSE BREAK LOOP
            IF (Z(3,T).LE.E) THEN
               WRITE(*,'(1X,2E15.8)') Z(1,T),Z(2,T)
               T = T - 1
            ELSE
               GOTO 80
            ENDIF
70       CONTINUE

C ***   REFINE SEARCH OF A STACKED SQUARE
80       R = Z(1,T)
         S = Z(2,T)
         C = Z(3,T) / 2
         M = 2
         T = T - 1
90       CONTINUE
         STOP
         END

C ******************************************************************
         SUBROUTINE FUNCT1(X,Y,F)
C ******************************************************************

C ***   SUBROUTINE TO EVALUATE FIRST FUNCTION
         F = X * X + Y * Y - 4
         RETURN
         END

C ******************************************************************
         SUBROUTINE FUNCT2(X,Y,G)
C ******************************************************************

C ***   SUBROUTINE TO EVALUATE SECOND FUNCTION
         G = X - Y
         RETURN
         END
```

For the listed functions, a circle and a line that intersect at $(\sqrt{2},\ \sqrt{2})$ and $(-\sqrt{2},\ -\sqrt{2})$, the CROSS program gives the following results:

```
SIMULTANEOUS CROSSING OF TWO CURVES
ENTER SEARCH LOWER LEFT COORDINATES
X = ?-2
Y = ?-2
ENTER SEARCH DIMENSION?4
ENTER NUMBER OF GRID DIVISIONS?10
ENTER REQUIRED PRECISION?0.0001
CROSSING COORDINATES ARE
X                    Y

1.41416016           1.41425781
1.41425781           1.41416016
1.41416016           1.41416016
-1.41425781          -1.41425781
-1.41435547          -1.41425781
-1.41425781          -1.41435547
```

In this example, it turns out that, for each intersection, three of the smallest squares, with dimensions 0.0001 or less, each have crossings of both curves along their borders. The CROSS program prints the coordinates of such squares.

An easy way to examine the operation of the stack is to insert the instructions

```
801 PRINT "...................."
802 FOR I = 1 TO T
803 PRINT Z(1,T+1-I), Z(2,T+1-I), Z(3,T+1-I)
804 NEXT I

850 PRINT "...................."
```

These print the contents of the stack, top item first, each time an item can be put on the stack.

4.5.3 Positioning a Robot Arm

Figure 4-10(a) shows a two-link robot arm. One end of the arm is fixed at the origin of the coordinate system, and the other end is at coordinate (x, y). The arm's positioning actuators control the link angles θ and ϕ, as shown. In terms of θ and ϕ, the coordinates of the joint between links are (a, b), where

$$a = 0.5 \cos \theta$$

$$b = 0.5 \sin \theta$$

The coordinates (x, y) are given by

$$x - a = 0.35 \cos (\phi + \theta)$$

$$y - b = 0.35 \sin (\phi + \theta)$$

or

$$x = 0.5 \cos \theta + 0.35 \cos (\phi + \theta)$$

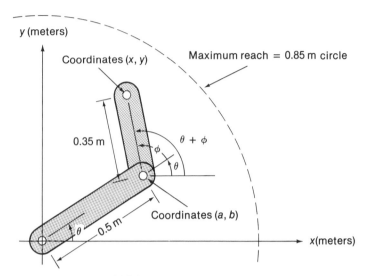

(a) Robot arm geometry

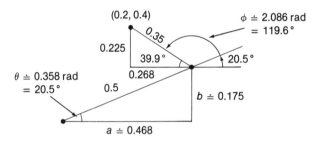

(b) First solution

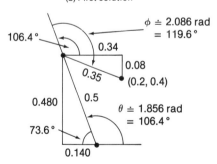

(c) Second solution

Figure 4-10   Positioning a robot arm.

$$y = 0.5 \sin \theta + 0.35 \sin (\phi + \theta)$$

The angles θ and ϕ that will position the end of the arm at coordinates (0.2, 0.4) are given by simultaneous solution of

$$0.5 \cos \theta + 0.35 \cos (\phi + \theta) - 0.2 = 0$$

$$0.5 \sin \theta + 0.35 \sin (\phi + \theta) - 0.4 = 0$$

Using the CROSS program with

```
1010 F = 0.5 * COS (X) + 0.35 * COS (X+Y) - 0.2
2010 G = 0.5 * SIN (X) + 0.35 * SIN (X+Y) - 0.4
```

gives

```
SIMULTANEOUS CROSSING OF TWO CURVES
ENTER SEARCH LOWER LEFT COORDINATES
X = ?-3.14
Y = ?-3.14
ENTER SEARCH DIMENSION ?6.28
ENTER NUMBER OF GRID DIVISIONS ?10
ENTER REQUIRED PRECISION?0.0001
CROSSING COORDINATES ARE
X                       Y
 .35800293              2.08615283
 .35800293              2.08607617
1.85617236             -2.08615283
1.85617236             -2.08622949
```

There are two sets of link angles that will position the arm as desired, $\theta \approx 0.358$, $\phi \approx 2.086$ and $\theta \approx 1.856$, $\phi \approx -2.086$. The geometry of these solutions is shown in Figure 4-10(b) and (c). The existence of multiple solutions, as in this case, is one of the complications of robotics. Some decision making is generally necessary, even at this fundamental level.

4.6 NEWTON–RAPHSON METHODS

For a nonlinear function of two variables $f(x, y)$, the Newton–Raphson zero location algorithm involves approximating the function by a two-variable Taylor series about the most recent trial point (x_i, y_i):

$$f(x, y) = f(x_i, y_i) + \frac{1}{1!} \left[\frac{\partial f(x_i, y_i)}{\partial x} (x - x_i) + \frac{\partial f(x_i, y_i)}{\partial y} (y - y_i) \right]$$

$$+ \frac{1}{2!} \left[\frac{\partial^2 f(x_i, y_i)}{\partial x^2} (x - x_i)^2 + 2 \frac{\partial^2 f(x_i, y_i)}{\partial x \, \partial y} (x - x_i)(y - y_i) \right.$$

$$+ \left. \frac{\partial^2 f(x_i, y_i)}{\partial y^2} (y - y_i)^2 \right] + \cdots$$

If the point (x_{i+1}, y_{i+1}) is a zero of $f(x, y)$, then

$$f(x_{i+1}, y_{i+1}) = f(x_i, y_i)$$

$$+ \frac{1}{1!} \left[\frac{\partial f(x_i, y_i)}{\partial x} (x_{i+1} - x_i) + \frac{\partial f(x_i, y_i)}{\partial y} (y_{i+1} - y_i) \right] + \cdots = 0$$

Approximating the above by just the constant and linear terms, the next trial point (x_{i+1}, y_{i+1}) is given by

$$f(x_i, y_i) + \frac{\partial f(x_i, y_i)}{\partial x} (x_{i+1} - x_i) + \frac{\partial f(x_i, y_i)}{\partial y} (y_{i+1} - y_i) = 0$$

or

$$\frac{\partial f(x_i, y_i)}{\partial x} x_{i+1} + \frac{\partial f(x_i, y_i)}{\partial y} y_{i+1} = \frac{\partial f(x_i, y_i)}{\partial x} x_i \qquad (4\text{-}1)$$

$$+ \frac{\partial f(x_i, y_i)}{\partial y} y_i - f(x_i, y_i)$$

Equation (4-1) is a linear equation in the variables x_{i+1} and y_{i+1} to be satisfied by the next trial point. Because the series was truncated, it is unlikely that any choice of (x_{i+1}, y_{i+1}) satisfying the equation will be the exact coordinates of a zero of $f(x, y)$, but to the extent that (x_i, y_i) is close to a zero, the approximation is a good one. Taking x_{i+1} and y_{i+1} to be variables, the straight line defined by (4-1) is an approximation to a segment of the zero-crossing curve of $f(x, y)$.

The Newton–Raphson algorithm in two variables has the same shortcomings as the one-dimensional version of the algorithm discussed in Section 3.4.2. It is used in some situations when it is known that the trial point (x_i, y_i) is close to a zero location. In general, it is unreliable.

For simultaneous zero crossings of two functions, $f(x, y)$ and $g(x, y)$, the Newton–Raphson conditions are two simultaneous linear algebraic equations in the two unknown coordinates of the next trial point (x_{i+1}, y_{i+1}):

$$\frac{\partial f(x_i, y_i)}{\partial x} x_{i+1} + \frac{\partial f(x_i, y_i)}{\partial y} y_{i+1} = \frac{\partial f(x_i, y_i)}{\partial x} x_i + \frac{\partial f(x_i, y_i)}{\partial y} y_i - f(x_i, y_i)$$

$$\frac{\partial g(x_i, y_i)}{\partial x} x_{i+1} + \frac{\partial g(x_i, y_i)}{\partial y} y_{i+1} = \frac{\partial g(x_i, y_i)}{\partial x} x_i + \frac{\partial g(x_i, y_i)}{\partial y} y_i - g(x_i, y_i)$$

(4-2)

The simultaneous solution of (4-2) gives a unique next trial point. For example, for the functions

$$f(x, y) = x^2 + y^2 - 4 \qquad g(x, y) = x - y$$

the intersecting circle and straight line,

$$\frac{\partial f}{\partial x} = 2x, \qquad \frac{\partial f}{\partial y} = 2y \qquad \frac{\partial g}{\partial x} = 1, \qquad \frac{\partial g}{\partial y} = -1$$

If the previous trial point was at coordinates

$$(x_i, y_i) = (2, -3)$$

the Newton–Raphson conditions for the coordinates of the next trial point are

$$\frac{\partial f(x_i, y_i)}{\partial x} = 4, \quad \frac{\partial f(x_i, y_i)}{\partial y} = -6 \quad \frac{\partial g(x_i, y_i)}{\partial x} = 1, \quad \frac{\partial g(x_i, y_i)}{\partial y} = -1$$

$$f(x_i, y_i) = 9, \quad g(x_i, y_i) = 5$$

$$4x_{i+1} - 6y_{i+1} = (4)(2) + (-6)(-3) - 9 = 17$$

$$x_{i+1} - y_{i+1} = (1)(2) + (-1)(-3) - 5 = 0$$

or

$$(x_{i+1}, y_{i+1}) = (-8.5, -8.5)$$

Beginning with a trial point nearer one of the intersections, which are at $(\sqrt{2}, \sqrt{2})$ and $(-\sqrt{2}, -\sqrt{2})$, say

$$(x_i, y_i) = (1, 1)$$

gives a much better next trial point:

$$(x_{i+1}, y_{i+1}) = (1.5, 1.5)$$

When one must find the simultaneous zero crossings of many functions of many variables, the Newton–Raphson method with the appropriate decision making to restart a search that is diverging is one of a very few practical alternatives. The conditions on the next trial point consist of a single linear algebraic equation for each function, so the determination of the next trial point simply involves solving a set of simultaneous linear algebraic equations involving the partial derivatives of the functions. The method is also called *linearization about an operating point* in some applications.

4.7 GRADIENT SEARCHES FOR MULTIDIMENSIONAL EXTREMA

Finding the zero crossings of nonlinear functions is to find the solutions of the corresponding nonlinear equations. Another important operation is the location of the relative extrema of a function. In Chapter 3 we developed methods for finding the extrema of functions of a single variable and now we extend those methods to functions of more than one variable, emphasizing the two variable case for clarity and simplicity. As in Chapter 3, we concentrate on finding relative minima; finding relative maxima simply involves finding the minima of the negative of a function.

In one dimension a numerical search for a relative minimum involves only the decision of whether to increase or decrease the variable. But in two or more dimensions, one must decide on a two- or more-dimensional direction to proceed.

4.7.1 Properties of Multidimensional Extrema

A continuous function of two variables, $f(x, y)$ has critical points where

$$\frac{\partial f}{\partial x} = 0 \quad \text{and} \quad \frac{\partial f}{\partial y} = 0$$

simultaneously. For a function for which the partial derivatives exist, a two-dimensional Taylor series expansion of $f(x, y)$ about a point (x_c, y_c) is

$$f(x, y) = f(x_c, y_c) + \frac{1}{1!} \left[\frac{\partial f(x_c, y_c)}{\partial x} (x - x_c) + \frac{\partial f(x_c, y_c)}{\partial y} (y - y_c) \right]$$

$$+ \frac{1}{2!} \left[\frac{\partial^2 f(x_c, y_c)}{\partial x^2} (x - x_c)^2 + 2 \frac{\partial^2 f(x_c, y_c)}{\partial x \, \partial y} (x - x_c)(y - y_c) \right.$$

$$\left. + \frac{\partial^2 f(x_c, y_c)}{\partial y^2} (y - y_c)^2 \right] + \cdots$$

If (x_c, y_c) is a critical point of f, then

$$\frac{\partial f(x_c, y_c)}{\partial x} = 0 \quad \text{and} \quad \frac{\partial f(x_c, y_c)}{\partial y} = 0$$

For points very near the critical point,

$$f(x, y) \approx f(x_c, y_c) + \frac{1}{2!} \left[\frac{\partial^2 f(x_c, y_c)}{\partial x^2} (x - x_c)^2 \right.$$

$$\left. + 2 \frac{\partial^2 f(x_c, y_c)}{\partial x \, \partial y} (x - x_c)(y - y_c) + \frac{\partial^2 f(x_c, y_c)}{\partial y^2} (y - y_c)^2 \right]$$

The critical point is a relative maximum if the term in brackets is negative for all (x, y) near the critical point. It is a relative minimum if the term in brackets is positive for all (x, y) near the critical point. If the bracketed term is positive for some (x, y) and negative for other (x, y) near the critical point, the critical point is neither a relative maximum nor a relative minimum. If the bracketed term $= 0$, then higher-order terms in the Taylor series expansion must be examined to determine the nature of the critical point.

The term in brackets is a *quadratic form* in the variables $(x - x_c)$ and $(y - y_c)$. The corresponding term in the Taylor series expansion of a function of more than two variables is also a quadratic form, involving more than two variables. We will examine the properties of quadratic forms in detail in Chapter 8. For the two-variable case, it can be shown that, defining

$$J = \begin{vmatrix} \dfrac{\partial^2 f(x_c, y_c)}{\partial x^2} & \dfrac{\partial^2 f(x_c, y_c)}{\partial x \, \partial y} \\ \dfrac{\partial^2 f(x_c, y_c)}{\partial x \, \partial y} & \dfrac{\partial^2 f(x_c, y_c)}{\partial y^2} \end{vmatrix}$$

$$= \frac{\partial^2 f(x_c, y_c)}{\partial x^2} \frac{\partial^2 f(x_c, y_c)}{\partial y^2} - \left[\frac{\partial^2 f(x_c, y_c)}{\partial x \, \partial y} \right]^2$$

if $J > 0$ and $\partial^2 f/\partial x^2 < 0$, the critical point is a relative maximum. If $J > 0$ and $\partial^2 f/\partial x^2 > 0$, the critical point is a relative minimum, and if $J < 0$, the critical point is not a relative extremum of the function.

When the needed partial derivatives are not available or are inconvenient to program, numerical methods that search directly for relative extrema are an important tool. These are the subject of the sections to follow. As in the case of functions of a single variable, we concentrate here on finding relative minima; relative maxima are found in similar ways or by locating the relative minima of the negative of the function.

Candidate points for the global maxima and minima of a function within a region include the relative maxima and minima and all points on the boundary of the region. In a single dimension, if global extrema are to be found over an interval of the variable, the endpoints of that interval could be maximum or minimum. In two or more dimensions, any of the points on the boundary of the region could be extrema. When the function has two variables and the two-dimensional boundaries are straight lines, it is simple to perform one-dimensional searches for extrema along them. When the boundaries are more complicated, a search for extrema along them can be more complicated.

4.7.2 Two-Dimensional Gradient Search

Imagining a two-dimensional function to be analogous to the elevation of land as a function of position, it is easy to see how relative minima can be located by traveling in the direction of greatest downhill slope. Of course, there is the remote possibility of ending exactly on a saddle point, and one might never find the entrance to a mine shaft (the bottom of which is a minimum) in this way, but usually the method is effective.

Figure 4-11 shows a contour map of a function of two variables together with several steps of a downhill search for the function's minimum. The direction and amount of greatest uphill slope is the *gradient* of the function. The gradient is the direction having components in each direction that are partial derivatives of the function in those directions. The direction and amount of greatest downhill slope is the negative of the gradient. The gradient is always perpendicular to the contour lines of a contour map.

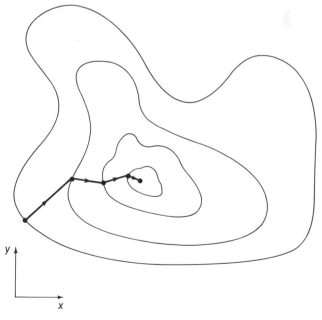

Figure 4-11   Gradient search for a two-dimensional relative minimum. The contours are evenly spaced values of the function but are drawn through the search points.

For a function $f(x, y)$, a gradient search for relative minima proceeds in steps with direction proportional to $-\partial f/\partial x$ in the x direction and $-\partial f/\partial y$ in the y direction. It is essentially Newton's method (section 3.4.2) in two dimensions. How large should the steps be? As with Newton's method, there are many ways to choose and vary the step size, none of which is entirely satisfactory.

The simplest choice is to take steps of constant proportionality to the negative gradient, that is, $-c(\partial f/\partial x)$ in the x direction and $-c(\partial f/\partial y)$ in the y direction, where c is a positive *convergence constant*. A good choice for c usually depends on some knowledge of the nature of the function. If c is too large, a minimum will be repeatedly overshot, with the search making large excursions back and forth past the minimum. The search points can even get increasingly remote from the minimum instead of converging on it, a situation called *divergence* or instability. If c is too small, convergence to a relative minimum will be exceedingly slow. A value of c that is too small for one function can be too large for another function. More sophisticated gradient searches monitor the sizes of the steps, attempt to identify divergence and slow convergence when it is occurring, and change the value of c to improve the situation.

4.7.3 Gradient Program

A two-dimensional gradient algorithm is given in Table 4-9, and the corresponding computer program, called GRADIENT, is given in Table 4-10. The user is to change the listed parameters C and E and the function and its derivatives as needed, of course. The program is written to loop back for a new set of search starting coordinates so that multiple minima can be more easily located.

TABLE 4-9 Algorithm for Two-Dimensional Relative Minimum Location Using the Gradient

DEFINITIONS

A relative minimum of the function $f(x, y)$ with known partial derivatives

$$g = \frac{\partial f}{\partial x} \qquad h = \frac{\partial f}{\partial y}$$

is to be located, if possible, starting the search at a user-selected set of coordinates (x, y). The relative minimum is to be located with a precision such that the magnitude of the gradient is less than a user-specified amount ϵ. The user also selects the convergence constant c such that the next point tested is c times the gradient away from the previous tested point.

ALGORITHM

1. Enter the starting coordinates (x, y) of the search.
2. If the magnitude of the gradient at (x, y) is less than the required precision ϵ, print the approximate coordinates (x, y) of the minimum and the value of the function at (x, y) and go to step 1 for another set of starting coordinates (lines 210–260 and 310–360).
3. Increment x by $-cg$ and y by $-ch$,

$$x \leftarrow x - cg$$
$$y \leftarrow y - ch$$

and go to step 2 (lines 270–300).

TABLE 4-10

Computer Program in BASIC to Find the Two-Dimensional Relative Minima Using the Gradient (GRADIENT)

VARIABLES USED

C = convergence constant

E = required precision

X = x-direction search variable

Y = y-direction search variable

F = value of the function

G = value of the partial derivative of the function with respect to X

H = value of the partial derivative of the function with respect to Y

LISTING

```
100     PRINT "RELATIVE MINIMUM USING THE GRADIENT"
110   REM SET CONVERGENCE CONSTANT AND PRECISION
120     C =.25
130     E =.00001
140   REM ENTER INITIAL COORDINATES
150     PRINT "ENTER THE INITIAL X COORDINATE";
160     INPUT X
170     PRINT "ENTER THE INITIAL Y COORDINATE";
180     INPUT Y
190   REM PRINT HEADING
200     PRINT "COORDINATES OF MINIMUM"
210   REM EVALUATE PARTIAL DERIVATIVES
220     GOSUB 2000
230     GOSUB 3000
240   REM PRINT RESULT AND GO ON IF GRADIENT
250   REM MAGNITUDE IS LESS THAN PRECISION
260     IF SQR(G*G +H*H)< E THEN GOTO 310
270   REM INCREMENT VARIABLES AND CONTINUE SEARCH
280     X =X -C*G
290     Y =Y -C*H
300     GOTO 210
310   REM PRINT RESULT AND GO ON
320     PRINT "X =";X
330     PRINT "Y =";Y
340     GOSUB 1000
350     PRINT "FUNCTION HAS VALUE";F
360     GOTO 140
370     END
1000  REM SUBROUTINE TO EVALUATE FUNCTION
1010    F =X*X -2*X +2*Y*Y -8*Y +12
1020    RETURN
2000  REM SUBROUTINE TO EVALUATE PARTIAL DERIVATIVE
2010  REM WITH RESPECT TO X
2020    G =2*X -2
2030    RETURN
3000  REM SUBROUTINE TO EVALUATE PARTIAL DERIVATIVE
3010  REM WITH RESPECT TO Y
3020    H =4*Y -8
3030    RETURN
```

(continued)

Computer Program in FORTRAN to Find the Two-Dimensional Relative Minima Using the Gradient (GRADIENT)

```
C ********************************************************************
C PROGRAM GRADIENT -- RELATIVE MINIMUM USING THE GRADIENT
C ********************************************************************
      WRITE(*,'(1X,''RELATIVE MINIMUM USING THE GRADIENT'')')

C *** SET CONVERGENCE CONSTANT AND PRECISION
      C = 0.25
      E = 0.00001

C *** ENTER INITIAL COORDINATES
      DO 10 I=1, 100
          WRITE(*,'(1X,''ENTER THE INITIAL X COORDINATE '',\)')
          READ(*,*) X
          WRITE(*,'(1X,''ENTER THE INITIAL Y COORDINATE '',\)')
          READ(*,*) Y

C *** PRINT HEADING
          WRITE(*,'(1X,''COORDINATES OF MINIMUM'')')

C *** EVALUATE PARTIAL DERIVATIVES, MAXIMUM OF 100 ITERATIONS
          DO 20 J=1, 100
              CALL DERIV1(X,G)
              CALL DERIV2(Y,H)

C *** PRINT RESULT AND GO ON IF GRADIENT
C *** MAGNITUDE IS LESS THAN PRECISION
              IF (SQRT(G*G + H*H).LT.E) GOTO 30

C *** INCREMENT VARIABLES AND CONTINUE SEARCH
              X =X -C*G
              Y =Y -C*H
 20       CONTINUE

C *** PRINT RESULT AND GO ON
 30       WRITE(*,'(1X,''X = '',E15.8)') X
          WRITE(*,'(1X,''Y = '',E15.8)') Y

          CALL FUNCT(X,Y,F)
          WRITE(*,'(1X,''FUNCTION HAS VALUE = '',E15.8)') F
 10   CONTINUE

      STOP
      END

C ********************************************************************
      SUBROUTINE FUNCT(X,Y,F)
C ********************************************************************

C *** SUBROUTINE TO EVALUATE FUNCTION
      F =X*X -2*X +2*Y*Y -8*Y +12
      RETURN
      END

C ********************************************************************
      SUBROUTINE DERIV1(X,G)
C ********************************************************************

C *** SUBROUTINE TO EVALUATE PARTIAL DERIVATIVE
C *** WITH RESPECT TO X
      G =2*X -2
      RETURN
      END
```

```
C ****************************************************************
      SUBROUTINE DERIV2(Y,H)
C ****************************************************************

C *** SUBROUTINE TO EVALUATE PARTIAL DERIVATIVE
C *** WITH RESPECT TO Y
      H =4*Y -8
      RETURN
      END
```

For the function listed,

$$f(x, y) = (x - 1)^2 + 2(y - 2)^2 + 3 = x^2 - 2x + 2y^2 - 8y + 12$$

which is minimum at $x = 1$, $y = 2$, the GRADIENT program gives

```
RELATIVE MINIMUM USING THE GRADIENT
ENTER THE INITIAL X COORDINATE?0
ENTER THE INITIAL Y COORDINATE?0
COORDINATES OF MINIMUM
X = .999996185
Y = 2
FUNCTION HAS VALUE 3
```

Any starting point will give this result because of the simple ''cup'' shape of this function.

4.7.4 Using Numerical Derivatives

A program that uses numerical approximation to the partial derivatives needed for a gradient search is listed in Table 4-11. It is called NUMGRAD and differs from the

TABLE 4-11

Computer Program in BASIC for a Two-Dimensional Gradient Search Using Numerical Derivatives (NUMGRAD)

VARIABLES USED

Same as for the GRADIENT Program (Table 4-10) and, in addition,

    U = numerical derivative increment
    S = stored value of the function

LISTING

```
100      PRINT "RELATIVE MINIMA USING NUMERICAL GRADIENTS"
110    REM SET PARAMETERS
120      C = .25
121      U = .001
130      E = .001
140    REM ENTER INITIAL COORDINATES
150      PRINT "ENTER THE INITIAL X COORDINATES";
160      INPUT X
170      PRINT "ENTER THE INITIAL Y COORDINATES";
180      INPUT Y
190    REM PRINT HEADING
```

(continued)

```
200        PRINT "COORDINATES OF MINIMUM"
210     REM EVALUATE PARTIAL DERIVATIVES
220        GOSUB 2000
240     REM PRINT RESULT AND GO ON IF GRADIENT
250     REM MAGNITUDE IS LESS THAN PRECISION
260        IF SQR(G*G +H*H)< E THEN GOTO 310
270     REM INCREMENT VARIABLES AND CONTINUE SEARCH
280        X = X-C*G
290        Y = Y-C*H
300        GOTO 210
310     REM PRINT RESULT AND GO ON
320        PRINT "X ="; X
330        PRINT "Y =";Y
340        GOSUB 1000
350        PRINT "FUNCTION HAS VALUE"; F
360        GOTO 140
370        END
1000    REM SUBROUTINE TO EVALUATE FUNCTION
1010       F =X*X -2*X +2*Y*Y-8*Y +12
1020       RETURN
2000    REM SUBROUTINE TO EVALUATE PARTIAL DERIVATIVES
2010       GOSUB 1000
2020       S =F
2030       X = X+U
2040       GOSUB 1000
2050       G =(F -S)/U
2060       X =X -U
2070       Y =Y +U
2080       GOSUB 1000
2090       H= (F -S)/U
2100       Y =Y-U
2110       RETURN
```

Computer Program in FORTRAN for a Two-Dimensional Gradient Search Using Numerical Derivatives (NUMGRAD)

```
C *****************************************************************
C PROGRAM NUMGRAD -- RELATIVE MINIMA USING NUMERICAL GRADIENTS
C *****************************************************************

      WRITE(*,'(1X,''RELATIVE MINIMA USING NUMERICAL GRADIENTS'')')

C *** SET PARAMETERS
      C = 0.25
      U = 0.001
      E = 0.001

C *** ENTER INITIAL COORDINATES, MAXIMUM OF 100 LOOPS
      DO 10 J=1, 100
         WRITE(*,'(1X,''ENTER THE INITIAL X COORDINATES '',\)')
         READ(*,*) X
         WRITE(*,'(1X,''ENTER THE INITIAL Y COORDINATES '',\)')
         READ(*,*) Y

C *** PRINT HEADING
         WRITE(*,'(1X,''COORDINATES OF MINIMUM'')')

C *** EVALUATE PARTIAL DERIVATIVES, MAXIMUM OF 1000 ITERATIONS
         DO 20 I=1, 1000
            CALL FUNCT2(F,U,S,X,Y,H)

C *** PRINT RESULT AND GO ON IF GRADIENT
C *** MAGNITUDE IS LESS THAN PRECISION
            IF (SQRT(G*G + H*H).LT.E) GOTO 30
```

```
C *** INCREMENT VARIABLES AND CONTINUE SEARCH
            X = X - C*G
            Y = Y - C*H
   20       CONTINUE

C *** PRINT RESULT AND GO ON
   30       WRITE(*,'(1X,''X = '',E15.8)') X
            WRITE(*,'(1X,''Y = '',E15.8)') Y
            CALL FUNCT1(X,Y,F)
            WRITE(*,'(1X,''FUNCTION HAS VALUE '',E15.8)') F
   10    CONTINUE
         STOP
         END

C ******************************************************************
         SUBROUTINE FUNCT1(X,Y,F)
C ******************************************************************

C *** SUBROUTINE TO EVALUATE FUNCTION
         F = X*X - 2.0*X + 2.0*Y*Y - 8.0*Y + 12.0
         RETURN
         END

C ******************************************************************
         SUBROUTINE FUNCT2(F,U,S,X,Y,H)
C ******************************************************************

C *** SUBROUTINE TO EVALUATE PARTIAL DERIVATIVES
         CALL FUNCT1(X,Y,F)
         S = F
         X = X + U
         CALL FUNCT1(X,Y,F)
         G = (F - S)/U
         X = X - U
         Y = Y + U
         CALL FUNCT1(X,Y,F)
         H = (F - S)/U
         Y = Y - U
         RETURN
         END
```

GRADIENT program only in that subroutines 2000 and 3000, which evaluate $g = \partial f/\partial x$ and $h = \partial f/\partial y$, are replaced by a new subroutine 2000 that approximates the partial derivatives by

$$G = \frac{f(X + U,Y) - f(X,Y)}{U}$$

and

$$H = \frac{f(X,Y + U) - f(X,Y)}{U}$$

where u is a small number selected by the user.

When run with the listed function and starting coordinates (0, 0), the results are

```
RELATIVE MINIMA USING NUMERICAL GRADIENTS
ENTER THE INITIAL X COORDINATE?0
ENTER THE INITIAL Y COORDINATE?0
COORDINATES OF MINIMUM
X = .99901203
Y = 1.99950021
FUNCTION HAS VALUE 3.000000148
```

which compares well with the minimum found for the same function with the GRADIENT program. As the numerical derivatives are approximated over a span of U, the obtainable precision of the result is likely to be only on the order of U.

4.8 STEEPEST DESCENT

One of the best ways of controlling the step size in a gradient search is the method of steepest descent. Instead of taking steps with size proportional to the magnitude of the gradient, the search is composed of a series of smaller steps of controlled size. The result is performance that is in many ways similar to a one-dimensional bisection search.

4.8.1 Steepest Descent Method

The steepest descent method is illustrated in Figure 4-12. Starting from an initial point (x_1, y_1), relatively small steps of equal size are taken in the direction of the negative gradient at (x_1, y_1) as long as the value of the function at each step continues to be smaller. When the next such step results in an increase in the function [at (x_2, y_2) in the

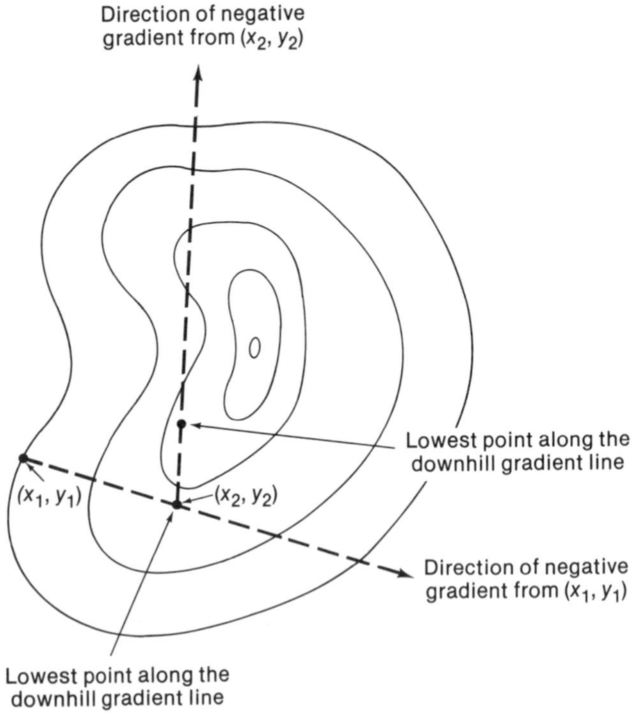

Direction of negative
gradient from (x_2, y_2)

Lowest point along the
downhill gradient line

(x_1, y_1)

(x_2, y_2)

Direction of negative
gradient from (x_1, y_1)

Lowest point along the
downhill gradient line

Figure 4-12 Steepest descent method.

figure] the gradient is recalculated, then small steps are taken in the direction of the new gradient, and so on. Each time the gradient is recalculated, the step size is reduced so that, eventually, the steps become very small.

A steepest descent algorithm is given in Table 4-12. It involves most of the procedures of the gradient algorithm but takes small steps of controlled size until a step

TABLE 4-12   Steepest Descent Minimum-Locating Algorithm

DEFINITIONS
A relative minimum of the function $f(x, y)$ with known partial derivatives

$$g = \frac{\partial f}{\partial x} \qquad h = \frac{\partial f}{\partial y}$$

is to be located, if possible, starting the search at a user-selected set of coordinates (x, y). The relative minimum is to be located with a precision such that the magnitude of the gradient is less than a user-specified amount ϵ. The user also selects an initial step size c.

ALGORITHM

1. Set the step size d to the initial step size c (lines 140–150).

$$d \leftarrow c$$

2. Enter the starting coordinates (x, y) of the search (lines 160–200).
3. If the magnitude of the gradient $\sqrt{g^2 + h^2}$ at (x, y) is less than the required precision ϵ, print the approximate coordinates (x, y) of the minimum and the value of the function at (x, y) and go to step 2 for another set of starting coordinates (lines 230–290 and 420–470).
4. Evaluate the function at (x, y):

$$p = f(x, y)$$

(lines 300–330).

5. Increment the variables with a step size d in the direction of the negative gradient,

$$x \leftarrow x - \frac{dg}{\sqrt{g^2 + h^2}}$$

$$y \leftarrow y - \frac{dh}{\sqrt{g^2 + h^2}}$$

and evaluate the function at the new coordinates

$$f \leftarrow f(x, y)$$

(lines 340–370).

6. If $f < p$, go to step 5 to take another step of length d in the direction of the negative gradient. If $f \geq p$, a relative minimum is nearby. Halve the step size,

$$d \leftarrow \frac{d}{2}$$

and go to step 3 to recalculate the gradient (lines 380–410).

results in an increase in the value of the function. Each time the gradient is recalculated, the size of these smaller steps is halved.

4.8.2 Steepest Descent Program

A computer program using this algorithm, STEEPEST, is given in Table 4-13. Lines 160–290 and 420–480 and the subroutines are identical to the instructions in the GRADIENT program except that the magnitude of the gradient is assigned to the variable S in line 280 because S will also be used for another purpose elsewhere. Steps 300–410 constitute a search with step size D for a relative minimum in the direction of the negative gradient. Each time the search for a minimum is completed, the step size for the next such search is halved.

TABLE 4-13

Computer Program in BASIC for Two-Dimensional Minimum Location by Steepest Descent (STEEPEST)

VARIABLES USED

> C = initial step size
> E = required precision
> D = step size
> X = x-direction search variable
> Y = y-direction search variable
> F = value of the function
> P = previous value of the function
> G = value of the partial derivative with respect to x
> H = value of the partial derivative with respect to y
> S = magnitude of the gradient of the function

LISTING

```
100      PRINT "STEEPEST DESCENT TO FIND MINIMA"
110    REM   SET INITIAL INCREMENT AND PRECISION
120      C = 1.0
130      E = .00001
140    REM   SET INCREMENT TO INITIAL INCREMENT
150      D = C
160    REM   ENTER INITIAL COORDINATES
170      PRINT "ENTER THE INITIAL X COORDINATE";
180      INPUT X
190      PRINT "ENTER THE INITIAL Y COORDINATE";
200      INPUT Y
210    REM   PRINT HEADING
220      PRINT "COORDINATES OF MINIMUM"
230    REM   EVALUATE PARTIAL DERIVATIVES
240      GOSUB 2000
250      GOSUB 3000
260    REM   PRINT RESULT AND GO ON IF GRADIENT
270    REM   MAGNITUDE IS LESS THAN PRECISION
280      S =  SQR (G * G + H * H)
290      IF S < E THEN GOTO 420
300    REM   EVALUATE FUNCTION AT (X,Y)
310      GOSUB 1000
```

```
320   REM  SET P TO VALUE OF FUNCTION AT (X,Y)
330     P = F
340   REM  INCREMENT VARIABLES AND EVALUATE FUNCTION
350     X = X - D * G / S
360     Y = Y - D * H / S
370     GOSUB 1000
380   REM  IF GOING DOWNHILL, CONTINUE SEARCH
390     IF F < P THEN GOTO 330
400     D = D / 2
410     GOTO 230
420   REM  PRINT RESULT AND GO ON
430     PRINT "X = ";X
440     PRINT "Y = ";Y
450     GOSUB 1000
460     PRINT "FUNCTION HAS VALUE ";F
470     GOTO 160
480     END
1000  REM  SUBROUTINE TO EVALUATE FUNCTION
1010    F = X * X - 2 * X + 2 * Y * Y - 8 * Y + 12
1020    RETURN
2000  REM  SUBROUTINE TO EVALUATE PARTIAL DERIVATIVE
2010  REM  WITH RESPECT TO X
2020    G = 2 * X - 2
2030    RETURN
3000  REM  SUBROUTINE TO EVALUATE PARTIAL DERIVATIVE
3010  REM  WITH RESPECT TO Y
3020    H = 4 * Y - 8
3030    RETURN
```

Computer Program in FORTRAN for Two-Dimensional Minimum Location by Steepest Descent (STEEPEST)

```
C ********************************************************************
C PROGRAM STEEPEST - STEEPEST DESCENT TO FIND MINIMA
C ********************************************************************

      WRITE(*,'(1X,''STEEPEST DESCENT TO FIND MINIMA'')')

C *** SET INITIAL INCREMENT AND PRECISION
      C = 1.0
      E = 0.00001

C *** SET INCREMENT TO INITIAL INCREMENT
      D = C

C *** ENTER INITIAL COORDINATES, MAXIMUM OF 100 LOOPS
      DO 10 K=1, 100
          WRITE(*,'(1X,''ENTER THE INITIAL X COORDINATE '',\)')
          READ(*,*) X
          WRITE(*,'(1X,''ENTER THE INITIAL Y COORDINATE '',\)')
          READ(*,*) Y

C *** PRINT HEADING
          WRITE(*,'(1X,''COORDINATES OF MINIMUM'')')

C *** EVALUATE PARTIAL DERIVATIVES, MAXIMUM OF 1000 ITERATIONS
          DO 20 J=1, 1000
              CALL PARTX(X,G)
              CALL PARTY(Y,H)

C *** PRINT RESULT AND GO ON IF GRADIENT
C *** MAGNITUDE IS LESS THAN PRECISION
              S = SQRT(G*G + H*H)
              IF (S.LT.E) THEN
```

(continued)

```
C *** PRINT RESULT AND GO ON
                WRITE(*,'(1X,''X = '',E15.8)') X
                WRITE(*,'(1X,''Y = '',E15.8)') Y
                CALL FUNCT(X,Y,F)
                WRITE(*,'(1X,''FUNCTION HAS VALUE '',E15.8)') F
                STOP

           ELSE

C *** EVALUATE FUNCTION AT (X,Y)
                CALL FUNCT(X,Y,F)

C *** SET P TO VALUE OF FUNCTION AT (X,Y), ITERATE MAXIMUM
C *** OF 1000 TIMES
                DO 30 I=1, 100
                   P = F

C *** INCREMENT VARIABLES AND EVALUATE FUNCTION
                X = X - D*G/S
                Y = Y - D*H/S
                CALL FUNCT(X,Y,F)

C *** IF GOING DOWNHILL, CONTINUE SEARCH
                IF (F.GE.P) GOTO 40
 30             CONTINUE
 40             CONTINUE
                D = D/2.0
           ENDIF
 20     CONTINUE
 10   CONTINUE
      STOP
      END

C ********************************************************************
      SUBROUTINE FUNCT(X,Y,F)
C ********************************************************************

C *** SUBROUTINE TO EVALUATE FUNCTION
      F = X*X - 2.0*X + 2.0*Y*Y - 8.0*Y + 12.0
      RETURN
      END

C ********************************************************************
      SUBROUTINE PARTX(X,G)
C ********************************************************************

C *** SUBROUTINE TO EVALUATE PARTIAL DERIVATIVE
C *** WITH RESPECT TO X
      G = 2.0*X - 2.0
      RETURN
      END

C ********************************************************************
      SUBROUTINE PARTY(Y,H)
C ********************************************************************

C *** SUBROUTINE TO EVALUATE PARTIAL DERIVATIVE
C *** WITH RESPECT TO Y
      H = 4.0*Y - 8.0
      RETURN
      END
```

When run with the listed parameters and function, the STEEPEST program very quickly gives

```
STEEPEST DESCENT TO FIND MINIMA
ENTER THE INITIAL X COORDINATE?0
ENTER THE INITIAL Y COORDINATE?0
COORDINATES OF MINIMUM
X = 1
Y = 1.9999985
FUNCTION HAS VALUE 3
```

which is correct. For an initial point, instead, at (100, 200) convergence is exceedingly slow. With this initial point and

$$120 \ C = 10$$

convergence is again fast, giving

```
ENTER THE INITIAL X COORDINATE?100
ENTER THE INITIAL Y COORDINATE?200
COORDINATES OF MINIMUM
X = 1
Y = 2.00000145
FUNCTION HAS VALUE 3
```

The user is, of course, to change the function and its partial derivatives and adjust the parameters as needed for a particular application.

An enhanced routine might count the number of steps for each location of a minimum and halve the step size only if a relatively small number of steps, say 4–10, were needed. It might also increase the step size if the number of steps to a minimum is large.

4.8.3 Optimizing Production

Before beginning production, a manufacturing company will usually evaluate the costs of production and the market in some detail. When the company makes competing products and when its manufacturing capacity is limited, there is usually an optimum product mix that maximizes the total profit on those items. Figure 4-13(a) shows the per-item material and labor costs for two models of pocket calculators for timing and evaluating the performance of swimmers in athletic events. Each cost is of the form

$$c(n) = a_1 + \frac{a_2}{n}$$

where n is the number of units manufactured in a month and where a_1 and a_2 are constants. The cost is very high for small n and levels out to a minimum amount for large n. This is a good model for a range of n that is not too small. The more expensive of the two calculators has a per-item cost

$$c_1(n_1) = 6 + \frac{2500}{n_1}$$

while the cheaper model's cost is

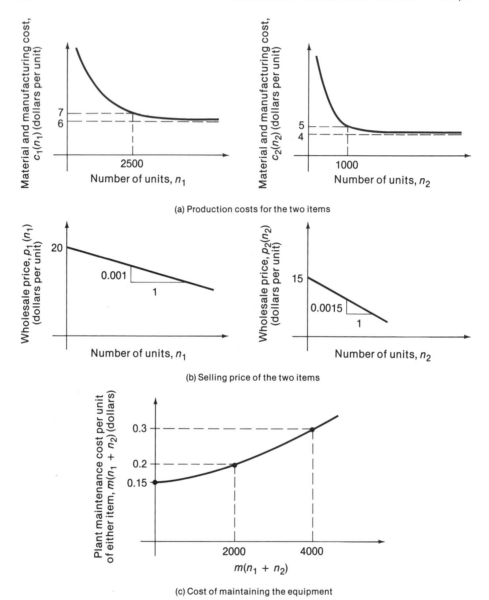

(a) Production costs for the two items

(b) Selling price of the two items

(c) Cost of maintaining the equipment

Figure 4-13   Costs and prices for two competing products made by the same company.

$$c_2(n_2) = 4 + \frac{1000}{n_2}$$

The predicted wholesale selling prices of the two items are shown in Figure 4-13(b). Each price is modeled linearly, with the price dropping, the more items that are produced. The per-item prices are

$$p_1(n_1) = 20 - 0.001n_1$$

and

$$p_2(n_2) = 15 - 0.0015n_2$$

which are valid for numbers of items produced per month, n_1 less than about 10,000 and n_2 less than about 5000.

The equipment used in manufacture requires maintenance and repair that only depends on the total number of items produced, not their mix. Based on past experience, the per-item maintenance cost is related to the total number of items produced per month as shown in Figure 4-13(c). If the curve is quadratic, of the form

$$m(n_1 + n_2) = \beta_1 + \beta_2(n_1 + n_2) + \beta_3(n_1 + n_2)^2$$

where the β's are constants, then

$$\beta_1 = 0.15$$

and β_2 and β_3 satisfy

$$0.15 + 2000\beta_2 + (2000)^2\beta_3 = 0.2$$

$$0.15 + 4000\beta_2 + (4000)^2\beta_3 = 0.3$$

Scaling these equations by letting

$$b_2 = 1000\beta_2 \qquad b_3 = 10^6\beta_3$$

they become

$$0.15 + 2b_2 + 4b_3 = 0.2$$

$$0.15 + 4b_2 + 16b_3 = 0.3$$

which has solution, using the SOLVE program,

$$b_2 = 0.0125 \qquad b_3 = 0.00625$$

or

$$\beta_2 = 1.25 \times 10^{-5} \qquad \beta_3 = 6.25 \times 10^{-9}$$

Subtracting the costs from the selling prices, the total month's profit on these two items as a function of the numbers produced is expected to be

$$P(n_1, n_2) = n_1 p_1(n_1) + n_2 p_2(n_2) - n_1 c_1(n_1)$$

$$- n_2 c_2(n_2) - (n_1 + n_2)m(n_1 + n_2)$$

$$= n_1\left(20 - 0.001n_1 - 6 - \frac{2500}{n_1}\right) + n_2\left(15 - 0.0015n_2 - 4 - \frac{1000}{n_2}\right)$$

$$- (n_1 + n_2)[0.15 + 1.25 \times 10^{-5}(n_1 + n_2)$$

$$+ 6.25 \times 10^{-9}(n_1 + n_2)^2]$$

Its partial derivatives with respect to n_1 and n_2 are

$$\frac{\partial P}{\partial n_1} = 13.85 - 0.002n_1 - 2.5 \times 10^{-5}(n_1 + n_2) - 1.875 \times 10^{-8}(n_1 + n_2)^2$$

$$\frac{\partial P}{\partial n_2} = 10.85 - 0.003n_2 - 2.5 \times 10^{-5}(n_1 + n_2) - 1.875 \times 10^{-8}(n_1 + n_2)^2$$

When the STEEPEST program (which finds the *minimum*) is run with

```
120   C = 100
130   E = 0.001
1010  F = -14*X + 0.001*X*X + 3500 - 11*Y + 0.0015*Y*Y +
          0.15*(X+Y) + 1.25E-5*(X+Y)*(X+Y) + 6.25E-9
          *(X+Y)*(X+Y)*(X+Y)
2020  G = -13.85 + 0.002*X + 2.5E-5*(X+Y) + 1.875E-8
          *(X+Y) *(X+Y)
3020  H = -10.85 + 0.003*Y + 2.5E-5*(X+Y) + 1.875E-8
          *(X+Y)*(X+Y)
```

the extremum found is

$$X = 6040.72; \qquad Y = 3027.87$$

for which

$$F = -57084.86$$

meaning that profit will be maximum and about \$57,085 when $n_1 = 6041$ and $n_2 = 3028$.

PROBLEMS

4-1. Use the CURVE program to find zero-crossing points, then zero-crossing curves of the following functions in the regions indicated.
   (a) $f(x, y) = 3x^2 + 2y^2 - 4$ for $-2 \le x \le 2$ and $-2 \le y \le 2$
   (b) $f(x, y) = 2x^2 + 2y^2 - 4x + 8y - 8$ for $-4 \le x \le 2$ and $-5 \le y \le 1$

4-2. Add graphics to the CURVE program so that it produces a plot on a screen, printer, or plotter of the zero-crossing points located.

4-3. Carefully describe a generalization of the CURVE algorithm for locating points on the zero-crossing *surface* of a function of three variables.

4-4. Use the TRACK program to find closely spaced points on the zero-crossing curves of the following functions. Then sketch the curves in the region indicated.
   (a) $f(x, y) = x^2 - 4x + 3y^2 + 6y - 2$ for $-1 \le x \le 5$ and $-4 \le y \le 1$
   (b) $f(x, y) = x^2 + 2y - 2x - 2$ for $-1 \le x \le 3$ and $-1 \le y \le 2$

4-5. Add graphics to the TRACK program so that it produces a plot on a screen, printer, or plotter of the zero-crossing points located. Connect the points with short straight lines to form approximate curves.

4-6. Carefully explain how the TRACK algorithm might be applied to the automatic steering of a vehicle on a single-lane, one-way road. Suppose that the vehicle has a movable optical sensor that can detect the left side, right side, and center line of the road.

4-7. Use the SCAN program to find information about the zero-crossing curve of the following function. Then sketch the curves in the indicated region.

$$f(x, y) = \sin 3x + \cos y \qquad \text{for } 0 \le x \le 2 \quad \text{and} \quad 0 \le y \le 2$$

4-8. A visitor from the planet Zorn is content with using a square search region for a grid search for zero crossings (as in the SCAN algorithm), but feels that it is our cultural bias that makes us tend to subdivide that into squares. The visitor's natural inclination is subdivision into triangles. Describe such an algorithm.

4-9. Modify the SCAN program so that it performs a refined grid search for the relative minima of a two-dimensional function, as described in Section 4.4.3.

4-10. Use the SCAN2 program to locate the vicinities of simultaneous zero crossings of the following pairs of functions within the indicated region. Locate the simultaneous zero crossing to within a distance of 0.0001 unit by running more crude searches, then refining them.

(a) $f(x, y) = 2x^2 + 2y^2 - 4x + 8y - 8$
$\quad g(x, y) = -x + y$ $\qquad$ for $-2 \le x \le 1$ and $-2 \le y \le 1$

(b) $f(x, y) = 2x^2 - y^2 - 4$
$\quad g(x, y) = x^2 + y^2 - 4$ $\qquad$ for $-3 \le x \le 3$ and $-2 \le y \le 2$

4-11. Sketch contour maps for the following functions by using either the CURVE or the TRACK program or both.

(a) $f(x, y) = 2x^2 + y^2 - 4$ $\quad$ for $-2 \le x \le 2$ and $-3 \le y \le 3$

(b) $f(x, y) = y^2 - 4x - 6y + 1$ $\quad$ for $-3 \le x \le 3$ and $-2 \le y \le 6$

4-12. Use the CROSS program to locate the simultaneous zero crossings of the following pairs of functions within the regions indicated. Locate the simultaneous zero crossings within a distance of 0.001.

(a) $f(x, y) = 0.135 \cos x + 0.0945 \cos (x + y) - 0.054$

$\qquad\qquad\qquad\qquad\qquad$ for $-4 \le x \le 4$ and $-4 \le y \le 4$

$\quad g(x, y) = 0.135 \sin x + 0.0945 \sin (x + y) - 0.108$

(b) $f(x, y) = 2x^2 - y^2 - 4$
$\quad g(x, y) = x^2 + y^2 - 4$ $\qquad$ for $-3 \le x \le 3$ and $-2 \le y \le 2$

4-13. The stack used in connection with the CROSS program is last in, first out. There are a variety of applications in which other stack arrangements are more desirable. Carefully describe how each of the following kinds of stack can be implemented in software.

(a) In waiting on customers, processing orders, assigning university class schedules, and similar applications, a first-in, first-out stack is desirable.

(b) In responding to business and mail, and in controlling industrial plants (especially during emergencies), it is important to associate a priority with each stacked item and to deal with the highest-priority item first.

4-14. For the robot arm of Section 4.5.3, find all pairs of link angles θ and ϕ that bring the arm's end to the following coordinates.

(a) $(0.1, -0.25)$ $\qquad$ (b) $(-0.3, -0.1)$

4-15. For the robot arm of Section 4.5.3, suppose that the link angles are limited to the values $0 \le \theta \le 180$ and $-90 \le \phi \le 90$. What range of coordinates (x, y) can be reached by the end of the arm?

4-16. A third link of length 0.25 m is added to the robot arm of Section 4.5.3, and β is the counterclockwise angle of the new link with respect to the second link. Find formulas for the coordinates of the arm's end in terms of the angle θ, ϕ, and β. Then find all combinations of the angles for which the end of the arm is at the following coordinates.

(a) $(-0.5, 0)$ $\qquad$ (b) $(0.4, 0.3)$

4-17. For the robot arm of the Section 4.5.3, design an algorithm for deciding which link angles should be used (when there is more than one possibility) in quickly moving the arm's end from a position (x_1, y_1) which link angles (θ_1, ϕ_1) to a new position (x_2, y_2).

4-18. Write, test, and debug a program that uses the Newton–Raphson method of finding the simultaneous zero crossings of two functions of two variables. Then use the program to find the following simultaneous zero crossings.

(a) $f(x, y) = 2x^2 + 2y^2 - 4x + 8y - 8$ for $-2 \le x \le 1$ and $-2 \le y \le 1$
    $g(x, y) = -x + y$

(b) $f(x, y) = 2x^2 - y^2 - 4$ for $-3 \le x \le 3$ and $-2 \le y \le 2$
    $g(x, y) = x^2 + y^2 - 4$

4-19. Use one or more computer programs to find the relative extremum of

$$f(x, y, z) = x^2 + y^2 + z^2 - 2x + 4y - 2z + 12$$

4-20. Use the CROSS program to locate simultaneous zero crossings of the first partial derivatives of

$$f(x, y) = x^3 + 3xy - 3y$$

for all x and y. Using this information, determine the relative maxima and relative minima of f. Also determine the global maximum and global minimum, if they exist.

4-21. Use the GRADIENT program to find the relative minima of the following functions.

(a) $f(x, y) = x^2 + y^2 - 4x + 6y + 25$ for $0 \le x \le 3$ and $-4 \le y \le 0$
(b) $f(x, y) = x^3 + y^3 - 3xy$ for $-2 \le x \le 2$ and $-2 \le y \le 2$

4-22. Use the GRADIENT program to find the relative extrema (both maxima and minima) of the following functions.

(a) $f(x, y) = x^3 + y^3 + 3xy$
(b) $f(x, y) = y^3 - 2x^3 + 2x^2y - xy^2 + 3x - 3y$

4-23. Use the NUMGARD program to find the relative maxima of the following functions.

(a) $f(x, y) = -x^2 - y^2 + 2x + 4y - 3$
(b) $f(x, y) = 2x^2y + 4y^2x + xy$

4-24. Use the STEEPEST program to find the relative minima of the following functions.

(a) $f(x, y) = x^2 + y^2 - 4x + 6y + 25$ for $0 \le x \le 3$ and $-4 \le y \le 0$
(b) $f(x, y) = x^3 + y^3 - 3xy$ for $-2 \le x \le 2$ and $-2 \le y \le 2$

4-25. Write, test, and debug a computer program to approximately locate the relative minima of a function $f(x, y)$ with a grid search as described in Section 4.4.3.

4-26. Write, test, and debug a three-dimensional gradient program for finding the relative minima of functions $f(x, y, z)$ of three variables. Use the program to find the minima of

$$f(x, y, z) = x^2 + y^2 + z^2 - 2x - 4y - 4z - 16$$

4-27. Improve the STEEPEST program by incorporating the suggestions made at the end of Section 4.7.2.

4-28. In the production optimization problem of Section 4.8.3, let

$$c_1(n_1) = 7 + \frac{1500}{n_1} \qquad c_2(n_2) = 5 + \frac{1000}{n_2}$$

$$P_1(n_1) = 20 - 0.0005n_1 \qquad P_2(n_2) = 17 - 0.001n_2$$

instead. Find the optimum product mix that maximizes the total profit on the two items.

4-29. In the production optimization problem of Section 4.8.3, let the coefficients β_1, β_2, and β_3 in

$$m(n_1 + n_2) = \beta_1 + \beta_2(n_1 + n_2) + \beta_3(n_1 + n_2)^2$$

be selected, instead, from

$$m(0) = 0.20 \qquad m(1000) = 0.22 \qquad m(3000) = 0.30$$

Find the optimum product mix.

5

POLYNOMIALS AND FACTORING

5.1 PREVIEW

Polynomials are special kinds of nonlinear functions. We devote a chapter to their study because of their importance to a wide variety of applications. Perhaps the most important approximations to other functions are derived from Taylor series expansions

$$f(x) = f(a) + \frac{\dot{f}(a)(x-a)}{1!} + \frac{\ddot{f}(a)(x-a)^2}{2!} + \cdots + \frac{f^{[n]}(\zeta)(x-a)^n}{n!}$$

where

$$\dot{f} = \frac{df}{dt}, \quad \ddot{f} = \frac{d^2f}{dt^2}, \quad \ldots, \quad f^{[n]} = \frac{d^nf}{dt^n}$$

Polynomials are also fundamental to problems in linear algebra and in the solution of differential equations.

The chapter begins with a discussion of some of the properties of polynomials, including properties and transformations of their roots. Then, in Section 5.3, a series of polynomial manipulation programs is developed. Algorithms for polynomial division, axis shifting, and evaluation at a complex value of the variable are programmed and tested. These are each used later as parts of other programs.

Polynomial factorization is fundamental to many important problems in analysis and computation, so much of the remainder of this chapter is concerned with factoring. Several approaches, their advantages and disadvantages, are discussed in Section 5.4. A complete, straightforward factoring program using real-part and imaginary-part root search is developed in Section 5.5.

The Routh–Hurwitz test is a powerful tabular procedure for determining, without factoring, the number of roots of a polynomial with positive real parts. It is the subject of Section 5.6, where computer programs for its implementation are given. In Section 5.7 the Routh–Hurwitz test and polynomial axis shifting are used to design a second factoring program. In it, the search for root real parts and imaginary parts is reduced to single-dimensional bisection searches instead of a two-dimensional one, giving excellent speed and dependability.

5.2 POLYNOMIAL PROPERTIES

A polynomial is a function of the form

$$p(s) = \alpha_n s^n + \alpha_{n-1} s^{n-1} + \cdots + \alpha_1 s + \alpha_0$$

The variable is often allowed to be complex:

$$s = x + iy$$

The $n + 1$ α's are real numbers that are the *coefficients* of the polynomial and assuming that $\alpha_n \neq 0$, n is the *degree* of the polynomial.

The *zeros* or *roots* of a polynomial are the solutions of

$$p(s) = \alpha_n s^n + \alpha_{n-1} s^{n-1} + \cdots + \alpha_1 s + \alpha_0 = 0 \qquad (5\text{-}1)$$

An nth-degree polynomial has n zeros $s_1, s_2, \ldots, s_n$. It can be factored in terms of its zeros as

$$p(s) = \alpha_n (s - s_1)(s - s_2) \cdots (s - s_n) \qquad (5\text{-}2)$$

Some of the zeros might be the same number, in which case the polynomial is said to have one or more *repeated roots*. If all the zeros of a polynomial are different numbers, the polynomial has *distinct roots*. For a polynomial with real coefficients, as we assume here, if a root is complex, its complex conjugate is also a root. A polynomial of odd degree then necessarily has at least one real root.

The two roots of a second-degree polynomial,

$$p(s) = \alpha_2 s^2 + \alpha_1 s + \alpha_0$$

are given by the quadratic formula,

$$s_1, s_2 = \frac{-\alpha_1 \pm \sqrt{\alpha_1^2 - 4\alpha_2\alpha_0}}{2\alpha_2}$$

There are complicated formulas for the roots of third- and fourth-degree polynomials, but none exist (or can exist) for polynomials of higher degree. Thus polynomials factoring is *necessarily* a numerical procedure for polynomials of degree 5 and higher. As a practical matter, factoring most third- and fourth-degree polynomials is often best done numerically, too.

Comparing equations (5-1) and (5-2), the roots and the coefficients of a polynomial are related in the following ways:

$$\frac{\alpha_0}{\alpha_n} = (-1)^n \text{ (product of the roots)} = (-1)^n s_1 s_2 \cdots s_n$$

$$\frac{\alpha_1}{\alpha_n} = (-1)^{n-1} \text{ (sum of all products of } n - 1 \text{ of the roots)}$$

$$= (-1)^{n-1} (s_2 s_3 s_4 \cdots s_n + s_1 s_3 s_4 \cdots s_n + s_1 s_2 s_3 \cdots s_{n-1})$$

$$\frac{\alpha_2}{\alpha_n} = (-1)^{n-2} \text{ (sum of all products of } n - 2 \text{ of the roots)}$$

$$\vdots$$

$$\frac{\alpha_{n-1}}{\alpha_n} = (-1)(\text{sum of all roots}) = (-1)(s_1 + s_2 + \cdots + s_n)$$

From these, various properties of the roots of a specific polynomial can be determined, probably the most useful of which is the Routh–Hurwitz test described in Section 5.6.

Sometimes it is helpful to transform a polynomial in some way that results in another, simpler polynomial with roots that are closely related to the original polynomial. For example, dividing a polynomial

$$p(s) = \alpha_n s^n + \alpha_{n-1} s^{n-1} + \cdots + \alpha_1 s + \alpha_0$$

by its leading coefficient, α_n, does not affect the polynomial's roots:

$$g(s) = \frac{p(s)}{\alpha_n} = s^n + \frac{\alpha_{n-1}}{\alpha_n} s^{n-1} + \cdots + \frac{\alpha_1}{\alpha_n} s + \frac{\alpha_0}{\alpha_n}$$

Another useful transformation is to substitute

$$s' = ks$$

where k is a nonzero constant, and multiply the resulting nth-degree polynomial by k^n, giving

$$p'(s') = k^n p\left(s \leftarrow \frac{s'}{k}\right) = k^n\left[\alpha_n \left(\frac{s'}{k}\right)^n + \alpha_{n-1}\left(\frac{s'}{k}\right)^{n-1} + \cdots + \alpha_1\left(\frac{s'}{k}\right) + \alpha_0\right]$$

$$= \alpha_n s'^n + k\alpha_{n-1} s'^{n-1} + k^2\alpha_{n-2} s'^{n-2} + \cdots + k^{n-1}\alpha_1 s' + k^n\alpha_0$$

which is a polynomial of the same degree with roots that are k times the roots of the original polynomial:

$$s_i' = ks_i \qquad i = 1, 2, \ldots, n$$

When the scaling constant k is chosen so that the constant term is $p'(s')$ equals the leading coefficient,

$$k^n\alpha_0 = \alpha_n$$

then

$$q(s') = \alpha_n\left(s'^n + \frac{k\alpha_{n-1}}{\alpha_n}s'^{n-1} + \cdots + \frac{k^{n-1}\alpha_1}{\alpha_n}s' + 1\right)$$

and the product of the roots of the scaled polynomial is $(-1)^n$. Some of these roots will probably have magnitude less that unity, and some will have magnitude greater than unity. To locate the roots for which $|s_i'| \le 0$, we need only search the region inside the unit circle on the complex plane, as indicated in Figure 5-1.

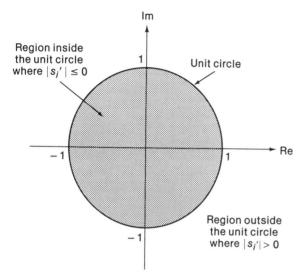

Im

Region inside
the unit circle
where $|s_i'| \le 0$

1

Unit circle

−1

1

Re

Region outside
the unit circle
where $|s_i'| > 0$

−1

Figure 5-1 Roots inside and outside the unit circle on the complex plane.

When $\alpha_0 \ne 0$, the substitution

$$s'' = \frac{1}{s}$$

and multiplication of an nth-degree polynomial by s''^n simply reverses the order of the polynomial coefficients:

$$r(s'') = s''^n p\left(s \leftarrow \frac{1}{s''}\right) = s''^n \left[\alpha_n\left(\frac{1}{s''}\right)^n + \alpha_{n-1}\left(\frac{1}{s''}\right)^{n-1} + \cdots + \alpha_1\left(\frac{1}{s''}\right) + \alpha_0\right]$$

$$= \alpha_0 s''^n + \alpha_1 s''^{n-1} + \cdots + \alpha_{n-1} s'' + \alpha_n$$

The transformed polynomial $r(s'')$ has roots that are the inverses of the roots of the original polynomial:

$$s_i'' = \frac{1}{s_i} \qquad i = 1, 2, \ldots, n$$

When this additional transformation is performed on the scaled polynomial $q(s')$, the roots of $q(s')$ that are outside the unit circle on the complex plane are transformed to inside the unit circle, and vice versa. Thus a search for polynomial roots, no matter what their location, can be reduced to finding the roots of $p'(s')$ inside the unit circle, and finding the roots of $p''(s'')$ inside the unit circle.

5.3 POLYNOMIAL MANIPULATION

5.3.1 Polynomial Division

In factoring polynomials, when a real root of a polynomial has been found, it is often desirable to divide by the known factor to give a one-degree-lower quotient polynomial having the remaining roots. An algorithm for polynomial division is given in Table 5-1.

TABLE 5-1   Algorithm for First-Degree Polynomial Division

DEFINITIONS

The dividend polynomial is

$$p(s) = \alpha_n s^n + \alpha_{n-1} s^{n-1} + \cdots + \alpha_1 s + \alpha_0$$

The divisor polynomial is

$$g(s) = cs + d$$

The quotient to be found is

$$q(s) = \beta_{n-1} s^{n-1} + \beta_{n-2} s^{n-2} + \cdots + \beta_1 s + \beta_0$$

and the remainder is r.

ALGORITHM

For n steps, beginning with $i = 1$ and ending with $i = n$:

$$\beta_{n-i} \leftarrow \frac{\alpha_{n-i+1}}{c}$$

$$\alpha_{n-i} \leftarrow \alpha_{n-i} - \frac{d\alpha_{n-i+1}}{c}$$

The coefficient α_0 becomes the remainder r.

The division begins with

$$\frac{\alpha_n}{c} s^{n-1}$$

$$cs + d \overline{\smash{\big)}\ \alpha_n s^n + \alpha_{n-1} s^{n-1} + \alpha_{n-2} s^{n-2} + \cdots + \alpha_1 s + \alpha_0}$$

$$\alpha_n s^n + \frac{d\alpha_n}{c} s^{n-1}$$

$$\left(\alpha_{n-1} - \frac{d\alpha_n}{c} \right) s^{n-1} + \alpha_{n-2} s^{n-2} + \cdots + \alpha_1 s + \alpha_0$$

Denoting the partial remainder coefficient of s^{n-1} by α'_{n-1}, the next step is

$$cs + d \enclose{longdiv}{} \quad \begin{array}{l} \dfrac{\alpha_n}{c} s^{n-1} + \dfrac{\alpha'_{n-1}}{c} s^{n-2} \\[2mm] \overline{\alpha_n s^n + \alpha_{n-1} s^{n-1} + \alpha_{n-2} s^{n-2} + \cdots + \alpha_1 s + \alpha_0} \\[2mm] \alpha_n s^n + \dfrac{d\alpha_n}{c} s^{n-1} \\[2mm] \hline \alpha'_{n-1} s^{n-1} + \alpha_{n-2} s^{n-2} + \cdots + \alpha_1 s + \alpha_0 \\[2mm] \alpha'_{n-1} s^{n-1} + \dfrac{d\alpha'_{n-1}}{c} s^{n-2} \\[2mm] \hline \left(\alpha_{n-2} - \dfrac{d\alpha'_{n-1}}{c}\right) s^{n-2} + \cdots + \alpha_1 s + \alpha_0 \end{array}$$

Proceeding in this way, the last step is

$$cs + d \enclose{longdiv}{} \quad \begin{array}{l} \dfrac{\alpha_n}{c} s^{n-1} + \dfrac{\alpha'_{n-1}}{c} s^{n-2} + \dfrac{\alpha'_{n-2}}{c} s^{n-3} + \cdots + \dfrac{\alpha'_1}{c} \\[2mm] \overline{\alpha_n s^n + \alpha_{n-1} s^{n-1} + \alpha_{n-2} s^{n-2} + \cdots + \alpha_1 s + \alpha_0} \\[2mm] \vdots \\[2mm] \hline \alpha'_1 s + \alpha_0 \\[2mm] \alpha'_1 s + \dfrac{d\alpha'_1}{c} \\[2mm] \hline \alpha_0 - \dfrac{d\alpha'_1}{c} \end{array}$$

where the remainder is

$$\alpha'_0 = \alpha_0 - \frac{d\alpha'_1}{c}$$

A program in BASIC/FORTRAN, Called DIVIDE1, using this algorithm to compute the quotient and remainder of a first-degree polynomial division, is given in Table 5-2. For the dividend polynomial,

$$P(s) = s^4 - 3s^3 + 4s^2 + 2s - 2$$

and divisor

$$g(s) = 3s - 1$$

TABLE 5-2

Computer Program in BASIC for First-Degree Polynomial Division (DIVIDE1)

VARIABLES USED

N = degree of dividend polynomial

$P(20)$ = coefficients of the dividend polynomial

$P(N)$ is the leading coefficient and $P(0)$ is the constant term

$Q(19)$ = coefficients of the $(N - 1)$th-degree quotient polynomial; $Q(N - 1)$ is the leading coefficient and $Q(0)$ is the constant term

C = coefficient of s in the divisor polynomial

D = constant coefficient in the divisor polynomial

I = coefficient index

LISTING

```
100     PRINT "FIRST-DEGREE POLYNOMIAL DIVISION"
110     DIM P(20), Q(19)
120   REM ENTER DIVIDEND POLYNOMIAL AND DEGREE
130     PRINT "PLEASE ENTER DEGREE OF DIVIDEND";
140     INPUT N
150     FOR I = 0 TO N
160     PRINT "ENTER POWER ";N-I;" COEFFICIENT";
170     INPUT P(N - I)
180     NEXT I
190   REM ENTER FIRST-DEGREE DIVISOR POLYNOMIAL
200     PRINT "ENTER THE DIVISOR'S LEADING COEFFICIENT";
210     INPUT C
220     IF C = 0 THEN GOTO 190
230     PRINT "ENTER THE DIVISOR'S CONSTANT COEFFICIENT";
240     INPUT D
250   REM COMPUTE QUOTIENT COEFFICIENTS
260   REM P(0) BECOMES THE REMAINDER
270     FOR I = 1 TO N
280     Q(N - I) = P(N - I + 1) / C
290     P(N - I) = P(N - I) - D *  P(N - I + 1) / C
300     NEXT I
310   REM PRINT QUOTIENT AND REMAINDER
320     PRINT "DIVIDEND POLYNOMIAL IS"
330     FOR I = 1 TO N
340     PRINT Q(N - I),"S**";N - I
350     NEXT I
360     PRINT "REMAINDER IS ";P(0)
370     END
```

Computer Program in FORTRAN for First-Degree Polynomial Division (DIVIDE1)

```
C ******************************************************************
C PROGRAM DIVIDE1 -- FIRST-DEGREE POLYNOMIAL DIVISION
C ******************************************************************
      REAL P(20), Q(19)

      WRITE(*,'(1X,''FIRST-DEGREE POLYNOMIAL DIVISION'')')

C *** ENTER DIVIDEND POLYNOMIAL AND DEGREE
      WRITE(*,'(1X,''PLEASE ENTER DEGREE OF DIVIDEND '',\)')
      READ(*,*) N
```

```
        DO 10 I=1, N+1
          WRITE(*,'(1X,''ENTER POWER '',I3,'' COEFFICIENT '',\)') N-I+1
          READ(*,*) P(N-I+2)
 10     CONTINUE

C *** ENTER FIRST-DEGREE DIVISOR POLYNOMIAL
 20     WRITE(*,'(1X,''ENTER THE DIVISORS LEADING COEFFICIENT '',\)')
        READ(*,*) C
        IF (C.EQ.0.0) GOTO 20
        WRITE(*,'(1X,''ENTER THE DIVISORS CONSTANT COEFFICIENT '',\)')
        READ(*,*) D

C *** COMPUTE QUOTIENT COEFFICIENTS
C *** P(1) BECOMES THE REMAINDER
        DO 30 I=1, N
          Q(N-I+1) = P(N-I+2) / C
          P(N-I+1) = P(N-I+1) - D * P(N-I+2) / C
 30     CONTINUE

C *** PRINT QUOTIENT AND REMAINDER
        WRITE(*,'(1X,''DIVIDEND POLYNOMIAL IS'')')
        DO 40 I=1, N
          WRITE(*,'(1X,E15.8,''  S** '',I5)') Q(N-I+1), N-I
 40     CONTINUE
        WRITE(*,'(1X,''REMAINDER IS '',E15.8)') P(1)
        STOP
        END
```

for example, DIVIDE1 gives

```
        FIRST DEGREE POLYNOMIAL DIVISION
        PLEASE ENTER DEGREE OF DIVIDEND? 4
        ENTER POWER 4 COEFFICIENT ? 1
        ENTER POWER 3 COEFFICIENT ?-3
        ENTER POWER 2 COEFFICIENT ? 4
        ENTER POWER 1 COEFFICIENT ? 2
        ENTER POWER 0 COEFFICIENT ?-2
        ENTER THE DIVISOR'S LEADING COEFFICIENT ?3
        ENTER THE DIVISOR'S CONSTANT COEFFICIENT ?-1
        DIVIDEND POLYNOMIAL IS
        .333333333     S**3
        -.888888889     S**2
        1.03703704     S**1
        1.01234568     S**0
        REMAINDER IS -.987654321
```

To remove a pair of complex conjugate roots from a polynomial, division by a second-degree divisor is used. The first two steps of a second-degree division are as follows:

$$cs^2 + ds + e \overline{\smash{\big)}\ \alpha_n s^n + \alpha_{n-1}s^{n-1} + \alpha_{n-2}s^{n-2} + \alpha_{n-3}s^{n-3} + \cdots + \alpha_1 s + \alpha_0}$$

with quotient terms

$$\frac{\alpha_n}{c}s^{n-2} + \frac{\alpha'_{n-1}}{c}s^{n-3}$$

$$\alpha_n s^n + \frac{d\alpha_n}{c} s^{n-1} + \frac{e\alpha_n}{c} s^{n-2}$$

$$\alpha'_{n-1} s^{n-1} + \alpha'_{n-2} s^{n-2} + \alpha_{n-3} s^{n-3} + \cdots + \alpha_1 s + \alpha_0$$

$$\alpha'_{n-1} s^{n-1} + \frac{d\alpha'_{n-1}}{c} s^{n-2} + \frac{e\alpha'_{n-1}}{c} s^{n-3}$$

$$\alpha''_{n-2} s^{n-2} + \alpha'_{n-3} s^{n-3} + \cdots + \alpha_1 s + \alpha_0$$

where

$$\alpha'_{n-1} = \alpha_{n-1} - \frac{d\alpha_n}{c}$$

$$\alpha'_{n-2} = \alpha_{n-2} - \frac{e\alpha_n}{c}$$

and

$$\alpha''_{n-2} = \alpha'_{n-2} - \frac{d\alpha'_{n-1}}{c}$$

$$\alpha'_{n-3} = \alpha_{n-3} - \frac{e\alpha'_{n-1}}{c}$$

Proceeding in this way, the last step is

$$\frac{\alpha_n}{c} s^{n-2} + \frac{\alpha'_{n-1}}{c} s^{n-3} + \cdots + \frac{\alpha''_2}{c}$$

$$cs^2 + ds + e \enspace\bigg|\enspace \alpha_n s^n + \alpha_{n-1} s^{n-1} + \cdots + \alpha_2 s^2 + \alpha_1 s + \alpha_0$$

$$\vdots$$

$$\alpha''_2 s^2 + \alpha'_1 s + \alpha_0$$

$$\alpha''_2 s^2 + \frac{d\alpha''_2}{c} s + \frac{e\alpha''_2}{c}$$

$$\left(\alpha'_1 - \frac{d\alpha''_2}{c} \right) s + \left(\alpha_0 - \frac{e\alpha''_2}{c} \right)$$

and the remainder is

$$\alpha''_1 s + \alpha'_0 = \left(\alpha'_1 - \frac{d\alpha''_2}{c} \right) s + \left(\alpha_0 - \frac{e\alpha''_2}{c} \right)$$

The corresponding algorithm for polynomial division is given in Table 5-3.

    A computer program in BASIC/FORTRAN, DIVIDE2, that uses this algorithm to perform polynomial division is listed in Table 5-4. Steps 120–180 for dividend polynomial degree and coefficient entry are the same as for DIVIDE1. The remaining steps are similar

TABLE 5-3   Algorithm for Second-Degree Polynomial Division

DEFINITIONS

The dividend polynomial is

$$p(s) = \alpha_n s^n + \alpha_{n-1} s^{n-1} + \cdots + \alpha_1 s + \alpha_0$$

The divisor polynomial is

$$g(s) = cs^2 + ds + e$$

The quotient to be found is

$$q(s) = \beta_{n-2} s^{n-2} + \beta_{n-3} s^{n-3} + \cdots + \beta_1 s + \beta_0$$

and the remainder is

$$r(s) = \gamma_1 s + \gamma_0$$

ALGORITHM

For $n - 1$ steps, beginning with $i = 1$ and ending with $i = n - 1$,

$$\beta_{n-i-1} \leftarrow \frac{\alpha_{n-i+1}}{c}$$

$$\alpha_{n-i} \leftarrow \alpha_{n-i} - \frac{d\alpha_{n-i+1}}{c}$$

$$\alpha_{n-i-1} \leftarrow \alpha_{n-i-1} - \frac{e\alpha_{n-i+1}}{c}$$

TABLE 5-4

Computer Program in BASIC for Second-Degree Polynomial Division (DIVIDE2)

VARIABLES USED

N = degree in dividend polynomial

$P(20)$ = coefficients of the dividend polynomial

$P(N)$ is the leading coefficient and $P(0)$ is the constant term

$Q(19)$ = coefficients of the $(N - 1)$-degree quotient polynomial; $Q(N - 1)$ is the leading coefficient and $Q(0)$ is the constant term

C = coefficient of s^2 in the divisor polynomial

D = coefficient of s in the divisor polynomial

E = constant coefficient in the divisor polynomial

I = coefficient index

LISTING

```
100     PRINT "SECOND DEGREE POLYNOMIAL DIVISION"
110     DIM P(20),Q(19)
120   REM ENTER DIVIDEND POLYNOMIAL AND DEGREE";
130     PRINT "PLEASE ENTER DEGREE OF DIVIDEND";
140     INPUT N
```

(continued)

```
150      FOR I = 0 TO N
160      PRINT "ENTER POWER ";N -I;" COEFFICIENT";
170      INPUT P(N -I)
180      NEXT I
190   REM ENTER SECOND DEGREE DIVISOR POLYNOMIAL
200      PRINT "ENTER THE DIVISOR'S LEADING COEFFICIENT";
210      INPUT C
220      IF C =0 THEN GOTO 190
230      PRINT "ENTER THE DIVISOR'S MIDDLE COEFFICIENT";
240      INPUT D
250      PRINT "ENTER THE DIVISOR'S CONSTANT COEFFICIENT";
260      INPUT E
270   REM COMPUTE QUOTIENT COEFFICIENTS
280   REM P(1) AND P(0) BECOME REMAINDER COEFFICIENTS
290      FOR I = 1 TO N -1
300      Q(N -I -1) = P(N -I +1)/C
310      P(N -I) = P(N -I)- D*P(N -I +1)/C
320      P(N -I -1) = P(N - I -1)-E*P(N -I +1)/C
330      NEXT I
340   REM PRINT QUOTIENT AND REMAINDER
350      PRINT "DIVIDEND POLYNOMIAL IS"
360      FOR I =2 TO N
370      PRINT Q(N-I),"S**"; N-I
380      NEXT I
390      PRINT "REMAINDER IS"
400      PRINT P(1), "S"
410      PRINT "+";P(0)
420      END
```

Computer Program in FORTRAN for Second-Degree Polynomial Division (DIVIDE2)

```
C ********************************************************************
C PROGRAM DIVIDE2 -- SECOND DEGREE POLYNOMIAL DIVISION
C ********************************************************************
      REAL P(20), Q(19)

      WRITE(*,'(1X,''SECOND DEGREE POLYNOMIAL DIVISION'')')

C *** ENTER DIVIDEND POLYNOMIAL AND DEGREE
      WRITE(*,'(1X,''PLEASE ENTER DEGREE OF DIVIDEND '',\)')
      READ(*,*) N
      DO 10 I=1, N+1
         WRITE(*,'(1X,''ENTER POWER '',I3,'' COEFFICIENT '',\)') N-I+1
         READ(*,*) P(N-I+2)
 10   CONTINUE

C *** ENTER SECOND DEGREE DIVISOR POLYNOMIAL
 20   WRITE(*,'(1X,''ENTER THE DIVISORS LEADING COEFFICIENT '',\)')
      READ(*,*) C
      IF (C.EQ.0.0) GOTO 20
      WRITE(*,'(1X,''ENTER THE DIVISORS MIDDLE COEFFICIENT '',\)')
      READ(*,*) D
      WRITE(*,'(1X,''ENTER THE DIVISORS CONSTANT COEFFICIENT '',\)')
      READ(*,*) E

C *** COMPUTE QUOTIENT COEFFICIENTS
C *** P(2) AND P(1) BECOME REMAINDER COEFFICIENTS
      DO 30 I=1, N-1
         Q(N-I) = P(N-I+2)/C
         P(N-I+1) = P(N-I+1) - D*P(N-I+2)/C
         P(N-I) = P(N-I) - E*P(N-I+2)/C
 30   CONTINUE
```

```
C *** PRINT QUOTIENT AND REMAINDER
      WRITE(*,'(1X,''DIVIDEND POLYNOMIAL IS'')')
      DO 40 I=2, N
         WRITE(*,'(1X,E15.8,''  S** '',I5)') Q(N-I+1), N-I
 40   CONTINUE
      WRITE(*,'(1X,''REMAINDER IS '',E15.8,'' S'')') P(2)
      WRITE(*,'(1X,''              + '',E15.8)') P(1)
      STOP
      END
```

but involve a divisor of second degree and a first-degree remainder. When the polynomial

$$p(s) = s^4 + 4s^3 - 2s^2 - 3s + 2$$

is divided by

$$g(s) = 3s^2 - s + 2$$

for example, DIVIDE2 gives the following result:

```
DIVIDEND POLYNOMIAL IS
 .333333333          S**2
1.44444444           S**1
 -.407407408         S**0
REMAINDER IS
-6.2962963           S
+2.81481482
```

5.3.2 Polynomial Evaluation

Evaluation of a polynomial

$$p(s) = \alpha_n s^n + \alpha_{n-1} s^{n-1} + \cdots + \alpha_1 s + \alpha_0$$

of a possibly complex value of $s = x + iy$,

$$p(s = x + iy) = \alpha_n(x + iy)^n + \alpha_{n-1}(x + iy)^{n-1} + \cdots + \alpha_1(x + iy) + \alpha_0$$

can be done by computing and collecting various powers of $x + iy$ and combining them. An easier and more elegant polynomial evaluation method is first to form

$$\alpha_n(x + iy) + \alpha_{n-1}$$

then form

$$[\alpha_n(x + iy) + \alpha_{n-1}](x + iy) + \alpha_{n-2} = \alpha_n(x + iy)^2 + \alpha_{n-1}(x + iy) + \alpha_{n-2}$$

then

$$[\alpha_n(x + iy)^2 + \alpha_{n-1}(x + iy) + \alpha_{n-2}](x + iy) + \alpha_{n-3}$$

$$= \alpha_n(x + iy)^3 + \alpha_{n-1}(x + iy)^2 + \alpha_{n-2}(x + iy) + \alpha_{n-3}$$

until, after the nth such step, the result is

$$[\alpha_n(x + iy)^{n-1} + \alpha_{n-1}(x + iy)^{n-2} + \cdots + \alpha_1] (x + iy) + \alpha_0$$
$$= \alpha_n(x + iy)^n + \alpha_{n-1}(x + iy)^{n-1} + \cdots + \alpha_1(x + iy) + \alpha_0$$

This algorithm for polynomial evaluation is given in Table 5-5. First, a step-by-step description, similar to those used previously, is listed. Then an alternative, more compact description is given. Most readers will find the alternative statement of the algorithm easier to understand initially, but the step-by-step version easier to relate to the corresponding computer program. In defining an algorithm, one should use whatever description method or methods are best suited to the purpose at hand. Rote translation

TABLE 5-5   Algorithm for Polynomial Evaluation with a Complex Number

DEFINITIONS

The polynomial is

$$p(s) = \alpha_n s^n + \alpha_{n-1} s^{n-1} + \cdots + \alpha_1 s + \alpha_0$$

Evaluation is at

$$s = x + iy$$

and the result is

$$p(s = x + iy) = u + iv = \alpha_n(x + iy)^n$$
$$+ \alpha_{n-1}(x + iy)^{n-1} + \cdots + \alpha_1(x + iy) + \alpha_0$$

ALGORITHM

1. Set u to the original polynomial's leading coefficient

$$u \leftarrow \alpha_n$$

and set v to zero:

$$v \leftarrow 0$$

(lines 2010–2020).

2. Perform the following n steps, beginning with $k = 1$ and ending with $k = n$ (lines 2030–2070).
   a. Set

$$w \leftarrow ux - vy + \alpha_{n-k}$$

The variable w is the real part of the product $(x + iy)(u + iv) + \alpha_{n-k}$ (line 2040).
   b. Set

$$v \leftarrow vx + uy$$

The variable v is the imaginary part of the product $(x + iy)(u + iv) + \alpha_{n-k}$ (line 2050).
   c. Transfer

$$u \leftarrow w$$

This now makes u the real part of the product $(x + iy)(u + iv) + \alpha_{n-k}$. This real part was computed first as w so that u would not be changed before its use in step b (line 2060).

ALTERNATIVE STATEMENT OF THE ALGORITHM

$$u \leftarrow \alpha_n$$

$$v \leftarrow 0$$

$$u + \underline{i}v \leftarrow (x + \underline{i}y)(u + \underline{i}v) + \alpha_{n-1}$$

$$= \alpha_n(x + \underline{i}y) + \alpha_{n-1}$$

$$u + \underline{i}v \leftarrow (x + \underline{i}y)(u + \underline{i}v) + \alpha_{n-2}$$

$$= \alpha_n(x + \underline{i}y)^2 + \alpha_{n-1}(x + \underline{i}y) + \alpha_{n-2}$$

$$\vdots$$

$$u + \underline{i}v \leftarrow (x + \underline{i}y)(u + \underline{i}v) + \alpha_0$$

$$= \alpha_n(x + \underline{i}y)^n + \alpha_{n-1}(x + \underline{i}y)^{n-1} + \cdots + \alpha_1(x + \underline{i}y) + \alpha_0$$

of computer code to a diagram or listing of steps is seldom very useful to anyone. This polynomial evaluation algorithm is used for the EVAL program in BASIC/FORTRAN, listed in Table 5-6. The EVAL algorithm is written as a subroutine to emphasize its simplicity and because it will be used later as part of a larger program. For the polynomial

$$p(s) = (s + 2 + \underline{i}3)(s + 2 - \underline{i}3)(s + 1)(s + 4)$$

$$= s^4 + 9s^3 + 37s^2 + 81s + 52$$

for example, EVAL gives

```
RESULTING VALUE IS
0 + I 0
```

verifying that $-2 + \underline{i}3$ is a root:

$$p(s = -2 + \underline{i}3) = 0 + \underline{i}0$$

TABLE 5-6

Computer Program in BASIC for Polynomial Evaluation with a Complex Number (EVAL)

VARIABLES USED

N = polynomial degree

P(20) = original polynomial coefficients; P(N) is the coefficient of the highest-order term and p(0) is the coefficient of the constant term

X = real part of the variable

Y = imaginary part of the variable

U = real part of the result

V = imaginary part of the result

W = temporary real part

I = polynomial coefficient index in main program

K = polynomial coefficient index in subroutine

(continued)

LISTING

```
100      PRINT "POLYNOMIAL EVALUATION"
110      DIM P(20)
120    REM ZERO POLYNOMIAL ARRAY
130      FOR I = 0 TO 20
140      P(I)=0
150      NEXT I
160    REM ENTER POLYNOMIAL DEGREE AND COEFFICIENTS
170      PRINT "PLEASE ENTER POLYNOMIAL DEGREE"
180      INPUT N
190      FOR I = 0 TO N
200      PRINT "ENTER POWER ";N-I;" COEFFICIENTS";
210      INPUT P(N-I)
220      NEXT I
230    REM ENTER THE COMPLEX VARIABLE
240      PRINT "ENTER THE VARIABLE REAL PART";
250      INPUT X
260      PRINT "ENTER THE VARIABLE IMAGINARY PART";
270      INPUT Y
280    REM EVALUATE POLYNOMIAL
290      GOSUB 2000
300    REM PRINT RESULT
310      PRINT "RESULTING VALUE IS"
320      PRINT U;" +I ";V
330      END
2000   REM SUBROUTINE TO EVALUATE A POLYNOMIAL
2010     U = P(N)
2020     V = 0
2030     FOR K = 1 TO N
2040     W = U * X - V * Y +P(N-K)
2050     V = V * X + U * Y
2060     U = W
2070     NEXT K
2080     RETURN
```

Computer Program in FORTRAN for Polynomial Evaluation with a Complex Number (EVAL)

```
C ***********************************************************************
C PROGRAM EVAL -- POLYNOMIAL EVALUATION
C ***********************************************************************
      REAL P(21)

      WRITE(*,'(1X,''POLYNOMIAL EVALUATION'')')

C *** ZERO POLYNOMIAL ARRAY
      DO 10 I=1, 21
         P(I)=0.0
 10   CONTINUE

C *** ENTER POLYNOMIAL DEGREE AND COEFFICIENTS
      WRITE(*,'(1X,''PLEASE ENTER POLYNOMIAL DEGREE '',\)')
      READ(*,*) N
      DO 20 I=1, N+1
         WRITE(*,'(1X,''ENTER POWER '',I3,'' COEFFICIENT '',\)') N-I+1
         READ(*,*) P(N-I+2)
 20   CONTINUE

C *** ENTER THE COMPLEX VARIABLE
      WRITE(*,'(1X,''ENTER THE VARIABLE REAL PART '',\)')
      READ(*,*) X
```

```
          WRITE(*,'(1X,''ENTER THE VARIABLE IMAGINARY PART '',\)')
          READ(*,*) Y

C *** EVALUATE POLYNOMIAL
          CALL FUNCT(P,N,X,Y,U,V)

C *** PRINT RESULT
          WRITE(*,'(1X,''RESULTING VALUE IS'')')
          WRITE(*,'(1X,E15.8,'' +I '',E15.8)') U,V
          STOP
          END

C *******************************************************************
          SUBROUTINE FUNCT(P,N,X,Y,U,V)
C *******************************************************************
          REAL P(21)

C *** SUBROUTINE TO EVALUATE A POLYNOMIAL
          U = P(N+1)
          V = 0.0
          DO 10 K=1, N
             W = U*X - V*Y + P(N-K+1)
             V = V*X + U*Y
             U = W
10        CONTINUE
          RETURN
          END
```

5.3.3 Axis Shifting

Another important polynomial operation is that of axis shifting. For a polynomial

$$q(s) = \beta_n s^n + \beta_{n-1} s^{n-1} + \cdots + \beta_1 s + \beta_0$$

an axis shift by the real number c gives a new polynomial,

$$p(s) = q(s + c) = \beta_n(s + c)^n + \beta_{n-1}(s + c)^{n-1} + \cdots + \beta_1(s + c) + \beta_0$$

$$= \alpha_n s^n + \alpha_{n-1} s^{n-1} + \cdots + \alpha_1 s + \alpha_0$$

An efficient method of finding the new polynomial coefficients from the old ones is the following. First, form

$$p_1(s) = \beta_n(s + c) + \beta_{n-1}$$

then form

$$p_2(s) = [\beta_n(s + c) + \beta_{n-1}](s + c) + \beta_{n-2} = p_1(s)(s + c) + \beta_{n-2}$$

$$= \beta_n(s + c)^2 + \beta_{n-1}(s + c) + \beta_{n-2}$$

and

$$p_3(s) = [\beta_n(s + c)^2 + \beta_{n-1}(s + c) + \beta_{n-2}](s + c) + \beta_{n-3}$$

$$= p_2(s)(s + c) + \beta_{n-3} = \beta_n(s + c)^3$$

$$+ \beta_{n-1}(s + c)^2 + \beta_{n-2}(s + c) + \beta_{n-3}$$

After the nth such step, there results

$$p_n(s) = [\beta_n(s + c)^{n-1} + \beta_{n-1}(s + c)^{n-2} + \cdots + \beta_1](s + c) + \beta_0$$

$$= \beta_n(s + c)^n + \beta_{n-1}(s + c)^{n-1} + \cdots + \beta_1(s + c) + \beta_0$$

Beginning with

$$\alpha_0 = \beta_n$$

the first intermediate polynomial is

$$p_1(s) = \beta_n(s + c) + \beta_{n-1} = \alpha_0(s + c) + \beta_{n-1}$$

so that the replacements

$$\alpha_1 \leftarrow \alpha_0$$

$$\alpha_0 \leftarrow c\alpha_0 + \beta_{n-1}$$

give its coefficients:

$$p_1(s) = \alpha_1 s + \alpha_0$$

The second intermediate polynomial is

$$p_2(s) = p_1(s)(s + c) + \beta_{n-2} = (\alpha_1 s + \alpha_0)(s + c) + \beta_{n-2}$$

so that the replacements

$$\alpha_2 \leftarrow \alpha_1$$

$$\alpha_1 \leftarrow c\alpha_1 + \alpha_0$$

$$\alpha_0 \leftarrow c\alpha_0 + \beta_{n-2}$$

give its coefficients:

$$p_2(s) = \alpha_2 s^2 + \alpha_1 s + \alpha_0$$

Similarly, the coefficients of the third intermediate polynomial,

$$p_3(s) = p_2(s)(s + c) + \beta_{n-3}$$

are obtained from the coefficients of the second with the replacements

$$\alpha_3 \leftarrow \alpha_2$$

$$\alpha_2 \leftarrow c\alpha_2 + \alpha_1$$

$$\alpha_1 \leftarrow c\alpha_1 + \alpha_0$$

$$\alpha_0 \leftarrow c\alpha_0 + \beta_{n-3}$$

and so on. At the nth iteration, the coefficients of the shifted polynomial are formed with the replacements

$$\alpha_n \leftarrow \alpha_{n-1}$$

$$\alpha_{n-1} \leftarrow c\alpha_{n-1} + \alpha_{n-2}$$

$$\vdots$$

$$\alpha_1 \leftarrow c\alpha_1 + \alpha_0$$

$$\alpha_0 \leftarrow c\alpha_0 + \beta_0$$

Using this result, an algorithm for finding axis-shifted polynomial coefficients is given in Table 5-7. The corresponding computer program in BASIC/FORTRAN is listed in Table 5-8. The SHIFT program is written as a subroutine because, like EVAL, it too will be used later as part of a larger program. For the polynomial

$$q(s) = (s + 3)^4$$

for example, the SHIFT program gives

```
POLYNOMIAL AXIS SHIFT
PLEASE ENTER POLYNOMIAL DEGREE?4
ENTER POWER 4 COEFFICIENT?1
ENTER POWER 3 COEFFICIENT?12
ENTER POWER 2 COEFFICIENT?54
ENTER POWER 1 COEFFICIENT?108
ENTER POWER 0 COEFFICIENT?81
ENTER THE AXIS SHIFT AMOUNT?-3
SHIFTED POLYNOMIAL IS
1     S**4
0     S**3
0     S**2
0     S**1
0     S**0
```

verifying that

$$p(s) = q(s - 3) = s^4$$

TABLE 5-7   Algorithm for Polynomial Axis Shifting

DEFINITIONS

The original polynomial is

$$q(s) = \beta_n s^n + \beta_{n-1} s^{n-1} + \cdots + \beta_1 s + \beta_0$$

The amount of right shift of the imaginary axis is c, and the intermediate polynomials are

$$p_i(s) = \alpha_n s^i + \alpha_{n-1} s^{i-1} + \cdots + \alpha_{n-i+1} s + \alpha_{n-i}$$

(continued)

The resulting polynomial is

$$p(s) = p_n(s) = \beta_n(s + c)^n + \beta_{n-1}(s + c)^{n-1} + \cdots + \beta_1(s + c) + \beta_0$$

$$= \alpha_n s^n + \alpha_{n-1} s^{n-1} + \cdots + \alpha_1 s + \alpha_0$$

ALGORITHM

1. Form the first intermediate polynomial, $p_1(s)$:

$$\alpha_1 \leftarrow \beta_n$$

$$\alpha_0 \leftarrow c*\beta_n + \beta_{n-1}$$

(lines 1010–1030).

2. Form the remaining $n - 1$ intermediate polynomials, $p_i(s)$, beginning with $i = 2$ and ending with $i = n$.

a. Set the leading coefficient,

$$\alpha_i \leftarrow \alpha_{i-1}$$

(lines 1060–1070).

b. Form the middle $i - 1$ coefficients. Beginning with $j = 1$ and ending with $j = i - 1$, set

$$\alpha_{i-j} = \alpha_{i-j-1} + c\alpha_{i-j} \qquad j = 1, 2, \ldots, i - 1$$

(lines 1080–1110).

c. Form the last coefficient

$$\alpha_0 \leftarrow c\alpha_0 + \beta_{n-i}$$

(lines 1120–1130).

TABLE 5-8

Computer Program in BASIC for Polynomial Axis Shifting (SHIFT)

VARIABLES USED

N	=	polynomial degree
Q(20)	=	original polynomial coefficients; Q(N) is the coefficient of the highest-order term and Q(0) is the coefficient of the constant term
P(20)	=	resulting polynomial coefficients
C	=	amount of axis shift
I	=	polynomial coefficient index; intermediate polynomial number index
J	=	intermediate polynomial coefficient index

LISTING

```
100     PRINT "POLYNOMIAL AXIS SHIFT"
110     DIM P(20),Q(20)
120   REM ENTER POLYNOMIAL DEGREE AND COEFFICIENTS
130     PRINT "PLEASE ENTER POLYNOMIAL DEGREE";
140     INPUT N
150     FOR I = 0 TO N
160     PRINT "ENTER POWER ";N -I;"COEFFICIENT";
170     INPUT Q(N -I)
```

```
180      NEXT I
190   REM ENTER THE AMOUNT OF AXIS SHIFT
200      PRINT "ENTER THE AXIS SHIFT AMOUNT";
210      INPUT C
220   REM SHIFT POLYNOMIAL
230      GOSUB 1000
240   REM PRINT RESULT
250      PRINT "SHIFTED POLYNOMIAL IS"
260      FOR I=0 TO N
270      PRINT P(N -I), "S**";N-I
280      NEXT I
290      END
1000  REM SUBROUTINE TO AXIS SHIFT A POLYNOMIAL
1010  REM FORM FIRST INTERMEDIATE POLYNOMIAL
1020     P(1) = Q(N)
1030     P(0) = C*Q(N) + Q(N -1)
1040  REM FORM REMAINING INTERMEDIATE POLYNOMIALS
1050     FOR I = 2 TO N
1060  REM SET LEADING COEFFICIENTS
1070     P(I) = P(I-1)
1080  REM FORM MIDDLE COEFFICIENTS
1090     FOR J = 1 TO I -1
1100     P(I -J) = P(I -J -1) + C*P(I -J)
1110     NEXT J
1120  REM FORM LAST COEFFICIENT
1130     P(0)=C * P(0)+ Q(N -I)
1140     NEXT I
1150     RETURN
```

Computer Program in FORTRAN for Polynomial Axis Shifting (SHIFT)

```
C ********************************************************************
C PROGRAM SHIFT -- POLYNOMIAL AXIS SHIFT
C ********************************************************************
      REAL P(20), Q(20)

      WRITE(*,'(1X,''POLYNOMIAL AXIS SHIFT'')')

C *** ENTER POLYNOMIAL DEGREE AND COEFFICIENTS
      WRITE(*,'(1X,''PLEASE ENTER POLYNOMIAL DEGREE '',\)')
      READ(*,*) N
      DO 10 I=1, N+1
         WRITE(*,'(1X,''ENTER POWER '',I2,'' COEFFICIENT '',\)') N-I+1
         READ(*,*) Q(N-I+2)
  10  CONTINUE

C *** ENTER THE AMOUNT OF AXIS SHIFT
      WRITE(*,'(1X,''ENTER THE AXIS SHIFT AMOUNT '',\)')
      READ(*,*) C

C *** SHIFT POLYNOMIAL
      CALL AXISFT(Q,N,C,P)

C *** PRINT RESULT
      WRITE(*,'(1X,''SHIFTED POLYNOMIAL IS'')')
      DO 20 I=1, N+1
         WRITE(*,'(1X,E15.8,'' S** '',I2)') P(N-I+2), N-I+1
  20  CONTINUE
      STOP
      END

C ********************************************************************
      SUBROUTINE AXISFT(Q,N,C,P)
```

(continued)

```
C  ********************************************************************
      REAL P(20), Q(20)
C *** SUBROUTINE TO AXIS SHIFT A POLYNOMIAL
C *** FORM FIRST INTERMEDIATE POLYNOMIAL
      P(2) = Q(N+1)
      P(1) = C*Q(N+1) + Q(N)
C *** FORM REMAINING INTERMEDIATE POLYNOMIALS
      DO 10 I=2, N
C *** SET LEADING COEFFICIENTS
         P(I+1) = P(I)
C *** FORM MIDDLE COEFFICIENTS
         DO 20 J=1, I-1
            P(I-J+1) = P(I-J) + C*P(I-J+1)
  20     CONTINUE
C *** FORM LAST COEFFICIENT
         P(1) = C*P(1)+Q(N-I+1)
  10  CONTINUE
      RETURN
      END
```

5.4 SOME FACTORING METHODS

5.4.1 Finding and Removing Real Roots

Often, the real roots of odd multiplicity are factored from a polynomial before a search for complex roots is begun. One way to find the real, distinct roots of a polynomial is to perform a search for real zeros such as BISECT (Table 3-3). Another, similar way is to consider the function tested to be the remainder after division by a first-order root term. Like the value of the polynomial itself, the remainder changes algebraic sign whenever the trial real root location passes a root of odd multiplicity. Trial division has the advantage of automatically producing the quotient polynomial, reducing the degree of the polynomial to be tested for other roots.

For example, for the polynomial

$$p(s) = 3s^3 + 10s^2 + 21s + 6$$

division by a first-degree factor with a root at $s = 1$ gives a remainder of 40:

$$
\begin{array}{r}
3s^2 + 13s + 34 \\
\hline
s - 1 \,\big)\, 3s^3 + 10s^2 + 21s + 6 \\
3s^3 - 3s^2 \\
\hline
13s^2 + 21s + 6 \\
13s^2 - 13s \\
\hline
34s + 6 \\
34s - 34 \\
\hline
40
\end{array}
$$

Division by a factor with a root at $s = -1$ gives a remainder of -8:

$$
\begin{array}{r}
3s^2 + 7s + 14 \\
\hline
s + 1 \enclose{longdiv}{3s^3 + 10s^2 + 21s + 6} \\
3s^3 + 3s^2 \\
\hline
7s^2 + 21s + 6 \\
7s^2 + 7s \\
\hline
14s + 6 \\
14s + 14 \\
\hline
-8
\end{array}
$$

The change in algebraic sign of the remainder indicates that this polynomial has a root between $s = -1$ and $s = 1$. The results of other divisions by first-degree root terms are tabulated below, where it is seen that it is easy to quickly narrow down the root location, which happens to be at $s = \frac{1}{3}$.

ROOT	REMAINDER
1	40
−1	−8
0	6
−0.5	−2.375
−0.25	1.328
−0.375	−0.627

Once a root has been found with sufficient accuracy, the lower-degree quotient polynomial can be tested for other real roots.

In this process, as in the bisection search procedure to be developed, the remainder after first-degree division is considered to be a function of the trial divisor root location. Figure 5-2 illustrates the idea. This function $f(x)$, given by an algorithm instead of as an explicit formula, is a continuous function of the variable x that has zero crossings at the values of x for which $(s - x)$ is a factor of the dividend polynomial.

Table 5-9 lists a program called REALFACT that uses a bisection search on a first-order trial division remainder to locate real roots. When a real root is found with the desired accuracy, it is printed and the search for the remaining roots continues using the quotient, which is the polynomial with the known root factor removed. When no more real roots in the search interval can be found, the program prints the remaining quotient, to be processed further.

The REALFACT routine uses major parts of the DIVIDE1 and BISECT programs. First, the polynomial to be tested, stored as the array P( ), and the search parameters are entered by the user. A bisection search begins at the left endpoint of the interval tested and proceeds far enough past the right endpoint (B + 2 * C in line 260) to ensure

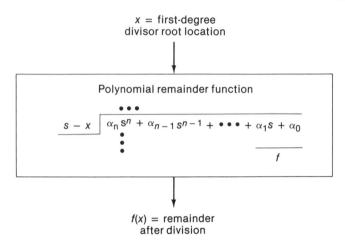

Figure 5-2   Viewing a first-degree division remainder as a function of the divisor coefficient.

TABLE 5-9

Program in BASIC to Find and Remove Real Roots from a Polynomial (REALFACT)

VARIABLES USED

N = polynomial degree

$P(20)$ = original polynomial coefficients; quotient polynomial at each step as a root is removed; $P(N)$ is the leading coefficient and $P(0)$ is the constant term

$Q(19)$ = quotient polynomial

$R(20)$ = copy of $P(\cdot)$ used for division and changed by the division process; $R(0)$ is the remainder of the division

A = left search boundary

B = right search boundary

C = initial search increment

D = present increment of the variable

E = smallest increment of the variable

X = search variable

F = value of the remainder; value of the function at the left test point

G = value of the function at the right test point

I = polynomial coefficient index

LISTING

```
100     PRINT "POLYNOMIAL REAL ROOT LOCATION AND DIVISION"
110     DIM P(20),Q(19),R(20)
120   REM   ENTER POLYNOMIAL AND DEGREE
130     PRINT "PLEASE ENTER POLYNOMIAL DEGREE";
140     INPUT N
150     FOR I = 0 TO N
160     PRINT "ENTER POWER ";N - I;"COEFFICIENT";
170     INPUT P(N - I)
180     NEXT I
```

```
190    REM   ENTER BOUNDARIES, INCREMENT AND PRECISION
200      PRINT "PLEASE ENTER LEFT SEARCH BOUNDARY";
210      INPUT A
220      PRINT "ENTER RIGHT SEARCH BOUNDARY";
230      INPUT B
240      PRINT "ENTER INITIAL STEP SIZE";
250      INPUT C
260      B = B + 2 * C
270      PRINT "ENTER DESIRED PRECISION";
280      INPUT E
290    REM   PRINT HEADING
300      PRINT "REAL ROOT","REMAINDER"
310    REM   SET INCREMENT TO INITIAL INCREMENT
320      D = C
330    REM   EVALUATE FUNCTION AT LEFT BOUNDARY
340      X = A
350      GOSUB 1000
360    REM   SET G TO LEFT VALUE OF FUNCTION
370      G = F
380    REM   INCREMENT VARIABLE. PRINT QUOTIENT IF RIGHT
390    REM   BOUNDARY EXCEEDED; OTHERWISE EVALUATE FUNCTION
400      X = X + D
410      IF X > B THEN   GOTO 650
420      GOSUB 1000
430    REM   IF NO FUNCTION SIGN CHANGE, CONTINUE SEARCH
440      IF G * F > 0 THEN   GOTO 370
450    REM   PRINT ROOT IF INCREMENT IS LESS THAN E;
460    REM   OTHERWISE MOVE X BACK, HALVE INCREMENT, AND
470    REM   REEVALUATE
480      IF D < E THEN   GOTO 520
490      X = X - D
500      D = D / 2
510      GOTO 380
520    REM   PRINT ROOT AND SEARCH DIVIDEND
530      PRINT X,F
540    REM   TRANSFER QUOTIENT Q TO P
550      FOR I = 0 TO N - 1
560      P(I) = Q(I)
570      NEXT I
580    REM   RESET DEGREE N; PRINT QUOTIENT IF QUOTIENT
590    REM   IS CONSTANT
600      N = N - 1
610      IF N = 0 THEN   GOTO 650
620    REM   RESET LEFT BOUNDARY AND CONTINUE SEARCH
630      A = X + C
640      GOTO 310
650    REM   PRINT QUOTIENT AND END
660      PRINT "QUOTIENT POLYNOMIAL IS"
670      FOR I = 0 TO N
680      PRINT P(N - I),"S**";N - I
690      NEXT I
700      END
1000   REM   SUBROUTINE FOR FIRST DEGREE POLY DIVISION
1010   REM   TRANSFER P TO R
1020     FOR I = 0 TO N
1030     R(I) = P(I)
1040     NEXT I
1050     FOR I = 1 TO N
1060     Q(N - I) = R(N - I + 1)
1070     R(N - I) = R(N - I) + X * R(N - I + 1)
1080     NEXT I
1090     F = R(0)
1100     RETURN
```

(continued)

Program in FORTRAN to Find and Remove Real Roots from a Polynomial (REALFACT)

```
C ******************************************************************
C PROGRAM REALFACT -- POLYNOMIAL REAL ROOT LOCATION AND DIVISION
C ******************************************************************
      REAL P(20), Q(19), X, F
      INTEGER N

      WRITE(*,'(1X,''POLYNOMIAL REAL ROOT LOCATION AND DIVISION'')')

C *** ENTER POLYNOMIAL AND DEGREE
      WRITE(*,'(1X,''PLEASE ENTER POLYNOMIAL DEGREE '',\)')
      READ(*,*) N
      DO 10 I=1, N+1
         WRITE(*,'(1X,''ENTER POWER '',I2,'' COEFFICIENT '',\)') N-I+1
         READ(*,*) P(N-I+2)
   10 CONTINUE

'C *** ENTER BOUNDARIES, INCREMENT AND PRECISION
      WRITE(*,'(1X,''PLEASE ENTER LEFT SEARCH BOUNDARY '',\)')
      READ(*,*) A
      WRITE(*,'(1X,''ENTER RIGHT SEARCH BOUNDARY '',\)')
      READ(*,*) B
      WRITE(*,'(1X,''ENTER INITIAL STEP SIZE '',\)')
      READ(*,*) C
      B = B + 2*C
      WRITE(*,'(1X,''ENTER DESIRED PRECISION '',\)')
      READ(*,*) E

C *** PRINT HEADING
      WRITE(*,'(1X,''    REAL ROOT        REMAINDER'')')

C *** LOOP MAXIMUM OF 1000 TIMES
C *** SET INCREMENT TO INITIAL INCREMENT
      DO 20 II=1, 1000
         D = C

C *** EVALUATE FUNCTION AT LEFT BOUNDARY
         X = A
         CALL POLDIV(P,N,X,F,Q)

C *** SET G TO LEFT VALUE OF FUNCTION
         G = F

C *** INCREMENT VARIABLE (MAXIMUM OF 1000 ITERATIONS)
C *** PRINT QUOTIENT IF RIGHT BOUNDARY EXCEEDED;
C *** OTHERWISE EVALUATE FUNCTION
         DO 30 JJ=1, 1000
            X = X + D
            IF (X.GT.B) GOTO 40
            CALL POLDIV(P,N,X,F,Q)

C *** IF NO FUNCTION SIGN CHANGE, CONTINUE SEARCH
            IF (G*F.GT.0.0) THEN
               G = F
            ELSE

C *** PRINT ROOT IF INCREMENT IS LESS THAN E;
C *** OTHERWISE MOVE X BACK, HALVE INCREMENT, AND
C *** REEVALUATE
               IF (D.LT.E) GOTO 50
               X = X - D
               D = D/2
```

```
          ENDIF
30        CONTINUE

C *** PRINT ROOT AND SEARCH DIVIDEND
50        WRITE(*,'(1X,E15.8,1X,E15.8)') X, F

C *** TRANSFER QUOTIENT Q TO P
          DO 60 I=1, N
              P(I) = Q(I)
60        CONTINUE

C *** RESET DEGREE N; PRINT QUOTIENT IF QUOTIENT
C *** IS CONSTANT
          N = N - 1
          IF (N.EQ.0) GOTO 40

C *** RESET LEFT BOUNDARY AND CONTINUE SEARCH
          A = X + C
20        CONTINUE

C *** PRINT QUOTIENT AND END
40     WRITE(*,'(1X,''QUOTIENT POLYNOMIAL IS'')')
          DO 70 I=1, N+1
              WRITE(*,'(1X,E15.8,'' S** '',I2)') P(N-I+2), N-I+1
70        CONTINUE

       STOP
       END

C ********************************************************************
       SUBROUTINE POLDIV(P,N,X,F,Q)
C ********************************************************************
       REAL P(20), Q(19), R(20), X, F
       INTEGER N

C *** SUBROUTINE FOR FIRST DEGREE POLY DIVISION
C *** TRANSFER P TO R
       DO 10 I=1, N+1
           R(I) = P(I)
10     CONTINUE
       DO 20 I=1, N
           Q(N-I+1) = R(N-I+2)
           R(N-I+1) = R(N-I+1) + X*R(N-I+2)
20     CONTINUE
       F = R(1)
       RETURN
       END
```

that a root at the endpoint will be located. The BISECT program did not evaluate the function outside the search boundaries because of the possibility that the function used would be infinite or undefined there. There is no such problem with a polynomial function, however. The bisection search is line-for-line identical to that in the BISECT program, with the function searched given by subroutine 1000. The subroutine computes the quotient $Q(\cdot)$ and the remainder F of the division of the polynomial P by the first-order divisor. When the root X is found with sufficient accuracy, the remainder F is very nearly zero, and the root found and actual value of F are printed. The polynomial $P(\cdot)$ is then replaced by the quotient $Q(\cdot)$, and the search continues. When all roots are found or when the right endpoint is reached, the search ends and the quotient is printed.

For the polynomial

$$p(s) = s^4 + s^3 + 6s^2 - 14s - 20$$

the REALFACT program gives

```
POLYNOMIAL REAL ROOT LOCATION AND DIVISION
PLEASE ENTER POLYNOMIAL DEGREE ? 4
ENTER POWER 4 COEFFICIENT ? 1
ENTER POWER 3 COEFFICIENT ? 1
ENTER POWER 2 COEFFICIENT ? 6
ENTER POWER 1 COEFFICIENT ? -14
ENTER POWER 0 COEFFICIENT ? -20
PLEASE ENTER LEFT SEARCH BOUNDARY ? -5
ENTER RIGHT SEARCH BOUNDARY ? 5
ENTER INITIAL STEP SIZE ? 0.1
ENTER DESIRED PRECISION ? 0.0001
REAL ROOT      REMAINDER
 -.999902354   -2.63635287E-03
1.99999999      8.78655323E-04
QUOTIENT POLYNOMIAL IS
1              S**2
2.00009764     S**1
10.0000976     S**0
```

It correctly locates the real roots at $s = -1$ and $s = 2$ and quotient

$$q(s) = s^2 + 2s + 10$$

which contains a complex conjugate pair of root. For the polynomial

$$p(s) = (s - 3)^3 = s^3 - 9s^2 + 27s - 27$$

REALFACT finds only one of the three repeated roots because the quotient after one $(s - 3)$ factor is removed has a second-order ("grazing") root at $s = 3$ which the bisection process does not identify:

```
REAL ROOT          REMAINDER
3.00058593         0
QUOTIENT POLYNOMIAL IS
1                  S**2
 -5.99941408       S**1
9                  S**0
```

5.4.2 Trial Quadratic Division

One common way of finding complex conjugate pairs of polynomial roots is to divide the polynomial by a trial quadratic divisor and search the first-degree remainder polynomial for those conjugate root locations for which both remainder coefficients are simultaneously zero. For example, for the polynomial

$$p(s) = s^3 - s^2 + 3s + 5 = (s + 1)(s - 1 - \underline{i}2)(s - 1 + \underline{i}2)$$

$$= (s + 1)(s^2 - 2s + 5)$$

division by a quadratic divisor

$$(s - x - iy)(s - x + iy) = s^2 - 2xs + x^2 + y^2$$

gives

$$
\require{enclose}
\begin{array}{r}
s + (-1 + 2x) \\
\hline
s^2 - 2xs + (x^2 + y^2) \enclose{longdiv}{s^3 - \quad s^2 + \qquad\qquad 3s + 5}
\end{array}
$$

$$s^3 - 2xs^2 + (x^2 + y^2)s$$
$$\overline{(-1 + 2x)s^2 + (3 - x^2 - y^2)s + 5}$$
$$(-1 + 2x)s^2 - 2x(-1 + 2x)s + (x^2 + y^2)(-1 + 2x)$$
$$\overline{(3x^2 - y^2 - 2x + 3)s + (5 + x^2 + y^2 - 2x^3 - 2xy^2)}$$

so that the coefficients of the first-degree remainder polynomial are, in terms of the trial root location $x + iy$,

$$u(x, y) = 3x^2 - y^2 - 2x + 3$$
$$v(x, y) = 5 + x^2 + y^2 - 2x^3 - 2xy^2$$

Using these functions in the CROSS program (Table 4-8)

```
1010 F = 3*X*X - Y*Y - 2*X + 3
2010 G = 5 + X*X + Y*Y - 2*X*X*X - 2*X*Y*Y
```

gives

```
SIMULTANEOUS CROSSING OF TWO CURVES
ENTER SEARCH LOWER LEFT COORDINATES
X = ?0
Y = ?0
ENTER SEARCH DIMENSIONS? 3
ENTER NUMBER OF GRID DIVISIONS? 10
ENTER REQUIRED PRECISION? 0.0001
CROSSING COORDINATES ARE
X              Y
.999975586    2.00002441
.999975586    1.99995117
```

so that the complex conjugate root pair is located correctly.

A factoring program using this method might first remove all real roots of odd multiplicity. It would then use a quadratic polynomial division routine such as DIVIDE2 to compute the first-degree remainder coefficients for each trial X and Y, and to remove root factor terms from the polynomial, as they are found. Trial quadratic division is generally a good polynomial factoring method. It is straightforward to use the CROSS (Table 4-8) and DIVIDE2 (Table 5-4) routines to assemble a polynomial factoring program of this kind. The search region is then over the possible range of divisor coefficients.

5.4.3 Bairstow's Method

Bairstow's method is a commonly used polynomial factoring procedure that is essentially trial quadratic division with a two-dimensional Newton–Raphson search for the divisor coefficients that result in a zero remainder. Generally, a Newton–Raphson search trades dependability for convergence speed, and this is the case with Bairstow's method. Sometimes there is failure to converge and, additionally, special consideration should be given to situations involving repeated root and to even polynomials, polynomials containing only even terms.

The first-degree remainder

$$\gamma_1 s + \gamma_0$$

of a second-degree trial divisor

$$s^2 + ds + e$$

of the polynomial to be factored is computed and the coefficients, d and e, of the divisor are changed at each step in an attempt to find those coefficients for which the remainder is zero. In this search, the two remainder coefficients are considered to be functions of the two divisor coefficients

$$\delta_1(d, e) \qquad \text{and} \qquad \delta_0(d, e)$$

These functions are fairly complicated, nonlinear ones, so the increments of d and e to be made at the next step are somewhat involved to calculate. In the end, they require a second quadratic division and the solution of two simultaneous linear algebraic equations.

We will not pursue Bairstow's method further here because there are other polynomial factoring methods that are more sure and just as fast. When programmed to accommodate the various special cases, such as even polynomials, and to restart the search automatically when it is not converging, Bairstow's method is adequate.

5.5 FACTORING BY REAL AND IMAGINARY PART SEARCH

Because of the method's dependability and relative simplicity, we now consider real-part and imaginary-part search to locate the roots of a polynomial. A program to remove real and complex conjugate pairs of roots from the polynomial, as they are found, will be developed. The EVAL algorithm (Table 5-5) will be used to find the real and imaginary parts of the polynomial for various complex values of the variable s. The CROSS algorithm (Table 4-7) will be used to find those values of s for which the polynomial's real and imaginary parts are simultaneously zero. A general polynomial factoring program such as the one to be developed here is fairly extensive and sophisticated, so, along the way, there is an opportunity to learn more about the process of program design.

5.5.1 Search Method

Another good way of locating a root, possibly complex,

$$s = x + iy$$

of a polynomial $p(s)$ is to express the polynomial in terms of x and y, find the real and imaginary parts of the result, and equate to zero:

$$\text{Re}[p(s = x + iy)] = u(x, y) = 0$$

$$\text{Im}[p(s = x + iy)] = u(x, y) = 0$$

A number s, possibly complex, is a root of the polynomial if, when substituted, it simultaneously makes the real and imaginary parts of the polynomial zero. One can then use a two-dimensional simultaneous zero-crossing program to find x and y, the real and imaginary parts of the root. If the polynomial has only complex root pairs (including possibly repeated roots of even multiplicity), then once any such root is located, it and the complex conjugate can be factored from the polynomial:

$$\frac{p(s)}{(s - x - iy)(s - x + iy)} = \frac{p(s)}{s^2 - (2x)s + (x^2 + y^2)} = q(s)$$

The lower-degree remainder polynomial $q(s)$ can then be partially factored in the same way, and so on until all of the root pairs are found. As a simple numerical example, the polynomial

$$p(s) = s^3 - s^2 + 3s + 5 = (s + 1)(s - 1 - i2)(s - 1 + i2)$$

has, for $s = x + iy$, real and imaginary parts given by

$$p(s = x + iy) = (x + iy)^3 - (x + iy)^2 + 3(x + iy) + 5$$

$$= x^3 + 3ix^2y - 3xy^2 - iy^3 - x^2 - 2ixy + y^2 + 3x + 3iy + 5$$

$$= (x^3 - 3xy^2 - x^2 + y^2 + 3x + 5) + i(3x^2y - y^3 - 2xy + 3y)$$

or

$$\text{Re}[p(s = x + iy)] = u(x, y) = x^3 - 3xy^2 - x^2 + y^2 + 3x + 5$$

$$\text{Im}[p(s = x + iy)] = v(x, y) = 3x^2y - y^3 - 2xy + 3y$$

The CROSS program when used to find the coordinates of

$$u(x, y) = 0$$

$$v(x, y) = 0$$

with

```
1010 F = X*X*X - 3*X*Y*Y - X*X + Y*Y + 3*X + 5
2010 G = 3*X*X*Y - Y*Y*Y - 2*X*Y + 3*Y
```

gives, for example,

```
SIMULTANEOUS CROSSING OF TWO CURVES
ENTER SEARCH LOWER LEFT COORDINATES
X = ? - 2
Y = ? - 2
ENTER SEARCH DIMENSION?5
```

```
ENTER NUMBER OF GRID DIVISIONS?10
ENTER REQUIRED PRECISION?0.0001
CROSSING COORDINATES ARE
X                          Y
1                          2
.999938965                 2
1                          1.999938965
.999938965                 1.999938965
-1                         0
-1.00006104                0
-1                         -6.10351563EE-05
-1.00006104                -6.10351563EE-05
1                          -2
.999938965                 -2
```

which correctly locates the roots in the vicinity of $s = 1 \pm i2$ and $s = -1$.

In locating polynomial roots in this way, we are considering the two remainder coefficients of a quadratic polynomial division to be functions of the divisor root locations. Figure 5-3 shows the idea. These functions are continuous and have simultaneous zeros when $x + iy$ and $x - iy$ are roots of the dividend polynomial.

A factoring program using this method would use a polynomial evaluation routine such as the EVAL program to evaluate the real and imaginary parts of the polynomial. To speed operation and simplify matters in the case of repeated roots, it would remove root factor terms from the polynomial as they are found, using routines such as DIVIDE1 [for real roots (Table 5-2)] and DIVIDE2 [for complex conjugate root pairs (Table 5-4)].

When the polynomial has repeated roots, there are multiple intersections of the $u(x, y) = 0$ and $v(x, y) = 0$ curves; that is, for a double root, two $u(x, y) = 0$ and two $v(x, y) = 0$ curves intersect one another at the root location, as in Figure 5-4. There is no particular difficulty in locating repeated roots with this method except that the complicated nature of the $u(x, y) = 0$ and $v(x, y) = 0$ curves in the vicinity of a root might require a relatively small initial grid spacing so that the program does not skip over a zero crossing.

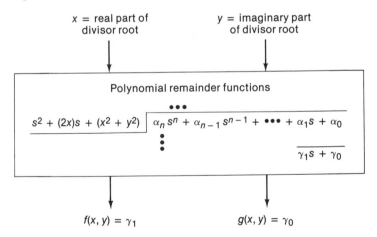

Figure 5-3 Polynomial remainder coefficients as functions of the real and imaginary parts of a quadratic divisor.

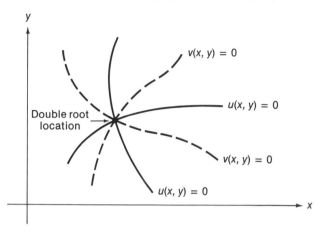

Figure 5-4 Nature of the $u(x, y) = 0$ and $v(x, y) = 0$ curves in the vicinity of a double root of the polynomial.

5.5.2 Removing Located Roots

To develop a general factoring program using real part and imaginary part searches, we will first design and test a main program that removes first-degree and second-degree factors from the polynomial as they are found. Locating the roots will be done by subroutines that are to be developed later using the CROSS and EVAL algorithms. An algorithm for this main program is given in Table 5-10. It is expressed in terms of a *flow diagram*, a form that is well suited to starting and understanding moderate degrees of decision making, as are involved here.

The corresponding main program, POLYROOT, is listed in Table 5-11. The search dimension D that is entered by the user defines the square search region for a roof $s = x + iy$ shown in Figure 5-5. This region is divided into 100 squares, 10 on a side, for further search for a simultaneous crossing of the real-part-equals-zero and the imaginary-part-equals-zero curves of the polynomial. When a root is found, it is determined whether or not the root is real. If real, the first-degree factor is removed from the polynomial before searching for other roots. If a complex root is found, both it and its complex conjugate are removed from the polynomial.

Especially with a program like this involved, it is a very good idea to test portions of the instructions to see that they are behaving as expected before proceeding further. One simple test for these instructions is to simulate the operation of the subroutines. For example, the temporary additional instructions

```
1000 FOR I = 0 TO N
1010 PRINT P(N - I)
1020 NEXT I
1030 X = 3
1040 Y = 0
1050 IF N = 2 THEN END
1060 RETURN
```

print the polynomial coefficients, simulate the subroutine's location of a real root at (3, 0), and print the coefficients of the quotient polynomial, which has become the new

TABLE 5-10   Algorithm to Factor a Polynomial, Using Real Part and Imaginary Part Search

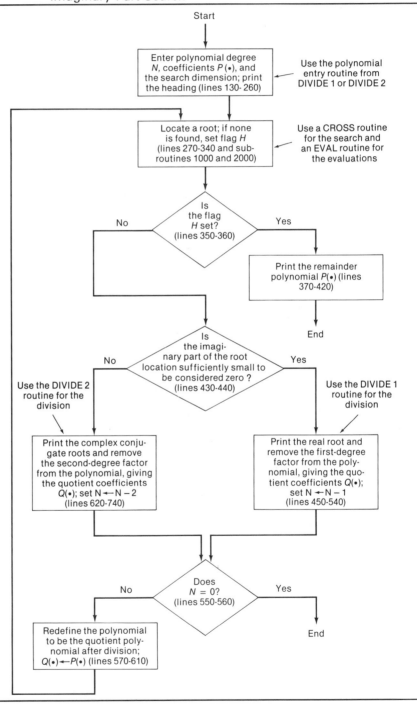

TABLE 5-11

Portion of a Computer Program in BASIC for Polynomial Factoring (Main Part of POLYROOT)

VARIABLES USED

N = degree of the polynomial tested

$P(20)$ = coefficients of the tested polynomial; $P(N)$ is the leading coefficient and $P(0)$ is the constant term

D = initial search dimension, as in Figure 5-3

E = required precision; ultimately, the two zero-crossing curves must be within E of one another to be considered to have crossed

R = x coordinate of the lower left corner of a square

S = y coordinate of the left corner of a square

C = dimension of the smaller squares searched

M = number of smaller squares on each side of the search region

H = flag to indicate that no simultaneous zero crossing has been found: $F = 0$ if a crossing is found; otherwise, $H = 1$

X = x coordinate of a square

Y = y coordinate of a square

$Q(20)$ = coefficients of the quotient polynomial, after division by a first- or second-degree divisor

I = polynomial coefficient index; array index

J = array index

LISTING

```
010    REM POLYROOT

100        PRINT "FACTORING WITH RE & IM PART SEARCH"
110        DIM P(20),Q(20),A(12,12),B(12,12),L(12,12)
120        DIM K(12,12), Z(3,50)
130    REM   ENTER POLYNOMIAL, DIMENSION, AND PRECISION
140        PRINT "PLEASE ENTER DEGREE OF POLYNOMIAL";
150        INPUT N
160        FOR I = 0 TO N
170        PRINT "ENTER POWER ";N - I;"COEFFICIENT";
180        INPUT P(N - I)
190        NEXT I
200        PRINT "ENTER SEARCH DIMENSION";
210        INPUT D
220        PRINT "ENTER THE REQUIRED PRECISION";
230        INPUT E
240    REM   PRINT HEADING
250        PRINT "POLYNOMIAL ROOTS ARE"
260        PRINT "REAL","IMAGINARY"
270    REM   FIND A ROOT
280    REM   SET SEARCH PARAMETERS AND FLAG
290        R = - D
300        S = - D / 100
310        C = D / 5
320        M = 10
330        H = 0
340        GOSUB 1000
350    REM   PRINT MESSAGE AND END IF NO ROOT FOUND
360        IF H = 0 THEN GOTO   430
```

(continued)

```
370     PRINT "NO MORE ROOTS FOUND IN SEARCH REGION"
380     PRINT "REMAINING POLYNOMIAL IS"
390     FOR I = 0 TO N
400     PRINT P(N - I),"S**";N - I
410     NEXT I
420     GOTO 750
430  REM   TEST IF THE ROOT IS REAL
440     IF Y > E THEN   GOTO 620
450  REM   FOR A REAL ROOT...
460  REM   PRINT THE ROOT
470     PRINT X,0
480  REM   REMOVE THE FIRST-DEGREE TERM FROM P()
490     FOR I = 1 TO N
500     Q(N - I) = P(N - I + 1)
510     P(N - I) = P(N - I) + X * P(N - I + 1)
520     NEXT I
530  REM   REDUCE THE POLYNOMIAL DEGREE BY ONE
540     N = N - 1
550  REM   DETERMINE IF ALL ROOTS FOUND
560     IF N = 0 THEN   GOTO 750
570  REM REPLACE P() BY Q()
580     FOR I = 0 TO N
590     P(I) = Q(I)
600     NEXT I
610     GOTO 270
620  REM   FOR A COMPLEX ROOT...
630  REM   PRINT THE CONJUGATE ROOTS
640     PRINT X,Y
650     PRINT X, - Y
660  REM   REMOVE THE SECOND-DEGREE TERM FROM P()
670     FOR I = 1 TO N - 1
680     Q(N - I - 1) = P(N - I + 1)
690     P(N - I) = P(N - I) + 2 * X * P(N - I + 1)
700     P(N - I - 1) = P(N - I - 1) - (X * X + Y * Y) * P(N - I + 1)
710     NEXT I
720  REM   REDUCE THE POLYNOMIAL DEGREE BY TWO
730     N = N - 2
740     GOTO 550
750     END
```

Portion of a Computer Program in FORTRAN for Polynomial Factoring (Main Part of POLYROOT)

```
C ******************************************************************
C PROGRAM POLYROOT -- FACTORING WITH REAL AND IMAGINARY PART SEARCH
C ******************************************************************
      REAL    P(20),    Q(20)
      INTEGER H
      DATA    ZERO/0.0/

      WRITE(*,'(1X,''FACTORING WITH RE & IM PART SEARCH'')')

C *** ENTER POLYNOMIAL, DIMENSION, AND PRECISION
      WRITE(*,'(1X,''PLEASE ENTER DEGREE OF POLYNOMIAL '',\)')
      READ(*,*) N
      DO 10 I=1, N+1
         WRITE(*,'(1X,''ENTER POWER '',I2,'' COEFFICIENT '',\)') N-I+1
         READ(*,*) P(N-I+2)
   10 CONTINUE
      WRITE(*,'(1X,''ENTER SEARCH DIMENSION '',\)')
      READ(*,*) D
      WRITE(*,'(1X,''ENTER THE REQUIRED PRECISION '',\)')
      READ(*,*) E
```

```
C *** PRINT HEADING
      WRITE(*,'(1X,''POLYNOMIAL ROOTS ARE'')')
      WRITE(*,'(1X,''      REAL          IMAGINARY'')')

C *** FIND A ROOT
C *** SET SEARCH PARAMETERS AND FLAG, MAXIMUM OF 100 LOOPS
      DO 20 II=1, 100
         R = -D
         S = -D/100.0
         C = D/5.0
         M = 10.0
         H = 0
         CALL ZEROS(P,N,R,S,C,M,E,X,Y,H)

C *** PRINT MESSAGE AND END IF NO ROOT FOUND
         IF (H.NE.0) THEN
            WRITE(*,'(1X,''NO MORE ROOTS FOUND IN SEARCH REGION'')')
            WRITE(*,'(1X,''REMAINING POLYNOMIAL IS'')')
            DO 30 I=1, N+1
               WRITE(*,'(1X,E15.8,'' S** '',I2)') P(N-I+2), N-I+1
 30         CONTINUE
            STOP
         ENDIF

C *** TEST IF THE ROOT IS REAL
         IF (Y.LE.E) THEN

C *** FOR A REAL ROOT...
C *** PRINT THE ROOT
            WRITE(*,'(1X,E15.8,1X,E15.8)') X, ZERO

C *** REMOVE THE FIRST-DEGREE TERM FROM P()
            DO 50 I=1, N
               Q(N-I+1) = P(N-I+2)
               P(N-I+1) = P(N-I+1) + X*P(N-I+2)
 50         CONTINUE

C *** REDUCE THE POLYNOMIAL DEGREE BY ONE
            N = N-1
C *** DETERMINE IF ALL ROOTS FOUND
            IF (N.EQ.0) STOP

C *** REPLACE P() BY Q()
            DO 60 I=1, N+1
               P(I) = Q(I)
 60         CONTINUE

         ELSE

C *** FOR A COMPLEX ROOT...
C *** PRINT THE CONJUGATE ROOTS

            WRITE(*,'(1X,E15.8,1X,E15.8)') X,  Y
            WRITE(*,'(1X,E15.8,1X,E15.8)') X, -Y

C *** REMOVE THE SECOND-DEGREE TERM FROM P()
            DO 80 I=1, N-1
               Q(N-I)   = P(N-I+2)
               P(N-I+1) = P(N-I+1) + 2*X*P(N-I+2)
               P(N-I)   = P(N-I) - (X*X + Y*Y)*P(N-I+2)
 80         CONTINUE

C *** REDUCE THE POLYNOMIAL DEGREE BY TWO
            N = N-2
```

(continued)

```
C *** DETERMINE IF ALL ROOTS FOUND
            IF (N.EQ.0) STOP

C *** REPLACE P() BY Q()
            DO 90 I=1, N+1
                P(I) = Q(I)
   90       CONTINUE
          ENDIF
   20   CONTINUE

        STOP
        END
```

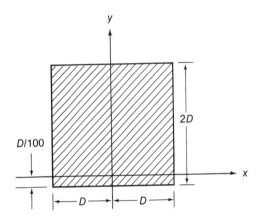

Figure 5-5 Search area of the POLY-ROOT program.

polynomial to be tested. For an entered polynomial

$$p(s) = 2s^3 - 10s^2 + 22s - 30 = 2(s - 3)(s^2 - 2s + 5)$$

this test shows the root and quotient polynomials to be computed correctly.

```
POLYNOMIAL ROOTS ARE
   REAL      IMAGINARY

    2                        ORIGINAL POLYNOMIAL
  -10                        COEFFICIENTS
   22
  -30


    3            0

    2
   -4                        NEW POLYNOMIAL COEFFICIENTS
   10
```

To test how these instructions handle the location of a complex factor, the additional instructions

```
1000 FOR I = 0 TO N
1010 PRINT P(N - 1)
1020 NEXT I
1030 X = 1
1040 Y = 2
1050 IF N = 1 THEN END
1060 RETURN
```

can be added instead. For the same polynomial, the correct factors are printed, as is the correct quotient polynomial:

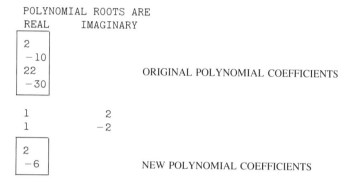

```
POLYNOMIAL ROOTS ARE
REAL      IMAGINARY
```
```
 2
-10
22                              ORIGINAL POLYNOMIAL COEFFICIENTS
-30
```
```
 1              2
 1             -2
```
```
 2
-6                              NEW POLYNOMIAL COEFFICIENTS
```

To test the operation of the flag, setting the flag is simulated by the temporary subroutine

```
1000 FOR I = 0 TO N
1010 PRINT P(N - I)
1020 NEXT I
1030 H = 1
1040 RETURN
```

The correct result,

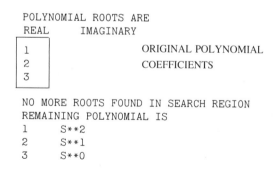

```
POLYNOMIAL ROOTS ARE
REAL      IMAGINARY
```
```
1                               ORIGINAL POLYNOMIAL
2                               COEFFICIENTS
3
```
```
NO MORE ROOTS FOUND IN SEARCH REGION
REMAINING POLYNOMIAL IS
1       S**2
2       S**1
3       S**0
```

is obtained.

When this set of instructions was first written, there *was* a mistake. The algebraic sign in the expression on line 500 was a minus instead of the plus it should have been. The first of these tests showed that for a first-degree divisor, the quotient polynomial was not computed correctly. The error was thus quickly located and corrected.

5.5.3 Factoring Program

When subroutines for simultaneous zero-crossing location and for polynomial evaluation are added to the foregoing main program, the result is a general polynomial factoring program. The simultaneous zero-crossing routine is listed in Table 5-12. It is the heart of the CROSS program, renumbered and with a few minor changes. First, only a single

TABLE 5-12

Zero-Crossing Subroutine in BASIC for the POLYROOT Factoring Program

ADDITIONAL VARIABLES USED

$$T \;=\; \text{stack pointer}$$
$$A(12,12) \;=\; \text{array of values of the function F}$$
$$B(12,12) \;=\; \text{array of values of the function G}$$
$$F \;=\; \text{real part of the evaluated polynomial}$$
$$G \;=\; \text{imaginary part of the evaluated polynomial}$$
$$L(12,12) \;=\; \text{array of squares containing zero crossings of the function F, as a function of grid indices}$$
$$K(12,12) \;=\; \text{array of squares containing zero crossings of the function G, as a function of grid indices}$$
$$Z(3,50) \;=\; \text{stack of coordinates and dimensions of squares to be searched further}$$

LISTING

```
1000  REM   SUBROUTINE TO LOCATE SIMULTANEOUS ZEROS
1010  REM
1020  REM   ROUTINE ACCEPTS LOWER LEFT SEARCH COORDI-
1030  REM   NATES(R,S), SEARCH SQUARE DIMENSION C,
1040  REM   NUMBER OF DIVISIONS M, AND REQUIRED PRE-
1050  REM   CISION E.  IT RETURNS THE COORDINATES
1060  REM   (X,Y) FOR THE FIRST CROSSING FOUND.  IF
1070  REM   NO CROSSING IS FOUND, THE L FLAG IS SET
1080  REM   TO 1.  SUBROUTINE 2000 IS USED TO EVALUATE
1090  REM   THE TWO FUNCTIONS F AND G.
1100  REM
1110  REM   ZERO THE STACK POINTER
1120     T = 0
1130  REM   STORE GRID POINT VALUES OF FUNCTIONS AND
1140  REM   ZERO THE L AND K ARRAYS
1150     FOR J = 1 TO M + 1
1160     FOR I = 1 TO M + 1
1170     X = R + (I - 1) * C
1180     Y = S + (J - 1) * C
1190     GOSUB 2000
1200     A(I,J) = F
1210     B(I,J) = G
1220     L(I,J) = 0
1230     K(I,J) = 0
1240     NEXT I
1250     NEXT J
1260  REM   IDENTIFY SQUARES WITH ZERO CROSSINGS
1270     FOR J = 1 TO M + 1
1280     FOR I = 1 TO M + 1
1290     IF A(I,J) <  > 0 THEN GOTO 1340
1300     L(I,J) = 1
1310     L(I - 1,J) = 1
1320     L(I - 1,J - 1) = 1
```

```
1330    L(I,J - 1) = 1
1340    IF B(I,J) <  > 0 THEN GOTO 1390
1350    K(I,J) = 1
1360    K(I - 1,J) = 1
1370    K(I - 1,J - 1) = 1
1380    K(I,J - 1) = 1
1390    IF A(I,J) * A(I + 1,J) > = 0 THEN  GOTO 1420
1400    L(I,J) = 1
1410    L(I,J - 1) = 1
1420    IF B(I,J) * B(I + 1,J) > = 0 THEN  GOTO 1450
1430    K(I,J) = 1
1440    K(I,J - 1) = 1
1450    IF A(I,J) * A(I,J + 1) > = 0 THEN  GOTO 1480
1460    L(I,J) = 1
1470    L(I - 1,J) = 1
1480    IF B(I,J) * B(I,J + 1) > = 0 THEN  GOTO 1510
1490    K(I,J) = 1
1500    K(I - 1,J) = 1
1510    NEXT I
1520    NEXT J
1530    REM   STACK COORDINATES AND DIMENSION
1540    FOR J = 1 TO M
1550    FOR I = 1 TO M
1560    IF L(I,J) * K(I,J) = 0 THEN   GOTO 1610
1570    T = T + 1
1580    Z(1,T) = R + (I - 1) * C
1590    Z(2,T) = S + (J - 1) * C
1600    Z(3,T) = C
1610    NEXT I
1620    NEXT J
1630    REM  IF STACK EMPTY, SET FLAG AND RETURN
1640    IF T = 0 THEN H = 1
1650    IF T = 0 THEN   GOTO 1780
1660    REM  IF SUFFICIENT PRECISION, RETURN
1670    IF Z(3,T) > E THEN   GOTO 1710
1680    X = Z(1,T)
1690    Y = Z(2,T)
1700    GOTO 1780
1710    REM   REFINE SEARCH OF A STACKED SQUARE
1720    R = Z(1,T)
1730    S = Z(2,T)
1740    C = Z(3,T) / 2
1750    M = 2
1760    T = T - 1
1770    GOTO 1130
1780    RETURN
```

Zero-Crossing Subroutine in FORTRAN for the POLYROOT Factoring Program

```
C ********************************************************************
      SUBROUTINE ZEROS(P,N,R,S,C,M,E,X,Y,H)
C ********************************************************************
      REAL    A(12,12), B(12,12), Z(3,50), P(20)
      INTEGER L(12,12), K(12,12), H

C *** SUBROUTINE TO LOCATE SIMULTANEOUS ZEROS

C *** ROUTINE ACCEPTS LOWER LEFT SEARCH COORDINATES (R,S), SEARCH
C *** SQUARE DIMENSION C, NUMBER OF DIVISIONS M, AND REQUIRED
C *** PRECISION E.  IT RETURNS THE COORDINATES (X,Y) FOR THE FIRST
C *** CROSSING FOUND.  IF NO CROSSING IS FOUND, THE H FLAG IS SET
```

(continued)

```
C *** TO 1.  SUBROUTINE FUNCT IS USED TO EVALUATE THE TWO FUNCTIONS
C *** F AND G.

C *** ZERO THE STACK POINTER
      T = 0

C *** MAIN LOOP, MAXIMUM OF 1000 ITERATIONS
      DO 5 II=1, 1000

C *** STORE GRID POINT VALUES OF FUNCTIONS AND
C *** ZERO THE L AND K ARRAYS
          DO 10 J=1, M+1
              DO 20 I=1, M+1
                  X = R + (I-1)*C
                  Y = S + (J-1)*C
                  CALL FUNCT(P,N,X,Y,F,G)
                  A(I,J) = F
                  B(I,J) = G
                  L(I,J) = 0
                  K(I,J) = 0
 20           CONTINUE
 10       CONTINUE

C *** IDENTIFY SQUARES WITH ZERO CROSSINGS
          DO 30 J=1, M+1
              DO 40 I=1, M+1
                  IF (A(I,J).EQ.0.0) THEN
                      L(I+1,J+1) = 1
                      L(I,J+1) = 1
                      L(I,J) = 1
                      L(I+1,J) = 1
                  ENDIF
                  IF (B(I,J).EQ.0.0) THEN
                      K(I+1,J+1) = 1
                      K(I,J+1) = 1
                      K(I,J) = 1
                      K(I+1,J) = 1
                  ENDIF
                  IF (A(I,J)*A(I+1,J).LT.0.0) THEN
                      L(I+1,J+1) = 1
                      L(I+1,J) = 1
                  ENDIF
                  IF (B(I,J)*B(I+1,J).LT.0.0) THEN
                      K(I+1,J+1) = 1
                      K(I+1,J) = 1
                  ENDIF
                  IF (A(I,J)*A(I,J+1).LT.0.0) THEN
                      L(I+1,J+1) = 1
                      L(I,J+1) = 1
                  ENDIF
                  IF (B(I,J)*B(I,J+1).LT.0.0) THEN
                      K(I+1,J+1) = 1
                      K(I,J+1) = 1
                  ENDIF
 40           CONTINUE
 30       CONTINUE

C *** STACK COORDINATES AND DIMENSION
          DO 50 J=1, M
              DO 60 I=1, M
                  IF (L(I+1,J+1)*K(I+1,J+1).NE.0.0) THEN
                      T = T+1
                      Z(1,T) = R + (I-1)*C
                      Z(2,T) = S + (J-1)*C
```

```
                        Z(3,T) = C
                      ENDIF
   60            CONTINUE
   50         CONTINUE

C *** IF STACK EMPTY, SET FLAG AND RETURN
              IF (T.EQ.0) THEN
                 H=1
                 RETURN
              ENDIF

C *** IF SUFFICIENT PRECISION, RETURN
              IF (Z(3,T).LE.E) THEN
                 X = Z(1,T)
                 Y = Z(2,T)
                 RETURN
              ENDIF

C *** REFINE SEARCH OF A STACKED SQUARE
              R = Z(1,T)
              S = Z(2,T)
              C = Z(3,T)/2
              M = 2
              T = T-1
   5       CONTINUE
           RETURN
           END
```

subroutine call is made to obtain both the real-part and the imaginary-part functions. In CROSS, the two functions were computed by separate subroutines.

Second, when the stack is empty, meaning that no square contains a potential simultaneous zero crossing, the flag H is set to 1 from zero, and control is returned to the main program. In CROSS, there was no flag; a message was printed then program execution ended. Third, when a square is found with simultaneous zero crossings on its boundaries having dimension E or less, the search ends. Instead of printing the zero-crossing location, and looking for other such squares, the variables (x, y) are set to the coordinates of the square found and control is returned to the main program.

The subroutine for polynomial evaluation is listed in Table 5-13. It is the EVAL evaluation routine, using F and G for the real and imaginary parts of the result instead

TABLE 5-13

Polynomial Evaluation Subroutine in BASIC for the POLYROOT Factoring Program

ADDITIONAL VARIABLES USED

F = real parts of the intermediate polynomials; real part of the evaluated polynomial

G = imaginary parts of the intermediate polynomials; imaginary part of the evaluated polynomial

W = temporary storage of F

V = index

(continued)

LISTING

```
2000   REM  SUBROUTINE TO EVALUATE A POLYNOMIAL
2010     F = P(N)
2020     G = 0
2030     FOR V = 1 TO N
2040     W = F * X - G * Y + P(N - V)
2050     G = G * X + F * Y
2060     F = W
2070     NEXT V
2080     RETURN
```

Polynomial Evaluation Subroutine in FORTRAN for the POLYROOT Factoring Program

```
C  *****************************************************************
       SUBROUTINE FUNCT(P,N,X,Y,F,G)
C  *****************************************************************
       REAL P(20)

C *** SUBROUTINE TO EVALUATE A POLYNOMIAL
       F = P(N+1)
       G = 0.0
       DO 10 I=1, N
         W = F*X - G*Y + P(N-I+1)
         G = G*X + F*Y
         F = W
10     CONTINUE
       RETURN
       END
```

of U and V. Because the variables I and J are being used as indices when this subroutine is called, the variable V is used in the FOR . . . NEXT loop.

To test the zero-crossing location subroutine, the temporary instructions

```
100 DIM A(21,21), B(21,21), L(21,21), K(21,21), Z(3,50)
110 R = 0
120 S = 0
130 C = 0.3
140 M = 10
150 E = 0.0001
160 GOSUB 1000
170 PRINT X,Y
180 END
```

and the temporary subroutine

```
2000 REM SUBROUTINE TO EVALUATE TWO FUNCTIONS
2010 F = X * X + Y * Y - 4
2020 G = X - Y
2030 RETURN
```

defining the circle and line curves used previously, give the correct solution,

$$1.41416016 \qquad 1.4142334$$

The temporary instructions

```
100 DIM P(20)
110 N = 4
120 P(4) = 1
130 P(3) = 9
140 P(2) = 37
150 P(1) = 81
160 P(0) = 52
170 X = -2
180 Y = 3
190 GOSUB 2000
200 PRINT F,G
210 END
```

followed by the evaluation subroutine, give the correct evaluation of the polynomial

$$p(s) = s^4 + 9s^3 + 37s^2 + 81s + 52$$

as $s = -2 \pm i3$, which is

$$0 \qquad 0$$

Testing both subroutines together with, instead, the temporary instructions

```
100 DIM A(21,21), B(21,21), L(21,21), K(21,21), Z(3,50)
110 DIM P(20)
120 N = 3
130 P(3) = 1
140 P(2) = -1
150 P(1) = 3
160 P(0) = 5
170 R = 0
180 S = 0
190 C = 0.3
200 M = 10
210 E = 0.0001
220 GOSUB 1000
230 PRINT X,Y
240 END
```

correctly locates the root of the polynomial

$$p(s) = s^3 - s^2 + 3s + 5$$

at $s = 1 + i2$:

$$.999975596 \qquad 2.00002441$$

When subroutines 1000 and 2000 in Tables 5-12 and 5-13 are added to the POLYROOT main program of Table 5-11, a general polynomial factoring program results. For the polynomial

$$p(s) = s^3 - s^2 + 3s + 5 = (s + 1)(s^2 - 2s + 5)$$

which has roots at $s = -1$ and $s = 1 \pm i2$, the results are

REAL	IMAGINARY
.999975587	1.99998047
.999975587	-1.99998047
-.999975586	0

For the polynomial

$$p(s) = s^4 - 2s^3 + 3s^2 + 4s - 5$$

the roots found are these:

REAL	IMAGINARY
1.16411133	1.85730469
1.16411133	-1.85730469
.869018555	0
-1.19729004	0

5.6 ROUTH–HURWITZ TEST

An outstanding numerical polynomial factoring method uses the Routh–Hurwitz test to locate the real parts of the roots and another method to find their imaginary parts. The method thus replaces the usual two-dimensional search for a root with two one-variable searches, a large saving in computational effort. The Routh–Hurwitz test is a numerical procedure for determining the number of roots of a polynomial that are in the right half (to the right of the imaginary axis) of the complex plane [right half plane (RHP)]. By doing polynomial axis shifts in a bisection pattern and comparing the number of RHP roots of the axis-shifted polynomials, the real parts of the roots can be located to a desired precision. The test also gives the polynomial satisfied by the imaginary parts of the located roots, and these are found with a second one-variable search.

5.6.1 Fundamental Test

An example of the Routh–Hurwitz test of the polynomial

$$p(s) = 2s^4 + 3s^3 + 5s^2 + 2s + 6$$

is as follows. First, the initial part of the array is formed. The powers of s are written to the left and the polynomial coefficients are alternated between the first and second rows, as shown. It is helpful to imagine the rows to continue to the right with entries of zeros.

$$
\begin{array}{c|ccc}
s^4 & 2 & 5 & 6 \\
s^3 & 3 & 2 & \\
s^2 & & & \\
s^1 & & & \\
s^0 & & &
\end{array}
$$

The array is completed by proceeding, row by row, calculating the elements of the next row. Each element calculated is derived from four elements in the two rows above, two of them at the left column and two in the column to the right of the element being calculated. In each case, the calculated element is the negative of the determinant of the four elements above, divided by the lower left element above. For the example, the first element of the s^2 row is

$$-\frac{\begin{vmatrix} 2 & 5 \\ 3 & 2 \end{vmatrix}}{3} = \frac{11}{3}$$

the second element of the s^2 row is

$$-\frac{\begin{vmatrix} 2 & 6 \\ 3 & 0 \end{vmatrix}}{3} = 6$$

the first element of the s^1 row is

$$-\frac{\begin{vmatrix} 3 & 2 \\ \frac{11}{3} & 6 \end{vmatrix}}{\frac{11}{3}} = -\frac{32}{11}$$

and so on.

s^4	2	5	6		s^4	2	5	6		s^4	2	5	6
s^3	3	2			s^3	3	2	0		s^3	3	2	
s^2	$\frac{11}{3}$				s^2	$\frac{11}{3}$	6			s^2	$\frac{11}{3}$	6	
s^1					s^1					s^1	$-\frac{32}{11}$		
s^0					s^0					s^0			

The completed Routh–Hurwitz array is shown below. The number of RHP roots of $p(s)$ is the number of algebraic sign changes in the elements of the left column of the array, proceeding from top to bottom. For this example, there are two sign changes in the left column, as indicated with the arrows; therefore, $p(s)$ has two RHP roots:

s^4	2	5	6
s^3	3	2	
s^2	$\frac{11}{3}$	6	
s^1	$-\frac{32}{11}$		
s^0	6		

5.6.2 Left-Column Zeros

It sometimes happens that the polynomial coefficients are such that a zero occurs in the left column of the array, so that the array cannot be completed. The situation where there is a zero at the left of a row, but the entire row does not consist of zeros, is termed a *left-column zero*. For example, the polynomial

$$p(s) = 3s^4 + 6s^3 + 2s^2 + 4s + 5$$

has array that begins as follows:

s^4	3	2	5
s^3	6	4	
s^2	0	5	
s^1			
s^0			

The array cannot be completed in the usual way, because of the necessity to divide by zero.

When a left-column zero occurs, it is easiest to form a new polynomial with an additional known root, increasing its order but changing the coefficients so that a left-column zero does not occur. For example, adding an additional LHP root at $s = -1$ to the previous polynomial gives

$$p'(s) = (s + 1)(3s^4 + 6s^3 + 2s^2 + 4s + 5)$$

$$= 3s^5 + 9s^4 + 8s^3 + 6s^2 + 9s + 5$$

s^5	3	8	9
s^4	9	6	5
s^3	6	$\frac{22}{3}$	
s^2	-5	5	
s^1	$\frac{40}{3}$		
s^0	5		

This polynomial has two RHP roots, so the original polynomial $p(s)$ has two RHP roots.

5.6.3 Premature Termination

The situation where an entire row of zeros occurs in an array is termed a *premature termination*. For example, the Routh–Hurwitz test of the polynomial

$$p(s) = s^5 + 2s^4 + 8s^3 + 11s^2 + 16s + 12$$

terminates prematurely at the s^1 row:

s^5	1	8	16
s^4	2	11	12
s^3	$\frac{5}{2}$	10	
s^2	3	12	
s^1	0	0	
s^0			

Premature termination occurs whenever there is an even or odd polynomial divisor of the original polynomial. Even and odd polynomials have root locations that are symmetric about the imaginary axis. The coefficients of the even or odd divisor polynomial are those given in the row above the row of zeros:

$$p_{\text{divisor}}(s) = 3s^2 + 12 = 3(s^2 + 4) = 3(s + 2\underline{i})(s - 2\underline{i})$$

The tested polynomial thus has two imaginary axis roots, at $s = \pm 2\underline{i}$. To complete the array, the row of zeros is replaced by the coefficients of the derivative of the divisor polynomial:

$$\frac{dp_{\text{divisor}}(s)}{ds} = 6s$$

s^5	1	8	16
s^4	2	11	12
s^3	$\frac{5}{2}$	10	
s^2	3	12	
s^1	6		
s^0	12		

There are no algebraic sign changes in the left-column elements, so this polynomial has no RHP roots.

If a polynomial has imaginary axis (IA) roots, it has an even or odd polynomial divisor and its Routh–Hurwitz test will always terminate prematurely. Thus if there is no premature termination, the tested polynomial has no IA roots. However, there will be an even or odd polynomial divisor and premature termination, too, for any set of polynomial roots that are symmetric about the imaginary axis.

Another example is given below, where a premature termination occurred below the s^4 row:

$$p(s) = s^6 + s^5 + 2s^4 - s^3 - 5s^2 - 2s - 6$$

s^6	1	2	-5	-6
s^5	1	-1	-2	
s^4	3	-3	-6	
s^3	12	-6		
s^2	$-\frac{3}{2}$	-6		
s^1	-54			
s^0	-6			

This polynomial has one RHP root. Its fourth-degree polynomial divisor's Routh–Hurwitz test is that below the dashed line. It has one RHP root (i.e., the original polynomial's RHP root is in the even divisor) and since an even or odd polynomial's roots are symmetric about the imaginary axis, the divisor must have exactly one root in the left-half of the complex plane (LHP). The other two divisor roots are then on the imaginary axis. The number of roots of the original polynomial in the three regions are then

$$\text{RHP} = 1 \qquad \text{IA} = 2 \qquad \text{LHP} = 3$$

It is always possible to determine the separate numbers of RHP, IA, and LHP roots of a polynomial in this way from a Routh–Hurwitz test because the test of any even or odd polynomial divisor is included in the test of the original polynomial.

5.6.4 Array-Completion Program

A computer program called ROUTH for Routh–Hurwitz testing is listed in Table 5-14. The polynomial to be tested is entered into the array $P(\cdot)$, in the usual way, in lines 120–230. The program is written with the array completion itself a subroutine so that this portion of it can easily be used later as part of larger programs. The number of columns m of the array is determined in line 2030, and the first two rows of the array are transferred from the polynomial coefficients in lines 2110–2140. Then, if there is no zero in the left column, the array is completed in lines 2210–2260.

TABLE 5-14

Computer Program in BASIC for Fundamental Routh–Hurwitz Testing (ROUTH)

This program terminates if a zero occurs in the left column of the array.

VARIABLES USED

N	=	polynomial degree
P(20)	=	polynomial coefficients; P(N) is the coefficient of the highest-order term and P(0) is the coefficient of the constant term
R(21,12)	=	Routh–Hurwitz array as a function of row and column

F = flag to indicate if a zero has occurred in the left column of the array; if none has occurred, F $= -1$; if one has occurred, F $= 1$

M = number of columns of the Routh–Hurwitz array

I = polynomial coefficient index; array column index

J = array row index

LISTING

```
100      PRINT "FUNDAMENTAL ROUTH-HURWITZ TEST"
110      DIM P(20), R(21,12)
120   REM ENTER POLYNOMIAL DEGREE AND COEFFICIENTS
130      PRINT "PLEASE ENTER POLYNOMIAL DEGREE";
140      INPUT N
150   REM ENTER THE POLYNOMIAL COEFFICIENTS
160      FOR I = 0 TO N
170      PRINT "ENTER POWER"; N-I;" COEFFICIENT";
180      INPUT P(N-I)
190      NEXT I
200   REM FORM THE ROUTH-HURWITZ ARRAY
210      GOSUB 2000
220   REM PRINT ARRAY
230      PRINT "ROUTH-HURWITZ ARRAY IS"
240      FOR J= 1 TO M
250      PRINT "COLUMN ";J
260      FOR I= 1 TO N+1
270      PRINT R(I,J)
280      NEXT I
290      NEXT J
300   REM NOTIFY IF ZERO IN LEFT COLUMN
310      IF F> 0 THEN PRINT "ARRAY HAS ZERO IN LEFT COLUMN"
320      END
2000  REM SUBROUTINE FOR BASIC ROUTH-HURWITZ TEST
2010  REM SET FLAG AND NO. COLUMNS OF ARRAY
2020     F =-1
2030     M= INT (N/2+1)
2040  REM ZERO ARRAY R
2050     FOR I =1 TO 21
2060     FOR J =1 TO 12
2070     R(I,J) =0
2080     NEXT J
2090     NEXT I
2100  REM FROM FIRST TWO LINES OF ARRAY
2110     FOR I =1 TO M
2120     R(1,I) =P(N -2*I +2)
2130     IF N -2*I +1 < 0 THEN GOTO 2150
2140     R(2,I) =P(N -2*I +1)
2150     NEXT I
2160  REM COMPLETE ARRAY
2170     FOR I =1 TO N-1
2180  REM CHECK FOR ZERO IN LEFT COLUMN
2190     IF R(I+1,1) =0 THEN F= 1
2200     IF R(I+1,1) =0 THEN GOTO 2270
2210  REM COMPLETE EACH ROW
2220     FOR J= 1 TO M-1
2230     R(I + 2,J)= R(I+ 1,1)*R(I,J+1)-R(I,1)*R(I+1,J+1)
2240     R(I+ 2,J)= R(I+2,J)/R(I+1,1)
2250     NEXT J
2260     NEXT I
2270  .  RETURN
```

(continued)

Computer Program in FORTRAN for Fundamental Routh–Hurwitz Testing (ROUTH)

```
C ********************************************************************
C PROGRAM ROUTH -- FUNDAMENTAL ROUTH-HURWITZ TEST
C ********************************************************************
      REAL P(20), R(21,12)
      INTEGER F

      WRITE(*,'(1X,''FUNDAMENTAL ROUTH-HURWITZ TEST'')')

C *** ENTER POLYNOMIAL DEGREE AND COEFFICIENTS
      WRITE(*,'(1X,''PLEASE ENTER POLYNOMIAL DEGREE '',\)')
      READ(*,*) N

C *** ENTER THE POLYNOMIAL COEFFICIENTS
      DO 10 I=1, N+1
         WRITE(*,'(1X,''ENTER POWER '',I2,'' COEFFICIENT '',\)') N-I+1
         READ(*,*) P(N-I+2)
  10  CONTINUE

C *** FORM THE ROUTH-HURWITZ ARRAY
      CALL ROUTH(P,N,R,F,M)

C *** PRINT ARRAY
      WRITE(*,'(1X,''ROUTH-HURWITZ ARRAY IS'')')
      DO 20 J=1, M
         WRITE(*,'(1X,''COLUMN '',I2)') J
         DO 30 I=1, N+1
            WRITE(*,'(1X,E15.8)') R(I,J)
  30     CONTINUE
  20  CONTINUE

C *** NOTIFY IF ZERO IN LEFT COLUMN
      IF (F.GT.0) WRITE(*,'(1X,''ARRAY HAS ZERO IN LEFT COLUMN'')')
      STOP
      END

C ********************************************************************
      SUBROUTINE ROUTH(P,N,R,F,M)
C ********************************************************************
      REAL P(20), R(21,12)
      INTEGER F

C *** SUBROUTINE FOR BASIC ROUTH-HURWITZ TEST
C *** SET FLAG AND NO. COLUMNS OF ARRAY
      F = -1
      M =  INT(N/2+1)

C *** ZERO ARRAY R
      DO 10 I=1, 21
         DO 20 J=1, 12
            R(I,J) = 0.0
  20     CONTINUE
  10  CONTINUE

C *** FROM FIRST TWO LINES OF ARRAY
      DO 30 I=1, M
         R(1,I) = P(N - 2*I + 3)
         IF ((N - 2*I +1).GE.0) THEN
            R(2,I) = P(N - 2*I + 2)
         ENDIF
  30  CONTINUE
```

```
C *** COMPLETE ARRAY
      DO 40 I=1, N-1

C *** CHECK FOR ZERO IN LEFT COLUMN
        IF (R(I+1,1).EQ.0.0) THEN
          F=1
          RETURN
        ENDIF

C *** COMPLETE EACH ROW
        DO 50 J=1, M-1
          R(I+2,J) = R(I+1,1)*R(I,J+1)  - R(I,1)*R(I+1,J+1)
          R(I+2,J)= R(I+2,J)/R(I+1,1)
   50   CONTINUE
   40 CONTINUE
      RETURN
      END
```

The variable F is used as a flag to indicate whether the array computed has a zero in the left column. If it does, F is changed from a value of -1 to 1, the subroutine is exited, and the main program prints a message. The array $R(\cdot,\cdot)$ is set to zero in the subroutine (in steps 2040–2090) rather than in the main program, so that there will be no difficulties with leftover nonzero elements if the subroutine is called more than once.

For the polynomial

$$p(s) = s^4 - 6s^3 - 6s^2 + 22s - 15$$

the Routh program gives

```
           FUNDAMENTAL ROUTH-HURWITZ TEST
           PLEASE ENTER POLYNOMIAL DEGREE?4
           ENTER POWER 4 COEFFICIENT?1
           ENTER POWER 3 COEFFICIENT?-6
           ENTER POWER 2 COEFFICIENT?-6
           ENTER POWER 1 COEFFICIENT?22
           ENTER POWER 0 COEFFICIENT?-15
           ROUTH-HURWITZ ARRAY IS
           COLUMN 1
           1
           -6
           -2.33333333
           60.5714286
           -15
           COLUMN 2
           -6
           22
           -15
           0
           0
           COLUMN 3
           -15
           0
           0
           0
           0
```

or

$$
\begin{array}{c|ccc}
s^4 & 1 & -6 & -15 \\
s^3 & -6 & 22 & 0 \\
s^2 & -2.33 & -15 & 0 \\
s^1 & 60.57 & 0 & 0 \\
s^0 & -15 & 0 & 0
\end{array}
$$

which shows the three RHP roots.

5.6.5 General Test Program

An expanded version of the ROUTH program, called RHTEST, that does accommodate left-column zeros and premature terminations is listed in Table 5-15. The flag F now indicates special situations as follows:

$F = -1$ denotes no left-column zero and no premature termination of the array

$F = 0$ signals that a left-column zero has occurred

$F > 0$ indicates that a premature termination has occurred

The value of F is the integer array row number that gives the coefficients of the divisor polynomial. No accounting is made for the unlikely occurrence of both a left-column zero and a premature termination or multiples of either or both. The routine does handle these events, however. Using only a single flag F, the flag simply records the most recent event, if any.

TABLE 5-15

Computer Program in BASIC for General Routh–Hurwitz Testing (RHTEST)

This program accommodates left-column zeros and premature termination.

VARIABLES USED

N	=	polynomial degree
P(20)	=	polynomial coefficients; P(N) is the coefficient of the highest-order term and P(0) is the coefficient of the constant term
R(21,12)	=	Routh–Hurwitz array as a function of row and column
F	=	flag for left-column zero and premature termination in the array; if neither has occurred, $F = -1$; if the original array has a left-column zero, $F = 0$; if a premature termination of the array has occurred, $F > 0$ is the row number of the divisor polynomial
M	=	number of columns of the Routh–Hurwitz array
L	=	number of left-column sign changes in the array
Z	=	random number between 0 and 1
I	=	polynomial coefficient index; array column index
J	=	array row index

LISTING

```
100       PRINT "GENERAL ROUTH-HURWITZ TEST"
110       DIM P(20),R(21,12)
120   REM   ENTER POLYNOMIAL DEGREE AND COEFFICIENTS
130       PRINT "PLEASE ENTER POLYNOMIAL DEGREE";
140       INPUT N
150   REM   INPUT THE POLYNOMIAL COEFFICIENTS
160       FOR I = 0 TO N
170       PRINT "ENTER POWER ";N - I;" COEFFICIENT";
180       INPUT P(N - I)
190       NEXT I
200   REM   FORM THE ROUTH-HURWITZ ARRAY
210       GOSUB 2000
220   REM   PRINT ARRAY
230       PRINT "ROUTH-HURWITZ ARRAY IS"
240       FOR J = 1 TO M
250       PRINT "COLUMN ";J
260       FOR I = 1 TO N + 1
270       PRINT R(I,J)
280       NEXT I
290       NEXT J
300       PRINT "NUMBER OF RHP ROOTS IS ";L
310   REM   NOTIFY IF LEFT-COLUMN ZERO OR PREMATURE TERMINATION
320       IF F < 0 THEN PRINT "NO LEFT-COLUMN ZEROS OR PREMATURE TERMINATION"
330       IF F = 0 THEN PRINT "ARRAY HAS LEFT-COLUMN ZERO"
340       IF F > 0 THEN PRINT "ARRAY HAS PREMATURE TERMINATION IN ROW ";F
350       END
2000  REM   SUBROUTINE FOR ROUTH-HURWITZ TEST
2010  REM   SET FLAG AND NO. COLUMNS OF ARRAY
2020      F = - 1
2030      M = INT (N / 2 + 1)
2040  REM   ZERO ARRAY R
2050      FOR I = 1 TO 21
2060      FOR J = 1 TO 12
2070      R(I,J) = 0
2080      NEXT J
2090      NEXT I
2100  REM   FORM FIRST TWO LINES OF ARRAY
2110      FOR I = 1 TO M
2120      R(1,I) = P(N - 2 * I + 2)
2130      IF N - 2 * I + 1 < 0 THEN GOTO 2150
2140      R(2,I) = P(N - 2 * I + 1)
2150      NEXT I
2160  REM   COMPLETE ARRAY
2170      FOR I = 1 TO N - 1
2180  REM   CHECK FOR ZERO IN LEFT COLUMN
2190      IF R(I + 1,1) < > 0 THEN GOTO 2310
2200  REM   IF ZERO IN LEFT COLUMN, LOOK FOR PREMATURE TERMINATION
2210      FOR J = 2 TO M
2220      IF R(I + 1,J) < > 0 THEN F = 0
2230      NEXT J
2240      IF F = 0 THEN GOTO 2430
2250  REM   SET F TO ROW NUMBER OF PREMATURE TERMINATION
2260      F = I
2270  REM   REPLACE ROW OF ZEROS BY DERIVATIVE COEFFICIENT
2280      FOR J = 1 TO M
2290      R(I + 1,J) = (N - I - 2 * J + 3) * R(I,J)
2300      NEXT J
2310  REM   COMPLETE EACH ROW
2320      FOR J = 1 TO M - 1
2330      R(I + 2,J) = R(I + 1,1) * R(I,J + 1) - R(I,1) * R(I + 1,J + 1)
2340      R(I + 2,J) = R(I + 2,J) / R(I + 1,1)
2350      NEXT J
```

(continued)

```
2360    NEXT I
2370  REM   COUNT LEFT COLUMN SIGN CHANGES
2380    L = 0
2390    FOR I = 1 TO N
2400    IF SGN (R(I,1)) < > SGN (R(I + 1,1)) THEN L = L + 1
2410    NEXT I
2420    GOTO 2520
2430  REM   WHEN LEFT COLUMN ZERO, INTRODUCE A NEW LHP ROOT
2440    Z = .5 + RND (1)
2450    P(N + 1) = P(N)
2460    FOR I = 0 TO N - 1
2470    P(N - I) = P(N - I - 1) + Z * P(N - I)
2480    NEXT I
2490    P(0) = Z * P(0)
2500    N = N + 1
2510    GOTO 2030
2520    RETURN
```

Computer Program in FORTRAN for General Routh–Hurwitz Testing (RHTEST)

```
C **********************************************************************
C PROGRAM RHTEST -- GENERAL ROUTH-HURWITZ TEST
C **********************************************************************
      REAL P(20), R(21,12)
      INTEGER F

      WRITE(*,'(1X,''GENERAL ROUTH-HURWITZ TEST'')')

C ***   ENTER POLYNOMIAL DEGREE AND COEFFICIENTS
      WRITE(*,'(1X,''PLEASE ENTER POLYNOMIAL DEGREE '',\)')
      READ(*,*) N

C ***   INPUT THE POLYNOMIAL COEFFICIENTS
      DO 10 I=0, N
        WRITE(*,'(1X,''ENTER POWER '',I3,'' COEFFICIENT '',\)') N-I
        READ(*,*) P(N-I+1)
10    CONTINUE

C ***   FORM THE ROUTH-HURWITZ ARRAY
      CALL RH(N,P,R,L,F,M)

C ***   PRINT ARRAY
      WRITE(*,'(1X,''ROUTH-HURWITZ ARRAY IS'')')
      DO 20 J=1, M
        WRITE(*,'(1X,''COLUMN '',I3)') J
        DO 30 I=1, N+1
          WRITE(*,'(1X,E15.8)') R(I,J)
30      CONTINUE
20    CONTINUE
      WRITE(*,'(1X,''NUMBER OF RHP ROOTS IS '',I3)') L

C ***   NOTIFY IF LEFT-COLUMN ZERO OR PREMATURE TERMINATION
      IF(F.LT.0) WRITE(*,'(1X,
     *  ''NO LEFT-COLUMN ZEROS OR PREMATURE TERMINATION'')')
      IF(F.EQ.0) WRITE(*,'(1X,''ARRAY HAS LEFT-COLUMN ZERO'')')
      IF(F.GT.0) WRITE(*,'(1X,
     *  ''ARRAY HAS PREMATURE TERMINATION IN ROW '',I3)') F
      STOP
      END

C **********************************************************************
      SUBROUTINE RH(N,P,R,L,F,M)
C **********************************************************************
```

```
        REAL P(20), R(21,12)
        INTEGER F

C ***   SUBROUTINE FOR ROUTH-HURWITZ TEST
C ***   SET FLAG AND NO. COLUMNS OF ARRAY
        F = -1
 2030 M = INT(N/2.0 + 1.0)

C ***   ZERO ARRAY R
        DO 10 I=1, 21
           DO 20 J=1, 12
              R(I,J) = 0.0
 20        CONTINUE
 10     CONTINUE

C ***   FORM FIRST TWO LINES OF ARRAY
        DO 30 I=1, M
           R(1,I) = P(N-(2*I)+3)
           IF ((N-(2*I)+1).GE.0) THEN
              R(2,I) = P(N-(2*I)+2)
           ELSE
              GOTO 35
           ENDIF
 30     CONTINUE

C ***   COMPLETE ARRAY
 35     DO 40 I=1, N-1

C ***   CHECK FOR ZERO IN LEFT COLUMN
           IF (R(I+1,1).EQ.0.0) THEN

C ***   IF ZERO IN LEFT COLUMN, LOOK FOR PREMATURE TERMINATION
              DO 50 J=2, M
                 IF (R(I+1,J).NE.0) F=0
 50           CONTINUE
              IF (F.EQ.0) THEN

C ***   WHEN LEFT COLUMN ZERO, INTRODUCE A NEW LHP ROOT
                 CALL RANNUM(RAN)
                 Z = 0.5+RAN
                 P(N+2) = P(N+1)

                 DO 90 J=0, N-1
                    P(N-J+1) = P(N-J) + Z*P(N-J+1)
 90              CONTINUE

                 P(1) = Z * P(1)
                 N = N + 1
                 GOTO 2030
              ENDIF

C ***   SET F TO ROW NUMBER OF PREMATURE TERMINATION
              F=I

C ***   REPLACE ROW OF ZEROS BY DERIVATIVE COEFFICIENT
              DO 60 J=1, M
                 R(I+1,J)=(N-I-(2*J)+3)*R(I,J)
 60           CONTINUE
           ENDIF

C ***   COMPLETE EACH ROW
           DO 70 J=1, M-1
              R(I+2,J) = R(I+1,1)*R(I,J+1) - R(I,1)*R(I+1,J+1)
              R(I+2,J) = R(I+2,J)/R(I+1,1)
```

(continued)

```
70        CONTINUE
40     CONTINUE

C ***   COUNT LEFT COLUMN SIGN CHANGES
       L = 0
       DO 80 I=1, N
          IF (SIGN(1.0,R(I,1)).NE.SIGN(1.0,R(I+1,1))) L=L+1
80     CONTINUE

       RETURN
       END


C***************************************************************************
       SUBROUTINE RANNUM(RAN)
C***************************************************************************

C *** RETURNS PSEUDOGAUSSIAN RANDOM NUMBERS WITH UNIFORM DISTRIBUTION
C *** FROM 0 TO 1.  A SEED VALUE HAS BEEN SELECTED AS 32767.
       REAL X
       INTEGER*2 A(2), B(2)
       INTEGER*4 C

C *** MAKE B(1) AND C SHARE SAME MEMORY LOCATION -- B(1) BECOMES THE
C *** LAST TWO BYTES OF THE FOUR BYTE INTEGER C
       EQUIVALENCE (C,B(1))

C *** A(2) IS CONSTANT OF FORM (8T - 3)
C *** A(1) IS SEED CONSTANT (MUST BE ODD)
       DATA A(2)/32717/
       DATA A(1)/32767/

C *** GENERATE UNIFORM PSEUDORANDOM CONSTANTS BY TAKING
C *** THE LAST 2 BYTES OF A FOUR BYTE INTEGER AND
C *** MULTIPLYING THESE BYTES BY THE A(2) CONSTANT.
       C = A(1)*A(2)
       A(1) = B(1)

C *** NORMALIZE BETWEEN 0 AND 1
       X = (A(1) + 32767.0)/65534.0
       IF(X.LT.0.0) X = 0.0
       IF(X.GT.1.0) X = 1.0
       RAN = X

       RETURN
       END
```

If there is a left-column zero, the polynomial tested $p(s)$ is changed to

$$p(s) \leftarrow (s + z)p(s)$$

where z is a randomly generated number between 0.5 and 1.5. This additional root of the new polynomial is in the LHP and so does not affect the calculated number of RHP roots. The roots addition is very likely to remove the left-column zero, and if it does not, another LHP root is added, and so on. These steps are performed when necessary by lines 2420–2490.

If there is a premature termination, the row of zeros in the array are replaced by the derivative coefficients of the divisor polynomial, and the test continues. These steps are performed when necessary by lines 2240–2290. The number of algebraic sign changes of the left-column elements (which is the number of RHP roots of the polynomial) is found in steps 2360–2410 and is returned by the subroutine to the main program as the value of the variable L.

The RHTEST program, when applied to the polynomial

$$p(s) = s^4 + 2s^3 + 3s^2 + 6s - 1$$

gave

```
ROUTH-HURWITZ ARRAY IS
COLUMN 1
1
2.60311763
1.20623526
1.65805136
3.28916542
-.60311763
COLUMN 2
4.20623526
7.80935289
2.85039628
-.60311763
0
0
COLUMN 3
2.61870578
-.60311763
0
0
0
0

NUMBER OF RHP ROOTS IS 1
ARRAY HAS LEFT-COLUMN ZERO
```

during one run, or

s^5	1	4.206	2.619
s^4	2.603	7.809	-0.603
s^3	1.206	2.850	0
s^2	1.658	-0.603	0
s^1	3.289	0	0
s^0	-0.603	0	0

As the array for $p(s)$ has a left-column zero,

s^4	1	3	-1
s^3	2	6	
s^2	0	-1	
s^1			
s^0			

the program multiplied the original polynomial by the factor $(s + 0.603)$ and tested the resulting fifth-degree polynomial, which does not have a left-column zero. Since the number 0.603 was derived from a random number generator, different LHP factors will occur each time this program is run.

For the polynomial

$$p(s) = (s^2 + 1)(s - 2)(s + 3)$$
$$= s^4 + s^3 - 5s^2 + s - 6$$

the RHTEST program gives

```
ROUTH-HURWITZ ARRAY IS
COLUMN 1
1
1
-6
-12
-6
COLUMN 2
-5
1
-6
0
0
COLUMN 3
-6
0
0
0
0
NUMBER OF RHP ROOTS IS 1
ARRAY HAS PREMATURE TERMINATION IN ROW 3
```

$$
\begin{array}{c|ccc}
s^4 & 1 & -5 & -6 \\
s^3 & 1 & 1 & 0 \\
\hline
s^2 & -6 & -6 & 0 \\
s^1 & -12 & 0 & 0 \\
s^0 & -6 & 0 & 0
\end{array}
$$

which identifies the even polynomial divisor, $s^2 + 1$.

The RHTEST program can exhibit difficulties if numerical imprecision results in a left-column zero or a row of zeros being not quite zero numerically. An example of this will occur on some relatively short-word-length machines for the polynomial

$$p(s) = s^5 + 3s^4 + 4s^3 + 7s^2 + 4s + 2$$

for which the RHTEST program will give something similar to

```
ROUTH-HURWITZ ARRAY IS
COLUMN 1
1
3
1.66666667
.999999999
-2.79396773E-09
2
COLUMN 2
4
7
3.33333333
2
0
0
COLUMN 3
4
2
0
0
0
0
NUMBER OF RHP ROOTS IS 2
NO LEFT-COLUMN ZEROS OR PREMATURE TERMINATION
```

or

s^5	1	4	4
s^4	3	7	2
s^3	1.667	3.333	0
s^2	0.999	2	0
s^1	-2.79×10^{-9}	0	0
s^0	2	0	0

when there really should be a row of zeros in the s^1 row, indicative of a premature termination of the array. The polynomial does *not* have two RHP roots, but these are indicated because numerical error resulted in a small negative number (rather than a small positive one) in the left column.

One solution to this problem is to take to be zero any small entry that could, if zero, signal a left-column zero or a premature termination. This can be done by replacing lines 2180 and 2210 with

```
2180 IF ABS (R(I + 1,1)) > 1 E-6 THEN GOTO 2300
2210 IF ABS (R(I + 1,J)) > 1 E-6 THEN F = 0
```

If a polynomial truly has a small but truly nonzero left-column entry, no harm will be done by introducing an additional root before testing. However, it is possible that a premature termination will be detected when one really does not quite exist.

5.7 FACTORING USING ROUTH–HURWITZ METHODS

5.7.1 Axis Shifting and Routh–Hurwitz Testing

Probably the best polynomial factoring method uses the Routh–Hurwitz test and polynomial axis shifts to locate the real parts and the imaginary parts of the roots. The method described here, when automatic removal of located roots and bisection imaginary part search for repeated roots of any multiplicity are added, has been found to be fast as the fastest alternative methods. It is extremely dependable.

To shift the roots of a polynomial σ units to the right (shifting the imaginary axis σ units to the left), replace s in the original polynomial by $(s - \sigma)$. For example, the polynomial

$$p(s) = s^4 + 15s^3 + 84s^2 + 210s + 200$$

has all roots in the LHP, as can easily be verified with a Routh–Hurwitz test. Replacing s by $(s - 2)$ will shift the polynomial roots 2 units to the right, or the imaginary axis 2 units to the left:

$$p(s) = s\{s[s(s + 15) + 84] + 210\} + 200$$

$$p'(s) = (s - 2)\,\{(s - 2)\,[(s - 2)\,(s - 2 + 15) + 84] + 210\} + 200$$

$$= s^4 + 7s^3 + 18s^2 + 22s + 12$$

A Routh–Hurwitz test of the shifted polynomial indicates no RHP roots; hence the original polynomial has all roots to the left of $s = -2$ on the complex plane.

By repeated axis shifts and Routh–Hurwitz tests in a bisection pattern, the real parts of a polynomial's roots can be located. Figure 5-6 shows the character of the function involved. Further, once the real parts are known, the imaginary parts of the roots are easily found by searching for the purely imaginary values of the variable for which the Routh–Hurwitz test divisor polynomial is zero. Polynomial factoring is thereby reduced to two real-variable single-dimension searches instead of a two-dimensional search.

5.7.2 General Factoring Program

An algorithm for locating the real and imaginary roots of a polynomial using Routh–Hurwitz testing is given in Table 5-16 in flow diagram form. It begins with the polynomial degree h and its coefficients β_i, the dimensions of a search interval along the real axis from e to $b = -e$, and the smallest acceptable error in precision of the real-part locations, g. The search starts at the right boundary e of the real axis and performs Routh–Hurwitz tests in bisection pattern, moving left, to locate the real parts of the roots with the required precision.

If a single root's real part is located, it is printed and the search continues to the left. If the real part of a pair of roots is located, a premature termination of the Routh–Hurwitz array is presumed, and the imaginary parts of the roots are determined from the divisor polynomial of the array, as follows: Suppose that the x^2 row of the test has entries

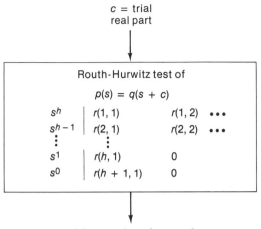

c = trial
real part

Routh-Hurwitz test of

$$p(s) = q(s + c)$$

s^h	$r(1, 1)$	$r(1, 2)$ •••
s^{h-1}	$r(2, 1)$	$r(2, 2)$ •••
⋮	⋮	
s^1	$r(h, 1)$	0
s^0	$r(h + 1, 1)$	0

$w(c)$ = number of roots of
the polynomial that are
to the right of c

(a) Routh-Hurwitz test result as a function
of the amount of polynomial axis shift

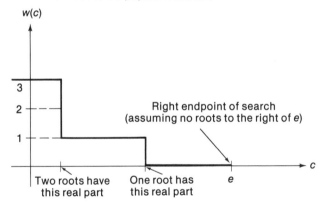

Figure 5-6   Viewing the Routh–Hurwitz
test as a function of axis shift.

(b) Nature of the number-of-roots function

s^2	$r(h - 1, 1)$ $r(h - 1, 2)$ 0 ···	
s^1		
s^0		

so that

$$r(h - 1, 1)s^2 + r(h - 1, 2)$$

is the even polynomial divisor of the axis-shifted polynomial. The imaginary parts of the roots are then given by $s = iy$, where

$$r(h - 1, 1)(iy)^2 + r(h - 1, 2) = 0$$

TABLE 5-16   Algorithm for Polynomial Factorization by Routh–Hurwitz Methods

DEFINITIONS

A search for root real parts of the polynomial

$$q(s) = \beta_h s^h + \beta_{h-1} s^{h-1} + \cdots + \beta_1 s + \beta_0$$

extends from $s = e$ at the right to $s = -e$ at the left. The variable c is used for the real-part search variable, u is the initial search increment, d is the present search increment, and g is the required precision, which is the largest increment to be used. The axis-shifted polynomial,

$$p(s) = q(s + c) = \alpha_h s^h + \alpha_{h-1} s^{h-1} + \cdots + \alpha_1 s + \alpha_0$$

is Routh–Hurwitz tested to determine the number of its RHP roots, w. The variable y is used to save the previous number of RHP roots found. The Routh–Hurwitz entries are

s^h	$r(1, 1)$	$r(1, 2)$	$r(1, 3)$	$\cdots$
s^{h-1}	$r(2, 1)$	$r(2, 2)$	$r(2, 3)$	$\cdots$
$\vdots$	$\vdots$			
s^2	$r(h - 1, 1)$	$r(h - 1, 2)$	0	$\cdots$
s^1	$r(h, 1)$	0	0	$\cdots$
s^0	$r(h + 1, 1)$	0	0	$\cdots$

and v is the imaginary part of a complex root.

ALGORITHM

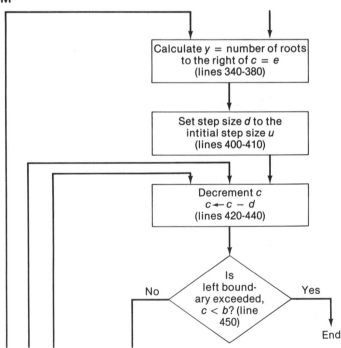

Calculate y = number of roots
to the right of $c = e$
(lines 340-380)

Set step size d to the
intitial step size u
(lines 400-410)

Decrement c
$c \leftarrow c - d$
(lines 420-440)

Is
left bound-
ary exceeded,
$c < b$? (line
450)

No   Yes

End

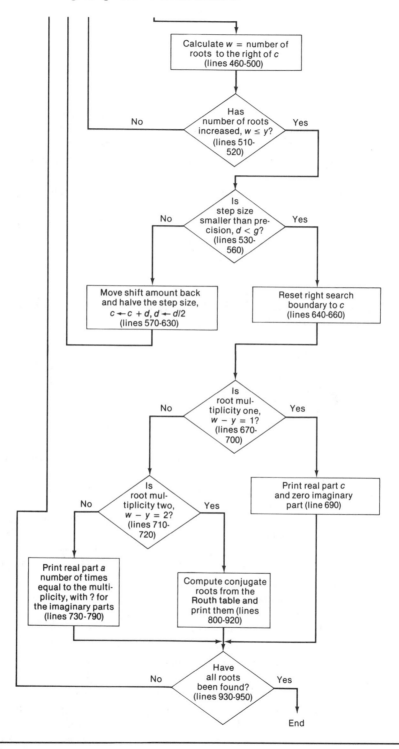

or

$$y = \pm \sqrt{\frac{r(h - 1, 2)}{r(h - 1, 1)}}$$

Of course, if low precision is required in the real-part location, imaginary-part location accuracy can be compromised. If more than two roots having the same real part are located, their real parts are printed but the imaginary parts are not determined here. It is a simple matter to do so, for instance by performing a single-variable bisection search of the divisor polynomial.

A computer program in BASIC/FORTRAN that uses this algorithm for factoring is given in Table 5-17. It uses, without change, the subroutine from the SHIFT axis-shifting program of Table 5-8 and, also without change, the subroutine for general Routh–Hurwitz testing form the RHTEST program of Table 5-15.

TABLE 5-17

Computer Program in BASIC for Polynomial Factorization by Routh–Hurwitz Methods (FACTOR)

VARIABLES USED

H	=	original polynomial degree
N	=	axis-shifted polynomial degree
Q(20)	=	original polynomial
P(20)	=	axis-shifted polynomial; its degree N can be increased form H if the Routh–Hurwitz test adds a root to the polynomial to accommodate a left-column zero
E	=	real-part search dimension; the real-part search extends between $-E$ and E
B	=	$-E$, the left boundary of the real-part search
C	=	temporary real-part search left boundary; C is moved from right to left, from $C = E$ toward $C = -E$
U	=	E/3 is the initial step size for the right-to-left search for the real parts of roots
G	=	user-entered real-part error bound; search for the real part of a root continues until the real part is located within an interval of G units
R(21,12)	=	Routh–Hurwitz array as a function of row and column
L	=	number of RHP roots found by the Routh–Hurwitz test
Y	=	previous number of RHP roots found
D	=	real-part search step size
W	=	present number of RHP roots found
V	=	imaginary part of a complex root
I	=	polynomial coefficient index; multiple root index; intermediate polynomial coefficient index; array column index
J	=	intermediate polynomial coefficient index; array row index
F	=	flag for left-column zero and premature termination in the array; if neither has occurred, $F = -1$; if the original array had a left-column zero, $F = 0$;

if a premature termination of the array has occurred, $F > 0$ is the row
number of the divisor polynomial

M = number of columns in the array in the Routh–Hurwitz test subroutine

Z = random number between 0 and 1 in the Routh–Hurwitz test subroutine

LISTING

```
100      PRINT "POLYNOMIAL FACTORIZATION"
110      DIM P(20),Q(20),R(21,12)
120    REM  ENTER POLYNOMIAL DEGREE AND COEFFICIENTS
130      PRINT "PLEASE ENTER POLYNOMIAL DEGREE";
140      INPUT H
150      FOR I = 0 TO H
160      PRINT "ENTER POWER ";H - I;"COEFFICIENT";
170      INPUT Q(H - I)
180      NEXT I
190    REM  ENTER SEARCH PARAMETERS.  B IS LEFT REAL PART
200    REM  BOUNDARY, E IS RIGHT REAL PART BOUNDARY, C IS
210    REM  INITIAL REAL PART LOCATION, U IS INITIAL REAL
220    REM  PART SEARCH STEP SIZE, AND D IS RUNNING STEP
230    REM  SIZE.  G IS LARGEST ACCEPTABLE ERROR OF THE
240    REM  RESULT.
250      PRINT "ENTER SEARCH DIMENSION";
260      INPUT E
270      B = - E
280      C = E
290      U = E / 3
300      PRINT"ENTER REAL PART ERROR BOUND";
310      INPUT G
320    REM  PRINT HEADING
330      PRINT "REAL PART","IMAG PART"
340    REM  CALCULATE Y=NO. ROOTS TO RIGHT OF C
350      N=H
360      GOSUB 1000
370      GOSUB 2000
380      Y = L
390    REM  BISECTION LOOP TO NARROW DOWN ROOT REAL PARTS
400    REM  SET STEP SIZE D TO INITIAL STEP SIZE
410      D = U
420    REM  SHIFT AXIS D UNITS TO THE LEFT; END IF
430    REM LEFT BOUNDARY EXCEEDED
440      C = C - D
450      IF C < B THEN  GOTO 960
460    REM CALCULATE W=NO. ROOT TO THE RIGHT OF C
470      N = H
480      GOSUB 1000
490      GOSUB 2000
500      W = L
510    REM  IF NO. OF ROOTS HAS NOT INCREASED, INCREMENT AGAIN
520      IF W < = Y THEN GOTO 420
530    REM  IF NO. OF ROOTS HAS INCREASED, SEE IF INCREMENT
540    REM  SIZE D IS SMALLER THAN REAL PART ERROR BOUND
550    REM  G.  IF SO, AXIS SHIFT HAS SUFFICIENT ACCURACY
560      IF D < G THEN  GOTO 640
570    REM  IF AXIS SHIFT DOES NOT YET HAVE SUFFICIENT
580    REM  ACCURACY, MOVE AXIS BACK TO THE RIGHT WHERE
590    REM  IT WAS ON THE PREVIOUS STEP, HALVE THE STEP SIZE,
600    REM  AND LOOP BACK TO INCREMENT AGAIN
610      C = C + D
620      D = D / 2
630      GOTO 420
640    REM  SET RIGHT SEARCH BOUNDARY E TO AMOUNT OF AXIS
650    REM  SHIFT C
```

(continued)

```
660     E = C
670     REM  MULTIPLICITY OF ROOTS IN VICINITY OF REAL PART
680     REM  C IS W-Y.  IF MULTIPLICITY IS ONE, PRINT ROOT
690       IF W - Y = 1 THEN   PRINT C,0
700       IF W - Y = 1 THEN   GOTO 930
710     REM  IF MULTIPLICITY IS TWO, ROOTS ARE CONJUGATE
720       IF W - Y = 2 THEN   GOTO 800
730     REM  IF MULTIPLICITY IS GREATER THAN TWO, PRINT REAL
740     REM  PART AND QUESTION MARK A NUMBER OF TIMES EQUAL TO
750     REM  THE MULTIPLICITY
760       FOR I = 1 TO W - Y
770       PRINT C," ?"
780       NEXT I
790       GOTO 930
800     REM  FOR QUADRATIC ROOTS, A PREMATURE TERMINATION
810     REM  SHOULD OCCUR (OR APPROXIMATELY SO) AT LINE
820     REM  (H-2) OF THE ROUTH-HURWITZ TABLE.  FIND THE
830     REM  IMAGINARY PART V (AND - V) FROM THESE TWO
840     REM  ENTRIES AND PRINT THE ROOT LOCATIONS
850     REM  IF THE QUOTIENT IS NEGATIVE, A SMALL NUMERICAL
860     REM  ERROR HAS OCCURRED AND THE ROOTS MUST BE
870     REM  VIRTUALLY ON THE REAL AXIS.
880       V = R(H - 1,2) / R(H - 1,1)
890       IF V < 0 THEN V = 0
900       V = SQR (V)
910       PRINT C,V
920       PRINT C, -V
930     REM  END IF ALL ROOTS PASSED, OTHERWISE CONTINUE SEARCH
940       IF W = H THEN   GOTO 960
950       GOTO 340
960       END
1000
```

```
                                    ┌─────────────────────────────────┐
◄───────────────────────────────────│ Subroutine to axis—shift a polynomial
                                    │ from the SHIFT program; TABLE 5-8│
                                    └─────────────────────────────────┘
1150
2000
                                    ┌──────────────────────────────────────┐
◄───────────────────────────────────│ Subroutine to Routh-Hurwitz—test a polynomial
                                    │ from the RHTEST program; TABLE 5-15│
                                    └──────────────────────────────────────┘
2520
```

Computer Program in FORTRAN for Polynomial Factorization by Routh–Hurwitz Methods (FACTOR)

```
C ********************************************************************
C PROGRAM FACTOR -- POLYNOMIAL FACTORIZATION
C ********************************************************************
      REAL P(20), Q(20), R(21,12)
      INTEGER F, H, I, N, Y, W
      DATA ZERO/0.0/

      WRITE(*,'(1X,''POLYNOMIAL FACTORIZATION'')')

C ***   ENTER POLYNOMIAL DEGREE AND COEFFICIENTS
      WRITE(*,'(1X,''PLEASE ENTER POLYNOMIAL DEGREE '',\)')
      READ(*,*) H
      DO 10 I=0, H
         WRITE(*,'(1X,''ENTER POWER '',I3,'' COEFFICIENT '',\)') H-I
         READ(*,*) Q(H-I+1)
   10 CONTINUE

C ***   ENTER SEARCH PARAMETERS.  B IS LEFT REAL PART
C ***   BOUNDARY, E IS RIGHT REAL PART BOUNDARY, C IS
C ***   INITIAL REAL PART LOCATION, U IS INITIAL REAL
```

```
C ***   PART SEARCH STEP SIZE, AND D IS RUNNING STEP
C ***   SIZE.  G IS LARGEST ACCEPTABLE ERROR OF THE
C ***   RESULT.
        WRITE(*,'(1X,''ENTER SEARCH DIMENSION '',\)')
        READ(*,*) E
        B = -E
        C =  E
        U =  E/3.0

        WRITE(*,'(1X,''ENTER REAL PART ERROR BOUND '',\)')
        READ(*,*) G

C ***   PRINT HEADING
        WRITE(*,'(1X,''    REAL PART        IMAG PART'')')

C ***   CALCULATE Y=NO. ROOTS TO RIGHT OF C; ITERATE A MAXIMUM
C ***   OF 100 TIMES
        DO 50 K=1, 100
          N=H
          CALL SHIFT(Q,N,C,P)
          CALL RH(N,P,R,L,F,M)
          Y = L

C ***   BISECTION LOOP TO NARROW DOWN ROOT REAL PARTS
C ***   SET STEP SIZE D TO INITIAL STEP SIZE
          D = U

C ***   SHIFT AXIS D UNITS TO THE LEFT; END IF
C ***   LEFT BOUNDARY EXCEEDED; ITERATE A MAXIMUM OF 1000 TIMES
          DO 20 I=1, 1000
            C = C - D
            IF (C.LT.B) STOP

C ***   CALCULATE W=NO. ROOT TO THE RIGHT OF C
            N = H
            CALL SHIFT(Q,N,C,P)
            CALL RH(N,P,R,L,F,M)
            W = L

C ***   IF NO. OF ROOTS HAS NOT INCREASED, INCREMENT AGAIN
            IF (W.LE.Y) GOTO 20

C ***   IF NO. OF ROOTS HAS INCREASED, SEE IF INCREMENT
C ***   SIZE D IS SMALLER THAN REAL PART ERROR BOUND
C ***   G.  IF SO, AXIS SHIFT HAS SUFFICIENT ACCURACY
            IF (D.LT.G) GOTO 30

C ***   IF AXIS SHIFT DOES NOT YET HAVE SUFFICIENT
C ***   ACCURACY, MOVE AXIS BACK TO THE RIGHT WHERE
C ***   IT WAS ON THE PREVIOUS STEP, HALVE THE STEP SIZE,
C ***   AND LOOP BACK TO INCREMENT AGAIN
            C = C + D
            D = D/2
 20       CONTINUE

C ***   SET RIGHT SEARCH BOUNDARY E TO AMOUNT OF AXIS
C ***   SHIFT C
 30       E = C

C ***   MULTIPLICITY OF ROOTS IN VICINITY OF REAL PART
C ***   C IS W-Y.  IF MULTIPLICITY IS ONE, PRINT ROOT
          IF (W-Y.EQ.1) THEN
            WRITE(*,'(1X,E15.8,1X,E15.8)') C,ZERO
```

(continued)

```
C ***   IF MULTIPLICITY IS TWO, ROOTS ARE CONJUGATE
          ELSEIF (W-Y.EQ.2) THEN

C ***   FOR QUADRATIC ROOTS, A PREMATURE TERMINATION
C ***   SHOULD OCCUR (OR APPROXIMATELY SO) AT LINE
C ***   (H-2) OF THE ROUTH-HURWITZ TABLE.  FIND THE
C ***   IMAGINARY PART V (AND - V) FROM THESE TWO
C ***   ENTRIES AND PRINT THE ROOT LOCATIONS
C ***   IF THE QUOTIENT IS NEGATIVE, A SMALL NUMERICAL
C ***   ERROR HAS OCCURRED AND THE ROOTS MUST BE
C ***   VIRTUALLY ON THE REAL AXIS.
            V = R(H-1,2)/R(H-1,1)
            IF (V.LT.0) V=0
            V = SQRT(V)
            WRITE(*,'(1X,E15.8,1X,E15.8)') C,  V
            WRITE(*,'(1X,E15.8,1X,E15.8)') C, -V

C ***   IF MULTIPLICITY IS GREATER THAN TWO, PRINT REAL
C ***   PART AND QUESTION MARK A NUMBER OF TIMES EQUAL TO
C ***   THE MULTIPLICITY
          ELSE
            DO 40 I=1, W-Y
              WRITE(*,'(1X,E15.8,'' ?'')') C
 40         CONTINUE
          ENDIF

C ***   END IF ALL ROOTS PASSED, OTHERWISE CONTINUE SEARCH
          IF (W.EQ.H) STOP
 50     CONTINUE

        STOP
        END
```

> Subroutine to axis—shift a polynomial from the SHIFT program; TABLE 5-8

> Subroutine to Routh-Hurwitz test a polynomial from the RHTEST program; TABLE 5-15

For the polynomial

$$q(s) = s^4 + 2s^3 - s^2 + 3s + 4$$

for example, the FACTOR program gives

```
POLYNOMIAL FACTORIZATION
PLEASE ENTER POLYNOMIAL DEGREE?4
ENTER POWER 4 COEFFICIENT?1
ENTER POWER 3 COEFFICIENT?2
ENTER POWER 2 COEFFICIENT?-1
ENTER POWER 1 COEFFICIENT?3
ENTER POWER 0 COEFFICIENT?4
ENTER SEARCH DIMENSION?10
ENTER REAL PART ERROR BOUND?0.0001
REAL PART        IMAG PART
.726928713       1.1288172
.72698713       -1.1288172
-.853271483       0
-2.60070801       0
```

which shows that $q(s)$ factors approximately as

$$q(s) = (s - 0.727 - \underline{i}1.129)(s - 0.727 + \underline{i}1.129)(s + 0.853)(s + 2.601)$$

5.7.3 Finding Control System Modes

Continuous-time automatic control systems are described by transfer functions consisting of ratios of polynomials. A transfer function, as a function of the variable s, is the ratio of the output signal to the input signal when both output and input are exponential, varying with time t as exp (st). Alternatively, a transfer function is the ratio of the Laplace transform of the output to the Laplace transform of the input when all initial conditions are zero.

The block diagram of Figure 5-7 describes the feedback control of submarine depth in terms of the individual transfer functions of the submarine dynamics and ballast controller. The overall transfer function of the system, relating the depth output to the desired depth input, is

$$T(s) = \frac{G_1(s)G_2(s)}{1 + G_1(s)G_2(s)} = \frac{5(4s^2 + 0.6s + 2)/s(s^3 + 4s^2 + 5s)}{1 + 5(4s^2 + 0.6s + 2)/s(s^3 + 4s^2 + 5s)}$$

$$= \frac{5(4s^2 + 0.6s + 2)}{s^4 + 4s^3 + 25s^2 + 3s + 10}$$

The denominator of the overall transfer function $T(s)$ describes the zero-input behavior of the overall system, that is, the kind of outputs the system can have when the input is zero. The zero-input component of $y(t)$ is of the form

$$y_{\text{zero input}}(t) = c_1 e^{s_1 t} + c_2 e^{s_2 t} + c_3 e^{s_3 t} + c_4 e^{s_4 t}$$

where s_1, s_2, s_3, and s_4 are the roots of the denominator polynomial of $T(s)$ called the system *poles*, and C_1, C_2, C_3, and C_4 are arbitrary constants that depend on the initial conditions.

If any of the denominator polynomial roots is positive or has a positive real part, the corresponding term in $y_{\text{zero input}}(t)$, called a *mode*, expands with time. The control system is then *unstable*, an undesirable situation. For the system of Figure 5-7, the poles are the roots of the polynomial

$$p(s) = s^4 + 4s^3 + 25s^2 + 3s + 10$$

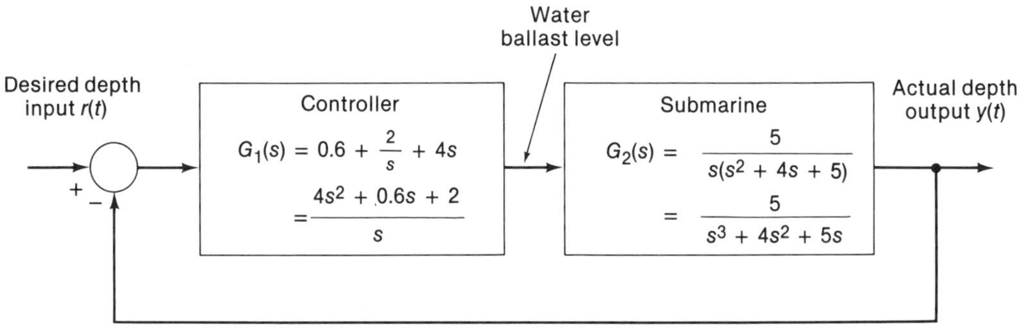

Figure 5-7   Block diagram of an automatic control system.

and are found by the FACTOR program to be

REAL PART	IMAG PART
− .0283823458	.6400015161
− .0283823458	− .6400015161
−1.97168986	4.52496458
−1.97168986	−4.52496458

The system is thus stable and has a zero-input response component with terms that decay to zero as $\exp(-0.0284t)$ or faster.

If, instead, the controller is made adjustable, with individual transfer function of the form

$$G_1(s) = 0.6 + \frac{K}{s} + 4s = \frac{4s^2 + 0.6s + K}{s}$$

then the overall transfer function of the control system is, in terms of the adjustable constant K,

$$T(s) = \frac{G_1(s)G_2(s)}{1 + G_1(s)G_2(s)}$$

$$= \frac{5(4s^2 + 0.6s + K)/s(s^3 + 4s^2 + 5s)}{1 + 5(4s^2 + 0.6s + K)/s(s^3 + 4s^2 + 5s)}$$

$$= \frac{5(4s^2 + 0.6s + K)}{s^4 + 4s^3 + 25s^2 + 3s + 5K}$$

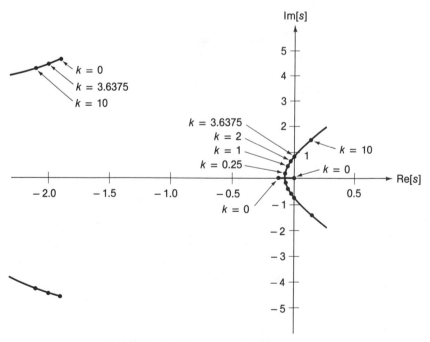

Figure 5-8 Root-locus plot for the control system.

Repeated factoring of the denominator polynomial gives the root locus plot of Figure 5-8, which shows the locations poles of $T(s)$ on the complex plane as a function of the adjustable constant K. For $K = 3.6375$ and larger, the overall system has poles with positive real parts and is unstable.

PROBLEMS

5-1. How does the transformation $s' = -s$ change the coefficients of a polynomial?

5-2. Scale the polynomial

$$p(s) = 3s^4 - 2s^3 + s^2 + 5s - 6$$

so that the magnitude of the product of its roots is unity. Show how the new and old roots are related to one another.

5-3. Scale the polynomial

$$p(s) = -2s^4 + 3s^3 - 5s^2 - 6s + 10$$

so that the sum of its roots is unity. Show how the new and old roots are related to one another.

5-4. Find the Taylor series expansion of the function

$$f(x) = \frac{e^x}{x^2 + 1} \qquad \text{about } x = -\frac{1}{2}$$

If the series is truncated after three terms, how do the value of $f(x)$ and the value of the series compare at $x = -\frac{1}{4}$?

5-5. Use the DIVIDE1 program to find the following quotients and their remainders. Check by dividing by hand.

(a) $\dfrac{s^4 - 4s^3 + 5s - 6}{-2s + 3}$ **(b)** $\dfrac{-3s^3 + 2s^2 - 4s + 7}{s}$

5-6. Use the DIVIDE2 program to find the following quotients and their remainders. Check by multiplying the quotient and divisor and adding the remainder.

(a) $\dfrac{4s^4 - 3s^3 + 2s^2 + s - 8}{s^2 + 3}$ **(b)** $\dfrac{6.7s^4 + 5.5}{1.2s^2 - s + 3.4}$

5-7. Use the EVAL program to evaluate the following polynomials at the indicated values of the variable s. Check the results by hand calculation.

(a) $p(s) = 2s^3 - 3s + 4; \quad s = -2 + i$
(b) $p(s) = s^4 + 2s^3 - 3s^2 + 5s + 7; \quad s = 3$

5-8. Expand the function

$$f(x) = xe^{-2x}$$

in a Taylor series about $x = 0$. If the Taylor series is truncated after m terms, find what m must be so that the function and its truncated Taylor series approximation differ in magnitude by no more than 0.001 at $x = 0.4$.

5-9. Use the EVAL program to determine if the given value of s is a root of the polynomial indicated.

(a) $p(s) = 2.1s^4 - 3.3s^3 + 6s^2 - 11.3; \quad s = -1.2 + i3.5$
(b) $p(s) = s^5 + 3s^4 + 9s^3 + 4s^2 - 30s; \quad s = -1 + i3$

5-10. Use the EVAL program and a trial-and-error search to find a real part of the variable x for which

$$s = x + \underline{i}2$$

gives a purely imaginary value of the polynomial

$$p(s) = s^3 - 2s^2 + 3s + 5$$

It will be helpful to modify the program so that the polynomial coefficients do not have to be reentered for each evaluation.

5-11. For the polynomial

$$p(s) = s^3 - 3s^2 + 2s + 1$$

use the SHIFT program to compute the following. Check the results by hand calculation.
(a) $p(s - 2)$ **(b)** $p(s + 3)$

5-12. When the axis of a polynomial is shifted to the location of a real root of that polynomial, the new polynomial will have s as a factor. Use the SHIFT program and trial and error to find a real root of

$$p(s) = s^3 + 4s^2 + s + 7$$

It will be helpful to modify the program so that the original polynomial coefficients do not have to be reentered for each axis shifting.

5-13. Develop an algorithm for scaling a polynomial with the transformation

$$q(s') = k^n p(s \leftarrow s')$$

where

$$s' = ks$$

and n is the degree of $p(s)$. Test the method on the polynomial

$$p(s) = 3s^3 + 2s^2 - 4s + 5$$

with $k = \frac{1}{2}$. Using the algorithm, write, test, and debug a computer program to perform the transformation on any user-entered polynomial of degree 30 or less.

5-14. Use the BISECT program of Chapter 3 to find the real roots of odd multiplicity of the following polynomials. Then use the DIVIDE1 program to remove these roots, resulting in a polynomial having the remaining roots,
(a) $p(s) = s^4 + 4s^3 + 12s^2 + 22s + 14$
(b) $p(s) = 0.7s^4 + 0.87s^3 - 1.66s^2 + 0.36s - 6.93$

5-15. Use the REALFACT program to find all the real roots of odd multiplicity of the following polynomials.
(a) $p(s) = s^4 - 5s^2 + 4$
(b) $p(s) = 3.14s^5 + 1.41s^4 + 6.07s^2 + 2.33s$
(c) $p(s) = s^4 + 100s^3 - 1000s^2 + 10^6$

5-16. Use the POLYROOT program to find the roots of the following polynomials, then check the results by multiplying the polynomial factors together to obtain the original polynomial.
(a) $p(s) = s^3 - 0.3s^2 + 0.44s + 1.17$
(b) $p(s) = 2s^4 + 6s^3 + 5s^2 + 4s + 1$

5-17. Use the POLYROOT program to find the roots of the following polynomials, then check the results with the EVAL program.
   (a) $p(s) = 3s^3 - 2s + 4$
   (b) $p(s) = s^3 + 2.1s^2 - s + 1.05$

5-18. Add to the POLYROOT program (Tables 5-11, 5-12, and 5-13) the ability to recognize when first- or second-degree polynomials remain to be factored toward the end of the process of factoring a higher-degree polynomial. Print the roots of these from formulas instead of doing a further real part and imaginary part search. Test the improved program with the following polynomials.
   (a) $p(s) = -6s + 2$
   (b) $p(s) = 4s^2 - 2s + 1$
   (c) $p(s) = 3s^2 + 7s + 2$
   (d) $p(s) = s^3 - 3s^2 + 2s + 7$
   (e) $p(s) = -s^4 + 6s^2 + 8s - 10$

5-19. For each of the following polynomials, determine how many roots are in the RHP, how many are in the LHP, and how many are on the imaginary axis.
   (a) $p(s) = s^3 + 2s^2 + 3s + 7$
   (b) $p(s) = 2s^5 + 6s^4 - 3s^3 + 5s^2 + 2s + 4$
   (c) $p(s) = s^4 + 3s^3 + 2s^2 + 6s + 5$
   (d) $p(s) = s^5 + 3s^3 + 2s^2 + 4s$
   (e) $p(s) = 3s^4 - 2s^2 - 1$
   (f) $p(s) = s^5 + s^4 + 4s^3 + 3s^2 + 3s$

5-20. Use an axis shift and a Routh–Hurwitz test to determine how many roots of the polynomial

$$p(s) = -s^4 - 4s^3 + 10s^2 + 20s + 50$$

are to the left of $s = -2$ on the complex plane.

5-21. Use the FACTOR program to find roots of the following polynomials.
   (a) $p(s) = s^4 - 3s^3 + 4s^2 - 10$
   (b) $p(s) = s^4 + 7.3s^3 - 6.2s^2 + 5.6s - 3.8$
   (c) $p(s) = 6s^5 - 3s^4 + 2s^3 + s^2 + s + 4$
   (d) $p(s) = s^5 + 1$

5-22. Modify the FACTOR program to also accommodate the case of a divisor of multiplicity 3 being found. This would mean that the polynomial tested has three roots with the same real part. Test the improved program with the following polynomials, each of which has three roots with the same real part.
   (a) $p(s) = (s + 2)^3 = s^3 + 6s^2 + 12s + 8$
   (b) $p(s) = s^4 + 5s^3 + 18s^2 + 34s + 20$

5-23. For the automatic submarine depth control system of Section 5.7.3, let

$$G_2(s) = \frac{-\frac{1}{2}s + 5}{s^3 + 4s^2 + 5s}$$

instead. Find the poles of $T(s)$ and the form of the zero-input component of $y(t)$.

5-24. In the feedback submarine depth system of Section 5.7.3, let the controller transfer function be

$$G_1(s) = K + \frac{2}{s} + 4s$$

where K is adjustable, $0 \leq k \leq 50$. Construct a root-locus plot showing the poles of $T(s)$ on the complex plane as a function of K.

6

MATRIX ALGEBRA

6.1 PREVIEW

The manipulation of several sets of simultaneous linear algebraic equations, as is often required in science and engineering, can be exceedingly cumbersome without the economy of notation offered by matrix algebra. Matrix methods also give a considerably more sophisticated and powerful viewpoint of linear algebra. For a while now, our attention will turn back to emphasize analysis more than computation, but as we shall see in Chapter 7, computation very soon again assumes a key role. Not only do numerical methods relieve us of the drudgery of hand calculation, but they are essential in all but the simplest situation for such a needed operation as polynomial factorization.

The basic concepts of matrix algebra, including the matrix inverse and adjugate, are presented in Section 6.2. In Section 6.3, matrix operations are expressed in terms of elementary row and column manipulations performed by matrix multiplications. Gauss–Jordan pivoting and related operations are placed in this form. Computer programs for matrix inversion and for simultaneous equation solution are developed in Section 6.4. For the latter, the determinant of the equations to be solved is used as an indicator of possible numerical difficulty. Later, a more dependable measure of the closeness to singularity of a set of linear equations, their singular values, will be employed. In Section 6.5, solution of linear algebraic equations and matrix inversion using the LU decomposition method are discussed. An important application of matrix methods is vector analysis. The matrix viewpoint of vector analysis is introduced in Section 6.6; vector spaces and changes of coordinates, particularly orthogonal coordinates, are discussed in Section 6.7.

6.2 BASIC MATRIX ALGEBRA

The manipulation of sets of linear algebraic equations and the determination of their properties are greatly enhanced by matrix notation. For many readers, it is likely that the material of this section is a review, in which case it serves as an outline of our major concerns at this point and to establish terminology and notation. For the reader to whom this section's material is new, the effort applied now to its mastery will pay dividends, not only in this course of study, but long into the future.

6.2.1 Matrix Methods

A matrix is a rectangular array of numbers or functions. Matrices are symbolized by letters, and an individual element of a matrix is denoted by a letter with subscripts, usually the lowercase form of the symbol for the matrix itself. The first subscript is the row of the element and the second subscript is the element's column. For example, for the matrix

$$\mathbf{A} = \begin{bmatrix} 1 & 0 & 1 \\ 0 & -2 & 2 \\ 6 & 0 & 3 \\ 1 & 0 & 5 \end{bmatrix}$$

the element in the second row and third column is

$$a_{23} = 2$$

A matrix with m rows and n columns is of dimension $m \times n$. The matrix $\mathbf{A}$ above, for example, is 4×3. A *square* matrix has an equal number of rows and columns. A *column vector*, or just *vector*, is a matrix with just one column, for example,

$$\mathbf{b} = \begin{bmatrix} 3 \\ -1 \\ 0 \\ 2 \end{bmatrix}$$

Vectors are usually denoted by lowercase symbols and their elements are signified by the symbol with a single subscript that is the element's row number. For the vector $\mathbf{b}$, $b_3 = 0$ is the element. A vector with m rows is called an m-vector. A matrix with all elements *zero* is a zero or *null* matrix. An $n \times m$ zero matrix is denoted by $\mathbf{0}_{mn}$ or by just $\mathbf{0}$ when its dimensions are obvious. A square matrix with all elements zero except possibly those on the diagonal is called a *diagonal* matrix, and a diagonal matrix with all elements unity is an *identity* matrix. An $n \times n$ identity matrix is denoted by $\mathbf{I}_n$ or by just $\mathbf{I}$. For example,

$$\mathbf{I}_3 = \begin{bmatrix} 1 & 0 & 0 \\ 0 & 1 & 0 \\ 0 & 0 & 1 \end{bmatrix}$$

Two matrices are equal,

$$\mathbf{A} = \mathbf{B}$$

only if they have the same dimensions and if all corresponding elements in the two matrices are equal. Matrices with the same dimensions are termed *conformable*. The product of a scalar k and a matrix $\mathbf{A}$ is another matrix

$$\mathbf{B} = k\mathbf{A}$$

of the same dimensions as $\mathbf{A}$, with each element k times the corresponding element of $\mathbf{A}$:

$$b_{ij} = ka_{ij}$$

The sum or difference of two matrices,

$$\mathbf{C} = \mathbf{A} \pm \mathbf{B}$$

makes sense only if $\mathbf{A}$ and $\mathbf{B}$ are conformable. The sum or the difference $\mathbf{C}$ then has the same dimensions as $\mathbf{A}$ and $\mathbf{B}$ and has elements that are the sums or differences of the corresponding elements of elements of $\mathbf{A}$ and $\mathbf{B}$:

$$c_{ij} = a_{ij} \pm b_{ij}$$

The order in which several conformable matrices are added or subtracted is of no consequence.

A square matrix $\mathbf{A}$ for which

$$a_{ij} = a_{ji}$$

is *symmetric*. If

$$a_{ij} = -a_{ji}$$

the matrix is *skew symmetric*. For example,

$$\mathbf{A} = \begin{bmatrix} 1 & -1 & 2 \\ -1 & 3 & 0 \\ 2 & 0 & 4 \end{bmatrix}$$

is symmetric, and

$$\mathbf{B} = \begin{bmatrix} 0 & 1 & 2 & -3 \\ -1 & 0 & 4 & 5 \\ -2 & -4 & 0 & 6 \\ 3 & -5 & -6 & 0 \end{bmatrix}$$

is skew symmetric. Any square matrix can be expressed as the sum of a symmetric matrix plus a skew symmetric matrix.

Interchange of the rows and columns of a matrix is denoted by the dagger symbol, as in $\mathbf{B} = \mathbf{A}^\dagger$. The matrix $\mathbf{B}$ is the transpose of the matrix $\mathbf{A}$. The transpose of a column vector is a *row vector*:

$$\mathbf{b}^\dagger = \begin{bmatrix} b_1 \\ b_2 \\ \vdots \\ b_n \end{bmatrix}^\dagger = [b_1 \quad b_2 \quad \cdots \quad b_n]$$

Usually, lowercase symbols are used to represent column vectors, then a row vector is written as the transpose of the corresponding column vector.

The product of two matrices,

$$C = AB$$

exists only if the number of columns of A equals the number of rows of B. The matrices are then conformable for multiplication. If A is $m \times n$ and B is $n \times p$, the product is $m \times p$:

$$\begin{array}{ccc} C & = & A \quad B \\ m \times p & & m \times n \quad n \times p \end{array}$$

and

$$c_{ij} = a_{i1}b_{1j} + a_{i2}b_{2j} + \cdots + a_{in}b_{nj} = \sum_{k=1}^{n} a_{ik}b_{kj}$$

Multiplication of a matrix A on the left, *premultiplication*, or on the right, *postmultiplication*, but an identity matrix of proper dimension leaves A unchanged:

$$IA = A \qquad AI = A$$

Generally, however, AB does not equal BA. For matrix products, the ordering of the product is important because matrix multiplication is associative, i.e.

$$ABC = A(BC) = (AB)C$$

Premultiplication and postmultiplication have the distributive property

$$A(B + C) = AB + AC \qquad (A + B)C = AC + BC$$

and

$$A0 = 0A = 0$$

but $AB = AC$ does not generally imply that $B = C$ and $AB = 0$ does not necessarily mean that $A = 0$ or $B = 0$.

Partitioning, indicated by dashed lines, indicates how a matrix is composed of smaller *submatrices*. For example,

$$A = \left[\begin{array}{ccc:cc} 1 & 2 & 0 & 3 & -4 \\ -1 & -3 & 0 & -2 & 0 \\ \hdashline 6 & 3 & 7 & 2 & 5 \\ -4 & 1 & 8 & 4 & 0 \end{array} \right] = \left[\begin{array}{c:c} A_{11} & A_{12} \\ \hdashline A_{21} & A_{22} \end{array} \right]$$

indicates how the matrix A is composed of the four submatrices

$$A_{11} = \begin{bmatrix} 1 & 2 & 0 \\ -1 & -3 & 0 \end{bmatrix} \qquad A_{12} = \begin{bmatrix} 3 & -4 \\ -2 & 0 \end{bmatrix}$$

$$A_{21} = \begin{bmatrix} 6 & 3 & 7 \\ -4 & 1 & 8 \end{bmatrix} \qquad A_{22} = \begin{bmatrix} 2 & 5 \\ 4 & 0 \end{bmatrix}$$

The partitioning

$$\mathbf{A} = [\mathbf{a} \mid \mathbf{b} \mid \cdots]$$

indicates that the columns of $\mathbf{A}$ are the vectors $\mathbf{a}$, $\mathbf{b}$, ... The partitioning

$$\mathbf{A} = \begin{bmatrix} \mathbf{x}^\dagger \\ ---- \\ \mathbf{y}^\dagger \\ ---- \\ \vdots \end{bmatrix}$$

indicates that the rows of $\mathbf{A}$ are the row vectors $\mathbf{x}^\dagger$, $\mathbf{y}^\dagger$, ...

The transpose of a matrix product is the product of the transposed matrices in reverse order:

$$(\mathbf{AB})^\dagger = \mathbf{B}^\dagger\mathbf{A}^\dagger$$

In general,

$$(\mathbf{AB} \cdots \mathbf{PQ})^\dagger = \mathbf{Q}^\dagger\mathbf{P}^\dagger \cdots \mathbf{B}^\dagger\mathbf{A}^\dagger$$

The *trace* of a square matrix $\mathbf{A}$ is the sum of the diagonal elements of the matrix:

$$\text{trace}(\mathbf{A}) = a_{11} + a_{22} + \cdots + a_{nn} = \sum_{i=1}^{n} a_{ii}$$

The complex conjugate of a matrix, denoted by $\mathbf{A}^*$, is the matrix composed of the complex conjugates of the elements of $\mathbf{A}$.

When a matrix $\mathbf{A}$ has functions as elements, the derivative of $\mathbf{A}$ with respect to a scalar is the matrix composed of elements that are the derivatives of the corresponding elements of $\mathbf{A}$. Higher derivatives, partial derivatives, and integrals are defined similarly, as matrices of the same dimension, composed of the corresponding derivatives or integrals of the original elements.

6.2.2 Matrix Product Program

Table 6-1 is a program in BASIC/FORTRAN that computes the product of two conformable matrices,

$$\mathbf{C} = \mathbf{AB}$$

The user enters the matrix dimensions and elements, and the program prints the elements of the product matrix. Because the format of side-by-side printing of several numbers is dependent on the column capacity of the printer or display used, printing matrices row by row is apt to be difficult to read if more than one line is needed for each matrix row. This program, and later ones involving matrices, accepts and prints matrices column by column. This uses a little more screen space or printer paper than row-by-row printing but emphasizes user convenience.

The matrix $\mathbf{A}$ has dimensions $M \times N$ and $\mathbf{B}$ is $N \times P$, so the product $\mathbf{C} = \mathbf{AB}$ is $M \times P$. To form an element of the product, the element of $\mathbf{C}$ is first set to zero (line

TABLE 6-1

Computer Program in BASIC for Matrix Entry, Multiplication, and Printing (MULT)

VARIABLES USED

M	= number of rows of matrix **A** = number of rows of matrix **C** = **AB**
N	= number of columns of matrix **A** = number of rows of matrix **B**
P	= number of columns of matrix **B** = number of columns of matrix **C** = **AB**
A(20,20)	= elements of matrix **A** as a function of row and column
B(20,20)	= elements of matrix **B**
C(20,20)	= elements of matrix **C**
I	= row index
J	= column index
K	= product term index

LISTING

```
100      PRINT "MATRIX MULTIPLICATION"
110      DIM A(20,20),B(20,20),C(20,20)
120   REM   ENTER DIMENSIONS AND ELEMENTS OF MATRIX A
130      PRINT "PLEASE ENTER NO. ROWS OF A";
140      INPUT M
150      PRINT "ENTER NO. COLUMNS OF A";
160      INPUT N
170      PRINT "ENTER MATRIX A, COLUMN BY COLUMN"
180      FOR J = 1 TO N
190      PRINT "COLUMN ";J
200      FOR I = 1 TO M
210      PRINT "ROW ";I;
220      INPUT A(I,J)
230      NEXT I
240      NEXT J
250   REM   ENTER DIMENSION AND ELEMENTS OF MATRIX B
260      PRINT "NO. ROWS OF B MUST BE ";N
270      PRINT "PLEASE ENTER NO. COLUMNS OF B";
280      INPUT P
290      PRINT "ENTER MATRIX B, COLUMN BY COLUMN"
300      FOR J = 1 TO P
310      PRINT "COLUMN ";J
320      FOR I = 1 TO N
330      PRINT "ROW ";I;
340      INPUT B(I,J)
350      NEXT I
360      NEXT J
370   REM   FORM THE PRODUCT MATRIX C
380   REM   FOR EACH ROW OF C
390      FOR I = 1 TO M
400   REM   FOR EACH COLUMN OF C
410      FOR J = 1 TO P
420      C(I,J) = 0
430      FOR K = 1 TO N
440      C(I,J) = C(I,J) + A(I,K) * B(K,J)
450      NEXT K
460      NEXT J
470      NEXT I
480   REM   PRINT MATRIX PRODUCT C=AB
490      PRINT "PRODUCT MATRIX COLUMNS ARE"
```

(continued)

```
500        FOR J = 1 TO P
510        PRINT "COLUMN ";J
520        FOR I = 1 TO M
530        PRINT C(I,J)
540        NEXT I
550        NEXT J
560        END
```

Computer Program in FORTRAN for Matrix Entry, Multiplication, and Printing (MULT)

```
C ***********************************************************************
C PROGRAM MULT -- MATRIX MULTIPLICATION
C ***********************************************************************
      REAL A(20,20), B(20,20), C(20,20)

      WRITE(*,'(1X,''MATRIX MULTIPLICATION'')')

C *** ENTER DIMENSIONS AND ELEMENTS OF MATRIX A
      WRITE(*,'(1X,''PLEASE ENTER NO. ROWS OF A '',\)')
      READ(*,*) M
      WRITE(*,'(1X,''ENTER NO. COLUMNS OF A '',\)')
      READ(*,*) N
      WRITE(*,'(1X,''ENTER MATRIX A, COLUMN BY COLUMN'')')
      DO 10 J=1, N
         WRITE(*,'(1X,''COLUMN '',I2)') J
         DO 20 I=1, M
            WRITE(*,'(1X,''ROW '',I2,'' = '',\)') I
            READ(*,*) A(I,J)
20       CONTINUE
10    CONTINUE

C ***  ENTER DIMENSION AND ELEMENTS OF MATRIX B
      WRITE(*,'(1X,''NO. ROWS OF B MUST BE '',I2)') N
      WRITE(*,'(1X,''PLEASE ENTER NO. COLUMNS OF B '',\)')
      READ(*,*) P
      WRITE(*,'(1X,''ENTER MATRIX B, COLUMN BY COLUMN'')')
      DO 30 J=1, P
         WRITE(*,'(1X,''COLUMN '',I2)') J
         DO 40 I=1, N
            WRITE(*,'(1X,''ROW '',I2,'' = '',\)') I
            READ(*,*) B(I,J)
40       CONTINUE
30    CONTINUE

C ***  FORM THE PRODUCT MATRIX C
C ***  FOR EACH ROW OF C
      DO 50 I=1, M

C ***  FOR EACH COLUMN OF C
         DO 60 J=1, P
            C(I,J) = 0.0
            DO 70 K=1, N
               C(I,J) = C(I,J) + A(I,K)*B(K,J)
70          CONTINUE
60       CONTINUE
50    CONTINUE

C *** PRINT MATRIX PRODUCT C=AB
      WRITE(*,'(1X,''PRODUCT MATRIX COLUMNS ARE'')')
      DO 80 J=1, P
         WRITE(*,'(1X,''COLUMN '',I2)') J
         DO 90 I=1, M
            WRITE(*,'(1X,E15.8)') C(I,J)
```

```
90       CONTINUE
80    CONTINUE

      STOP
      END
```

420), then the N product terms are summed, one at a time, to that element (lines 420–450). For the matrices

$$\mathbf{A} = \begin{bmatrix} 1 & -2 & 3 \\ 2 & 0 & 4 \end{bmatrix} \qquad \mathbf{B} = \begin{bmatrix} -1 & 5 \\ 6 & -4 \\ -3 & 7 \end{bmatrix}$$

the MULT program, for example, gives

```
MATRIX MULTIPLICATION
PLEASE ENTER NUMBER OF ROWS OF A?2
ENTER NUMBER OF COLUMNS OF A?3
ENTER MATRIX A, COLUMN BY COLUMN
COLUMN 1
ROW 1?1
ROW 2?2
COLUMN 2
ROW 1?-2
ROW 2?0
COLUMN 3
ROW 1?3
ROW 2?4
NUMBER OF ROWS OF B MUST BE 3
PLEASE ENTER NUMBER OF COLUMNS OF B?2
ENTER MATRIX B, COLUMN BY COLUMN
COLUMN 1
ROW 1?-1
ROW 2?6
ROW 3?-3
COLUMN 2
ROW 1?5
ROW 2?-4
ROW 3?7
PRODUCT MATRIX COLUMNS ARE
COLUMN 1
-22
-14
COLUMN 2
34
38
```

or

$$\mathbf{C} = \begin{bmatrix} -22 & 34 \\ -14 & 38 \end{bmatrix}$$

6.2.3 Linear Equations in Matrix Form

A set of m linear algebraic equations in n unknowns,

$$a_{11}x_1 + a_{12}x_2 + \cdots + a_{1n}x_n = b_1$$

$$a_{21}x_1 + a_{22}x_2 + \cdots + a_{2n}x_n = b_2$$

$$\vdots$$

$$a_{m1}x_1 + a_{m2}x_2 + \cdots + a_{mn}x_n = b_m$$

is

$$\mathbf{Ax = b}$$

in matrix notation, where the $m \times n$ matrix $\mathbf{A}$ is the array of coefficients

$$\mathbf{A} = \begin{bmatrix} a_{11} & a_{12} & \cdots & a_{1n} \\ a_{21} & a_{22} & \cdots & a_{2n} \\ \vdots & \vdots & \vdots & \vdots \\ a_{m1} & a_{m2} & \cdots & a_{mn} \end{bmatrix}$$

$\mathbf{x}$ is the vector of unknowns, and $\mathbf{b}$ is the vector of knowns:

$$\mathbf{x} = \begin{bmatrix} x_1 \\ x_2 \\ \vdots \\ x_n \end{bmatrix} \qquad \mathbf{b} = \begin{bmatrix} b_1 \\ b_2 \\ \vdots \\ b_m \end{bmatrix}$$

The matrix notation is useful because it saves a lot of writing and because it greatly simplifies the description of the equation manipulation. For example, the equations

$$\begin{aligned} 2x_1 + 6x_2 - 2x_3 &= 7 \\ -3x_1 - 3x_2 + x_3 &= 0 \\ + 4x_2 + 5x_3 &= -1 \end{aligned} \quad \text{are} \quad \begin{bmatrix} 2 & 6 & -2 \\ -3 & -3 & 1 \\ 0 & 4 & 5 \end{bmatrix} \begin{bmatrix} x_1 \\ x_2 \\ x_3 \end{bmatrix} = \begin{bmatrix} 7 \\ 0 \\ -1 \end{bmatrix} \quad (6\text{-}1)$$

in matrix form.

The matrix that is composed of the coefficients of the equations and the column of knowns,

$$\begin{bmatrix} a_{11} & a_{12} & \cdots & a_{1n} & b_1 \\ a_{21} & a_{22} & \cdots & a_{2n} & b_2 \\ \vdots & \vdots & \vdots & \vdots & \\ a_{m1} & a_{m2} & \cdots & a_{mm} & b_m \end{bmatrix} = [\mathbf{A} \vdots \mathbf{b}]$$

is called the *augmented matrix* for the set of equations. For the equations (6-1),

$$\begin{bmatrix} 2 & 6 & -2 & 7 \\ -3 & -3 & 1 & 0 \\ 0 & 4 & 5 & -1 \end{bmatrix}$$

is the augmented matrix.

6.2.4 Determinant, Inverse, and Adjugate

The determinant of a square matrix $\mathbf{A}$ is denoted by $|\mathbf{A}|$ and is uniquely defined by the three properties listed in Table 6-2. Further properties that follow from the defining ones are also listed in the table. The determinant of a 2×2 matrix is

$$\begin{vmatrix} a_{11} & a_{12} \\ a_{21} & a_{22} \end{vmatrix} = a_{11}a_{22} - a_{21}a_{12}$$

Determinants of square matrices of large dimension can be found by manipulating the array using the properties, or by expressing the determinant of a larger matrix in terms of determinants of smaller matrices with Laplace expansion, as in Chapter 2.

TABLE 6-2 Properties of the Determinant of a Square Matrix

DEFINING PROPERTIES

    (A) $|\mathbf{A}|$ is unchanged if the elements of any row (or column) are replaced by the sums of the elements of that row (column) and the corresponding ones of another row (column).

    (B) The value of the determinant is multiplied by k if all the elements of any row or column are each multiplied by a scalar k.

    (C) The determinant of an identity matrix is unity.

ADDITIONAL PROPERTIES

    (1) $|\mathbf{A}|$ is unchanged if the elements of any row (or column) are replaced by the sums of each of these elements with any number times the corresponding elements of another row (column).

    (2) The algebraic sign of $|\mathbf{A}|$ is reversed if any two rows or columns are interchanged.

    (3) $|\mathbf{A}| = 0$ if all elements of a row or column are zero or if the corresponding elements of any two rows or columns are identical or have a common ratio.

    (4) The determinant of a diagonal matrix is the product of the diagonal elements.

    (5) $|\mathbf{A}\dagger| = |\mathbf{A}|$.

    (6) The determinant of an upper or lower triangular matrix is the product of the diagonal elements.

    (7) $|\mathbf{AB}| = |\mathbf{A}|\,|\mathbf{B}|$, provided that $\mathbf{A}$ and $\mathbf{B}$ are square matrices. Applying this result repeatedly yields

$$|\mathbf{AB} \cdots \mathbf{PQ}| = |\mathbf{A}|\,|\mathbf{B}| \cdots |\mathbf{P}|\,|\mathbf{Q}|$$

The rank of a matrix, denoted by rank $(\mathbf{A})$, is the dimension of the largest square array within $\mathbf{A}$, formed by deleting rows and columns as necessary, having a nonzero determinant. Matrices having the largest possible rank for their dimension are termed *full rank*. A square matrix with zero determinants is not of full rank and is said to be *singular*.

The inverse of a square matrix $\mathbf{A}$, denoted by $\mathbf{A}^{-1}$, if it exists, satisfies

$$\mathbf{A}^{-1}\mathbf{A} = \mathbf{I}$$

It has the properties listed in Table 6-3. Cramer's rule can be used to solve for the inverse of a matrix in the following way. For the set of n equations in n variables,

$$\mathbf{Ax} = \mathbf{b}$$

TABLE 6-3   Properties of the Inverse of a Square Matrix

DEFINING PROPERTY

$\quad \mathbf{AA}^{-1} = \mathbf{I}$

FURTHER PROPERTIES

(1) $\mathbf{A}^{-1}\mathbf{A} = \mathbf{I}$.

(2) $\mathbf{I}^{-1} = \mathbf{I}$.

(3) $[\mathbf{A}^{-1}]^{-1} = \mathbf{A}$.

(4) $[\mathbf{A}^{-1}]^{\dagger} = [\mathbf{A}^{\dagger}]^{-1}$.

(5) The inverse of a nonsingular diagonal matrix $\mathbf{D}$ is diagonal. It has diagonal elements that are the reciprocals of the diagonal elements of $\mathbf{D}$.

(6) $(k\mathbf{A})^{-1} = (1/k)\mathbf{A}^{-1}$ for k a nonzero scalar.

(7) $(\mathbf{AB})^{-1} = \mathbf{B}^{-1}\mathbf{A}^{-1}$, provided that $\mathbf{A}$ and $\mathbf{B}$ are each square and nonsingular. Applying this result repeatedly yields

$$(\mathbf{AB} \cdots \mathbf{PQ})^{-1} = \mathbf{Q}^{-1}\mathbf{P}^{-1} \cdots \mathbf{B}^{-1}\mathbf{A}^{-1}$$

Multiplying on the left by $\mathbf{A}^{-1}$ gives

$$\mathbf{A}^{-1}\mathbf{Ax} = \mathbf{Ix} = \mathbf{A}^{-1}\mathbf{b}$$

Applying Cramer's rule, the solution for the ith variable is

$$x_i = \frac{|\mathbf{A}_i|}{|\mathbf{A}|}$$

Laplace expanding $|\mathbf{A}_i|$ along the ith column yields

$$\mathbf{x}_i = \frac{1}{|\mathbf{A}|}\left(|\mathbf{A}_{1i}|b_1 + |\mathbf{A}_{2i}|b_2 + \cdots + |\mathbf{A}_{ni}|b_n\right)$$

so

$$\mathbf{x} = \frac{1}{|\mathbf{A}|}\begin{bmatrix} |\mathbf{A}_{11}| & |\mathbf{A}_{21}| & \cdots & |\mathbf{A}_{n1}| \\ |\mathbf{A}_{12}| & |\mathbf{A}_{22}| & \cdots & |\mathbf{A}_{n2}| \\ \vdots & \vdots & \vdots & \vdots \\ |\mathbf{A}_{1n}| & |\mathbf{A}_{2n}| & \cdots & |\mathbf{A}_{nn}| \end{bmatrix}\mathbf{b} = \mathbf{A}^{-1}\mathbf{b}$$

provided that

$$|\mathbf{A}| \neq 0$$

If $|\mathbf{A}| = 0$, $\mathbf{A}^{-1}$ does not exist and the matrix $\mathbf{A}$ is said to be singular.

The matrix

$$\begin{bmatrix} |\mathbf{A}_{11}| & |\mathbf{A}_{21}| & \cdots & |\mathbf{A}_{n1}| \\ |\mathbf{A}_{12}| & |\mathbf{A}_{22}| & \cdots & |\mathbf{A}_{n2}| \\ \vdots & \vdots & \vdots & \vdots \\ |\mathbf{A}_{1n}| & |\mathbf{A}_{2n}| & \cdots & |\mathbf{A}_{nn}| \end{bmatrix}$$

is called the *adjugate* of the matrix $\mathbf{A}$ and is denoted by adj $(\mathbf{A})$. It is the transpose of the matrix formed by replacing each element of $\mathbf{A}$ with the cofactor of that element. Some properties of the adjugate are listed in Table 6-4.

An example of matrix inversion using the adjugate is as follows:

$$\mathbf{A} = \begin{bmatrix} 1 & -2 & 3 \\ 0 & 4 & -1 \\ 5 & 2 & -3 \end{bmatrix} \qquad (\text{adj } \mathbf{A})^{\dagger} = \begin{bmatrix} -10 & -5 & -20 \\ 0 & -18 & -12 \\ -10 & 1 & 4 \end{bmatrix}$$

$$|\mathbf{A}| = \begin{vmatrix} 4 & -1 \\ 2 & -3 \end{vmatrix} + 5 \begin{vmatrix} -2 & 3 \\ 4 & -1 \end{vmatrix} = -10 + 5(-10) = -60$$

$$\mathbf{A}^{-1} = \frac{\text{adj } \mathbf{A}}{|\mathbf{A}|} = \begin{bmatrix} \frac{10}{60} & 0 & \frac{10}{60} \\ \frac{5}{60} & \frac{18}{60} & -\frac{1}{60} \\ \frac{20}{60} & \frac{12}{60} & -\frac{4}{60} \end{bmatrix}$$

The solution of a set of n independent, consistent linear algebraic equations in n variables

$$\mathbf{A}\mathbf{x} = \mathbf{b}$$

TABLE 6-4 Properties of the Adjugate of a Square Matrix

DEFINING PROPERTY

For an $n \times n$ matrix $\mathbf{A}$

$$\text{adj } (\mathbf{A}) = \begin{bmatrix} |\mathbf{A}_{11}| & |\mathbf{A}_{21}| & \cdots & |\mathbf{A}_{n1}| \\ |\mathbf{A}_{12}| & |\mathbf{A}_{22}| & \cdots & |\mathbf{A}_{n2}| \\ \vdots & \vdots & \vdots & \vdots \\ |\mathbf{A}_{1n}| & |\mathbf{A}_{2n}| & \cdots & |\mathbf{A}_{nn}| \end{bmatrix}$$

FURTHER PROPERTIES

(1) $[\text{adj } (\mathbf{A})] \mathbf{A} = \mathbf{A} [\text{adj } (\mathbf{A})] = |\mathbf{A}| \mathbf{I}$

(2) $\text{adj } (\mathbf{A}) = |\mathbf{A}| \mathbf{A}^{-1}$

(3) For $\mathbf{A}$ and $\mathbf{B}$ each $n \times n$,

$$[\text{adj } (\mathbf{A})] [\text{adj } (\mathbf{B})] = \text{adj } (\mathbf{BA})$$

(4) If $\mathbf{A}$ is $n \times n$ and has rank $n - 1$, then

$$\text{rank } [\text{adj } (\mathbf{A})] = 1$$

(5) If $\mathbf{A}$ is $n \times n$ and has rank $n - 2$ or less, then

$$\text{adj } (\mathbf{A}) = 0$$

is given by

$$\mathbf{A}^{-1}\mathbf{A}\mathbf{x} = \mathbf{A}^{-1}\mathbf{b}$$

or

$$\mathbf{x} = \mathbf{A}^{-1}\mathbf{b}$$

In matrix algebra, there is no division operation, so the solution of a matrix equation such as

$$\mathbf{ABC} = \mathbf{D}$$

is generally more involved than it is when the quantities involved are scalars that can be "canceled." One tool in the solution of matrix equations such as the above is premultiplication and postmultiplication by matrix inverses when they exist. For example, to solve

$$\mathbf{ABC} = \mathbf{D}$$

for **B**, assuming that matrices **A** and **C** are square and nonsingular, premultiply by $\mathbf{A}^{-1}$ and postmultiply by $\mathbf{C}^{-1}$, giving

$$\mathbf{A}^{-1}\mathbf{ABCC}^{-1} = \mathbf{A}^{-1}\mathbf{DC}^{-1} \qquad \mathbf{B} = \mathbf{A}^{-1}\mathbf{DC}^{-1}$$

6.3 ELEMENTARY ROW AND COLUMN OPERATIONS

Simple operations on the rows and columns of a matrix are now expressed in terms of matrix multiplications. By doing so, we develop powerful methods for matrix operations, including matrix inversion.

6.3.1 Elementary Operations as Matrix Products

The operations involved in determinant evaluation and Gauss–Jordan pivoting can be expressed in terms of matrix multiplications. The following is a summary, in terms of 3×3 examples, of square matrix premultiplications that perform the basic operations of row interchange, row multiplication by a constant, and the addition of a row of a constant times another row. The simple types of matrices are called *elementary* matrices:

1. Row interchange

$$\begin{bmatrix} 0 & 1 & 0 \\ 1 & 0 & 0 \\ 0 & 0 & 1 \end{bmatrix} \begin{bmatrix} a_{11} & a_{12} & a_{13} \\ a_{21} & a_{22} & a_{23} \\ a_{31} & a_{32} & a_{33} \end{bmatrix} = \begin{bmatrix} a_{21} & a_{22} & a_{23} \\ a_{11} & a_{12} & a_{13} \\ a_{31} & a_{32} & a_{33} \end{bmatrix}$$

2. Row multiplication by a constant

$$\begin{bmatrix} 1 & 0 & 0 \\ 0 & k & 0 \\ 0 & 0 & 1 \end{bmatrix} \begin{bmatrix} a_{11} & a_{12} & a_{13} \\ a_{21} & a_{22} & a_{23} \\ a_{31} & a_{32} & a_{33} \end{bmatrix} = \begin{bmatrix} a_{11} & a_{12} & a_{13} \\ ka_{21} & ka_{22} & ka_{23} \\ a_{31} & a_{32} & a_{33} \end{bmatrix}$$

3. Addition to a row of a constant times another row

$$\begin{bmatrix} 1 & 0 & k \\ 0 & 1 & 0 \\ 0 & 0 & 1 \end{bmatrix} \begin{bmatrix} a_{11} & a_{12} & a_{13} \\ a_{21} & a_{22} & a_{23} \\ a_{31} & a_{32} & a_{33} \end{bmatrix} = \begin{bmatrix} a_{11} + ka_{31} & a_{12} + ka_{32} & a_{13} + ka_{33} \\ a_{21} & a_{22} & a_{23} \\ a_{31} & a_{32} & a_{33} \end{bmatrix}$$

A sequence of such row operations can be expressed as a product of these matrix pre-multiplications, which can be combined as premultiplication by a single, more complicated matrix if desired. Column operations can be performed by *post*multiplying by simple square matrices, as illustrated below. In terms of 3 × 3 matrices:

4. Column interchange

$$\begin{bmatrix} a_{11} & a_{12} & a_{13} \\ a_{21} & a_{22} & a_{23} \\ a_{31} & a_{32} & a_{33} \end{bmatrix} \begin{bmatrix} 0 & 1 & 0 \\ 1 & 0 & 0 \\ 0 & 0 & 1 \end{bmatrix} = \begin{bmatrix} a_{12} & a_{11} & a_{13} \\ a_{22} & a_{21} & a_{23} \\ a_{32} & a_{31} & a_{33} \end{bmatrix}$$

5. Column multiplication by a constant

$$\begin{bmatrix} a_{11} & a_{12} & a_{13} \\ a_{21} & a_{22} & a_{23} \\ a_{31} & a_{32} & a_{33} \end{bmatrix} \begin{bmatrix} 1 & 0 & 0 \\ 0 & 1 & 0 \\ 0 & 0 & k \end{bmatrix} = \begin{bmatrix} a_{11} & a_{12} & ka_{13} \\ a_{21} & a_{22} & ka_{23} \\ a_{31} & a_{32} & ka_{33} \end{bmatrix}$$

6. Addition of a constant times a column to another column

$$\begin{bmatrix} a_{11} & a_{12} & a_{13} \\ a_{21} & a_{22} & a_{23} \\ a_{31} & a_{32} & a_{33} \end{bmatrix} \begin{bmatrix} 1 & 0 & 0 \\ k & 1 & 0 \\ 0 & 0 & 1 \end{bmatrix} = \begin{bmatrix} a_{11} + ka_{12} & a_{12} & a_{13} \\ a_{21} + ka_{22} & a_{22} & a_{23} \\ a_{31} + ka_{32} & a_{32} & a_{33} \end{bmatrix}$$

As with the row operations, a sequence of column operations can be expressed as repeated matrix postmultiplications, which in turn can be written as one, more complicated matrix postmultiplication. Each of the elementary row and column operation matrices is non-singular provided, of course, that for row or column multiplication by k, $k \neq 0$.

6.3.2 Equation Solution

A series of row operations, in particular those for Gauss–Jordan pivoting, can be expressed as a product of the matrices for elementary row operations. This product of the simpler matrices can in turn be written as a single matrix premultiplication that expresses the conversion of the original equation matrix to the equivalent equation matrix for the fully pivoted equations. For example, for the equations

$$\begin{cases} 2x_1 - 3x_2 = -2 \\ x_1 + 4x_2 = 0 \end{cases}$$

the augmented matrix is

$$\begin{bmatrix} 2 & -3 & -2 \\ 1 & 4 & 0 \end{bmatrix}$$

The fully pivoted array can be obtained with the following sequence of row operations:

1. Normalize row one by dividing it by 2:

$$\begin{bmatrix} \frac{1}{2} & 0 \\ 0 & 1 \end{bmatrix} \begin{bmatrix} 2 & -3 & -2 \\ 1 & 4 & 0 \end{bmatrix} = \begin{bmatrix} 1 & -\frac{3}{2} & -1 \\ 1 & 4 & 0 \end{bmatrix}$$

2. Add -1 times row one to row two:

$$\begin{bmatrix} 1 & 0 \\ -1 & 1 \end{bmatrix} \begin{bmatrix} 1 & -\frac{3}{2} & -1 \\ 1 & 4 & 0 \end{bmatrix} = \begin{bmatrix} 1 & -\frac{3}{2} & -1 \\ 0 & \frac{11}{2} & 1 \end{bmatrix}$$

3. Normalize row two by dividing it by $\frac{11}{2}$:

$$\begin{bmatrix} 1 & 0 \\ 0 & \frac{2}{11} \end{bmatrix} \begin{bmatrix} 1 & -\frac{3}{2} & -1 \\ 0 & \frac{11}{2} & 1 \end{bmatrix} = \begin{bmatrix} 1 & -\frac{3}{2} & -1 \\ 0 & 1 & \frac{2}{11} \end{bmatrix}$$

4. Add $\frac{3}{2}$ times row two to row one:

$$\begin{bmatrix} 1 & \frac{3}{2} \\ 0 & 1 \end{bmatrix} \begin{bmatrix} 1 & -\frac{3}{2} & -1 \\ 0 & 1 & \frac{2}{11} \end{bmatrix} = \begin{bmatrix} 1 & 0 & -\frac{8}{11} \\ 0 & 1 & \frac{2}{11} \end{bmatrix}$$

Combining the operations gives the matrix that converts the original array to the fully pivoted array:

$$\begin{bmatrix} 1 & \frac{3}{2} \\ 0 & 1 \end{bmatrix} \begin{bmatrix} 1 & 0 \\ 0 & \frac{2}{11} \end{bmatrix} \begin{bmatrix} 1 & 0 \\ -1 & 1 \end{bmatrix} \begin{bmatrix} \frac{1}{2} & 0 \\ 0 & 1 \end{bmatrix} = \begin{bmatrix} \frac{4}{11} & \frac{3}{11} \\ -\frac{1}{11} & \frac{2}{11} \end{bmatrix}$$

and

$$\begin{bmatrix} \frac{4}{11} & \frac{3}{11} \\ -\frac{1}{11} & \frac{2}{11} \end{bmatrix} \begin{bmatrix} 2 & -3 & -2 \\ 1 & 4 & 0 \end{bmatrix} = \begin{bmatrix} 1 & 0 & -\frac{8}{11} \\ 0 & 1 & \frac{2}{11} \end{bmatrix}$$

For a set of linear algebraic equations

$$\mathbf{Ax} = \mathbf{b}$$

the relative ranks of the coefficient matrix $\mathbf{A}$ and the augmented matrix $[\mathbf{A} \mathbin{\vdots} \mathbf{b}]$ indicate the number of linearly independent equations and whether or not the equations are consistent. Of course,

$$\text{rank } ([\mathbf{A} \mathbin{\vdots} \mathbf{b}]) \geq \text{rank } (\mathbf{A})$$

since $\mathbf{A}$ is part of the augmented matrix. If

$$\text{rank } ([\mathbf{A} \mathbin{\vdots} \mathbf{b}]) = \text{rank } (\mathbf{A})$$

this rank is the number of linearly independent equations in the set. If there are one or more inconsistent equations,

$$\text{rank } [\mathbf{A} \mathrel{\vdots} \mathbf{b}] > \text{rank } (\mathbf{A})$$

as is easily verified by Gauss–Jordan pivoting.

6.3.3 Solution for the Inverse

For matrices of large dimension, the direct solution for the inverse,

$$\mathbf{A}^{-1} = \frac{\text{adj } (\mathbf{A})}{|\mathbf{A}|}$$

is very lengthy and tedious. An alternative method of computation of the inverse matrix is by means of the elementary transformations. A sequence of elementary row operations can always be found that will transform a *nonsingular* matrix to the identity matrix, since these operations simply represent a sequence of Gauss–Jordan pivoting steps, including normalization and possibly row interchange:

$$(\mathbf{P}_k \cdots \mathbf{P}_2 \mathbf{P}_1)\mathbf{A} = \mathbf{I}$$

The product of these elementary row operation matrices is the desired matrix inverse:

$$\mathbf{A}^{-1} = \mathbf{P}_k \cdots \mathbf{P}_2 \mathbf{P}_1$$

Similarly, a sequence of elementary column operations can always be found which will transform a nonsingular matrix to the identity matrix:

$$\mathbf{A}(\mathbf{Q}_1 \mathbf{Q}_2 \cdots \mathbf{Q}_j) = \mathbf{I}$$

The product of these elementary column operation matrices is also the desired matrix inverse:

$$\mathbf{A}^{-1} = \mathbf{Q}_1 \mathbf{Q}_2 \cdots \mathbf{Q}_j$$

If desired, combinations of elementary row and column operations can be used to transform a nonsingular matrix $\mathbf{A}$ to the identity matrix:

$$\mathbf{P}_i \cdots \mathbf{P}_2 \mathbf{P}_1 \mathbf{A} \mathbf{Q}_1 \mathbf{Q}_2 \cdots \mathbf{Q}_m = \mathbf{I}$$

Each of the elementary matrices is nonsingular, so

$$\mathbf{A} = \mathbf{P}_1^{-1} \mathbf{P}_2^{-1} \cdots \mathbf{P}_i^{-1} \mathbf{Q}_m^{-1} \cdots \mathbf{Q}_2^{-1} \mathbf{Q}_1^{-1}$$

Then

$$\mathbf{A}^{-1} = \mathbf{Q}_1 \mathbf{Q}_2 \cdots \mathbf{Q}_m \mathbf{P}_i \cdots \mathbf{P}_2 \mathbf{P}_1$$

The use of elementary operation matrices also offers a simple method of finding the determinant of a nonsingular matrix. If

$$\mathbf{P}_i \cdots \mathbf{P}_2 \mathbf{P}_1 \mathbf{A} \mathbf{Q}_1 \mathbf{Q}_2 \cdots \mathbf{Q}_m = \mathbf{I}$$

then

$$|\mathbf{P}_i| \cdots |\mathbf{P}_2| \, |\mathbf{P}_1| \, |\mathbf{A}| \, |\mathbf{Q}_1| \, |\mathbf{Q}_2| \cdots |\mathbf{Q}_m| = |\mathbf{I}|$$

and

$$|A| = \frac{1}{|P_i| \cdots |P_2| \, |P_1| \, |Q_1| \, |Q_2| \cdots |Q_m|}$$

The determinant of a row interchange matrix is -1, and that for multiplication of a row by the constant k is k. The determinant of a row operation matrix that adds k times a row to another row (Laplace expanding along the row containing a 1 and k) is unity. It is thus only necessary to keep track of the row multiplications and row interchanges.

As a numerical example of finding the inverse and determinant of a matrix in this way, the matrix

$$A = \begin{bmatrix} 2 & 4 \\ -1 & 3 \end{bmatrix}$$

can be transformed to an identity matrix with the following sequence of row operations:

1. Divide row one by 4:

$$\begin{bmatrix} \frac{1}{4} & 0 \\ 0 & 1 \end{bmatrix} \begin{bmatrix} 2 & 4 \\ -1 & 3 \end{bmatrix} = \begin{bmatrix} \frac{1}{2} & 1 \\ -1 & 3 \end{bmatrix}$$

2. Subtract 3 times row one from row two:

$$\begin{bmatrix} 1 & 0 \\ -3 & 1 \end{bmatrix} \begin{bmatrix} \frac{1}{2} & 1 \\ -1 & 3 \end{bmatrix} = \begin{bmatrix} \frac{1}{2} & 1 \\ -\frac{5}{2} & 0 \end{bmatrix}$$

3. Divide row two by $-\frac{5}{2}$:

$$\begin{bmatrix} 1 & 0 \\ 0 & -\frac{2}{5} \end{bmatrix} \begin{bmatrix} \frac{1}{2} & 1 \\ -\frac{5}{2} & 0 \end{bmatrix} = \begin{bmatrix} \frac{1}{2} & 1 \\ 1 & 0 \end{bmatrix}$$

4. Subtract $\frac{1}{2}$ times row two from row one:

$$\begin{bmatrix} 1 & -\frac{1}{2} \\ 0 & 1 \end{bmatrix} \begin{bmatrix} \frac{1}{2} & 1 \\ 1 & 0 \end{bmatrix} = \begin{bmatrix} 0 & 1 \\ 1 & 0 \end{bmatrix}$$

5. Interchange rows one and two:

$$\begin{bmatrix} 0 & 1 \\ 1 & 0 \end{bmatrix} \begin{bmatrix} 0 & 1 \\ 1 & 0 \end{bmatrix} = \begin{bmatrix} 1 & 0 \\ 0 & 1 \end{bmatrix}$$

Combining these operations,

$$\begin{bmatrix} 0 & 1 \\ 1 & 0 \end{bmatrix} \begin{bmatrix} 1 & -\frac{1}{2} \\ 0 & 1 \end{bmatrix} \begin{bmatrix} 1 & 0 \\ 0 & -\frac{2}{5} \end{bmatrix} \begin{bmatrix} 1 & 0 \\ -3 & 1 \end{bmatrix} \begin{bmatrix} \frac{1}{4} & 0 \\ 0 & 1 \end{bmatrix} = \begin{bmatrix} \frac{3}{10} & -\frac{2}{5} \\ \frac{1}{10} & \frac{1}{5} \end{bmatrix}$$

gives the inverse matrix:

$$\begin{bmatrix} \frac{3}{10} & -\frac{2}{5} \\ \frac{1}{10} & \frac{1}{5} \end{bmatrix} \begin{bmatrix} 2 & 4 \\ -1 & 3 \end{bmatrix} = \begin{bmatrix} 1 & 0 \\ 0 & 1 \end{bmatrix}$$

The determinant of **A** is the product

$$|\mathbf{A}| = (4)\left(-\tfrac{5}{2}\right)(-1) = \tfrac{1}{10}$$

This process can be mechanized somewhat by computing products of the elementary operations as the transformation of **A** to **I** proceeds. To keep the discussion simple, only the use of row operations will be discussed in detail, although column operations or a combination of row and column operations can be used if desired.

If

$$\mathbf{P}_k \cdots \mathbf{P}_2 \mathbf{P}_1 \mathbf{A} = \mathbf{I}$$

$$\mathbf{A}^{-1} = \mathbf{P}_k \cdots \mathbf{P}_2 \mathbf{P}_1 \mathbf{I}$$

To calculate the product of the elementary matrices as we proceed, start with the two arrays:

$$\mathbf{A} \qquad\qquad \mathbf{I}$$

Then, whatever elementary operations are performed upon **A**, perform the same operations upon **I**:

$$\begin{array}{ll} \mathbf{P}_1\mathbf{A} & \mathbf{P}_1\mathbf{I} \\ \mathbf{P}_2\mathbf{P}_1\mathbf{A} & \mathbf{P}_2\mathbf{P}_1\mathbf{I} \\ \vdots & \vdots \\ \mathbf{P}_k \cdots \mathbf{P}_2\mathbf{P}_1\mathbf{A} & \mathbf{P}_k \cdots \mathbf{P}_2\mathbf{P}_1\mathbf{I} \\ \mathbf{I} & \mathbf{A}^{-1} \end{array}$$

It is not necessary to deal with the elementary matrices at all, though. The same operations are to be performed on both **A** and **I**, but we need not determine specifically what the elementary matrices are. Simply Gauss–Jordan pivot the **A** array and, at the same time, do the same operations upon the **I** array. Additional operations on both arrays to normalize and order the rows of the **A** array to transform it to **I** results in **A**$^{-1}$ for the other array.

Before starting such a calculation, it might not be known whether **A** is singular, thus whether **A**$^{-1}$ exists. If **A** is singular, the pivoting operations will result in one or more zero rows of the pivoted **A** array, in which event transformation of the array to **I** is not possible. Thus singular matrices are easily identified.

6.3.4 Normal Form of a Matrix

Elementary row operations, $\mathbf{P}_1, \mathbf{P}_2, \ldots, \mathbf{P}_k$ and the elementary column operations, $\mathbf{Q}_1, \mathbf{Q}_2, \ldots, \mathbf{Q}_j$, all of which are nonsingular, can be combined into the nonsingular matrices **P** and **Q** to give

$$(\mathbf{P}_k \cdots \mathbf{P}_2\mathbf{P}_1)\mathbf{A}(\mathbf{Q}_1\mathbf{Q}_2 \cdots \mathbf{Q}_j) = \mathbf{PAQ} = \begin{bmatrix} \mathbf{I}_r & 0 \\ 0 & 0 \end{bmatrix}$$

for any matrix $\mathbf{A}$, square or not. The transformed matrix has all elements zeroes except the first r diagonal elements, which are 1's. The number of unity diagonal elements is the rank of $\mathbf{A}$. Any matrix $\mathbf{A}$ can thus be represented in the form

$$\mathbf{A} = \mathbf{P}^{-1}\begin{bmatrix} \mathbf{I}_r & \mathbf{0} \\ \mathbf{0} & \mathbf{0} \end{bmatrix}\mathbf{Q}^{-1}$$

where $\mathbf{P}$ and $\mathbf{Q}$ are nonsingular. Two matrices are *similar* if they have the same dimensions and the same rank. Since every matrix can be placed in a normal form with appropriate nonsingular premultiplication and postmultiplication matrices, if $\mathbf{A}$ and $\mathbf{B}$ are equivalent, each can be converted to the same normal form:

$$\mathbf{P}_A\mathbf{A}\mathbf{Q}_A = \begin{bmatrix} \mathbf{I}_r & \mathbf{0} \\ \mathbf{0} & \mathbf{0} \end{bmatrix} \qquad \mathbf{P}_B\mathbf{B}\mathbf{Q}_B = \begin{bmatrix} \mathbf{I}_r & \mathbf{0} \\ \mathbf{0} & \mathbf{0} \end{bmatrix}$$

Then

$$\mathbf{P}_A\mathbf{A}\mathbf{Q}_A = \mathbf{P}_B\mathbf{B}\mathbf{Q}_B$$

$$\mathbf{A} = (\mathbf{P}_A^{-1}\mathbf{P}_B)\mathbf{B}(\mathbf{Q}_B\mathbf{Q}_A^{-1})$$

and $\mathbf{A}$ can be obtained from $\mathbf{B}$ by appropriate nonsingular premultiplication and postmultiplication: If $\mathbf{A}$ and $\mathbf{B}$ are equivalent, there exist nonsingular matrices $\mathbf{P}$ and $\mathbf{Q}$ such that

$$\mathbf{A} = \mathbf{P}\mathbf{B}\mathbf{Q}$$

Similarly, $\mathbf{B}$ can be obtained from $\mathbf{A}$:

$$\mathbf{B} = (\mathbf{P}^{-1})\mathbf{A}(\mathbf{Q}^{-1})$$

A *nonsingular* matrix $\mathbf{A}$ can be represented in terms of its normal form as

$$\mathbf{A} = \mathbf{P}^{-1}\mathbf{I}\mathbf{Q}^{-1} = \mathbf{P}^{-1}\mathbf{Q}^{-1} = \mathbf{P}_1^{-1}\mathbf{P}_2^{-1}\cdots\mathbf{P}_k^{-1}\mathbf{Q}_j^{-1}\cdots\mathbf{Q}_2^{-1}\mathbf{Q}_1^{-1}$$

where the $\mathbf{P}_i$ and $\mathbf{Q}_i$ are elementary matrices. The inverse of an elementary matrix is another elementary matrix, as is easily verified, so any nonsingular matrix can be represented as the product of elementary matrices.

6.4 INVERSION PROGRAMS

6.4.1 Matrix Inversion by Pivoting Program

The pivoting method for matrix inversion described in the preceding section is expressed as an algorithm in Table 6-5. Row operations are used exclusively to transform the matrix $\mathbf{A}$ to the identity matrix. At the same time, the same operations on the matrix $\mathbf{B}$, which begins as the identity matrix, converts it to $\mathbf{A}^{-1}$. The method described is fundamentally different from the Gauss–Jordan pivoting program PIVOT (Table 2-4) in that the pivot element selected at each stage is the one with largest square in a *column* instead of a row. Had we elected to perform column operations instead, this algorithm would be more like the PIVOT algorithm.

TABLE 6-5   Algorithm for Matrix Inversion by Pivoting

DEFINITION

The inverse of the matrix **A**,

   $$\mathbf{B} = \mathbf{A}^{-1}$$

is to be calculated using elementary row operations of the **A** and the identity arrays.

ALGORITHM

1. Set the matrix **B** to the $n \times n$ identity matrix:

   $$\mathbf{B} \leftarrow \mathbf{I}$$

   (lines 230–310).

2. For each column starting with column $j = 1$ and ending with $j = n$ (lines 320–330 and 690).

   a. Find the element in column j and from row $i = j$ through row $i = n$ having the largest square. Let q be the row number of this element. If all such elements are zero, increment j, going to the next column (lines 340–410).

   b. If $q \neq j$, interchange rows j and q in both the matrix **A** and the matrix **B**. If $q = j$, no interchange is required (lines 420–520).

   c. Normalize the row j of **A** and of **B** by dividing by the pivot element.

   $$s \leftarrow a_{jj}$$

   $$a_{jk} \leftarrow \frac{a_{jk}}{s} \qquad k = 1, 2, \ldots, n$$

   $$b_{jk} \leftarrow \frac{b_{jk}}{s} \qquad k = 1, 2, \ldots, n$$

   (lines 530–580).

   d. Pivot on the elements a_{jj}. For each row i, starting with $i = 1$ and ending with row $i = n$, skipping row $i = j$,

   $$s \leftarrow a_{ij}$$
   $$a_{ik} \leftarrow a_{ik} - sa_{jk} \qquad k = 1, 2, \ldots, n$$
   $$b_{ik} \leftarrow b_{ik} - sb_{jk} \qquad k = 1, 2, \ldots, n$$

   (lines 590–690).

3. Print the matrix **B** (lines 700–770).

A computer program in BASIC/FORTRAN, called INVERSE, is listed in Table 6-6. The user enters the dimension and elements of the square matrix **A**, and the program computes and prints $\mathbf{A}^{-1}$. For the matrix

$$\mathbf{A} = \begin{bmatrix} 2 & 2 & -1 \\ -1 & 0 & 4 \\ 3 & -1 & -3 \end{bmatrix}$$

for example, INVERSE gives

```
MATRIX INVERSION BY PIVOTING
PLEASE ENTER MATRIX DIMENSION?3
ENTER MATRIX, COLUMN BY COLUMN
```

TABLE 6-6

Computer Program in BASIC for Matrix Inversion by Pivoting (INVERSE)

VARIABLES USED

N	= dimensions of matrices **A** and **B**
A(20,20)	= original matrix
B(20,20)	= matrix that becomes the inverse of **A** when the inverse exists
S	= squares of elements of the **A** matrix when searching for the element in a column with the largest square; an element of **A** stored temporarily when interchanging two rows of **A**; the pivot element when normalizing an equation; the element in the pivot column of an equation
P	= largest square of a set of elements in the **A** matrix
Q	= row numbers of elements with the largest square
T	= element of **B** stored temporarily when interchanging two rows of **B**
I	= row index
J	= column index
K	= row index

LISTING

```
100      PRINT "MATRIX INVERSION BY PIVOTING"
110      DIM A(20,20),B(20,20)
120   REM ENTER DIMENSION AND ELEMENTS OF A
130      PRINT "PLEASE ENTER MATRIX DIMENSION";
140      INPUT N
150      PRINT "ENTER MATRIX, COLUMN BY COLUMN"
160      FOR J = 1 TO N
170      PRINT "COLUMN ";J
180      FOR I = 1 TO N
190      PRINT "ROW";I;
200      INPUT A(I,J)
210      NEXT I
220      NEXT J
230   REM SET B TO THE IDENTITY MATRIX
240      FOR J = 1 TO N
250      FOR I = 1 TO N
260      B(I,J) = 0
270      NEXT I
280      NEXT J
290      FOR I = 1 TO N
300      B(I,I) = 1
310      NEXT I
320   REM FOR EACH COLUMN
330      FOR J = 1 TO N
340   REM FIND THE LARGEST MAGNITUDE ELEMENT IN
350   REM COLUMN J IN ROW J AND BELOW
360      P= 0
370      FOR I = J TO N
380      S = A(I,J) * A(I,J)
390      IF S > P THEN Q = I
400      IF S > P THEN P = S
410      NEXT I
420      IF P = 0 THEN GOTO 790
430   REM INTERCHANGE ROWS J AND Q
440      IF J = Q THEN GOTO 530
450      FOR K = 1 TO N
460      S = A(J,K)
```

```
470       T = B(J,K)
480       A(J,K) = A(Q,K)
490       B(J,K) = B(Q,K)
500       A(Q,K) = S
510       B(Q,K) = T
520      NEXT K
530   REM NORMALIZE ROW J
540      S = A(J,J)
550      FOR K = 1 TO N
560      A(J,K)=A(J,K)/S
570      B(J,K)=B(J,K) /S
580      NEXT K
590   REM PIVOT ON THE A(J,J) COEFFICIENT. FOR
600   REM EACH ROW EXCEPT ROW J
610      FOR I = 1 TO N
620      IF I = J THEN GOTO 680
630      S = A(I,J)
640      FOR K = 1 TO N
650      A(I,K)=A(I,K)-S*A(J,K)
660      B(I,K)=B(I,K)-S*B(J,K)
670      NEXT K
680      NEXT I
690      NEXT J
700   REM PRINT RESULT
710      PRINT "RESULTING MATRIX COLUMNS ARE"
720      FOR J = 1 TO N
730      PRINT "COLUMN ";J
740      FOR I = 1 TO N
750      PRINT B(I,J)
760      NEXT I
770      NEXT J
780      GOTO 810
790   REM PRINT SINGULARITY MESSAGE
800      PRINT "MATRIX IS SINGULAR:"
810      END
```

Computer Program in FORTRAN for Matrix Inversion by Pivoting (INVERSE)

```
C ********************************************************************
C PROGRAM INVERSE -- MATRIX INVERSION BY PIVOTING
C ********************************************************************
      REAL A(20,20), B(20,20)

      WRITE(*,'(1X,''MATRIX INVERSION BY PIVOTING'')')

C *** ENTER DIMENSION AND ELEMENTS OF A
      WRITE(*,'(1X,''PLEASE ENTER MATRIX DIMENSION '',\)')
      READ(*,*) N
      WRITE(*,'(1X,''ENTER MATRIX A, COLUMN BY COLUMN '')')
      DO 10 J=1, N
         WRITE(*,'(1X,''COLUMN '',I2)') J
         DO 20 I=1, N
            WRITE(*,'(1X,''ROW '',I2,'' = '',\)') I
            READ(*,*) A(I,J)
 20      CONTINUE
 10   CONTINUE

C *** SET B TO THE IDENTITY MATRIX
      DO 30 J=1, N
         DO 40 I=1, N
            B(I,J) = 0.0
 40      CONTINUE
```

(continued)

```
 30     CONTINUE
        DO 50 I=1, N
           B(I,I) = 1.0
 50     CONTINUE

C *** FOR EACH COLUMN
        DO 60 J=1, N

C *** FIND THE LARGEST MAGNITUDE ELEMENT IN
C *** COLUMN J IN ROW J AND BELOW
        P = 0.0
        DO 70 I=J, N
           S = A(I,J)*A(I,J)
           IF (S.GT.P) THEN
              Q=I
              P=S
           ENDIF
 70        CONTINUE

C *** PRINT SINGULARITY MESSAGE
        IF (P.EQ.0.0) THEN
           WRITE(*,'(1X,''MATRIX IS SINGULAR'')')
           STOP
        ENDIF

C *** INTERCHANGE ROWS J AND Q
        IF (J.NE.Q) THEN
           DO 80 K=1, N
              S = A(J,K)
              T = B(J,K)
              A(J,K) = A(Q,K)
              B(J,K) = B(Q,K)
              A(Q,K) = S
              B(Q,K) = T
 80           CONTINUE
        ENDIF

C *** NORMALIZE ROW J
        S = A(J,J)
        DO 90 K=1, N
           A(J,K)=A(J,K)/S
           B(J,K)=B(J,K) /S
 90     CONTINUE

C *** PIVOT ON THE A(J,J) COEFFICIENT. FOR
C *** EACH ROW EXCEPT ROW J
        DO 100 I=1, N
           IF (I.NE.J) THEN
              S = A(I,J)
              DO 110 K=1, N
                 A(I,K)=A(I,K)-S*A(J,K)
                 B(I,K)=B(I,K)-S*B(J,K)
 110             CONTINUE
           ENDIF
 100       CONTINUE
 60     CONTINUE

C *** PRINT RESULT
      WRITE(*,'(1X,''RESULTING MATRIX COLUMNS ARE'')')
      DO 120 J=1, N
         WRITE(*,'(1X,''COLUMN ''I2)') J
         DO 130 I=1, N
            WRITE(*,'(1X,E15.8)') B(I,J)
 130     CONTINUE
 120  CONTINUE

      STOP
      END
```

```
COLUMN 1
ROW 1?2
ROW 2?-1
ROW 3?3
COLUMN 2
ROW 1?2
ROW 2?0
ROW 3?-1
COLUMN 3
ROW 1?-1
ROW 2?4
ROW 3?-3
RESULTING MATRIX COLUMNS ARE
COLUMN 1
.16
.36
.04
COLUMN 2
.28
-.12
.32
COLUMN 3
.32
-.28
.08
```

or

$$\mathbf{A}^{-1} = \begin{bmatrix} 0.16 & 0.28 & 0.32 \\ 0.36 & -0.12 & -0.28 \\ 0.04 & 0.32 & 0.08 \end{bmatrix}$$

6.4.2 Equation Solution Program

Because it often happens that we wish to solve a set of n consistent, linearly independent linear algebraic equations in n unknowns, it is appropriate now to design a computer program for that purpose. The Gauss–Jordan pivoting program, PIVOT, of Table 2-4 can be used for this purpose, but it is really much more general. PIVOT accommodates a different number of equations than unknowns and identifies linearly dependent and inconsistent equations. But to make it possible to easily identify which equations are linearly dependent on or inconsistent with the others, their order is not rearranged so that a solution, when it exists, is not necessarily printed in the order of the variable subscripts.

The program called SOLVE in Table 6-7 is a modification of the Inverse program that gives the solution to n equations in n unknowns whenever it exists. It accepts the array of equation coefficients and computes its inverse as in the INVERSE program. At the same time, the determinant of the array is computed and printed so that the user can identify those situations where the determinant is extremely small but nonzero. Then the vector of knowns is entered and the solution vector computed and printed.

For the equations

$$\begin{cases} 2x_1 + x_2 - x_3 = -3 \\ -3x_1 - x_2 + 2x_3 = 5 \\ -x_1 + 4x_2 - 3x_3 = 0 \end{cases}$$

TABLE 6-7

Computer Program in BASIC to Solve *N* Equations in *N* Unknowns (SOLVE)

This is a modification of the INVERSE program of Table 6-6. The determinant of the equation coefficients is computed as **D** and printed. The vector of knowns is stored as the first row of the array **A**(1,20) after **A** has been used to compute the inverse matrix.

LISTING

```
100      PRINT "SIMULTANEOUS EQUATION SOLUTION"
110      DIM A(20,20),B(20,20)
120   REM ENTER DIMENSION AND ELEMENTS OF A
130      PRINT "ENTER NO. OF EQUATIONS AND UNKNOWNS";
140      INPUT N
150      PRINT "ENTER COEFFICIENTS, COLUMN BY COLUMN"
160      FOR J = 1 TO N
170      PRINT "COLUMN";J
180      FOR I = 1 TO N
190      PRINT "ROW";I;
200      INPUT A(I,J)
210      NEXT I
220      NEXT J
230   REM SET B TO THE IDENTITY MATRIX
240      FOR J = 1 TO N
250      FOR I = 1 TO N
260      B(I,J) = 0
270      NEXT I
280      NEXT J
290      FOR I = 1 TO N
300      B(I,I) = 1
310      NEXT I
311   REM SET D TO UNITY
312      D = 1
320   REM FOR EACH COLUMN
330      FOR J = 1 TO N
340   REM FIND THE LARGEST MAGNITUDE ELEMENT IN
350   REM COLUMN J IN ROW J AND BELOW
360      P =0
370      FOR I = J TO N
380      S = A(I,J)*A(I,J)
390      IF S > P THEN Q = I
400      IF S > P THEN P = S
410      NEXT I
420      IF P = 0 THEN GOTO 880
430   REM INTERCHANGE ROWS J AND Q
440      IF J = Q THEN GOTO 530
450      FOR K = 1 TO N
460      S = A(J,K)
470      T = B(J,K)
480      A(J,K)=A(Q,K)
490      B(J,K)=B(Q,K)
500      A(Q,K) =S
510      B(Q,K)=T
520      NEXT K
521      D = -D
530   REM NORMALIZE ROW J
540      S = A(J,J)
550      FOR K = 1 TO N
560      A(J,K)=A(J,K)/S
570      B(J,K)=B(J,K)/S
580      NEXT K
581      D = S*D
590   REM PIVOT ON THE A(J,J) COEFFICIENT FOR
600   REM EACH ROW EXCEPT ROW J
```

```
610        FOR I = 1 TO N
620        IF I = J THEN GOTO 680
630        S = A(I,J)
640        FOR K = 1 TO N
650        A(I,K)= A(I,K)-S*A(J,K)
660        B(I,K)= B(I,K)-S*B(J,K)
670        NEXT K
680        NEXT I
690        NEXT J
700     REM PRINT DETERMINANT
710        PRINT "DETERMINANT IS ";D
720     REM ENTER KNOWNS
730        PRINT "ENTER COLUMN OF KNOWNS"
740        FOR I = 1 TO N
750        PRINT "ROW";I;
760        INPUT A(1,I)
770        NEXT I
780     REM FIND AND PRINT SOLUTION
790        PRINT "SOLUTION VECTOR IS"
800        FOR I = 1 TO N
810        S=0
820        FOR J = 1 TO N
830        S = S+B(I,J) * A(1,J)
840        NEXT J
850        PRINT S
860        NEXT I
870        GOTO 900
880     REM PRINT NO SOLUTION MESSAGE
890        PRINT "COEFFICIENT ARRAY IS SINGULAR"
900        END
```

Computer Program in FORTRAN to Solve *N* Equations in *N* Unknowns (SOLVE)

```
C ********************************************************************
C PROGRAM SOLVE -- SIMULTANEOUS EQUATION SOLUTION
C ********************************************************************
      REAL A(20,20), B(20,20)

      WRITE(*,'(1X,''SIMULTANEOUS EQUATION SOLUTION'')')

C *** ENTER DIMENSION AND ELEMENTS OF A
      WRITE(*,'(1X,''ENTER NO. OF EQUATIONS AND UNKNOWNS '',\)')
      READ(*,*) N
      WRITE(*,'(1X,''ENTER COEFFICIENTS, COLUMN BY COLUMN'')')
      DO 10 J=1, N
         WRITE(*,'(1X,''COLUMN '',I2)') J
         DO 20 I=1, N
            WRITE(*,'(1X,''ROW '',I2,'' = '',\)') I
            READ(*,*) A(I,J)
20       CONTINUE
10    CONTINUE

C *** SET B TO THE IDENTITY MATRIX
      DO 30 J=1, N
         DO 40 I=1, N
            B(I,J) = 0.0
40       CONTINUE
30    CONTINUE
      DO 50 I=1, N
         B(I,I) = 1.0
50    CONTINUE

C *** SET D TO UNITY
      D = 1.0
```

(continued)

```
C *** FOR EACH COLUMN
      DO 60 J=1, N

C *** FIND THE LARGEST MAGNITUDE ELEMENT IN
C *** COLUMN J IN ROW J AND BELOW
          P = 0.0
          DO 70 I=J, N
             S = A(I,J)*A(I,J)
             IF (S.GT.P) THEN
                 Q = I
                 P = S
             ENDIF
   70     CONTINUE

C *** PRINT NO SOLUTION MESSAGE
          IF (P.EQ.0.0) THEN
             WRITE(*,'(1X,''COEFFICIENT ARRAY IS SINGULAR'')')
             STOP
          ENDIF

C *** INTERCHANGE ROWS J AND Q
          IF (J.NE.Q) THEN
             DO 80 K=1, N
                S = A(J,K)
                T = B(J,K)
                A(J,K)=A(Q,K)
                B(J,K)=B(Q,K)
                A(Q,K)=S
                B(Q,K)=T
   80        CONTINUE
             D = -D
          ENDIF

C *** NORMALIZE ROW J
          S = A(J,J)
          DO 90 K=1, N
             A(J,K)=A(J,K)/S
             B(J,K)=B(J,K)/S
   90     CONTINUE
          D = S*D

C *** PIVOT ON THE A(J,J) COEFFICIENT FOR
C *** EACH ROW EXCEPT ROW J
          DO 100 I=1, N
             IF (I.EQ.J) GOTO 100
             S = A(I,J)
             DO 110 K=1, N
                A(I,K)= A(I,K) - S*A(J,K)
                B(I,K)= B(I,K) - S*B(J,K)
  110        CONTINUE
  100     CONTINUE
   60 CONTINUE

C *** PRINT DETERMINANT
      WRITE(*,'(1X,''DETERMINANT IS '',E15.8)') D

C *** ENTER KNOWNS
      WRITE(*,'(1X,''ENTER COLUMN OF KNOWNS'')')
      DO 120 I=1, N
         WRITE(*,'(1X,''ROW '',I2,'' = '',\)') I
         READ(*,*) A(1,I)
  120 CONTINUE

C *** FIND AND PRINT SOLUTION
      WRITE(*,'(1X,''SOLUTION VECTOR IS'')')
```

```
      DO 130 I=1, N
         S = 0.0
         DO 140 J=1, N
            S = S + B(I,J)*A(1,J)
140      CONTINUE
         WRITE(*,'(1X,E15.8)') S
130   CONTINUE
      STOP
      END
```

the SOLVE program gives

```
SOLUTION VECTOR IS
-1.25
.25
.750000001
```

6.5 LU DECOMPOSITION METHOD

In the preceding section, solutions of linear algebraic equations and matrix inversion were obtained by means of elementary row or column interchanges performed by matrix multiplication. In this section we discuss a method that does not require any row or column interchanges and thus gives faster solutions than the methods discussed previously. This method is called the *LU decomposition*, the matrix factorization method, or the method of triangularization. It is popular and has been used extensively in those areas, such as structural analysis, where very large systems of equations are encountered.

6.5.1 The Method

Let **A** be a known $n \times n$ matrix of the form

$$\mathbf{A} = \begin{bmatrix} a_{11} & a_{12} & a_{13} & \cdots & a_{1n} \\ a_{21} & a_{22} & a_{23} & \cdots & a_{2n} \\ a_{31} & a_{32} & a_{33} & \cdots & a_{3n} \\ \vdots & \vdots & \vdots & \vdots & \vdots \\ a_{n1} & a_{n2} & a_{n3} & \cdots & a_{nn} \end{bmatrix}$$

It is desired to decompose **A**, if possible, into two matrices **L** and **U**, such that

$$\mathbf{A} = \mathbf{LU} \tag{6-2}$$

where **L** and **U** are lower and upper triangular matrices, which have the forms

$$\mathbf{L} = \begin{bmatrix} l_{11} & 0 & 0 & \cdots & 0 \\ l_{21} & l_{22} & 0 & \cdots & 0 \\ l_{31} & l_{32} & l_{33} & \cdots & 0 \\ \vdots & \vdots & \vdots & \vdots & \vdots \\ l_{n1} & l_{n2} & l_{n3} & \cdots & l_{nn} \end{bmatrix} \qquad \mathbf{U} = \begin{bmatrix} 1 & u_{12} & u_{13} & \cdots & u_{1n} \\ 0 & 1 & u_{23} & \cdots & u_{2n} \\ 0 & 0 & 1 & \cdots & u_{3n} \\ \vdots & \vdots & \vdots & \vdots & \vdots \\ 0 & 0 & 0 & \cdots & 1 \end{bmatrix} \tag{6-3}$$

In other words, we would like to determine the unknown elements of **L** and **U** such that equation (6-2) is satisfied. The unknown elements of **L** and **U** can be determined by directly multiplying $\mathbf{L}$ by $\mathbf{U}$ and equating the product to the corresponding elements of $\mathbf{A}$. The key idea behind the LU decomposition method is that such sequence of operations should be performed as in the following steps.

First, we determine the elements in the first column of $\mathbf{L}$. It is obviously by direct multiplication that the first column of $\mathbf{L}$ is identical to the first column of $\mathbf{A}$. Hence we can write

$$l_{i1} = a_{i1} \qquad i = 1, 2, \ldots, n$$

The second step is to determine the unknown elements in the first row of $\mathbf{U}$. Multiplying the first row of $\mathbf{L}$ by all the columns of $\mathbf{U}$ except the first and equating to the corresponding elements of $\mathbf{A}$, we note that

$$u_{1j} = \frac{a_{1j}}{l_{11}} \qquad j = 2, 3, \ldots, n$$

but since $l_{11} = a_{11}$, then

$$u_{1j} = \frac{a_{1j}}{a_{11}} \qquad j = 2, 3, \ldots, n$$

The third step is to solve for the elements in the second column of $\mathbf{L}$. Multiplying all the rows of $\mathbf{L}$ except the first by the second column of $\mathbf{U}$ and equating to the corresponding elements of $\mathbf{A}$ gives

$$l_{21}u_{12} + l_{22} = a_{22}$$
$$l_{31}u_{12} + l_{32} = a_{32}$$
$$\vdots$$
$$l_{n1}u_{12} + l_{n2} = a_{n2}$$

or

$$l_{i2} = a_{i2} - l_{i1}u_{12} \qquad i = 2, 3, \ldots, n$$

The fourth step is to compute the unknown elements in the second row of $\mathbf{U}$. Multiplying the second row of $\mathbf{L}$ by all the columns of $\mathbf{U}$ except the first two columns and equating to the corresponding elements of $\mathbf{A}$, we note that

$$l_{21}u_{13} + l_{22}u_{23} = a_{23}$$
$$l_{21}u_{14} + l_{22}u_{24} = a_{24}$$
$$\vdots$$
$$l_{21}u_{1n} + l_{22}u_{2n} = a_{2n}$$

or

$$u_{2j} = \frac{a_{2j} - l_{21}u_{1j}}{l_{22}} \qquad j = 3, 4, \ldots, n$$

Continuing this process by determining the third column of $\mathbf{L}$ followed by the third row of $\mathbf{U}$, the fourth column of $\mathbf{L}$ followed by the fourth row of $\mathbf{U}$, and so on until all the unknown elements of $\mathbf{L}$ and $\mathbf{U}$ are found.

A general algorithm for determining the elements of **L** and **U** is given in the following set of equations:

$$l_{i1} = a_{i1} \qquad\qquad\qquad (i = 1, 2, \ldots, n)$$

$$u_{1j} = \frac{a_{1j}}{a_{11}} \qquad\qquad\qquad (j = 2, 3, \ldots, n)$$

$$l_{ij} = a_{ij} - \sum_{k=1}^{j-1} l_{ik}u_{kj} \qquad \begin{cases} (j = 2, 3, \ldots, n) \\ (i = j, j+1, \ldots, n) \\ \text{incrementing } i \text{ for every } j \end{cases}$$

$$u_{ij} = \frac{1}{l_{ii}} \left(a_{ij} - \sum_{k=1}^{i-1} l_{ik}u_{kj} \right) \qquad \begin{cases} (i = 2, 3, \ldots, n) \\ (j = i+1, i+2, \ldots, n) \\ \text{incrementing } j \text{ for every } i \end{cases}$$

(6-4)

To demonstrate the steps involved in decomposing a matrix using the algorithm above, consider the following example. Suppose that it is desired to decompose **A** as shown:

$$\mathbf{A} = \begin{bmatrix} -1 & -2 & -4 & 1 \\ 2 & 7 & 14 & 4 \\ 1 & 4 & 9 & 6 \\ 4 & 10 & 17 & -5 \end{bmatrix} = \begin{bmatrix} l_{11} & 0 & 0 & 0 \\ l_{21} & l_{22} & 0 & 0 \\ l_{31} & l_{32} & l_{33} & 0 \\ l_{41} & l_{42} & l_{43} & l_{44} \end{bmatrix} \begin{bmatrix} 1 & u_{12} & u_{13} & u_{14} \\ 0 & 1 & u_{23} & u_{24} \\ 0 & 0 & 1 & u_{34} \\ 0 & 0 & 0 & 1 \end{bmatrix}$$

Applying the first equation of (6-4) over its range i gives

$$l_{11} = -1 \qquad l_{21} = 2 \qquad l_{31} = 1 \qquad l_{41} = 4$$

Applying the second equation of (6-4) over its range j gives

$$u_{12} = \frac{-2}{-1} = 2 \qquad u_{13} = 4 \qquad u_{14} = -1$$

Applying the third equation of (6-4) with $j = 2$ and i incremented from 2 to 4 gives

$$l_{22} = a_{22} - l_{21}u_{12} = 7 - (2)(2) = 3$$

$$l_{32} = a_{32} - l_{31}u_{12} = 4 - (1)(2) = 2$$

$$l_{42} = a_{42} - l_{41}u_{12} = 10 - (4)(2) = 2$$

Applying the fourth equation of (6-4) with $i = 2$ and j incremented from 3 to 4 gives

$$u_{23} = \frac{1}{l_{22}}(a_{23} - l_{21}u_{13}) = \tfrac{1}{3}(14 - (2)(4)) = 2$$

$$u_{24} = \frac{1}{l_{22}}(a_{24} - l_{21}u_{14}) = \tfrac{1}{3}(4 - (2)(-1)) = 2$$

Applying the third equation of (6-4) with $j = 3$ and i incremented from 3 to 4 gives

$$I_{33} = a_{33} - (l_{31}u_{13} + l_{32}u_{23}) = 9 - (4+4) = 1$$

$$l_{43} = a_{43} - (l_{41}u_{13} + l_{42}u_{23}) = 17 - (16+4) = -3$$

Applying the fourth equation of (6-4) with $i = 3$ and j incremented from 4 to 4 gives

$$u_{34} = \frac{1}{l_{33}} [a_{34} - (l_{31}u_{14} + l_{32}u_{24})] = 3$$

Finally, applying the third equation of (6-4) with $j = 4$ and i incremented from 4 to 4 gives

$$l_{44} = a_{44} - (l_{41}u_{14} + l_{42}u_{24} + l_{43}u_{34})$$

$$l_{44} = -5 - [(4)(-1) + (2)(2) + (-3)(3)] = 4$$

Examining the second and fourth equations of (6-4) shows that the decomposition of **A** is possibly only if $l_{ii} \neq 0$ for all i and $a_{11} \neq 0$. Using properties (6) and (7) of Table 6-2, the determinant of the matrix **A** as given by

$$|\mathbf{A}| = |\mathbf{L}| \, |\mathbf{U}|$$

equals the product of the diagonal elements of **L** times the product of the diagonal elements of **U**, that is,

$$|\mathbf{A}| = l_{11}l_{22}l_{33} \cdots l_{nn}$$

Therefore, if the matrix **A** is nonsingular, as is usually the case when solving an independent set of linear algebraic equations or inverting a matrix, all l_{ii} must be nonzero for all i. Furthermore, we have to assure that $a_{11} \neq 0$. If $a_{11} = 0$, we can interchange rows or columns of **A** so that $a_{11} \neq 0$. For example, the matrix

$$\mathbf{A} = \begin{bmatrix} 0 & 2 \\ 3 & 1 \end{bmatrix}$$

is nonsingular, but the decomposition

$$\mathbf{LU} = \begin{bmatrix} l_{11} & 0 \\ l_{21} & l_{22} \end{bmatrix} \begin{bmatrix} 1 & u_{12} \\ 0 & 1 \end{bmatrix}$$

implies that $l_{11} = 0$. Interchanging the two rows of **A** gives

$$\mathbf{A}' = \begin{bmatrix} 3 & 1 \\ 0 & 2 \end{bmatrix}$$

from which we obtain

$$\mathbf{L}'\mathbf{U}' = \begin{bmatrix} 3 & 0 \\ 0 & 2 \end{bmatrix} \begin{bmatrix} 1 & \frac{1}{3} \\ 0 & 1 \end{bmatrix}$$

Using the elementary operations of row interchange, therefore,

$$\mathbf{A} = \begin{bmatrix} 0 & 1 \\ 1 & 0 \end{bmatrix} \begin{bmatrix} 3 & 0 \\ 0 & 2 \end{bmatrix} \begin{bmatrix} 1 & \frac{1}{3} \\ 0 & 1 \end{bmatrix}$$

6.5.2 Solution of Linear Algebraic Equations

The LU decomposition method of finding solutions of a set of linear algebraic equations involves replacing the original set of equations with two new sets that give the same solution but are simpler to solve than the original set. Consider a set of n simultaneous linear independent algebraic equations in n variables that has the form

$$\mathbf{Ax} = \mathbf{b}$$

If the matrix $\mathbf{A}$ is decomposed into lower and upper triangular matrices such that

$$\mathbf{A} = \mathbf{LU}$$

where $\mathbf{L}$ and $\mathbf{U}$ are given by (6-3), then the solution of the equations

$$\mathbf{Ax} = \mathbf{L(Ux)} = \mathbf{b}$$

can be obtained by solving the two simpler sets of equations

$$\mathbf{Ly} = \mathbf{b} \tag{6-5a}$$

$$\mathbf{Ux} = \mathbf{y} \tag{6-5b}$$

The system of equations in (6-5a) is written as

$$\begin{bmatrix} l_{11} & 0 & 0 & \cdots & 0 \\ l_{21} & l_{22} & 0 & \cdots & 0 \\ l_{31} & l_{32} & l_{33} & \cdots & 0 \\ \vdots & \vdots & \vdots & \vdots & \vdots \\ l_{n1} & l_{n2} & l_{n3} & \cdots & l_{nn} \end{bmatrix} \begin{bmatrix} y_1 \\ y_2 \\ y_3 \\ \vdots \\ y_n \end{bmatrix} = \begin{bmatrix} b_1 \\ b_2 \\ b_3 \\ \vdots \\ b_n \end{bmatrix}$$

and can easily be solved for the y unknowns by forward substitution:

$$y_1 = \frac{b_1}{l_{11}}$$

$$y_2 = \frac{1}{l_{22}} (b_2 - l_{21} y_1)$$

$$y_3 = \frac{1}{l_{33}} (b_3 - l_{31} y_1 - l_{32} y_2)$$

$$\vdots$$

$$y_n = \frac{1}{l_{nn}} (b_n - l_{n1} y_1 - l_{n2} y_2 - \cdots - l_{n-1} y_{n-1})$$

or

$$y_1 = \frac{b_1}{l_{11}} \qquad\qquad (i = 1)$$

$$y_i = \frac{1}{l_{ii}} \left[b_i - \sum_{j=1}^{i-1} l_{ij} y_j \right] \qquad\qquad (i = 2, 3, \ldots, n)$$

Substituting these values of y in (6-5b) gives the system of equations

$$\begin{bmatrix} 1 & u_{12} & u_{13} & \cdots & \cdots & u_{1n} \\ 0 & 1 & u_{23} & \cdots & \cdots & u_{2n} \\ 0 & 0 & 1 & u_{34} & \cdots & u_{3n} \\ \vdots & \vdots & \vdots & \vdots & \vdots & \vdots \\ \cdots & \cdots & \cdots & \cdots & 1 & u_{(n-1,n)} \\ 0 & 0 & 0 & \cdots & \cdots & 1 \end{bmatrix} \begin{bmatrix} x_1 \\ x_2 \\ x_3 \\ \vdots \\ \\ x_n \end{bmatrix} = \begin{bmatrix} y_1 \\ y_2 \\ y_3 \\ \vdots \\ \\ y_n \end{bmatrix}$$

This system of equations can be solved for the x unknowns by simple backward substitution:

$$x_n = y_n$$

$$x_{n-1} = y_{n-1} - u_{(n-1,n)}x_n$$

$$\vdots$$

$$x_1 = y_1 - u_{12}x_2 - u_{13}x_3 - \cdots - u_{1n}x_n$$

or, in compact form,

$$x_n = y_n \qquad\qquad (i = n)$$

$$x_i = y_i - \sum_{j=i+1}^{n} u_{ij}x_j \qquad (i = n-1, n-2, \ldots, 2, 1)$$

As a numerical example, consider the system of equations

$$\begin{aligned} -x_1 - 2x_2 - 4x_3 + x_4 &= -10 \\ 2x_1 + 7x_2 + 14x_3 + 4x_4 &= 26 \\ x_1 + 4x_2 + 9x_3 + 6x_4 &= 13 \\ 4x_1 + 10x_2 + 17x_3 - 5x_4 &= 43 \end{aligned}$$

which has the matrix representation

$$\begin{bmatrix} -1 & -2 & -4 & 1 \\ 2 & 7 & 14 & 4 \\ 1 & 4 & 9 & 6 \\ 4 & 10 & 17 & -5 \end{bmatrix} \begin{bmatrix} x_1 \\ x_2 \\ x_3 \\ x_4 \end{bmatrix} = \begin{bmatrix} -10 \\ 26 \\ 13 \\ 43 \end{bmatrix}$$

Decomposing the coefficient matrix $\mathbf{A}$ into upper and lower triangular matrices as in the preceding example and solving equations (6-5a) and (6-5b), respectively, gives

$$\begin{bmatrix} -1 & 0 & 0 & 0 \\ 2 & 3 & 0 & 0 \\ 1 & 2 & 1 & 0 \\ 4 & 2 & -3 & 4 \end{bmatrix} \begin{bmatrix} y_1 \\ y_2 \\ y_3 \\ y_4 \end{bmatrix} = \begin{bmatrix} -10 \\ 26 \\ 13 \\ 43 \end{bmatrix}$$

$$y_1 = 10 \qquad y_2 = 2 \qquad y_3 = -1 \qquad y_4 = -1$$

Substituting these values in (6-5b), then, gives

$$\begin{bmatrix} 1 & 2 & 4 & -1 \\ 0 & 1 & 2 & 2 \\ 0 & 0 & 1 & 3 \\ 0 & 0 & 0 & 1 \end{bmatrix} \begin{bmatrix} x_1 \\ x_2 \\ x_3 \\ x_4 \end{bmatrix} = \begin{bmatrix} 10 \\ 2 \\ -1 \\ -1 \end{bmatrix}$$

and therefore,

$$x_4 = -1 \qquad x_3 = 2 \qquad x_2 = 0 \qquad x_1 = 1$$

in this order.

From a numerical standpoint, a more efficient way to solve a set of n simultaneous linear independent algebraic equations in n variables using the LU decomposition method is to combine the set of equations

$$\mathbf{A} = \mathbf{LU}$$

and

$$\mathbf{Ly} = \mathbf{b}$$

into a single matrix equation:

$$\begin{bmatrix} l_{11} & 0 & 0 & \cdots & 0 \\ l_{21} & l_{22} & 0 & \cdots & 0 \\ l_{31} & l_{32} & l_{33} & \cdots & 0 \\ \vdots & \vdots & \vdots & \vdots & \vdots \\ l_{n1} & l_{n2} & l_{n3} & \cdots & l_{nn} \end{bmatrix} \begin{bmatrix} 1 & u_{12} & u_{13} & \cdots & u_{1n} & y_1 \\ 0 & 1 & u_{23} & \cdots & u_{2n} & y_2 \\ 0 & 0 & 1 & \cdots & u_{3n} & y_3 \\ \vdots & \vdots & \vdots & \vdots & \vdots & \vdots \\ 0 & 0 & 0 & \cdots & 1 & y_n \end{bmatrix}$$

$$= \begin{bmatrix} a_{11} & a_{12} & a_{13} & \cdots & a_{1n} & b_1 \\ a_{21} & a_{22} & a_{23} & \cdots & a_{2n} & b_2 \\ a_{31} & a_{32} & a_{33} & \cdots & a_{3n} & b_3 \\ \vdots & \vdots & \vdots & \vdots & \vdots & \vdots \\ a_{n1} & a_{n2} & a_{n3} & \cdots & a_{nn} & b_n \end{bmatrix}$$

(6-6)

If, for convenience, we rename the columns

$$\begin{bmatrix} y_1 \\ y_2 \\ y_3 \\ \vdots \\ y_n \end{bmatrix} \quad \text{and} \quad \begin{bmatrix} b_1 \\ b_2 \\ b_3 \\ \vdots \\ b_n \end{bmatrix}$$

as

$$\begin{bmatrix} u_{(1,n+1)} \\ u_{(2,n+1)} \\ u_{(3,n+1)} \\ \vdots \\ u_{(n,n+1)} \end{bmatrix} \quad \text{and} \quad \begin{bmatrix} a_{(1,n+1)} \\ a_{(2,n+1)} \\ a_{(3,n+1)} \\ \vdots \\ a_{(n,n+1)} \end{bmatrix}$$

respectively, then

$$
\begin{bmatrix}
l_{11} & 0 & 0 & \cdots & 0 \\
l_{21} & l_{22} & 0 & \cdots & 0 \\
l_{31} & l_{32} & l_{33} & \cdots & 0 \\
\vdots & \vdots & \vdots & \vdots & \vdots \\
l_{n1} & l_{n2} & l_{n3} & \cdots & l_{nn}
\end{bmatrix}
\begin{bmatrix}
1 & u_{12} & u_{13} & \cdots & u_{1n} & u_{(1,n+1)} \\
0 & 1 & u_{23} & \cdots & u_{2n} & u_{(2,n+1)} \\
0 & 0 & 1 & \cdots & u_{3n} & u_{(3,n+1)} \\
\vdots & \vdots & \vdots & \vdots & \vdots & \vdots \\
0 & 0 & 0 & \cdots & 1 & u_{(n,n+1)}
\end{bmatrix}
$$

$$
=
\begin{bmatrix}
a_{11} & a_{12} & a_{13} & \cdots & a_{1n} & a_{(1,n+1)} \\
a_{21} & a_{22} & a_{23} & \cdots & a_{2n} & a_{(2,n+1)} \\
a_{31} & a_{32} & a_{33} & \cdots & a_{3n} & a_{(3,n+1)} \\
\vdots & \vdots & \vdots & \vdots & \vdots & \vdots \\
a_{n1} & a_{n2} & a_{n3} & \cdots & a_{nn} & a_{(n,n+1)}
\end{bmatrix}
$$

Because we are only interested in the values of x, there is no need to store the matrices **L** and **U** separately in the computer. Rather, we let **L** and **U** overwrite the matrix **A** by first storing the elements a_{ij} of the augmented **A** matrix and then replacing those elements with the computed elements of the **L** and **U** matrices. The algorithm in (6-4) is, therefore, modified as follows:

$$
a_{1j} = \frac{a_{1j}}{a_{11}} \qquad\qquad (j = 2, 3, \ldots, n + 1)
$$

$$
a_{ij} = a_{ij} - \sum_{k=1}^{j-1} a_{ik}a_{kj} \qquad
\begin{cases}
(j = 2, 3, \ldots, n) \\
(i = j, j + 1, \ldots, n) \\
\text{incrementing } i \text{ for every } j
\end{cases} \qquad (6\text{-}7)
$$

$$
a_{ij} = \frac{1}{a_{ii}}\left(a_{ij} - \sum_{k=1}^{i-1} a_{ik}a_{kj} \right) \qquad
\begin{cases}
(i = 2, 3, \ldots, n) \\
(j = i + 1, i + 2, \ldots, n + 1) \\
\text{incrementing } j \text{ for every } i
\end{cases}
$$

The third equation in (6-7) is all that is needed to solve for the x unknowns because it represents the augmented **U** matrix, which contains both the u elements and the y elements. Making sure that the augmented **U** matrix is the last to overwrite the augmented **A** matrix, the x unknowns are easily determined using the backward substitution formula,

$$
x_n = a_{(n,n+1)}
$$

$$
x_i = x_{(i,n+1)} - \sum_{j=i+1}^{n} a_{ij}x_j \qquad (i = n - 1, n - 2, \ldots, 1)
$$

A computer program that uses this algorithm for solving n consistent linear, independent algebraic equations in n variables is left as an exercise.

6.5.3 Matrix Inversion

The LU decomposition method may also be used to compute the inverse of a nonsingular matrix. As in the case of solving linear algebraic equations, the procedure of finding the inverse of a nonsingular matrix involves replacing the original matrix with lower and upper triangular matrices, which are simpler to invert than the original matrix. Consider a nonsingular matrix **A**. If **A** is decomposed into lower and upper triangular matrices such that

$$\mathbf{A} = \mathbf{LU}$$

where **L** and **U** have the forms given in (6-3), then the inverse of **A** is

$$\mathbf{A}^{-1} = (\mathbf{LU})^{-1} = \mathbf{U}^{-1}\mathbf{L}^{-1}.$$

Inverses of triangular matrices are easy to compute, because the inverse of a lower triangular matrix is itself lower triangular, and the inverse of an upper triangular matrix is also an upper triangular matrix. To show this, we use the defining property of matrix inversion. For the matrix **L**,

$$\begin{bmatrix} l_{11} & 0 & 0 & \cdots & 0 \\ l_{21} & l_{22} & 0 & \cdots & 0 \\ l_{31} & l_{32} & l_{33} & \cdots & 0 \\ \vdots & \vdots & \vdots & \vdots & \vdots \\ l_{n1} & l_{n2} & l_{n3} & \cdots & l_{nn} \end{bmatrix} \begin{bmatrix} z_{11} & 0 & 0 & \cdots & 0 \\ z_{21} & z_{22} & 0 & \cdots & 0 \\ z_{31} & z_{32} & z_{33} & \cdots & 0 \\ \vdots & \vdots & \vdots & \vdots & \vdots \\ z_{n1} & z_{n2} & z_{n3} & \cdots & z_{nn} \end{bmatrix}$$

$$= \begin{bmatrix} 1 & 0 & 0 & \cdots & 0 \\ 0 & 1 & 0 & \cdots & 0 \\ 0 & 0 & 1 & \cdots & 0 \\ \vdots & \vdots & \vdots & \vdots & \vdots \\ 0 & 0 & 0 & \cdots & 1 \end{bmatrix}$$

where the z symbols are elements of the **Z** matrix which is the inverse of **L**. The z unknowns can be determined by direct multiplication. Multiplying all the rows of **L** by the first column of **Z** and equating to the first column of the identity matrix gives

$$l_{11}z_{11} = 1$$

$$l_{21}z_{11} + l_{22}z_{21} = 0$$

$$l_{31}z_{11} + l_{32}z_{21} + l_{33}z_{31} = 0$$

$$\vdots$$

$$l_{n1}z_{11} + l_{n2}z_{21} + 1_{n3}z_{31} + \cdots + l_{nn}z_{n1} = 0$$

Multiplying all the rows of **L** by the second column of **Z** and equating the product to the second column of the identity matrix, we note that

$$l_{22}z_{22} = 1$$

$$l_{32}z_{22} + l_{33}z_{32} = 0$$

$$\vdots$$

$$l_{n2}z_{22} + l_{n3}z_{32} + \cdots + l_{nn}z_{n2} = 0$$

Continuing this process, then,

$$l_{33}z_{33} = 1$$

$$l_{43}z_{33} + l_{44}z_{43} = 0$$

$$\vdots$$

$$l_{n3}z_{33} + l_{n4}z_{43} + \cdots + l_{nn}z_{n3} = 0$$

and finally,

$$l_{nn}z_{nn} = 1$$

Examining these equations, the elements of the matrix

$$\mathbf{Z} = \mathbf{L}^{-1}$$

can therefore be determined recursively:

$$z_{ii} = \frac{1}{l_{ii}} \qquad\qquad (i = 1, 2, \ldots, n)$$

$$z_{ij} = -\frac{\sum\limits_{k=j}^{i-1} l_{ik}z_{kj}}{l_{ii}} \qquad \begin{cases} (j = 1, 2, \ldots, n-1) \\ (i = j+1, j+2, \ldots, n) \\ \text{incrementing } i \text{ for every } j \end{cases} \qquad (6\text{-}8)$$

$$z_{ij} = 0 \qquad\qquad (i < j)$$

provided that $l_{ii} \neq 0$ for all i. We showed in the preceding section that if $\mathbf{A}$ is nonsingular, all the diagonal elements l_{ii} of the $\mathbf{L}$ matrix are nonzero. Thus the division by l_{ii} is valid.

Similarly, the inverse of the upper triangular matrix $\mathbf{U}$ is also an upper triangular matrix, $\mathbf{V}$. Using the defining property of matrix inversion, the elements of $\mathbf{V}$ are determined by direct multiplication of $\mathbf{U}$ and $\mathbf{V}$ and equating the product to the identity matrix:

$$
\begin{bmatrix}
1 & u_{12} & u_{13} & \cdots & u_{1n} \\
0 & 1 & u_{23} & \cdots & u_{2n} \\
0 & 0 & 1 & \cdots & u_{3n} \\
\vdots & \vdots & \vdots & \vdots & \vdots \\
0 & 0 & 0 & \cdots & 1
\end{bmatrix}
\begin{bmatrix}
v_{11} & v_{12} & v_{13} & \cdots & v_{1n} \\
0 & v_{22} & v_{23} & \cdots & v_{2n} \\
0 & 0 & v_{33} & \cdots & v_{3n} \\
\vdots & \vdots & \vdots & \vdots & \vdots \\
0 & 0 & 0 & \cdots & v_{nn}
\end{bmatrix}
$$

$$
=
\begin{bmatrix}
1 & 0 & 0 & \cdots & 0 \\
0 & 1 & 0 & \cdots & 0 \\
0 & 0 & 1 & \cdots & 0 \\
\vdots & \vdots & \vdots & \vdots & \vdots \\
0 & 0 & 0 & \cdots & 1
\end{bmatrix}
$$

The result is the following recursive set of equations:

$$v_{ii} = 1 \qquad\qquad (i = 1, 2, \ldots, n)$$

$$v_{ij} = - \sum_{k=i+1}^{j} u_{ik} v_{kj} \qquad \begin{cases} (i = 1, 2, \ldots, n - 1) \\ (j = i + 1, i + 2, \ldots, n) \\ \text{incrementing } i \text{ for every } j \end{cases} \qquad (6\text{-}9)$$

$$v_{ij} = 0 \qquad\qquad (i > j)$$

An example of matrix inversion using the LU decomposition method is as follows:

$$\mathbf{A} = \begin{bmatrix} -1 & -2 & -4 & 1 \\ 2 & 7 & 14 & 4 \\ 1 & 4 & 9 & 6 \\ 4 & 10 & 17 & -5 \end{bmatrix}$$

As in the preceding example, the matrix $\mathbf{A}$ is decomposed into the following $\mathbf{L}$ and $\mathbf{U}$ matrices:

$$\mathbf{L} = \begin{bmatrix} -1 & 0 & 0 & 0 \\ 2 & 3 & 0 & 0 \\ 1 & 2 & 1 & 0 \\ 4 & 2 & -3 & 4 \end{bmatrix} \qquad \mathbf{U} = \begin{bmatrix} 1 & 2 & 4 & -1 \\ 0 & 1 & 2 & 2 \\ 0 & 0 & 1 & 3 \\ 0 & 0 & 0 & 1 \end{bmatrix}$$

To determine $\mathbf{A}^{-1}$, we need $\mathbf{L}^{-1}$ and $\mathbf{U}^{-1}$.

The elements of $\mathbf{L}^{-1}$ are determined using the set of equations in (6-8). First, we determine the diagonal elements $z_{11}, z_{22}, z_{33}, z_{44}$. Second, starting with $j = 1$ and i ranging from 2 to 4 in the second equation, we compute z_{21}, z_{31}, and z_{41}. Next, the second equation is reapplied with $j = 2$ and i incremented from 3 to 4 to compute z_{32}, and z_{42}. Finally, with $j = 3$ and $i = 4$, the second equation gives z_{43}. The remaining elements of the $\mathbf{Z}$ matrix are set to zero using the third equation of (6-8).

In a similar fashion, the elements of $\mathbf{U}^{-1}$ are determined using the set of equations in (6-9). First, we set each of the diagonal elements of $\mathbf{V}$ to 1. In the second equation, incrementing i from 1 to 3 and $j = i + 1$ for every i simultaneously, we compute v_{12}, v_{23}, v_{34}. The second equation is reapplied with i ranging from 1 to 3 and $j = i + 2$ (not to exceed $n = 4$) for every i, we compute v_{13} and v_{24}. Finally, the second equation is reapplied with i ranging from 1 to 3 and $j = i + 3$ (not to exceed $n = 4$), we compute v_{14}. The remaining elements of the $\mathbf{V}$ matrix are set to zero using the third equation of (6-9).

Based on the discussion above,

$$\mathbf{L}^{-1} = \begin{bmatrix} -1 & 0 & 0 & 0 \\ \frac{2}{3} & \frac{1}{3} & 0 & 0 \\ -\frac{1}{3} & -\frac{2}{3} & 1 & 0 \\ \frac{5}{12} & -\frac{2}{3} & \frac{3}{4} & \frac{1}{4} \end{bmatrix} \qquad \mathbf{U}^{-1} = \begin{bmatrix} 1 & -2 & 0 & 5 \\ 0 & 1 & -2 & 4 \\ 0 & 0 & 1 & -3 \\ 0 & 0 & 0 & 1 \end{bmatrix}$$

and therefore,

$$
\mathbf{A}^{-1} = \mathbf{U}^{-1}\mathbf{L}^{-1} =
\begin{bmatrix}
-\frac{1}{4} & -4 & \frac{15}{4} & \frac{5}{4} \\
3 & -1 & 1 & 1 \\
-\frac{19}{12} & \frac{4}{3} & -\frac{5}{4} & -\frac{3}{4} \\
\frac{5}{12} & -\frac{2}{3} & \frac{3}{4} & \frac{1}{4}
\end{bmatrix}
$$

6.6 VECTORS

Some basic ideas of vector analysis are now compared with matrix concepts. The result is a formulation of vector analysis in terms of matrices and considerable analytical power. Vector length, dot product, and orthogonality are expressed in terms of single-column matrices and are generalized where appropriate. Then linear independence, vector bases, and change of basis are expressed in terms of matrices.

6.6.1 Matrix Viewpoint of Vector Analysis

Vector analysis is concerned with the description of quantities, called *vectors*, that have magnitude and direction. A vector can be visualized as a directed line segment, an "arrow," as in Figure 6-1(a). The length of the line segment represents the magnitude

(a) Directed line segment

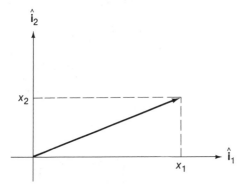

(b) Numerical components of a vector in an orthogonal coordinate system

Figure 6-1 Representing a vector.

of the vector quantity, and the direction of the line segment (in the sense of the arrow) represents the direction of the vector quantity. A numerical representation of a vector is obtained by specifying the coordinates of the top of the directed line segment in an orthogonal coordinate system having the directed line segment's base at its origin, as in Figure 6-1(b).

A vector can always be written as a unique linear combination of unit vectors in the coordinate directions in the following way:

$$\mathbf{x} = \begin{bmatrix} x_1 \\ x_2 \\ \vdots \\ x_n \end{bmatrix} = x_1 \begin{bmatrix} 1 \\ 0 \\ \vdots \\ 0 \\ 0 \end{bmatrix} + x_2 \begin{bmatrix} 0 \\ 1 \\ \vdots \\ 0 \\ 0 \end{bmatrix} + \cdots + x_n \begin{bmatrix} 0 \\ 0 \\ \vdots \\ 0 \\ 1 \end{bmatrix}$$

$$= x_1 \hat{\mathbf{i}}_1 + x_2 \hat{\mathbf{i}}_2 + \cdots + x_n \hat{\mathbf{i}}_n$$

The *unit coordinate vectors* are given the symbols $\hat{\mathbf{i}}_1, \hat{\mathbf{i}}_2, \ldots, \hat{\mathbf{i}}_n$.

The magnitude or length of a vector is, in terms of an orthogonal coordinate system,

$$|\mathbf{x}| = \sqrt{x_1^2 + x_2^2 + \cdots + x_n^2}$$

A more general assignment of a scalar size to each vector is the vector *norm*, which is required to have the properties listed in Table 6-8. The most common norm is the *Euclidean norm*, which for real vector elements $x_1, x_2, \ldots, x_n$ is

$$\|\mathbf{x}\| = \sqrt{x_1^2 + x_2^2 + \cdots + x_n^2}$$

which is the ordinary vector magnitude. If the elements of $\mathbf{x}$ are complex, the Euclidean norm is

$$\|\mathbf{x}\| = \sqrt{|x_1|^2 + |x_2|^2 + \cdots + |x_n|^2} = \sqrt{\mathbf{x}^\dagger \mathbf{x}}$$

where $|\mathbf{x}_i|$ denotes the complex magnitude of $\mathbf{x}_i$.

Norms other than the Euclidean norm find occasional uses. One possibility is the sum of the magnitudes of the elements of $\mathbf{x}$:

$$\|\mathbf{x}\| = |x_1| + |x_2| + \cdots + |x_n|$$

Another possibility is

$$\|\mathbf{x}\| = \sqrt{\mathbf{x}^\dagger \mathbf{Q}^\dagger \mathbf{Q} \mathbf{x}}$$

TABLE 6-8   Requirements of a Vector Norm

(1) $\|\mathbf{x}\| > 0$,   for $x \neq \mathbf{0}$
(2) $\|\mathbf{x}\| = 0$,   for $x = \mathbf{0}$
(3) $\|k\mathbf{x}\| = |k| \|\mathbf{x}\|$,   for any scalar k
(4) $\|\mathbf{x}\| + \|\mathbf{y}\| \geq \|\mathbf{x} + \mathbf{y}\|$ (triangle inequality)

where $\mathbf{Q}$ is some specific matrix. Yet another possibility is the maximum of the magnitude of the component of $\mathbf{x}$:

$$\|\mathbf{x}\| = \max\,(|x_1|, |x_2|, \ldots, |x_n|)$$

Here *norm* will mean the Euclidean norm unless otherwise noted.

     If

$$\|\mathbf{x}\| = 1$$

$\mathbf{x}$ is said to be a *unit vector*. A unit vector in the direction of a vector that is not necessarily unit can be constructed by dividing each element of the vector by the vector norm,

$$\hat{\mathbf{x}} = \frac{\mathbf{x}}{\|\mathbf{x}\|}$$

provided that $\|\mathbf{x}\| \neq 0$. A "hat" (e.g., $\hat{\mathbf{x}}$) is commonly used to denote a unit vector.

     The dot product of two vectors, $\mathbf{x}$ and $\mathbf{y}$, is

$$\mathbf{x} \cdot \mathbf{y} = |\mathbf{x}||\mathbf{y}|\,\cos\,\theta$$

where θ is the smallest angle between $\mathbf{x}$ and $\mathbf{y}$. In terms of unit coordinate vectors, where

$$\mathbf{x} = x_1\hat{\mathbf{i}}_1 + x_2\hat{\mathbf{i}}_2 + \cdots + x_n\hat{\mathbf{i}}_n$$

$$\mathbf{y} = y_1\hat{\mathbf{i}}_1 + y_2\hat{\mathbf{i}}_2 + \cdots + y_n\hat{\mathbf{i}}_n$$

then

$$\mathbf{x} \cdot \mathbf{y} = x_1y_1(\hat{\mathbf{i}}_1 \cdot \hat{\mathbf{i}}_1) + x_1y_2(\hat{\mathbf{i}}_1 \cdot \hat{\mathbf{i}}_2) + \cdots + x_1y_n(\hat{\mathbf{i}}_1 \cdot \hat{\mathbf{i}}_n)$$

$$+ x_2y_1(\hat{\mathbf{i}}_2 \cdot \hat{\mathbf{i}}_1) + x_2y_2(\hat{\mathbf{i}}_2 \cdot \hat{\mathbf{i}}_2) + \cdots + x_2y_n(\hat{\mathbf{i}}_2 \cdot \hat{\mathbf{i}}_n) + \cdots + x_ny_1(\hat{\mathbf{i}}_n \cdot \hat{\mathbf{i}}_1)$$

$$+ x_ny_2(\hat{\mathbf{i}}_n \cdot \hat{\mathbf{i}}_2) + \cdots + x_ny_n(\hat{\mathbf{i}}_n \cdot \hat{\mathbf{i}}_n)$$

$$= x_1y_1 + x_2y_2 + \cdots + x_ny_n$$

     A more general assignment of a scalar measure of the closeness in direction of two vectors of the same dimension is called the *inner product*. An inner product of two vectors $\mathbf{x}$ and $\mathbf{y}$ is denoted by $<\mathbf{x}, \mathbf{y}>$, a scalar function of the elements of the two vectors with the properties listed in Table 6-9.

TABLE 6-9   Requirement of a Vector Inner Product

(1) $<\mathbf{y}, \mathbf{x}> = <\mathbf{x}, \mathbf{y}>^*$ where $*$ denotes the complex conjugate

(2) $<k\mathbf{x}, \mathbf{y}> = k^* <\mathbf{x}, \mathbf{y}> = <\mathbf{x}, k^*\mathbf{y}>$, where k is a scalar

(3) $<\mathbf{x}, \mathbf{x}> > 0$ for $\mathbf{x} \neq 0$

(4) $<\mathbf{x} + \mathbf{u}, \mathbf{y} + \mathbf{v}> = <\mathbf{x}, \mathbf{y}> + <\mathbf{x}, \mathbf{v}> + <\mathbf{u}, \mathbf{y}> + <\mathbf{u}, \mathbf{v}>$

The most commonly used inner product is the *Euclidean inner product*, which for real vector elements is

$$<\mathbf{x}, \mathbf{y}> = \mathbf{x}^\dagger\mathbf{y} = x_1y_1 + x_2y_2 + \cdots + x_ny_n$$

which is the same as the dot product when the vector elements are real numbers. If the elements of x and y can be complex, the Euclidean inner product is

$$<\mathbf{x}, \mathbf{y}> = \mathbf{x}^\dagger\mathbf{y}^* = x_1y_1^* + x_2y_2^* + \cdots + x_ny_n^*$$

Other inner products find occasional uses, but here inner product will mean the Euclidean inner product unless otherwise noted.

The cross product of two vectors, $\mathbf{x}$ and $\mathbf{y}$, is defined only for 3-vectors. It is another vector

$$\mathbf{z} = \mathbf{x} \times \mathbf{y}$$

that is perpendicular to $\mathbf{x}$ and to $\mathbf{y}$ in the sense of the right-hand rule. The cross product has magnitude

$$|\mathbf{z}| = |\mathbf{x}|\,|\mathbf{y}|\,\sin\theta$$

where θ is the angle between $\mathbf{x}$ and $\mathbf{y}$. In terms of an orthogonal coordinate system with unit vectors $\hat{\mathbf{i}}_1$, $\hat{\mathbf{i}}_2$, and $\hat{\mathbf{i}}_3$,

$$\mathbf{x} = x_1\hat{\mathbf{i}}_1 + x_2\hat{\mathbf{i}}_2 + x_3\hat{\mathbf{i}}_3$$

$$\mathbf{y} = y_1\hat{\mathbf{i}}_1 + y_2\hat{\mathbf{i}}_2 + y_3\hat{\mathbf{i}}_3$$

and the cross product is

$$\mathbf{x} \times \mathbf{y} = x_1y_1(\hat{\mathbf{i}}_1 \times \hat{\mathbf{i}}_1) + x_1y_2(\hat{\mathbf{i}}_1 \times \hat{\mathbf{i}}_2) + x_1y_3(\hat{\mathbf{i}}_1 \times \hat{\mathbf{i}}_3) + x_2y_1(\hat{\mathbf{i}}_2 \times \hat{\mathbf{i}}_1)$$

$$+ x_2y_2(\hat{\mathbf{i}}_2 \times \hat{\mathbf{i}}_2) + x_2y_3(\hat{\mathbf{i}}_2 \times \hat{\mathbf{i}}_3) + x_3y_1(\hat{\mathbf{i}}_3 \times \hat{\mathbf{i}}_1) + x_3y_2(\hat{\mathbf{i}}_3 \times \hat{\mathbf{i}}_2)$$

$$+ x_3y_3(\hat{\mathbf{i}}_3 \times \hat{\mathbf{i}}_3)$$

$$= (x_2y_3 - x_3y_2)\hat{\mathbf{i}}_1 + (x_3y_1 - x_1y_3)\hat{\mathbf{i}}_2 + (x_1y_2 - x_2y_1)\hat{\mathbf{i}}_3$$

using

$$\hat{\mathbf{i}}_1 \times \hat{\mathbf{i}}_1 = 0 \qquad \hat{\mathbf{i}}_1 \times \hat{\mathbf{i}}_2 = \hat{\mathbf{i}}_3 \qquad \hat{\mathbf{i}}_1 \times \hat{\mathbf{i}}_3 = -\hat{\mathbf{i}}_2$$

$$\hat{\mathbf{i}}_2 \times \hat{\mathbf{i}}_1 = -\hat{\mathbf{i}}_3 \qquad \hat{\mathbf{i}}_2 \times \hat{\mathbf{i}}_2 = 0 \qquad \hat{\mathbf{i}}_2 \times \hat{\mathbf{i}}_3 = \hat{\mathbf{i}}_1$$

$$\hat{\mathbf{i}}_3 \times \hat{\mathbf{i}}_1 = \hat{\mathbf{i}}_2 \qquad \hat{\mathbf{i}}_3 \times \hat{\mathbf{i}}_2 = -\hat{\mathbf{i}}_1 \qquad \hat{\mathbf{i}}_3 \times \hat{\mathbf{i}}_3 = 0$$

The *outer product* (or *dyadic product*) of two vectors is a matrix that has as elements all possible products of one element of one vector with one element of the other vector. For an m-vector $\mathbf{x}$ and n-vector $\mathbf{y}$, each with real elements,

$$
\mathbf{x} >< \mathbf{y} = \mathbf{xy}^{\dagger} = \begin{bmatrix} x_1 y_1 & x_1 y_2 & \cdots & x_1 y_n \\ x_2 y_1 & x_2 y_2 & \cdots & x_2 y_n \\ \vdots & \vdots & \vdots & \vdots \\ x_m y_1 & x_m y_2 & \cdots & x_m y_n \end{bmatrix}
$$

If the elements of the vectors can be complex.

$$
\mathbf{x} >< \mathbf{y} = \mathbf{xy}^{*\dagger}
$$

The Euclidean inner and outer products are related by

$$
<\mathbf{x}, \mathbf{y}> = \mathbf{x}^{\dagger}\mathbf{y} = \text{trace } (\mathbf{xy}^{\dagger})
$$

6.6.2 Orthogonality and Linear Independence

Two vectors of nonzero norm are *orthogonal* to one another if

$$
<\mathbf{x}, \mathbf{y}> = 0
$$

Orthogonality is the same as perpendicularity for vectors with real elements if the inner product used is the usual Euclidean one. A set of vectors is *mutually orthogonal* if each vector in the set has nonzero norm and is orthogonal to every other vector in the set. For example, the 3-vectors

$$
\mathbf{x} = \begin{bmatrix} 1 \\ -1 \\ 0 \end{bmatrix} \qquad \mathbf{y} = \begin{bmatrix} 1 \\ 1 \\ 0 \end{bmatrix} \qquad \mathbf{z} = \begin{bmatrix} 0 \\ 0 \\ 3 \end{bmatrix}
$$

are mutually orthogonal, since

$$
<\mathbf{x}, \mathbf{y}> = <\mathbf{x}, \mathbf{z}> = <\mathbf{y}, \mathbf{z}> = 0
$$

It is convenient now to introduce superscript indices on vector symbols. Since $\mathbf{x}_i$ is used to denote the ith element of the vector $\mathbf{x}$, a superscript, for example $\mathbf{x}^j$, will be used to indicate the jth vector in some set of vectors. There is no conflict with the notation for an exponent here because a vector is not conformable for multiplication with itself. A set of m vectors, $\mathbf{x}^1, \mathbf{x}^2, \ldots, \mathbf{x}^m$, is linearly *independent* of one another if there are no scalars $k_1, k_2, \ldots, k_m$ except all k's zero for which

$$
k_1 \mathbf{x}^1 + k_2 \mathbf{x}^2 + \cdots + k_m \mathbf{x}^m = 0
$$

Otherwise, the set of vectors is said to be *linearly dependent*. In the case of complex vectors, the scalars $k_1, k_2, \ldots, k_m$ can be complex numbers. Linear independent means that no vector in the set is a linear combination of the other vectors in the set. For example, the 3-vectors

$$
\begin{bmatrix} 1 \\ 2 \\ 3 \end{bmatrix}, \quad \begin{bmatrix} -1 \\ 2 \\ -3 \end{bmatrix}, \quad \text{and} \quad \begin{bmatrix} 3 \\ 2 \\ 2 \end{bmatrix}
$$

are linearly independent, as are the 4-vectors

$$\begin{bmatrix} 1 \\ 0 \\ 0 \\ 0 \end{bmatrix}, \quad \begin{bmatrix} 0 \\ 1 \\ 0 \\ 0 \end{bmatrix}, \quad \text{and} \quad \begin{bmatrix} 0 \\ 0 \\ 0 \\ 1 \end{bmatrix}$$

The vectors

$$\begin{bmatrix} 1 \\ 2 \\ 3 \end{bmatrix}, \quad \begin{bmatrix} -4 \\ 5 \\ 0 \end{bmatrix}, \quad \begin{bmatrix} -2 \\ 9 \\ 6 \end{bmatrix}$$

are linearly dependent.

One way to determine whether or not a set of m n-vectors is linearly independent is to arrange the vectors as the columns (or their transposes as rows) of a matrix:

$$\mathbf{X} = [\mathbf{x}^1 \quad \mathbf{x}^2 \quad \cdots \quad \mathbf{x}^m]$$

The vectors are linearly independent if and only if

$$\text{rank} \ (\mathbf{X}) = m$$

because the rank of $\mathbf{X}$ is the number of linearly independent columns (and rows) of $\mathbf{X}$. No more than n n-vectors with elements real or complex numbers can be linearly independent. And n linearly independent n-vectors can always be found: for example, the n unit coordinate vectors.

6.6.3 Vector Interpretation of Homogeneous Equations

For a set of m linear algebraic equations in n unknowns,

$$\mathbf{Ax} = \mathbf{b}$$

one can think of the rows of $\mathbf{A}$ as vectors:

$$\mathbf{A} = \begin{bmatrix} \mathbf{a}_1^\dagger \\ \mathbf{a}_2^\dagger \\ \vdots \\ \mathbf{a}_m^\dagger \end{bmatrix}$$

A solution vector $\mathbf{x}$ then has the property

$$(<\mathbf{a}_1, \mathbf{x}> = b_1)$$

$$(<\mathbf{a}_2, \mathbf{x}> = b_2)$$

$$\vdots$$

$$(<\mathbf{a}_m, \mathbf{x}> = b_m)$$

That is, the inner product of $\mathbf{x}$ with each of the vectors $\mathbf{a}_1, \mathbf{a}_2, \ldots, \mathbf{a}_m$ are $b_1, b_2, \ldots, b_m$, respectively. For a set of homogeneous equations

$$\mathbf{Ax} = \mathbf{0}$$

a nontrivial solution $\mathbf{x}$, if it exists, is orthogonal to each of the vectors $\mathbf{a}_1, \mathbf{a}_2, \ldots, \mathbf{a}_m$:

$$(<\mathbf{a}_1, \mathbf{x}> = 0)$$

$$(<\mathbf{a}_2, \mathbf{x}> = 0)$$

$$\vdots$$

$$(<\mathbf{a}_m, \mathbf{x}> = 0)$$

As no more than n n-vectors can be linearly independent, if there are n linearly independent vectors among the $\mathbf{a}$'s, that is, if the rank of $\mathbf{A}$ is n, the trivial solution

$$\mathbf{x} = \mathbf{0}$$

is the only possible solution. If there are less than n linearly independent n-vectors among the $\mathbf{a}$'s, then a nontrivial solution for $\mathbf{x}$ exists because it is always possible to find an n-vector that is orthogonal to less than n different n-vectors. The norm of any nontrivial solution is arbitrary since its only requirement is that it must be orthogonal to each of the $\mathbf{a}$'s. If only $n - p$ of the rows of $\mathbf{A}$ are linearly independent, that is, if $\mathbf{A}$ is of rank $n - p$, then p linearly independent n-vectors $\mathbf{x}$ can be found that are orthogonal to each of the rows of $\mathbf{A}$.

If a matrix $\mathbf{A}$, which need not be square, is premultiplied or postmultiplied by a square nonsingular matrix $\mathbf{Q}$, the rank of the product is the same as the rank of $\mathbf{A}$. If the matrix $\mathbf{A}$ is taken to be the coefficient matrix for a set of homogeneous equations,

$$\mathbf{Ax} = \mathbf{0}$$

the rank of $\mathbf{A}$ is the number of linearly independent equations in the set. Postmultiplication of $\mathbf{A}$ by a conformable, nonsingular matrix $\mathbf{Q}$ can be interpreted as a change of variables

$$\mathbf{x} = \mathbf{Qx'}$$

$$\mathbf{Ax} = (\mathbf{AQ})\mathbf{x'} = \mathbf{0}$$

in the equations. Since the old variables can be recovered from the new ones through

$$\mathbf{x'} = \mathbf{Q}^{-1}\mathbf{x}$$

postmultiplication by a nonsingular matrix cannot change the intrinsic number of linearly independent equations in the set. The similar result that premultiplication by a nonsingular matrix leaves the rank of the product unchanged can be obtained by taking $\mathbf{A}^\dagger$ to be the coefficient matrix for some set of homogeneous equations.

6.6.4 Vector Bases and Subspaces

Any set of n linearly independent n-vectors is termed a *basis* for the n-dimensional space of all n-vectors. Every n-vector can be expressed as a unique linear combination of basis vectors. The unit coordinate vectors are an important set of basis vectors, but it often happens that other bases are especially useful. In general, suppose that the n-vectors $\mathbf{b}^1$, $\mathbf{b}^2, \ldots, \mathbf{b}^n$ are linearly independent and so form a basis for n-vectors. Expressing another n-vector $\mathbf{x}$ as a linear combination of the basis vectors,

$$\mathbf{x} = c_1\mathbf{b}^1 + c_2\mathbf{b}^2 + \cdots + c_n\mathbf{b}^n$$

where the c's are unknown constants, constitutes a set of n linear algebraic equations in n unknowns. In matrix form,

$$[\mathbf{b}^1 \quad \mathbf{b}^2 \quad \cdots \quad \mathbf{b}^n] \begin{bmatrix} c_1 \\ c_2 \\ \vdots \\ c_n \end{bmatrix} = \mathbf{BC} = \mathbf{x}$$

Since the basis vectors, which are columns of the matrix $\mathbf{B}$, are linearly independent, $\mathbf{B}^{-1}$ exists and the solution $\mathbf{C}$ exists and is unique.

One can also speak of and use basis representations for vector subspaces. For example, the two 4-vectors

$$\begin{bmatrix} 1 \\ 0 \\ 0 \\ 0 \end{bmatrix} \quad \text{and} \quad \begin{bmatrix} 0 \\ 1 \\ 0 \\ 0 \end{bmatrix}$$

is a basis for all 4-vectors of the form

$$\mathbf{x} = \begin{bmatrix} x_1 \\ x_2 \\ 0 \\ 0 \end{bmatrix} \tag{6-10}$$

since every such vector can be expressed as

$$\mathbf{x} = c_1 \begin{bmatrix} 1 \\ 0 \\ 0 \\ 0 \end{bmatrix} + c_2 \begin{bmatrix} 0 \\ 1 \\ 0 \\ 0 \end{bmatrix}$$

for some constants c_1 and c_2. Such a basis is said to *span* the space or subspace of the vectors. There are, of course, many possibilities for basis vectors. For example, for vectors of the form (6-10),

$$\begin{bmatrix} 2 \\ 1 \\ 0 \\ 0 \end{bmatrix} \quad \text{and} \quad \begin{bmatrix} 1 \\ -1 \\ 0 \\ 0 \end{bmatrix}$$

are also a basis.

The *Gram–Schmidt orthogonalization* procedure is a systematic method of forming an orthogonal set of vectors $\mathbf{y}^1, \mathbf{y}^2, \ldots, \mathbf{y}^m$ from a linearly independent set $\mathbf{x}^1, \mathbf{x}^2, \ldots,$

$\mathbf{x}^m$. Whatever subspace is spanned by the $\mathbf{y}$ vectors is also spanned by the orthogonal $\mathbf{x}$ vectors. The first vectors in the two sets are the same,

$$\mathbf{y}^1 = \mathbf{x}^1$$

To form $\mathbf{y}^2$, any component of $\mathbf{x}^2$ in the direction of $\mathbf{y}^1$ is subtracted:

$$\mathbf{y}^2 = \mathbf{x}^2 - \frac{(<\mathbf{x}^2, \mathbf{y}^1>)}{(<\mathbf{y}^1, \mathbf{y}^1>)} \mathbf{y}^1$$

For $\mathbf{y}^3$, components of $\mathbf{x}^3$ in the $\mathbf{y}^1$ and $\mathbf{y}^2$ directions are subtracted,

$$\mathbf{y}^3 = \mathbf{x}^3 - \frac{(<\mathbf{x}^3, \mathbf{y}^1>)}{(<\mathbf{y}^1, \mathbf{y}^1>)} \mathbf{y}^1 - \frac{(<\mathbf{x}^3, \mathbf{y}^2>)}{(<\mathbf{y}^2, \mathbf{y}^2>)} \mathbf{y}^2$$

and so on. The unit vectors

$$\mathbf{z}^i = \frac{\mathbf{y}^i}{\|\mathbf{y}^i\|}$$

then form an ortho*normal* set.

6.7 LINEAR TRANSFORMATION OF VECTORS

The subject of linear transformation of vectors is important in the analysis of many physical systems because their descriptions are often greatly simplified by a judicious choice of coordinate system. We first interpret a set of n linear algebraic equations in n unknowns

$$\mathbf{y} = \mathbf{Ax}$$

to be a transformation of the vector $\mathbf{x}$ into the vector $\mathbf{y}$. Transformations that leave the norm of the vector unchanged are of special practical importance because rotations of vectors, or equivalently of coordinate systems are of this type.

6.7.1 Range and Null Spaces

The multiplication of a vector $\mathbf{x}$ by a matrix to form another vector $\mathbf{y}$,

$$\mathbf{y} = \mathbf{Ax}$$

can be thought of as a transformation that converts the vector $\mathbf{x}$ to the vector $\mathbf{y}$. The transformation is linear because it always maps any scaled vector $\mathbf{x}$ to a $\mathbf{y}$ vector that is scaled by the same factor:

$$k\mathbf{y} = \mathbf{A}(k\mathbf{x})$$

Also, sums of vectors map to the corresponding sums of individual transformations. If

$$\mathbf{y}^1 = \mathbf{Ax}^1 \qquad \text{and} \qquad \mathbf{y}^2 = \mathbf{Ax}^2$$

then

$$\mathbf{y}^1 + \mathbf{y}^2 = \mathbf{A}(\mathbf{x}^1 + \mathbf{x}^2)$$

If the matrix $\mathbf{A}$ is square and nonsingular, the vector $\mathbf{x}$ can be recovered from $\mathbf{y}$ through

$$\mathbf{x} = \mathbf{A}^{-1}\mathbf{y}$$

Each vector $\mathbf{x}$ is then transformed to a unique vector $\mathbf{y}$ and each $\mathbf{y}$ corresponds to a unique $\mathbf{x}$. Suppose, instead, that $\mathbf{A}$ is not necessarily square and not necessarily of full rank. Let $\mathbf{A}$ be $m \times n$ and of rank r. Then the space of possible $\mathbf{y}$ vectors, which is r-dimensional, is termed the *range space* of $\mathbf{A}$. A basis for the range space of $\mathbf{A}$ is a largest set of linearly independent columns (which are m-vectors) of $\mathbf{A}$,

$$\mathbf{A} = [\mathbf{a}_1 \quad \mathbf{a}_2 \quad \cdots \quad \mathbf{a}_n]$$

since a general vector $\mathbf{x}$ maps to

$$\mathbf{y} = x_1\mathbf{a}_1 + x_2\mathbf{a}_2 + \cdots + x_n\mathbf{a}_n$$

For the matrix

$$\mathbf{A} = \begin{bmatrix} 2 & -4 & 1 & 3 \\ -1 & 2 & 0 & -1 \\ 3 & -6 & -1 & 2 \end{bmatrix}$$

the null space of $\mathbf{A}$ is spanned by the 3-vectors

$$\begin{bmatrix} 2 \\ -1 \\ 3 \end{bmatrix} \quad \text{and} \quad \begin{bmatrix} 1 \\ 0 \\ -1 \end{bmatrix}$$

(There are other possibilities as well, of course.) That is, every vector $\mathbf{y}$ can be expressed as

$$\mathbf{y} = c_1 \begin{bmatrix} 2 \\ -1 \\ 3 \end{bmatrix} + c_2 \begin{bmatrix} 1 \\ 0 \\ -1 \end{bmatrix}$$

where c_1 and c_2 are some constants.

If the matrix $\mathbf{A}$ of a linear transformation

$$\mathbf{y} = \mathbf{A}\mathbf{x}$$

is not of full rank, different vectors $\mathbf{x}$ are transformed to the same vector $\mathbf{y}$. The space of vectors $\mathbf{x}^0$ for which

$$\mathbf{A}\mathbf{x}^0 = \mathbf{0} \tag{6-11}$$

is termed the *null space* of the matrix $\mathbf{A}$. Any vector $\mathbf{x}^0$ in this space can be added to any vector $\mathbf{x}$ with no effect on the resulting $\mathbf{y}$:

$$\mathbf{y} = \mathbf{A}(\mathbf{x} + \mathbf{x}^0) = \mathbf{A}\mathbf{x}$$

A basis for the null space of $\mathbf{A}$ is a largest set of linearly independent n-vectors $\mathbf{x}^0$ satisfying (6-11).

For example, for the matrix

$$\mathbf{A} = \begin{bmatrix} 2 & -4 & 1 & 3 \\ -1 & 2 & 0 & -1 \\ 3 & -6 & -1 & 2 \end{bmatrix}$$

the vectors $\mathbf{x}^0$ for which

$$\mathbf{A}\mathbf{x}^0 = \mathbf{0}$$

satisfy

$$\begin{cases} 2x_1^0 - 4x_2^0 + x_3^0 + 3x_4^0 = 0 \\ -x_1^0 + 2x_2^0 \qquad - x_4^0 = 0 \\ 3x_1^0 - 6x_2^0 - x_3^0 + 2x_4^0 = 0 \end{cases} \qquad \text{(6-12)}$$

Using hand calculation or the PIVOT program, a fully pivoted equivalent set to (6-12) are

$$\begin{cases} -\tfrac{1}{2}x_1^0 + x_2^0 \qquad - \tfrac{1}{2}x_4^0 = 0 \\ \\ \qquad\qquad x_3^0 + x_4^0 = 0 \\ \\ \qquad\qquad\qquad 0 = 0 \end{cases}$$

showing that the third equation of (6-12) is linearly dependent on the other two equations. In terms of x_1^0 and x_4^0, these are

$$\begin{cases} x_2^0 = \tfrac{1}{2}x_1^0 + \tfrac{1}{2}x_4^0 \\ \\ x_3^0 = \qquad\quad -x_4^0 \end{cases}$$

so that x^0 is any vector of the form

$$\mathbf{x}^0 = \begin{bmatrix} x_1^0 \\ x_2^0 \\ x_3^0 \\ x_4^0 \end{bmatrix} = c_1 \begin{bmatrix} 1 \\ \tfrac{1}{2} \\ 0 \\ 0 \end{bmatrix} + c_2 \begin{bmatrix} 0 \\ \tfrac{1}{2} \\ -1 \\ 1 \end{bmatrix}$$

where c_1 and c_2 are any constants. Thus the 4-vectors

$$\begin{bmatrix} 1 \\ \tfrac{1}{2} \\ 0 \\ 0 \end{bmatrix} \quad \text{and} \quad \begin{bmatrix} 0 \\ \tfrac{1}{2} \\ -1 \\ 1 \end{bmatrix}$$

(and many other possibilities, too, of course) are a basis for the null space of $\mathbf{A}$.

If an $n \times n$ matrix has rank r, the integer $n - r$ called the nullity of the matrix. A nonsingular matrix has nullity zero. If a square matrix has just one linearly independent row (and column), its nullity is 1, and so forth. The nullity of the product of two square matrices $\mathbf{A}$ and $\mathbf{B}$ must be at least the largest of their individual nullities and can be as large as the sum of their nullities. This result is known as *Sylvester's law of nullity* and can be easily shown by noting that the nullity of a square matrix is the dimension of its null space.

6.7.2 Orthogonal Transformations

A nonsingular transformation

$$\mathbf{y} = \mathbf{Ax}$$

for which

$$\mathbf{A}^{-1} = \mathbf{A}\dagger$$

is called an orthogonal transformation. The matrix of an orthogonal transformation is called an *orthogonal* matrix. The norm of a vector is unchanged by an orthogonal transformation:

$$\|\mathbf{Ax}\|^2 = <\mathbf{Ax}, \mathbf{Ax}> = (\mathbf{Ax})\dagger(\mathbf{Ax})$$

$$= \mathbf{x}\dagger\mathbf{A}\dagger\mathbf{Ax} = \mathbf{A}^{-1}\mathbf{Ax} = \mathbf{x}\dagger\mathbf{x}$$

$$= \|\mathbf{x}\|^2$$

There are basically two types of orthogonal transformations. One is a pure rotation of the vector $\mathbf{x}$ to produce $\mathbf{y}$. The other type involves not only a vector rotation, but also a reversal of the algebraic signs of one or more of the vector's elements. Since the norm of a vector involves the squares of the vector elements, reversal of the algebraic sign of an element does not affect the vector norm. For example, a 2×2 matrix of the form

$$\mathbf{A} = \begin{bmatrix} \cos\theta & \sin\theta \\ \sin\theta & -\cos\theta \end{bmatrix}$$

is orthogonal for any angle θ. It represents a pure rotation of the vector $\mathbf{x}$ through the angle θ to form the vector $\mathbf{y}$, as shown in Figure 6-2(a). For pure rotation, the components of $\mathbf{y}$ are related to those of $\mathbf{x}$ by

$$\begin{cases} y_1 = x_1 \cos\theta + x_2 \sin\theta \\ y_2 = x_1 \sin\theta - x_2 \cos\theta \end{cases}$$

so that

$$\mathbf{y} = \begin{bmatrix} y_1 \\ y_2 \end{bmatrix} = \begin{bmatrix} \cos\theta & \sin\theta \\ \sin\theta & -\cos\theta \end{bmatrix} \begin{bmatrix} x_1 \\ x_2 \end{bmatrix} = \mathbf{Ax}$$

$$= (x_1 \cos\theta + x_2 \sin\theta)\hat{\mathbf{i}}_1 + (x_1 \sin\theta - x_2 \cos\theta)\hat{\mathbf{i}}_2$$

A 2×2 matrix of the form

$$\mathbf{A} = \begin{bmatrix} \cos\theta & \sin\theta \\ -\sin\theta & \cos\theta \end{bmatrix}$$

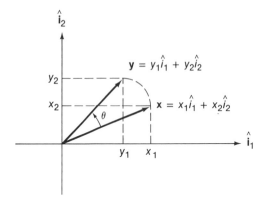

(a) Rotation of the vector **x** through the angle θ to produce **y**

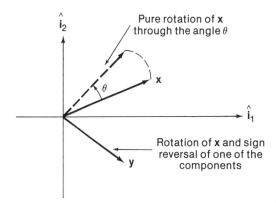

Figure 6-2   Orthogonal transformations of
(b) Rotation and sign reversal of a vector component   2-vectors.

is also orthogonal. It represents a pure rotation of the vector through the angle θ followed by an algebraic sign reversal of the second component of **y**, as shown in Figure 6-2(b):

$$\mathbf{y} = \begin{bmatrix} y_1 \\ y_2 \end{bmatrix} = \begin{bmatrix} \cos\theta & \sin\theta \\ -\sin\theta & \cos\theta \end{bmatrix} \begin{bmatrix} x_1 \\ x_2 \end{bmatrix} = \mathbf{Ax}$$

$$= (x_1 \cos\theta + x_2 \sin\theta)\hat{\mathbf{i}}_1 - (x_1 \sin\theta - x_2 \cos\theta)\hat{\mathbf{i}}_2$$

The 2×2 matrices

$$\begin{bmatrix} -\cos\theta & -\sin\theta \\ \sin\theta & -\cos\theta \end{bmatrix} \quad \text{and} \quad \begin{bmatrix} -\cos\theta & -\sin\theta \\ -\sin\theta & \cos\theta \end{bmatrix}$$

are also orthogonal and represent other combinations of pure rotation through the angle θ and algebraic sign reversal of one or both coordinates. Although rotation and component sign reversal in two dimensions is always equivalent to a rotation through some angle, that is not necessarily the case in three or more dimensions.

Table 6-10 lists properties of orthogonal matrices; property (1) is shown as follows: Since

$$|\mathbf{A}^\dagger| = |\mathbf{A}|$$

and

$$|\mathbf{A}^{-1}| = \frac{1}{|\mathbf{A}|}$$

Then

$$|\mathbf{A}| = \frac{1}{|\mathbf{A}|}$$

or

$$|\mathbf{A}|^2 = 1$$

It can be shown that $|\mathbf{A}| = (-1)^k$, where k is the number of vector component algebraic sign reversals which are made by the transformation. Thus an orthogonal transformation that is a pure rotation will have $|\mathbf{A}| = 1$, as will orthogonal transformations with an even number of component sign reversals. Property (2) follows by letting

$$\mathbf{A} = \begin{bmatrix} \mathbf{a}_1^\dagger \\ \mathbf{a}_2^\dagger \\ \vdots \\ \mathbf{a}_n^\dagger \end{bmatrix}$$

Since $\mathbf{A}\mathbf{A}^{-1} = \mathbf{A}\mathbf{A}^\dagger = \mathbf{I}$,

$$\begin{bmatrix} \mathbf{a}_1^\dagger \\ \mathbf{a}_2^\dagger \\ \vdots \\ \mathbf{a}_n^\dagger \end{bmatrix} [\mathbf{a}_1 \quad \mathbf{a}_2 \quad \cdots \quad \mathbf{a}_n] = \begin{bmatrix} 1 & 0 & \cdots & 0 \\ 0 & 1 & \cdots & 0 \\ \vdots & \vdots & \vdots & \vdots \\ 0 & 0 & \cdots & 1 \end{bmatrix}$$

Then

$$\mathbf{a}_i^\dagger \, \mathbf{a}_j^\dagger = \langle \mathbf{a}_i, \mathbf{a}_j \rangle = \begin{cases} (0, \ i \neq j) \\ (1, \ i = j) \end{cases}$$

The proof of (3) is similar.

TABLE 6-10   Properties of Orthogonal Matrices

DEFINING PROPERTY

  $\mathbf{A}^{-1} = \mathbf{A}^\dagger$

FURTHER PROPERTIES

  (1) $|\mathbf{A}| = \pm 1$.

  (2) The rows of an $n \times n$ orthogonal matrix are an orthonormal set of n-vectors.

  (3) The columns of an $n \times n$ orthogonal matrix are an orthonormal set of n-vectors.

6.7.3 Change of Coordinates

A linear transformation

$$\mathbf{x}' = \mathbf{Q}\mathbf{x}$$

can also be interpreted as being a transformation of the numerical representation of a vector $\mathbf{x}$ to a representation of the same vector in another coordinate system. Rotations of coordinate axes result in orthogonal transformation $\mathbf{Q}$. For example, a rotation of two-dimensional axes through the angle θ converts the components of the vector $\mathbf{x}$ to those of $\mathbf{x}'$ via the orthogonal transformation

$$\mathbf{x}' = \begin{bmatrix} x_1' \\ x_2' \end{bmatrix} = \mathbf{Q}\mathbf{x} = \begin{bmatrix} \cos\theta & \sin\theta \\ -\sin\theta & \cos\theta \end{bmatrix} \begin{bmatrix} x_1 \\ x_2 \end{bmatrix}$$

as shown in Figure 6-3. Since

$$x_1 = |\mathbf{x}| \cos\alpha$$

$$x_2 = |\mathbf{x}| \sin\alpha$$

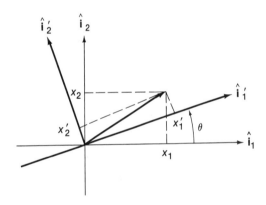

(a) Relation between the two coordinate systems

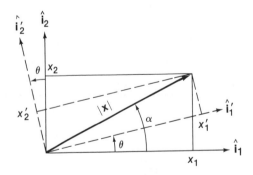

(b) Relations between the vector's components in the two coordinate systems

Figure 6-3 Rotation of coordinates in two dimensions.

then

$$x'_1 = |\mathbf{x}| \cos(\alpha - \theta) = |\mathbf{x}|(\cos\alpha\cos\theta + \sin\alpha\sin\theta) = x_1\cos\theta + x_2\sin\theta$$

$$x'_2 = |\mathbf{x}| \sin(\alpha - \theta) = |\mathbf{x}|(\sin\alpha\cos\theta - \cos\alpha\sin\theta) = -x_1\sin\theta + x_2\cos\theta$$

In three dimensions, general coordinate system rotation can be described by two angles, for instance the two *Euler angles* shown in Figure 6-4. Each of the individual rotations is described by an orthogonal change of coordinates. The rotation by angle θ in the (x_1, x_2)-plane gives

$$x' = \begin{bmatrix} x'_1 \\ x'_2 \\ x'_3 \end{bmatrix} = \begin{bmatrix} \cos\theta & \sin\theta & 0 \\ -\sin\theta & \cos\theta & 0 \\ 0 & 0 & 1 \end{bmatrix} \begin{bmatrix} x_1 \\ x_2 \\ x_3 \end{bmatrix} = \mathbf{Q}_1\mathbf{x}$$

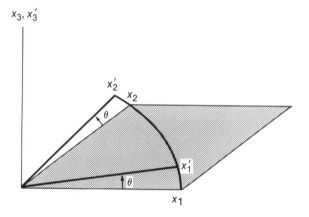

(a) Rotation in the (x_1, x_2)-plane by the angle θ

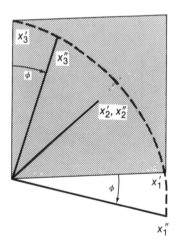

(b) Rotation in the (x'_1, x'_3)-plane by angle ϕ

Figure 6-4   General three-dimensional co-ordinate rotation as the plane rotation by angle θ followed by the plane rotation by angle ϕ.

and the rotation by angle ϕ in the (x_1', x_3')

$$\mathbf{x}'' = \begin{bmatrix} x_1'' \\ x_2'' \\ x_3'' \end{bmatrix} = \begin{bmatrix} \cos\phi & 0 & -\sin\phi \\ 0 & 1 & 0 \\ \sin\phi & 0 & \cos\phi \end{bmatrix} \begin{bmatrix} x_1' \\ x_2' \\ x_3' \end{bmatrix} = \mathbf{Q}_2\mathbf{x}'$$

The general rotation of coordinates is then the orthogonal transformation

$$\mathbf{x}'' = \begin{bmatrix} x_1'' \\ x_2'' \\ x_3'' \end{bmatrix} = \mathbf{Q}_2\mathbf{Q}_1\mathbf{x} = \begin{bmatrix} \cos\theta\cos\phi & \cos\phi\sin\theta & -\sin\phi \\ -\sin\theta & \cos\theta & 0 \\ \cos\theta\sin\phi & \sin\theta\sin\phi & \cos\phi \end{bmatrix} \begin{bmatrix} x_1 \\ x_2 \\ x_3 \end{bmatrix}$$

The order in which the individual rotations are done makes a difference, of course, since

$$\mathbf{Q}_2\mathbf{Q}_1 \neq \mathbf{Q}_1\mathbf{Q}_2.$$

PROBLEMS

6-1. Find the 3×4 matrix with elements

$$a_{ij} = i + 2^j$$

6-2. Explain how a triangular array such as

$$\begin{matrix} 1 & 2 & -3 \\ & -1 & 0 \\ & & 4 \end{matrix}$$

might be represented as a matrix.

6-3. What kind of matrix is both symmetric and skew-symmetric?

6-4. How many *different* elements can there be in an $n \times n$ symmetric matrix? In an $n \times n$ skew-symmetric matrix?

6-5. What effect does interchanging the rows with the columns (row i becomes column i) have for a symmetric matrix? For a skew-symmetric matrix?

6-6. Show that every square matrix can be written as a sum of a symmetric and a skew-symmetric matrix. Then decompose

$$\mathbf{A} = \begin{bmatrix} 1 & 0 & -8 & 9 \\ 5 & 2 & 1 & -1 \\ -2 & 3 & 3 & 0 \\ 7 & -6 & 6 & -4 \end{bmatrix}$$

into its symmetric and skew-symmetric parts.

6-7. Write, debug, and test computer programs that compute the following special kinds of matrix product.
(a) $\mathbf{c} = \mathbf{Ab}$, where $\mathbf{A}$ is $n \times n$ and $\mathbf{b}$ is an n-vector
(b) $\mathbf{c} = \mathbf{a}^\dagger\mathbf{b}$, where $\mathbf{a}$ and $\mathbf{b}$ are n-vectors
(c) $\mathbf{c} = \mathbf{ab}^\dagger$, where $\mathbf{a}$ and $\mathbf{b}$ are n-vectors

6-8. Find $\mathbf{A}^4 = \mathbf{AAAA}$, where

$$\mathbf{A} = \begin{bmatrix} 1 & 2 \\ -3 & 4 \end{bmatrix}$$

6-9. Find the following:

(a) Two square matrices **A** and **B** such that $\mathbf{AB} \neq \mathbf{BA}$.

(b) Two different nonzero square matrices **A** and **B** such that $\mathbf{AB} = \mathbf{BA}$.

(c) A set of matrices **A**, **B**, and **C** such that $\mathbf{AB} = \mathbf{AC}$ but $\mathbf{B} \neq \mathbf{C}$.

(d) Two nonzero 3×3 matrices such that $\mathbf{AB} = \mathbf{0}$.

(e) A 2×2 matrix **A** with real elements such that $\mathbf{A}^2 = \mathbf{AA} = -\mathbf{I}$. This matrix is analogous to $\underline{i} = \sqrt{-1}$.

6-10. Show that trace $(\mathbf{AB})$ = trace $(\mathbf{BA})$.

6-11. Indicate two different ways that the following matrices might be partitioned so that the submatrices are conformable for multiplication. Find the product using each of the partitionings.

$$\begin{bmatrix} 1 & 0 & 3 & -2 \\ -5 & 1 & -1 & 4 \\ 1 & -2 & 3 & 8 \\ -4 & -7 & 1 & -1 \end{bmatrix} \text{ and } \begin{bmatrix} -4 & -4 & 4 & 1 \\ -5 & -2 & 2 & -1 \\ 7 & 0 & -3 & -5 \\ -1 & 3 & 3 & -4 \end{bmatrix}$$

6-12. Show that if **A** and **B** are both $n \times n$,

$$\begin{bmatrix} \mathbf{A} & \text{any numbers} \\ \mathbf{0} & \mathbf{B} \end{bmatrix} = |\mathbf{A}| \, |\mathbf{B}|$$

6-13. A triangle in the x-y plane is defined by the three vertex points (x_1, y_1), (x_2, y_2), and (x_3, y_3). Show that the area of the triangle is

$$\pm \frac{1}{2} \begin{vmatrix} x_1 & y_1 & 1 \\ x_2 & y_2 & 1 \\ x_3 & y_3 & 1 \end{vmatrix}$$

The algebraic sign that applies in any given circumstance depends on the ordering of the rows of the determinant.

6-14. Three straight lines in the x-y plane are given by

$$y = m_1 x + b_1$$

$$y = m_2 x + b_2$$

$$y = m_3 x + b_3$$

where m_1, m_2, m_3, b_1, b_2, and b_3 are constants. If the three lines are to intersect at a common point, show that

$$\begin{vmatrix} m_1 & b_1 & 1 \\ m_2 & b_2 & 1 \\ m_3 & b_3 & 1 \end{vmatrix} = 0$$

6-15. Find the ranks of the following matrices.

(a) $\begin{bmatrix} 1 & 2 & 3 \\ 4 & 5 & 6 \\ 7 & 8 & 9 \end{bmatrix}$ (b) $\begin{bmatrix} 1 & 4 & -2 \\ 0 & 0 & -1 \\ 8 & 6 & -3 \\ 2 & 0 & 0 \\ 1 & 1 & 1 \end{bmatrix}$

6-16. What is the rank of a zero matrix? Of an $n \times n$ diagonal matrix?

6-17. Show that the rank of the product of two $n \times n$ matrices cannot be n unless the two matrices in the product have full rank, n.

6-18. Find the inverses, if they exist, of the following matrices, using

$$\mathbf{A}^{-1} = \frac{\text{adj } (\mathbf{A})}{|\mathbf{A}|}$$

(a) $\begin{bmatrix} 0 & 2 & 1 \\ -1 & 3 & 4 \\ -2 & 0 & 5 \end{bmatrix}$ (b) $\begin{bmatrix} 3 & 4 & 2 \\ -3 & 0 & 1 \\ -1 & -2 & 5 \end{bmatrix}$

6-19. Show that if $\mathbf{A}$ and $\mathbf{B}$ are each nonsingular,

$$\begin{bmatrix} \mathbf{A} & \mathbf{0} \\ \mathbf{0} & \mathbf{B} \end{bmatrix}^{-1} = \begin{bmatrix} \mathbf{A}^{-1} & \mathbf{0} \\ \mathbf{0} & \mathbf{B}-1 \end{bmatrix}$$

6-20. Show that
(a) $(\mathbf{A}^{-1})\dagger = (\mathbf{A}\dagger)^{-1}$
(b) $(\mathbf{A}^{-1})^{-1} = \mathbf{A}$

6-21. For the matrix

$$\mathbf{A} = \begin{bmatrix} 1 & -1 & 0 & 2 & -2 & 0 \\ 0 & 0 & -4 & 3 & -1 & 0 \\ 8 & -7 & 5 & 0 & 4 & 2 \\ 0 & 0 & 0 & -1 & -3 & 5 \end{bmatrix}$$

find a premultiplying matrix that adds twice row four to row one and three times row four to row two.

6-22. For the matrix

$$\mathbf{A} = \begin{bmatrix} 1 & -2 & 3 \\ 4 & -1 & 0 \\ 2 & 0 & -3 \end{bmatrix}$$

find a matrix $\mathbf{P}$ such that the product $\mathbf{PA}$ is a diagonal matrix.

6-23. For each of the matrices below, find matrices $\mathbf{P}$ and $\mathbf{Q}$ such that $\mathbf{PAQ}$ is a normal form.

(a) $\mathbf{A} = \begin{bmatrix} 2 & 0 & 5 \\ 3 & 0 & 2 \\ -1 & 0 & -2 \end{bmatrix}$ (b) $\mathbf{A} = \begin{bmatrix} 1 & 0 \\ 2 & 4 \\ 3 & -1 \end{bmatrix}$

6-24. Use row operations exclusively to evaluate the determination of the matrix

$$\begin{bmatrix} 1 & 1 & -1 & 1 \\ 1 & -1 & 0 & 2 \\ -2 & 0 & 1 & -1 \\ 0 & 1 & -1 & 0 \end{bmatrix}$$

6-25. Write, test, and debug a computer program to solve a set of linear, consistent, independent algebraic equations using the LU decomposition method.

6-26. Write, test, and debug a computer program to invert a nonsingular matrix using the LU decomposition method.

6-27. Find a general expression for the determinant of a 3×3 matrix in terms of the matrix elements.

6-28. Verify that

$$\|\mathbf{x} + \mathbf{y}\|^2 + \|\mathbf{x} - \mathbf{y}\|^2 = 2\|\mathbf{x}\|^2 + 2\|\mathbf{y}\|^2$$

6-29. Show that the norm

$$\|\mathbf{x}\| = \max(|x_1|, |x_2|, \ldots, |x_n|)$$

satisfies the four requirements for a norm listed in Table 6-8.

6-30. For the vectors

$$\mathbf{x} = \begin{bmatrix} 1 \\ 2 \\ 3 \end{bmatrix} \quad \text{and} \quad \mathbf{y} = \begin{bmatrix} 1 \\ 2 \\ -1 \end{bmatrix}$$

find the following:

(a) $\|\mathbf{x}\|$
(b) $\|\mathbf{y}\|$
(c) $<\mathbf{x}, \mathbf{y}>$
(d) A unit vector in the direction of $\mathbf{x}$
(e) A vector in the direction of $\mathbf{y}$ but with norm of 3
(f) A vector that is orthogonal to $\mathbf{x}$
(g) A unit vector that is orthogonal to both $\mathbf{x}$ and $\mathbf{y}$
(h) $\mathbf{x} > < \mathbf{y}$

6-31. Determine which of the following vectors are linearly independent of one another.

$$\begin{bmatrix} 1 \\ 2 \\ 1 \end{bmatrix}, \begin{bmatrix} 1 \\ 1 \\ 1 \end{bmatrix}, \begin{bmatrix} 0 \\ 1 \\ 1 \end{bmatrix}, \begin{bmatrix} 3 \\ 6 \\ 2 \end{bmatrix}, \begin{bmatrix} -1 \\ -2 \\ -1 \end{bmatrix}$$

6-32. Express the vector

$$\begin{bmatrix} 1 \\ 0 \\ -2 \\ 3 \end{bmatrix}$$

as a linear combination of the basis vectors

$$\begin{bmatrix} 1 \\ 0 \\ 1 \\ 0 \end{bmatrix}, \begin{bmatrix} 0 \\ 1 \\ 1 \\ 0 \end{bmatrix}, \begin{bmatrix} 0 \\ 2 \\ -1 \\ 0 \end{bmatrix}, \quad \text{and} \quad \begin{bmatrix} 2 \\ -2 \\ 0 \\ -1 \end{bmatrix}$$

6-33. Form an orthonormal basis for the subspace spanned by

$$\mathbf{x}^1 = \begin{bmatrix} 2 \\ -2 \\ 0 \end{bmatrix} \quad \text{and} \quad \mathbf{x}^2 = \begin{bmatrix} 1 \\ 0 \\ 1 \end{bmatrix}$$

6-34. Find an orthonormal basis, one vector in which has the same direction as (i.e., is proportional to) the vector.

$$\begin{bmatrix} 1 \\ -1 \\ -2 \end{bmatrix}$$

6-35. Find a vector that is orthogonal to both of the vectors

$$\begin{bmatrix} 1 \\ -2 \\ 3 \end{bmatrix} \quad \text{and} \quad \begin{bmatrix} 0 \\ -1 \\ 2 \end{bmatrix}$$

by solving a set of homogeneous algebraic equations.

6-36. Find a linear 3-vector transformation that transforms the vector

$$\begin{bmatrix} 1 \\ 0 \\ 0 \end{bmatrix} \quad \text{to} \quad \begin{bmatrix} -1 \\ 0 \\ 1 \end{bmatrix}$$

the vector

$$\begin{bmatrix} 0 \\ 1 \\ 0 \end{bmatrix} \quad \text{to} \quad \begin{bmatrix} 2 \\ 1 \\ 0 \end{bmatrix}$$

and the vector

$$\begin{bmatrix} 0 \\ 0 \\ 1 \end{bmatrix} \quad \text{to} \quad \begin{bmatrix} 1 \\ 1 \\ 1 \end{bmatrix}$$

6-37. Find a basis for the range space and a basis for the null space for the following matrices.

(a) $\begin{bmatrix} 2 & -2 & 3 \\ 1 & -1 & 1 \\ 1 & -1 & 2 \end{bmatrix}$ (b) $\begin{bmatrix} 1 & -2 & 3 & -1 \\ 2 & 2 & 0 & 4 \\ -1 & 1 & -2 & 0 \end{bmatrix}$

6-38. Find an *orthonormal* basis for the range space of

$$\begin{bmatrix} 1 & 0 & -1 \\ 2 & 1 & 1 \\ -1 & -1 & -2 \end{bmatrix}$$

6-39. Find two 3×3 matrices **A** and **B**, each of rank 2, for which **AB** has rank 1.

6-40. Determine whether or not the following are orthogonal matrices.

(a) $\begin{bmatrix} 0 & 0 & -1 \\ 0 & 1 & 0 \\ 1 & 0 & 0 \end{bmatrix}$ (b) $\begin{bmatrix} \dfrac{1}{2} & -\dfrac{1}{\sqrt{2}} & -\dfrac{1}{2} \\ \dfrac{1}{2} & \dfrac{1}{\sqrt{2}} & -\dfrac{1}{2} \\ \dfrac{1}{\sqrt{2}} & 0 & \dfrac{1}{\sqrt{2}} \end{bmatrix}$

6-41. Construct an orthogonal 3×3 matrix the first column of which is

$$\begin{bmatrix} \dfrac{1}{\sqrt{6}} \\ -\dfrac{2}{\sqrt{6}} \\ -\dfrac{1}{\sqrt{6}} \end{bmatrix}$$

6-42. Find the two-dimensional orthogonal transformation **A** in

$$\mathbf{y} = \mathbf{Ax}$$

for which the vector

$$\mathbf{x} = \begin{bmatrix} 2 \\ -3 \end{bmatrix}$$

is, in terms of the new coordinates, aligned with the first coordinate axis.

7

THE CHARACTERISTIC VALUE PROBLEM

7.1 PREVIEW

The characteristic value problem, the subject of this chapter, probably has the most far-reaching and penetrating applications of all of matrix analysis. To develop the basic problem, we examine those linear coordinate transformations of both the unknown vector and the vector of knowns that diagonalizes an $n \times n$ set of linear algebraic equations. Efficient methods for computing the quantities involved are then developed in later sections. It turns out that key parts of the problem involve polynomial factoring and the solutions of sets of simultaneous homogeneous linear algebraic equations.

The basic characteristic value problem is defined and solved in the next section. It is shown that inherent to an $n \times n$ matrix $\mathbf{A}$ is an nth-degree characteristic polynomial, the roots of which are called the *eigenvalues* of $\mathbf{A}$. Corresponding to each different eigenvalue is an *eigenvector* of $\mathbf{A}$. When the eigenvectors are assembled as columns of a transformation matrix, a *similarity transformation* of $\mathbf{A}$ to a diagonal matrix results.

The Cayley–Hamilton theorem follows from the results of the characteristic value problem and is this: Any square matrix satisfies its own characteristic equation. A constructive proof for matrices with distinct eigenvalues is given in Section 7.3. The Cayley–Hamilton theorem means that unlike functions of a scalar variable, functions of a square matrix that can be defined in terms of infinite power series can be expressed as an equivalent finite series. Such functions are the subject of Sections 7.4 and 7.5.

In Sections 7.6 and 7.7, powerful numerical algorithms are derived for determining the characteristic equation, determinant, adjugate, inverse, and eigenvectors of a square matrix. Computer programs for these computations are developed and demonstrated. In the final section, many of these ideas are brought together for the analysis of the vibrations of a spring–mass system.

7.2 TRANSFORMATION TO DIAGONAL FORM

7.2.1 Eigenvalues and Eigenvectors

The characteristic value problem of linear algebra addresses the following question: For the set of linear equations

$$\mathbf{Ax} = \mathbf{b}$$

where $\mathbf{A}$ is square, is there a nonsingular transformation of the solution vector $\mathbf{x}$ and the knowns vector $\mathbf{b}$

$$\mathbf{x} = \mathbf{Px}' \qquad \mathbf{x}' = \mathbf{P}^{-1}\mathbf{x}$$

$$\mathbf{b} = \mathbf{Pb}' \qquad \mathbf{b}' = \mathbf{P}^{-1}\mathbf{b}$$

such that in terms of the new coordinates, the equations are

$$\mathbf{\Lambda x}' = \mathbf{b}'$$

where $\mathbf{\Lambda}$ is a diagonal matrix? A transformation matrix $\mathbf{P}$ with this property is called a *modal matrix*, and the diagonal matrix $\mathbf{\Lambda}$ is called a *spectral matrix*. It is more convenient to define $\mathbf{P}$ as the transformation from $\mathbf{x}'$ to $\mathbf{x}$, as above, than as the transformation from $\mathbf{x}$ to $\mathbf{x}'$.

Substituting and premultiplying gives

$$\mathbf{Ax} = \mathbf{APx}' = \mathbf{Pb}'$$

$$(\mathbf{P}^{-1}\mathbf{AP})\mathbf{x}' = \mathbf{b}'$$

so that the spectral matrix is related to $\mathbf{A}$ by

$$\mathbf{\Lambda} = \mathbf{P}^{-1}\mathbf{AP} \tag{7-1}$$

In terms of the columns of the modal matrix, (7-1) is

$$\mathbf{AP} = \mathbf{P\Lambda} = [\mathbf{p}^1 \mid \mathbf{p}^2 \mid \cdots \mid \mathbf{p}^n]\begin{bmatrix} \lambda_1 & 0 & 0 & \cdots & 0 \\ 0 & \lambda_2 & 0 & \cdots & 0 \\ \vdots & \vdots & \vdots & \vdots & \vdots \\ 0 & 0 & 0 & \cdots & \lambda_n \end{bmatrix}$$

where the elements along the diagonal of the spectral matrix are $\lambda_1, \lambda_2, \ldots, \lambda_n$. Each column of the modal matrix $\mathbf{P}$ satisfies an equation of the form

$$\mathbf{Ap} = \lambda\mathbf{p}$$

or

$$(\lambda\mathbf{I} - \mathbf{A})\mathbf{p} = 0 \tag{7-2}$$

Of course, the trivial solution to (7-2), $\mathbf{p} = \mathbf{0}$, will be of no help, since a column of zeros would make $\mathbf{P}$ singular. The set of n homogeneous equations (7-2) in the n elements of $\mathbf{P}$ has a nontrivial solution only if their determinant is zero:

$$|\lambda\mathbf{I} - \mathbf{A}| = \mathbf{0}$$

This is an nth-degree polynomial equation in λ, termed the characteristic equation of the matrix $\mathbf{A}$:

$$\lambda^n + \alpha_{n-1}\lambda^{n-1} + \cdots + \alpha_1\lambda + \alpha_0 = 0$$

the n roots of the characteristic equation, $\lambda_1, \lambda_2, \ldots, \lambda_n$ are the *eigenvalues* (or *characteristic values* or *latent roots*) of the matrix $\mathbf{A}$. For example, for the 3×3 matrix

$$\mathbf{A} = \begin{bmatrix} 3 & 0 & -1 \\ -6 & 0 & 1 \\ 0 & -2 & -3 \end{bmatrix}$$

the characteristic equation is

$$|\lambda\mathbf{I} - \mathbf{A}| = \begin{vmatrix} \lambda - 3 & 0 & 1 \\ 6 & \lambda & -1 \\ 0 & 2 & \lambda + 3 \end{vmatrix}$$

$$= (\lambda - 3)(\lambda^2 + 3\lambda + 2) + 12$$

$$= \lambda^3 - 7\lambda + 6 = (\lambda - 1)(\lambda - 2)(\lambda + 3) = 0$$

The eigenvalues of $\mathbf{A}$ are thus

$$\lambda_1 = 1 \qquad \lambda_2 = 2 \qquad \lambda_3 = -3$$

An $n \times n$ matrix $\mathbf{A}$ always has an nth-degree characteristic equation and thus n eigenvalues. If the elements of $\mathbf{A}$ are real numbers, the coefficients of the characteristic equation are real numbers, so complex roots always occur in conjugate pairs. Table 7-1 lists some other important properties of the eigenvalues of a matrix.

Corresponding to each of the n eigenvalues λ_i of a matrix $\mathbf{A}$ is an eigenvector $\mathbf{p}$ that satisfies

$$\mathbf{A}\mathbf{p}^i = \lambda_i\mathbf{p}^i \tag{7-3}$$

The eigenvectors form the columns of the desired transformation matrix $\mathbf{P}$. If the eigenvalues λ_i are distinct, the eigenvectors can be found by determining nontrivial solutions $\mathbf{p}^i$ to (7-3), or

$$(\lambda_i\mathbf{I} - \mathbf{A})\mathbf{p}^i = 0 \tag{7-4}$$

Each involving a different eigenvalue λ_i. The sets of homogeneous equations (7-4) will each have a nontrivial solution because the λ_i are precisely the numbers that make $(\lambda_i\mathbf{I} - \mathbf{A})$ singular. Because the equations for the eigenvectors $\mathbf{p}^i$ are homogeneous, a nonzero constant times an eigenvector is also an eigenvector.

For example, for the matrix

$$\mathbf{A} = \begin{bmatrix} 3 & 0 & -1 \\ -6 & 0 & 1 \\ 0 & -2 & -3 \end{bmatrix}$$

it was found that the eigenvalues were $\lambda_1 = 1$, $\lambda_2 = 2$, and $\lambda_3 = -3$. The eigenvector $\mathbf{p}^1$ corresponding to λ_1 has components that satisfy

TABLE 7-1   Properties of the Eigenvalues of a Matrix

DEFINING PROPERTY

$$| \lambda \mathbf{I} - \mathbf{A} | = 0$$

FURTHER PROPERTIES

(1) $\lambda = 0$ is an eigenvalue of $\mathbf{A}$ if and only if $\mathbf{A}$ is singular. Since for $\lambda = 0$,

$$| \lambda \mathbf{I} - \mathbf{A} | = | -\mathbf{A} | = -| \mathbf{A} | = 0$$

(2) The eigenvalues of the matrix $(k\mathbf{A})$ are k times the eigenvalues of $\mathbf{A}$, for any scalar k because $| k\lambda \mathbf{I} - k\mathbf{A} | = 0$ for each value of λ for which

$$| \lambda \mathbf{I} - \mathbf{A} | = 0$$

(3) The eigenvalues of $\mathbf{A}^\dagger$ are the same as the eigenvalues of $\mathbf{A}$.

(4) The eigenvalues of $\mathbf{A}^{-1}$, provided that $\mathbf{A}^{-1}$ exists, are the inverses of the eigenvalues of $\mathbf{A}$:

$$| \lambda \mathbf{I} - \mathbf{A} | = | \lambda \mathbf{A}\mathbf{A}^{-1} - \mathbf{A} | = | \mathbf{A}(\lambda \mathbf{A}^{-1} - \mathbf{I}) | = -| \mathbf{A} | | \tfrac{1}{\lambda} \mathbf{I} - \mathbf{A}^{-1} |$$

(5) The eigenvalues of $\mathbf{A}^k$ (k an integer) are the eigenvalues of $\mathbf{A}$ raised to the kth power:

$$0 = | \lambda \mathbf{I} - \mathbf{A} | = | \lambda \mathbf{I} - \mathbf{A} | | \lambda \mathbf{I} + \mathbf{A} | = | (\lambda \mathbf{I} - \mathbf{A}) (\lambda \mathbf{I} + \mathbf{A}) | = | \lambda^2 \mathbf{I} - \mathbf{A}^2 |$$

(6) The eigenvalues of a diagonal matrix are the diagonal elements.

(7) The sum of the eigenvalues of an $n \times n$ matrix $\mathbf{A}$ with characteristic equation is trace $(\mathbf{A})$.

(8) The product of the eigenvalues of an $n \times n$ matrix $\mathbf{A}$ with characteristic equation is $| \mathbf{A} |$.

$$(\lambda_1 \mathbf{I} - \mathbf{A})\mathbf{p}^1 = \begin{bmatrix} -2 & 0 & 1 \\ 6 & 1 & -1 \\ 0 & 2 & 4 \end{bmatrix} \begin{bmatrix} p_{11} \\ p_{21} \\ p_{31} \end{bmatrix} = 0$$

or

$$\begin{aligned} -2p_{11} \qquad\quad + \quad p_{31} &= 0 \\ 6p_{11} + p_{21} - p_{31} &= 0 \\ 2p_{21} + 4p_{31} &= 0 \end{aligned}$$

These equations must be linearly dependent so that there is a nontrivial solution for $\mathbf{p}^1$. The second equation is a linear combination of the other two equations (it is -3 times the first equation plus $\tfrac{1}{2}$ times the third equation), so deleting it gives the equivalent set

$$\begin{aligned} -2p_{11} \qquad + \quad p_{31} &= 0 \\ 2p_{21} + 4p_{31} &= 0 \end{aligned}$$

Letting $p_{31} = 2$ for convenience, then,

$$\mathbf{p}^1 = \begin{bmatrix} p_{11} \\ p_{21} \\ p_{31} \end{bmatrix} = \begin{bmatrix} 1 \\ -4 \\ 2 \end{bmatrix}$$

Other choices of p_{31} simply scale the eigenvector, the norm of which is arbitrary.

For the eigenvector $\lambda_2 = 2$, the second eigenvector satisfies

$$(\lambda_2 \mathbf{I} - \mathbf{A})\mathbf{p}^2 = \begin{bmatrix} -1 & 0 & 1 \\ 6 & 2 & -1 \\ 0 & 2 & 5 \end{bmatrix} \begin{bmatrix} p_{12} \\ p_{22} \\ p_{32} \end{bmatrix} = 0$$

or

$$\begin{aligned} -p_{12} \quad\quad\quad + \quad p_{32} &= 0 \\ 6p_{12} + 2p_{22} - \quad p_{32} &= 0 \\ 2p_{22} + 5p_{32} &= 0 \end{aligned}$$

The second equation, being -6 times the first equation plus the third equation, is linearly dependent. Deleting it, an equivalent set is

$$\begin{aligned} -p_{12} \quad\quad\quad + \quad p_{32} &= 0 \\ 2p_{22} + 5p_{32} &= 0 \end{aligned}$$

Letting $p_{32} = 2$, then,

$$\mathbf{p}^2 = \begin{bmatrix} p_{12} \\ p_{22} \\ p_{32} \end{bmatrix} = \begin{bmatrix} 2 \\ -5 \\ 2 \end{bmatrix}$$

For $\lambda_3 = -3$,

$$(\lambda_3 \mathbf{I} - \mathbf{A})\mathbf{p}^3 = \begin{bmatrix} -6 & 0 & 1 \\ 6 & -3 & -1 \\ 0 & 2 & 0 \end{bmatrix} \begin{bmatrix} p_{13} \\ p_{23} \\ p_{33} \end{bmatrix} = 0$$

or

$$\begin{aligned} -6p_{13} \quad\quad\quad + \quad p_{33} &= 0 \\ 6p_{13} - 3p_{23} - p_{33} &= 0 \\ p_{23} \quad\quad\quad &= 0 \end{aligned}$$

Deleting the second, linearly dependent equation, an equivalent set is

$$\begin{aligned} -6p_{13} \quad\quad\quad + \quad p_{33} &= 0 \\ p_{23} \quad\quad\quad &= 0 \end{aligned}$$

In these $p_{23} = 0$, but either p_{13} or p_{33} is arbitrary. Choosing $p_{13} = 1$, then, an eigenvector corresponding to the eigenvalue is

$$\mathbf{p}^3 = \begin{bmatrix} p_{13} \\ p_{23} \\ p_{33} \end{bmatrix} = \begin{bmatrix} 1 \\ 0 \\ 6 \end{bmatrix}$$

Important properties of the eigenvectors of a matrix are summarized in Table 7-2. The linear independence of the eigenvectors corresponding to distinct eigenvalues can be shown as follows: Suppose that two eigenvectors, $\mathbf{p}^1$ and $\mathbf{p}^2$, corresponding to distinct

TABLE 7-2 Properties of the Eigenvectors of a Matrix

DEFINING PROPERTY

$$\mathbf{Ap} = \lambda\mathbf{p}$$

FURTHER PROPERTIES

(1) If two eigenvalues of a matrix are distinct, the corresponding eigenvectors are linearly independent.

(2) For a matrix $\mathbf{A}$ with real elements, the eigenvectors, if complex, can be expressed as complex conjugate pairs.

(3) The eigenvectors of a matrix $(k\mathbf{A})$ are identical to the eigenvectors of $\mathbf{A}$, for any scalar k. Since the eigenvalues of $(k\mathbf{A})$ are k times the eigenvalues of $\mathbf{A}$, if $\mathbf{Ap} = \lambda\mathbf{p}$, then $(k\mathbf{A})\mathbf{p} = (k\lambda)\mathbf{p}$.

(4) The eigenvectors of $\mathbf{A}^{-1}$ are the same as the eigenvectors of $\mathbf{A}$. Eigenvalues of $\mathbf{A}^{-1}$ are inverses of the eigenvalues of $\mathbf{A}$ and if $\mathbf{Ap} = \lambda\mathbf{p}$, then

$$\frac{1}{\lambda}\mathbf{A}^{-1}\mathbf{Ap} = \mathbf{A}^{-1}\mathbf{p}$$

or

$$\mathbf{A}^{-1}\mathbf{p} = \frac{1}{\lambda}\mathbf{p}$$

eigenvalues λ_1 and λ_2, are linearly dependent. This is equivalent to supposing that $\mathbf{p}^1 = \mathbf{p}^2$ because the scaling of an eigenvector is arbitrary. Then

$$(\lambda_1\mathbf{I} - \mathbf{A})\mathbf{p}^1 = (\lambda_2\mathbf{I} - \mathbf{A})\mathbf{p}^2 = (\lambda_2\mathbf{I} - \mathbf{A})\mathbf{p}^1 = 0$$

Subtracting gives

$$(\lambda_1 - \lambda_2)\mathbf{p}^1 = 0$$

which can only be true for $\mathbf{p}^1 = 0$ or $\lambda_1 = \lambda_2$. Thus $\mathbf{p}^1$ and $\mathbf{p}^2$ must be linearly independent. In a similar way, it can be shown that all eigenvectors of a matrix corresponding to distinct eigenvalues are pairwise linearly independent. This is sufficient to show the eigenvectors to be independent, since if all of a set of n vectors are pairwise linearly independent, they must span n dimensions.

7.2.2 Similarity Transformations

When a matrix $\mathbf{A}$ has distinct eigenvalues, its eigenvectors are linearly independent, so the modal matrix

$$\mathbf{P} = [\mathbf{p}^1 \mid \mathbf{p}^2 \mid \cdots \mid \mathbf{p}^n]$$

is nonsingular. The transformation

$$\mathbf{x} = \mathbf{Px}' \qquad \mathbf{x}' = \mathbf{P}^{-1}\mathbf{x}$$

$$\mathbf{b} = \mathbf{Pb}' \qquad \mathbf{b}' = \mathbf{P}^{-1}\mathbf{b}$$

when substituted into the equations

$$\mathbf{Ax} = \mathbf{b}$$

gives

$$\mathbf{APx'} = \mathbf{Pb'}$$

or

$$(\mathbf{P}^{-1}\mathbf{AP})\mathbf{x'} = \mathbf{\Lambda x'} = \mathbf{b'}$$

where the spectral matrix $\mathbf{\Lambda}$ is diagonal:

$$\mathbf{\Lambda} = \begin{bmatrix} \lambda_1 & 0 & 0 & \cdots & 0 \\ 0 & \lambda_2 & 0 & \cdots & 0 \\ \vdots & \vdots & \vdots & \vdots & \vdots \\ 0 & 0 & 0 & \cdots & \lambda_n \end{bmatrix}$$

The order that the eigenvalues appear along the diagonal of the spectral matrix is the order that the eigenvectors are placed as columns of the modal matrix $\mathbf{P}$. There are thus a number of different modal matrices, each yielding a different but closely related spectral matrix. Any column of a modal matrix can be multiplied by a nonzero constant, giving a new modal matrix, becuase the eigenvectors, being solutions of homogeneous equations, are of arbitrary norm. Eigenvector scaling changes in the modal matrix do not affect the spectral matrix.

Continuing with the example for which

$$\mathbf{A} = \begin{bmatrix} 3 & 0 & -1 \\ -6 & 0 & 1 \\ 0 & -2 & -3 \end{bmatrix}$$

$\lambda_1 = 1$, $\lambda_2 = 2$, $\lambda_3 = -3$, and

$$\mathbf{p}^1 \begin{bmatrix} 1 \\ -4 \\ 2 \end{bmatrix} \qquad \mathbf{p}^2 = \begin{bmatrix} 2 \\ -5 \\ 2 \end{bmatrix} \qquad \mathbf{p}^3 = \begin{bmatrix} 1 \\ 0 \\ 6 \end{bmatrix}$$

A modal matrix is

$$\mathbf{P} = \begin{bmatrix} 1 & 2 & 1 \\ -4 & -5 & 0 \\ 2 & 2 & 6 \end{bmatrix}$$

and

$$\mathbf{P}^{-1}\mathbf{AP} = \frac{1}{20} \begin{bmatrix} -30 & -10 & 5 \\ 24 & 4 & -4 \\ 2 & 2 & 3 \end{bmatrix} \begin{bmatrix} 3 & 0 & -1 \\ -6 & 0 & 1 \\ 0 & -2 & -3 \end{bmatrix} \begin{bmatrix} 1 & 2 & 1 \\ -4 & -5 & 0 \\ 2 & 2 & 6 \end{bmatrix}$$

$$= \begin{bmatrix} 1 & 0 & 0 \\ 0 & 2 & 0 \\ 0 & 0 & -3 \end{bmatrix}$$

If the matrix **A** has repeated eigenvalues, linearly independent eigenvectors corresponding to the same root must be found for each eigenvalue repetition, or else a nonsingular modal matrix **P** will not exist. There must be an additional linearly independent solution of

$$(\lambda_i \mathbf{I} - \mathbf{A})\mathbf{p}^i = 0$$

for each repetition of λ_i. This occurs only if

$$\text{rank } (\lambda_i \mathbf{I} - \mathbf{A}) = n - r$$

where r is the number of repetitions of the root λ_i, which is not likely to be the case.

Any nonsingular transformation of a square matrix **A** of the form $\mathbf{P}^{-1}\mathbf{AP}$, whether or not it results in a diagonal matrix, is termed a *similarity transformation*. Properties of similarity transformations are listed in Table 7-3. Two matrices, **A** and **B**, that are related by a similarity transformation are termed *similar*. Similar matrices have the same eigenvalues and thus the same characteristic equation.

TABLE 7-3   Properties of Similarity Transformations

A nonsingular transformation **P** of a matrix **A**, of the form

$$\mathbf{B} = \mathbf{P}^{-1}\mathbf{AP}$$

has the following properties:

(1) $\mathbf{P}^{-1}(\mathbf{A}^2)\mathbf{P} = \mathbf{B}^2$

(2) $\mathbf{P}^{-1}(\mathbf{A}^3)\mathbf{P} = \mathbf{B}^3$

and so on.

(3) Provided that $\mathbf{A}^{-1}$ exists,

$\quad$ $\mathbf{P}^{-1}(\mathbf{A}^{-1})\mathbf{P} = \mathbf{B}^{-1}$

(4) trace (**B**) = trace (**A**)

(5) $|\mathbf{B}| = |\mathbf{A}|$

(6) The eigenvalues of **B** are identical to those of **A**; they are unchanged by a similarity transformation.

7.2.3 Eigenvector Calculation

As the calculation of eigenvectors can be rather involved, we now give several more examples, using the aid of the PIVOT program of Table 2-4. A straightforward method of eigenvector computation for matrices with real distinct eigenvalues is summarized in the algorithm of Table 7-4.

For example, the matrix

$$\mathbf{A} = \begin{bmatrix} 3 & 1 & -1 \\ -22 & -10 & 32 \\ -2 & -1 & 3 \end{bmatrix}$$

has characteristic equation

$$\lambda^3 + 4\lambda^2 + \lambda - 6 = (\lambda - 1)(\lambda + 2)(\lambda + 3) = 0$$

TABLE 7-4 Algorithm for Finding Eigenvectors by Gauss–Jordan Pivoting

DEFINITION

It is desired to find an eigenvector **p** corresponding to the nonrepeated real eigenvalue λ of the $n \times n$ matrix **A**.

ALGORITHM

1. Form the array of coefficients of the n equations in the n elements of the eigenvector **p**,

$$(\lambda I - A)p = 0$$

by adding λ to each of the diagonal elements of $-A$:

$$
\begin{aligned}
(\lambda - a_{11})p_1 - a_{12}p_2 - \cdots - a_{1n}p_n &= 0 \\
-a_{21}p_1 + (\lambda - a_{22})p_2 - \cdots - a_{2n}p_n &= 0 \\
&\vdots \\
a_{n1}p_1 - a_{n2}p_2 - \cdots + (\lambda - a_{nn})p_n &= 0
\end{aligned}
\tag{7-5}
$$

2. Fully Gauss–Jordan pivot these equations using an algorithm such as that for the pivot program.

3. One of these equations will be linearly dependent on the others, indicated by a row of zeros in the fully pivoted array. Delete it from the set (7-5). There will remain $n - 1$ equations in the n variables $p_1, p_2, \ldots, p_n$. Assign a nonzero value, say unity, to any of the variables, place these knowns on the right side of the equations, and solve for the remaining $n - 1$ elements of the eigenvector **p**.

and eigenvalues

$$\lambda_1 = 1 \qquad \lambda_2 = -2 \qquad \lambda_3 = -3$$

For the first of these eigenvalues, the homogeneous equations to be solved for a corresponding eigenvector

$$
\mathbf{p}^1 = \begin{bmatrix} p_{11} \\ p_{21} \\ p_{31} \end{bmatrix}
$$

are

$$(1I - A)p^1 = 0$$

or

$$
\begin{aligned}
-2p_{11} - p_{21} + p_{31} &= 0 \\
22p_{11} + 11p_{21} - 32p_{31} &= 0 \\
2p_{11} + p_{21} - 2p_{31} &= 0
\end{aligned}
$$

The PIVOT program gives the equivalent fully pivoted equations as

```
FULLY PIVOTED EQUATIONS ARE
COEFFICIENTS OF EQUATION 1
1
.5
0
KNOWN IS 0
```

```
COEFFICIENTS OF EQUATION 2
0
0
1
KNOWN IS 0
COEFFICIENTS OF EQUATION 3
0
0
0
KNOWN IS 0
```

or

$$p_{11} + 0.5p_{21} = 0$$

$$p_{31} = 0$$

$$0 = 0$$

Letting $p_1^1 = 1$, an eigenvector is

$$\mathbf{p}^1 = \begin{bmatrix} 1 \\ -2 \\ 0 \end{bmatrix}$$

Note that in this example, one cannot just assign a nonzero value to any of the variables. p_{31} is necessarily zero here. Some decision logic is therefore necessary in using this method of eigenvector solution.

For a second eigenvalue, $\lambda_2 = -2$.

$$(-2\mathbf{I} - \mathbf{A})\mathbf{p}^2 = 0$$

gives

$$-5p_{12} - p_{22} + p_{32} = 0$$
$$22p_{12} + 8p_{22} - 32p_{32} = 0$$
$$2p_{12} + p_{22} - 5p_{32} = 0$$

Applying the PIVOT program to these, we have

```
FULLY PIVOTED EQUATIONS ARE
COEFFICIENTS OF EQUATION 1
1
.173913044
0
KNOWN IS 0
COEFFICIENTS OF EQUATION 2
0
-.130434783
1
KNOWN IS 0
COEFFICIENTS OF EQUATION 3
0
0
0
KNOWN IS 0
```

or

$$p_{12} + 0.174p_{22} \qquad \doteq 0$$
$$- 0.130p_{22} + p_{32} \doteq 0$$
$$0 \doteq 0$$

Letting $p_{22} = 1$ gives

$$\mathbf{p}^2 = \begin{bmatrix} -0.174 \\ 1 \\ 0.130 \end{bmatrix}$$

Similarly, for $\lambda_3 = -3$, the homogeneous equations for the eigenvector elements are

$$-6p_{13} - p_{23} + p_{33} = 0$$
$$22p_{13} + 7p_{23} - 32p_{33} = 0$$
$$2p_{13} + p_{23} - 6p_{33} = 0$$

Using PIVOT again gives the equivalent set,

```
FULLY PIVOTED EQUATIONS ARE
COEFFICIENTS OF EQUATION 1
1
.147058824
0
KNOWN IS 0
COEFFICIENTS OF EQUATION 2
0
-.117647059
1
KNOWN IS 0
COEFFICIENTS OF EQUATION 3
0
3.06499714E-10
0
KNOWN IS 0
```

or

$$p_{13} + 0.147p_{23} \qquad \doteq 0$$
$$- 0.118p_{23} + p_{33} \doteq 0$$
$$0 \doteq 0$$

Letting $p_{23} = 1$ yields

$$\mathbf{p}^3 = \begin{bmatrix} -0.147 \\ 1 \\ 0.118 \end{bmatrix}$$

For repeated eigenvalues, there may or may not be more than one linearly independent eigenvector. For example, the matrix

$$\mathbf{A} = \begin{bmatrix} -8 & 1 & 0 \\ -21 & 0 & 1 \\ -18 & 0 & 0 \end{bmatrix}$$

has characteristic equation

$$\lambda^3 + 8\lambda^2 + 21\lambda + 18 = (\lambda + 2)(\lambda + 3)^2 = 0$$

and thus a repeated $\lambda = -3$ eigenvalue. For the nonrepeated $\lambda_1 = -2$ eigenvalue, the homogeneous equations for the eigenvector

$$\mathbf{p}^1 = \begin{bmatrix} p_{11} \\ p_{21} \\ p_{31} \end{bmatrix}$$

are

$$(-2\mathbf{I} - \mathbf{A})\mathbf{p}^1 = 0$$

or

$$\begin{aligned} 6p_{11} - p_{21} &= 0 \\ 21p_{11} - 2p_{21} - p_{31} &= 0 \\ 18p_{11} - 2p_{31} &= 0 \end{aligned}$$

The PIVOT program gives the equivalent fully pivoted equations as

```
FULLY PIVOTED EQUATIONS ARE
COEFFICIENTS OF EQUATION 1
1
0
-.111111111
KNOWN IS 0
COEFFICIENTS OF EQUATION 2
0
1
-.666666667
KNOWN IS 0
COEFFICIENTS OF EQUATION 3
0
0
0
KNOWN IS 0
```

or

$$\begin{aligned} p_{11} - 0.111p_{31} &\doteq 0 \\ p_{21} - 0.667p_{31} &\doteq 0 \\ 0 &\doteq 0 \end{aligned}$$

Letting $p_{31} = 1$, this gives

$$\mathbf{p}^1 = \begin{bmatrix} 0.111 \\ 0.667 \\ 1 \end{bmatrix}$$

For the repeated eigenvalue $\lambda_2 = \lambda_3 = -3$, the homogeneous equations for the eigenvector(s) are

$$(-3\mathbf{I} - \mathbf{A})\mathbf{p}^2 = 0$$

or

$$
\begin{aligned}
5p_{12} - p_{22} &= 0 \\
21p_{12} - 3p_{22} - p_{32} &= 0 \\
18p_{12} \qquad\quad - 3p_{32} &= 0
\end{aligned}
$$

Using PIVOT, the result is the equivalent set of equations

```
FULLY PIVOTED EQUATIONS ARE
COEFFICIENTS OF EQUATION 1
1
0
-.166666667
KNOWN IS 0
COEFFICIENTS OF EQUATION 2
0
1
-.833333333
KNOWN IS 0
COEFFICIENTS OF EQUATION 3
0
0
0
KNOWN IS 0
```

or

$$
\begin{aligned}
p_{12} \qquad\quad - 0.167p_{32} &\doteq 0 \\
p_{22} - 0.833p_{32} &\doteq 0 \\
0 &\doteq 0
\end{aligned}
$$

Only one of the three variables is arbitrary, so there is only one linearly independent eigenvector. Choosing $p_{32} = 1$,

$$\mathbf{p}^2 = \begin{bmatrix} 0.167 \\ 0.833 \\ 1 \end{bmatrix}$$

For the matrix

$$\mathbf{A} = \begin{bmatrix} 1 & 0 & -1 \\ -1 & 2 & -1 \\ 0 & 0 & 2 \end{bmatrix} \tag{7-6}$$

however, even though the characteristic root $\lambda = 2$ is repeated, there are three linearly independent eigenvectors. For this matrix, the characteristic equation is

$$| \lambda \mathbf{I} - \mathbf{A} | = \lambda^3 - 5\lambda^2 + 8\lambda - 4 = (\lambda - 1)(\lambda - 2)^2 = 0$$

For the eigenvalue $\lambda = 1$, the homogeneous equations for the corresponding eigenvector are

$$
\begin{array}{ccc}
p_{31} = 0 & & p_{31} = 0 \\
p_{11} - p_{21} + p_{31} = 0 & \text{or} & p_{11} - p_{21} \quad = 0 \\
- p_{31} = 0 & & 0 = 0
\end{array}
$$

which is an equivalent fully pivoted set. The variable p_{31} cannot be chosen arbitrarily here. Choosing $p_{11} = 1$, an eigenvector is

$$\mathbf{p}^1 = \begin{bmatrix} 1 \\ 1 \\ 0 \end{bmatrix}$$

For the repeated $\lambda = 2$ eigenvalue, the homogeneous equations are

$$
\begin{array}{ccc}
p_{12} + p_{32} = 0 & & p_{12} + p_{32} = 0 \\
p_{12} + p_{32} = 0 & \text{or} & 0 = 0 \\
0 = 0 & & 0 = 0
\end{array}
$$

Here two of the three variables can be chosen arbitrarily. For $p_{12} = 0$, $p_{22} = 1$, then,

$$\mathbf{p}^2 = \begin{bmatrix} 0 \\ 1 \\ 0 \end{bmatrix}$$

For $p_{13} = 1$, $p_{23} = 0$,

$$\mathbf{p}^3 = \begin{bmatrix} 1 \\ 0 \\ -1 \end{bmatrix}$$

and these two eigenvectors corresponding to the same eigenvalue are linearly independent. A modal matrix for $\mathbf{A}$ of (7-6) is thus

$$\mathbf{P} = [\mathbf{p}^1 \mid \mathbf{p}^2 \mid \mathbf{p}^3] = \begin{bmatrix} 1 & 0 & 1 \\ 1 & 1 & 0 \\ 0 & 0 & -1 \end{bmatrix} \quad \text{and} \quad \mathbf{\Lambda} = \mathbf{P}^{-1}\mathbf{A}\mathbf{P} = \begin{bmatrix} 1 & 0 & 0 \\ 0 & 2 & 0 \\ 0 & 0 & 2 \end{bmatrix}$$

is diagonal, as expected.

If the eigenvalues of a matrix are complex, the corresponding eigenvectors will be complex. With proper scaling, the eigenvectors for complex conjugate eigenvalues will, themselves, be complex conjugates. For the matrix

$$\mathbf{A} = \begin{bmatrix} -2 & 1 \\ -5 & 0 \end{bmatrix}$$

for example, the characteristic equation is

$$| \lambda \mathbf{I} - \mathbf{A} | = \lambda^2 + 2\lambda + 5 = (\lambda + 1 + \underline{i}2)(\lambda + 1 - \underline{i}2) = 0$$

so the eigenvalues are

$$\lambda_1, \lambda_2 = -1 \pm \underline{i}2$$

For the first of these eigenvalues, the homogeneous equations for the eigenvectors are

$$[(-1 + \underline{i}2)\mathbf{I} - \mathbf{A}]\mathbf{p}^1 = \mathbf{0}$$

or

$$
\begin{aligned}
(1 + \underline{i}2)p_{11} - \quad\quad\quad\quad p_{21} &= 0 \\
5p_{11} + (-1 + \underline{i}2)p_{21} &= 0
\end{aligned}
\tag{7-7}
$$

Letting

$$p_{11} = e_{11} + \underline{i}f_{11} \quad\quad p_{21} = e_{21} + \underline{i}f_{21}$$

and separately equating the real and imaginary parts of each equation of (7-7), there results the four real equations

$$
\begin{aligned}
e_{11} - 2f_{11} - e_{21} \quad\quad\quad &= 0 \\
2e_{11} + f_{11} \quad\quad\quad - f_{21} &= 0 \\
5e_{11} \quad\quad\quad - e_{21} - 2f_{21} &= 0 \\
5f_{11} + 2e_{21} - f_{21} &= 0
\end{aligned}
$$

Using the PIVOT program, these equations have an equivalent fully pivoted set given by

```
FULLY PIVOTED EQUATIONS ARE
COEFFICIENTS OF EQUATION 1
0
1
.4
-.2
KNOWN IS 0
COEFFICIENTS OF EQUATION 2
1
0
-.2
-.4
KNOWN IS 0
COEFFICIENTS OF EQUATION 3
0
0
0
0
KNOWN IS 0
COEFFICIENTS OF EQUATION 4
0
0
0
0
KNOWN IS 0
```

or

$$f_{11} + 0.4e_{21} - 0.2f_{21} = 0$$
$$e_{11} \quad - 0.2e_{21} - 0.4f_{21} = 0$$
$$0 = 0$$
$$0 = 0$$

Choosing the first component of the eigenvector to be

$$e_{11} + \underline{i}f_{11} = 1 + \underline{i}0$$

then the second eigenvector component is given by

$$0.4e_{21} - 0.2f_{21} = \quad 0$$
$$-0.2e_{21} - 0.4f_{21} = -1$$

or

$$e_{21} + \underline{i}f_{21} = 1 + \underline{i}2$$

so that

$$\mathbf{P}^1 = \begin{bmatrix} 1 \\ 1 + \underline{i}2 \end{bmatrix}$$

The complex conjugate eigenvalue will have

$$\mathbf{p}^2 = \mathbf{p}^{1*} = \begin{bmatrix} 1 \\ 1 - \underline{i}2 \end{bmatrix}$$

as an associated eigenvector.

7.3 CAYLEY–HAMILTON THEOREM

Powers of a square matrix $\mathbf{A}$ involve simply multiplying $\mathbf{A}$ by itself:

$$\mathbf{A}^2 = \mathbf{AA} \qquad \mathbf{A}^3 = \mathbf{AAA}$$

and so on. With the additional definitions

$$\mathbf{A}^0 = \mathbf{I}$$

and

$$\mathbf{A}^{-2} = \mathbf{A}^{-1}\mathbf{A}^{-1} \qquad \mathbf{A}^{-3} = \mathbf{A}^{-1}\mathbf{A}^{-1}\mathbf{A}^{-1}$$

and so on, any integer power of a matrix is defined, and the law of exponents holds:

$$\mathbf{A}^m\mathbf{A}^n = \mathbf{A}^{(m+n)}$$

A similarity transformation that diagonalizes $\mathbf{A}$,

$$\mathbf{P}^{-1}\mathbf{AP} = \mathbf{\Lambda}$$

also diagonalizes any power of $\mathbf{A}$ since

$$\mathbf{P}^{-1}\mathbf{A}^2\mathbf{P} = \mathbf{P}^{-1}\mathbf{APP}^{-1}\mathbf{AP} = \mathbf{\Lambda}^2$$

$$\mathbf{P}^{-1}\mathbf{A}^3\mathbf{P} = \mathbf{P}^{-1}\mathbf{A}\mathbf{P}\mathbf{P}^{-1}\mathbf{A}\mathbf{P}\mathbf{P}^{-1}\mathbf{A}\mathbf{P} = \mathbf{\Lambda}^3$$

and so on. And if $\mathbf{A}$ is nonsingular (it then has no zero eigenvalues),

$$(\mathbf{P}^{-1}\mathbf{A}\mathbf{P})^{-1} = \mathbf{P}^{-1}\mathbf{A}^{-1}\mathbf{P} = \mathbf{\Lambda}^{-1}$$

$$\mathbf{P}^{-1}\mathbf{A}^{-2}\mathbf{P} = \mathbf{P}^{-1}\mathbf{A}^{-1}\mathbf{P}\mathbf{P}^{-1}\mathbf{A}^{-1}\mathbf{P} = \mathbf{\Lambda}^{-2}$$

and so on.

The Cayley–Hamilton theorem describes a remarkable property of every square matrix $\mathbf{A}$. If the characteristic equation of $\mathbf{A}$ is

$$\lambda^n + \alpha_{n-1}\lambda^{n-1} + \cdots + \alpha_1\lambda + \alpha_0 = 0$$

the matrix itself satisfies the same equation, namely

$$\mathbf{A}^n + \alpha_{n-1}\mathbf{A}^{n-1} + \cdots + \alpha_1\mathbf{A} + \alpha_0\mathbf{I} = 0 \qquad (7\text{-}8)$$

The result holds in general but will be shown now for matrices that can be diagonalized. For such a matrix

$$\mathbf{A} = \mathbf{P}\mathbf{\Lambda}\mathbf{P}^{-1}$$

where $\mathbf{P}$ is a modal matrix and $\mathbf{\Lambda}$ is the spectral matrix. Similarly,

$$\mathbf{A}^2 = \mathbf{P}^{-1}\mathbf{\Lambda}\mathbf{P}\mathbf{P}^{-1}\mathbf{\Lambda}\mathbf{P} = \mathbf{P}^{-1}\mathbf{\Lambda}^2\mathbf{P}$$

$$\mathbf{A}^3 = \mathbf{P}^{-1}\mathbf{\Lambda}^3\mathbf{P}$$

and so on. Then

$$\mathbf{A}^n + \alpha_{n-1}\mathbf{A}^{n-1} + \cdots + \alpha_1\mathbf{A} + \alpha_0\mathbf{I}$$

$$= \mathbf{P}[\mathbf{\Lambda}^n + \alpha_{n-1}\mathbf{\Lambda}^{n-1} + \cdots + \alpha_1\mathbf{\Lambda} + \alpha_0\mathbf{I}]\mathbf{p}^{-1}$$

$$= \mathbf{P}\begin{bmatrix} \lambda_1^n + \alpha_{n-1}\lambda_1^{n-1} + \cdots + \alpha_1\lambda_1 + \alpha_0 & & & 0 & \cdots \\ 0 & & & & \\ \vdots & & \lambda_2^n + \alpha_{n-1}\lambda_2^{n-1} + \cdots + \alpha_1\lambda_2 + \alpha_0 & & \cdots \\ 0 & & \vdots & & \\ & & 0 & & \cdots \end{bmatrix}\mathbf{P}^{-1} = 0$$

Using the Cayley–Hamilton theorem, the nth power of any $n \times n$ square matrix $\mathbf{A}$ can be expressed in terms of lesser powers of $\mathbf{A}$:

$$\mathbf{A}^n = -\alpha_{n-1}\mathbf{A}^{n-1} - \cdots - \alpha_1\mathbf{A} - \alpha_0\mathbf{I}$$

By repeatedly substituting for $\mathbf{A}^n$ any power of $\mathbf{A}$ can be expressed in terms of the $(n-1)$th and lower powers of $\mathbf{A}$, down to and including $\mathbf{A}^0 = \mathbf{I}$. For example, the 3×3 matrix

$$\mathbf{A} = \begin{bmatrix} 1 & -2 & 0 \\ 3 & -1 & 4 \\ 2 & 1 & 1 \end{bmatrix}$$

has characteristic equation

$$|\lambda\mathbf{I} - \mathbf{A}| = \begin{vmatrix} \lambda - 1 & 2 & 0 \\ -3 & \lambda + 1 & -4 \\ -2 & -1 & \lambda - 1 \end{vmatrix} = (\lambda - 1)(\lambda^2 - 5) - 2(-3\lambda - 5)$$

$$= \lambda^3 - \lambda^2 + \lambda + 15$$

Using the Cayley–Hamilton theorem,

$$\mathbf{A}^3 = \mathbf{A}^2 - \mathbf{A} - 15\mathbf{I}$$

so

$$\mathbf{A}^4 = \mathbf{A}(\mathbf{A}^2 - \mathbf{A} - 15\mathbf{I}) = \mathbf{A}^3 - \mathbf{A}^2 - 15\mathbf{A}$$

$$= \mathbf{A}^2 - \mathbf{A} - 15\mathbf{I} - \mathbf{A}^2 - 15\mathbf{A} = -16\mathbf{A} - 15\mathbf{I}$$

Multiplying each term in equation (7-8) by $\mathbf{A}^{-1}$, when $\mathbf{A}^{-1}$ exists and thus $\alpha_0 \neq 0$, gives an important relationship for the inverse of a matrix:

$$\mathbf{A}^{-1}(\mathbf{A}^n + \alpha_{n-1}\mathbf{A}^{n-1} + \cdots + \alpha_1\mathbf{A} + \alpha_0\mathbf{I})$$

$$= \mathbf{A}^{n-1} + \alpha_{n-1}\mathbf{A}^{n-2} + \cdots + \alpha_1\mathbf{I} + \alpha_0\mathbf{A}^{-1} = 0$$

or

$$\mathbf{A}^{-1} = \frac{1}{\alpha_0}(-\mathbf{A}^{n-1} - \alpha_{n-1}\mathbf{A}^{n-2} - \cdots - \alpha_1\mathbf{I}) \tag{7-9}$$

The inverse, too, can be expressed as a linear combination of $\mathbf{A}^{n-1}$ and lesser powers of $\mathbf{A}$.

7.4 FUNCTIONS OF A SQUARE MATRIX

7.4.1 Matrix Infinite Series

Certain functions of a square matrix occur commonly in the solution of practical problems and are defined by infinite geometric series,

$$\mathbf{F}(\mathbf{A}) = \sum_{i=0}^{\infty} k_i\mathbf{A}^i = k_0\mathbf{I} + k_1\mathbf{A} + k_2\mathbf{A}^2 + \cdots$$

similar to series for scalar functions. A matrix infinite series is said to be convergent if the resulting scalar series for each of the elements of the matrix function is convergent. Some important functions of a square matrix, defined in terms of infinite series matrix, are listed in Table 7-5. These series are convergent for all matrices $\mathbf{A}$.

Some important properties of the matrix exponential are listed in Table 7-6. Properties (1) through (3) can be shown by comparing series expansions. Property (4) is shown by applying (2) and (3), with $\mathbf{B} = \mathbf{A}^{-1}$.

TABLE 7-5 Common Matrix Functions Defined by Infinite Series

(1) Exponential

$$e^{\mathbf{A}} = \exp(\mathbf{A}) = \mathbf{I} + \mathbf{A} + \frac{\mathbf{A}^2}{2!} + \frac{\mathbf{A}^3}{3!} + \cdots$$

(2) Hyperbolic sine

$$\sinh \mathbf{A} = \frac{\exp(\mathbf{A}) - \exp(-\mathbf{A})}{2} = \mathbf{A} + \frac{\mathbf{A}^3}{3!} + \frac{\mathbf{A}^5}{5!} + \cdots$$

(3) Hyperbolic cosine

$$\cosh \mathbf{A} = \frac{\exp(\mathbf{A}) + \exp(-\mathbf{A})}{2} = \mathbf{I} + \frac{\mathbf{A}^2}{2!} + \frac{\mathbf{A}^4}{4!} + \cdots$$

(4) Sine

$$\sin \mathbf{A} = \frac{\exp(i\mathbf{A}) - \exp(-i\mathbf{A})}{2i} = \mathbf{A} - \frac{\mathbf{A}^3}{3!} + \frac{\mathbf{A}^5}{5!} - \cdots$$

(5) Cosine

$$\cos \mathbf{A} = \frac{\exp(i\mathbf{A}) + \exp(-i\mathbf{A})}{2} = \mathbf{I} - \frac{\mathbf{A}^2}{2!} + \frac{\mathbf{A}^4}{4!} - \cdots$$

TABLE 7-6 Properties of the Matrix Exponential

(1) $e^{\mathbf{A}}e^{\mathbf{B}} \neq e^{\mathbf{B}}e^{\mathbf{A}}$ in general

(2) If $\mathbf{BA} = \mathbf{AB}$, then $e^{\mathbf{A}}e^{\mathbf{B}} = e^{\mathbf{B}}e^{\mathbf{A}} = e^{(\mathbf{A}+\mathbf{B})}$

(3) $e^{0} = \mathbf{I}$

(4) $e^{-\mathbf{A}} = \mathbf{I} - \mathbf{A} + \frac{\mathbf{A}^2}{2!} - \frac{\mathbf{A}^3}{3!} + \cdots = (e^{\mathbf{A}})^{-1}$

7.4.2 Truncated Series Computation Program

One method of computing the matrix defined by an infinite series,

$$\mathbf{F}(\mathbf{A}) = k_0\mathbf{I} + k_1\mathbf{A} + k_2\mathbf{A}^2 + \cdots$$

is to truncate the series after a certain number of terms,

$$\mathbf{F}(\mathbf{A}) = k_0\mathbf{I} + k_1\mathbf{A} + k_2\mathbf{A}^2 + \cdots + k_m\mathbf{A}^m$$

A program in BASIC/FORTRAN, called SERIES and listed in Table 7-7, computes the sum of the first $M + 1$ terms of the series

$$e^{\mathbf{A}} = \mathbf{I} + \mathbf{A} + \frac{\mathbf{A}^2}{2!} + \cdots$$

The program is a little more clumsy for matrix exponentiation than it need be because it is desired to make the routine easy to modify so that it will compute other functions of $\mathbf{A}$. For matrix exponentiation only, it would be more efficient (and probably

TABLE 7-7

Computer Program in BASIC for Summing Matrix Series (SERIES)

VARIABLES USED

N = dimensions of all matrices

$A(20,20)$ = original matrix

$B(20,20)$ = power of the **A** matrix

$C(20,20)$ = temporary product of the **A** and **B** matrices

$D(20,20)$ = matrix series sum

M = power of the highest-degree term in the truncated matrix series

E = coefficient of a series term

I = row index

J = column index

K = row index in summation for matrix multiplication

L = term number index

LISTING

```
100      PRINT "MATRIX EXPONENTIATION"
110      DIM A(20,20),B(20,20),C(20,20),D(20,20)
120   REM ENTER DIMENSIONS AND ELEMENTS OF THE A MATRIX
130      PRINT "PLEASE ENTER MATRIX DIMENSION";
140      INPUT N
150      PRINT "ENTER MATRIX A, COLUMN BY COLUMN"
160      FOR J = 1 TO N
170      PRINT "COLUMN";J
180      FOR I = 1 TO N
190      PRINT "ROW";I;
200      INPUT A(I,J)
210      NEXT I
220      NEXT J
230   REM ENTER NO. TERMS
240      PRINT "ENTER POWER OF HIGHEST DEGREE TERM";
250      INPUT M
260   REM SET IDENTITY TERM COEFFICIENT (HERE, E=1)
270      E = 1
280   REM SET B=I AND D=E*I
290      FOR J = 1 TO N
300      FOR I = 1 TO N
310      B(I,J) = 0
320      D(I,J) = 0
330      NEXT I
340      NEXT J
350      FOR J = 1 TO N
360      B(J,J) = 1
370      D(J,J) = E
380      NEXT J
390   REM SUM THE NEXT M TERMS TO C
400      FOR L = 1 TO M
410   REM COMPUTE SERIES COEFFICIENT E AS A FUNCTION
420   REM OF DEGREE L (HERE, E=1/L!)
430      E = 1
440      FOR I = 1 TO L
450      E = E/I
460      NEXT I
470   REM MULTIPLY C=A*B
480      FOR I = 1 TO N
```

(continued)

```
490        FOR J = 1 TO N
500        C(I,J) = 0
510        FOR K = 1 TO N
520        C(I,J) = C(I,J) + A(I,K) * B(K,J)
530        NEXT K
540        NEXT J
550        NEXT I
560     REM ADD D=D+E*C AND TRANSFER C TO B
570        FOR I = 1 TO N
580        FOR J = 1 TO N
590        D(I,J) = D(I,J) + E*C(I,J)
600        B(I,J) = C(I,J)
610        NEXT J
620        NEXT I
630        NEXT L
640     REM PRINT TRUNCATED SERIES RESULT
650        PRINT "FUNCTION'S COLUMNS ARE"
660        FOR J = 1 TO N
670        PRINT "COLUMN ";J
680        FOR I = 1 TO N
690        PRINT D(I,J)
700        NEXT I
710        NEXT J
720        END
```

Computer Program in FORTRAN for Summing Matrix Series (SERIES)

```
C ************************************************************************
C PROGRAM SERIES -- MATRIX EXPONENTIATION
C ************************************************************************
      REAL A(20,20), B(20,20), C(20,20), D(20,20)

      WRITE(*,'(1X,''MATRIX EXPONENTIATION'')')

C *** ENTER DIMENSIONS AND ELEMENTS OF THE A MATRIX
      WRITE(*,'(1X,''PLEASE ENTER MATRIX DIMENSION '',\)')
      READ(*,*) N
      WRITE(*,'(1X,''ENTER MATRIX A, COLUMN BY COLUMN'')')
      DO 10 J=1, N
         WRITE(*,'(1X,''COLUMN '',I2)') J
         DO 20 I=1, N
            WRITE(*,'(1X,''ROW '',I2,'' = '',\)') I
            READ(*,*) A(I,J)
 20      CONTINUE
 10   CONTINUE

C *** ENTER NO. TERMS
      WRITE(*,'(1X,''ENTER POWER OF HIGHEST DEGREE TERM '',\)')
      READ(*,*) M

C *** SET IDENTITY TERM COEFFICIENT (HERE, E=1)
      E = 1.0

C *** SET B=I AND D=E*I
      DO 30 J=1, N
         DO 40 I=1, N
            B(I,J) = 0.0
            D(I,J) = 0.0
 40      CONTINUE
 30   CONTINUE
      DO 50 J=1, N
         B(J,J) = 1.0
         D(J,J) = E
 50   CONTINUE
```

```
C *** SUM THE NEXT M TERMS TO C
      DO 60 L=1, M

C *** COMPUTE SERIES COEFFICIENT E AS A FUNCTION
C *** OF DEGREE L (HERE, E=1/L!)
          E = 1.0
          DO 70 I=1, L
              E = E/I
 70       CONTINUE

C *** MULTIPLY C=A*B
          DO 80 I=1, N
              DO 90 J=1, N
                  C(I,J) = 0.0
                  DO 100 K=1, N
                      C(I,J) = C(I,J) + A(I,K)*B(K,J)
100               CONTINUE
 90           CONTINUE
 80       CONTINUE

C *** ADD D=D+E*C AND TRANSFER C TO B
          DO 110 I=1, N
              DO 120 J=1, N
                  D(I,J) = D(I,J) + E*C(I,J)
                  B(I,J) = C(I,J)
120           CONTINUE
110       CONTINUE
 60   CONTINUE

C *** PRINT TRUNCATED SERIES RESULT
      WRITE(*,'(1X,''FUNCTIONS COLUMNS ARE'')')
      DO 130 J=1, N
          WRITE(*,'(1X,''COLUMN '',I2)') J
          DO 140 I=1, N
              WRITE(*,'(1X,E15.8)') D(I,J)
140       CONTINUE
130   CONTINUE
      STOP
      END
```

better from the standpoint of numerical error) to divide the new term to be added by **L** at each step instead of computing

$$\mathbf{E} = \frac{1}{\mathbf{L}!}$$

every step.

For the matrix

$$\mathbf{A} = \begin{bmatrix} -0.4 & 0.4 \\ 0.2 & -0.6 \end{bmatrix}$$

the SERIES program gives

```
FUNCTION'S COLUMNS ARE
COLUMN 1
.695596824
.123133929
```

COLUMN 2
.246267858
.572462895

when 11 terms are summed. Convergence is rapid, as can be seen from the results of earlier truncations of the series:

$$\mathbf{D} = \mathbf{I} + \mathbf{A} = \begin{bmatrix} 0.6 & 0.4 \\ 0.2 & 0.4 \end{bmatrix} \qquad \mathbf{D} = \mathbf{I} + \mathbf{A} + \tfrac{1}{2}\mathbf{A}^2 = \begin{bmatrix} 0.72 & 0.2 \\ 0.1 & 0.62 \end{bmatrix}$$

$$\mathbf{D} = \mathbf{I} + \mathbf{A} + \tfrac{1}{2}\mathbf{A}^2 + \tfrac{1}{6}\mathbf{A}^3 = \begin{bmatrix} 0.690666667 & 0.256 \\ 0.128 & 0.562666667 \end{bmatrix}$$

$$\mathbf{D} = \mathbf{I} + \mathbf{A} + \tfrac{1}{2}\mathbf{A}^2 + \tfrac{1}{6}\mathbf{A}^3 + \tfrac{1}{24}\mathbf{A}^4 = \begin{bmatrix} 0.6964 & 0.24466666 \\ 0.122333333 & 0.57406667 \end{bmatrix}$$

One could automate the selection of the number M of series terms to include by ending when each of the elements of the term added has magnitude less than a certain small amount.

7.5 FINDING MATRIX FUNCTIONS

7.5.1 Evaluation by Diagonalization

Another method of finding a matrix defined by an infinite series is to use diagonalization. For a diagonal matrix,

$$\mathbf{A} = \begin{bmatrix} \lambda_1 & 0 & 0 & \cdots & 0 \\ 0 & \lambda_2 & 0 & \cdots & 0 \\ \vdots & \vdots & \vdots & \vdots & \vdots \\ 0 & 0 & 0 & \cdots & \lambda_n \end{bmatrix}$$

the function

$$\mathbf{F}(\mathbf{\Lambda}) = k_0\mathbf{I} + k_1\mathbf{\Lambda} + k_2\mathbf{\Lambda}^2 + \cdots$$

is

$$\mathbf{F}(\mathbf{\Lambda}) = \begin{bmatrix} k_0 + k_1\lambda_1 + k_2\lambda_1^2 + \cdots & 0 & \cdots \\ 0 & k_0 + k_1\lambda_2 + k_2\lambda_2^2 + \cdots & \cdots \\ \vdots & \vdots & \cdots \end{bmatrix}$$

$$= \begin{bmatrix} F(\lambda_1) & 0 & 0 & \cdots & 0 \\ 0 & F(\lambda_2) & 0 & \cdots & 0 \\ \vdots & \vdots & \vdots & \vdots & \vdots \\ 0 & 0 & 0 & \cdots & F(\lambda_n) \end{bmatrix}$$

If a matrix $\mathbf{A}$ involved in an infinite series can be diagonalized,

$$\mathbf{P}^{-1}\mathbf{AP} = \mathbf{\Lambda} \qquad \mathbf{A} = \mathbf{P}\mathbf{\Lambda}\mathbf{P}^{-1}$$

then

$$\mathbf{F}(\mathbf{A}) = k_0\mathbf{I} + k_1\mathbf{A} + k_2\mathbf{A}^2 + \cdots$$

$$= k_0 \mathbf{PIP}^{-1} + k_1 \mathbf{PP}^{-1}\mathbf{APP}^{-1} + k_2 \mathbf{PP}^{-1}\mathbf{A}^2\mathbf{PP}^{-1} + \cdots$$

$$= \mathbf{P}[k_0 \mathbf{I} + k_1 \mathbf{\Lambda} + k_2 \mathbf{\Lambda}^2 + \cdots]\mathbf{P}^{-1} = \mathbf{P}F(\mathbf{\Lambda})\mathbf{P}^{-1}$$

$$= \mathbf{P}\begin{bmatrix} F(\lambda_1) & 0 & 0 & \cdots & 0 \\ 0 & F(\lambda_2) & 0 & \cdots & 0 \\ \vdots & \vdots & \vdots & \vdots & \vdots \\ 0 & 0 & 0 & \cdots & F(\lambda_n) \end{bmatrix}\mathbf{P}^{-1}$$

As a numerical example, consider

$$\mathbf{A} = \begin{bmatrix} 3 & -4 \\ 0 & 2 \end{bmatrix}$$

The similarity transformation $\mathbf{P}$ given below diagonalizes $\mathbf{A}$:

$$\mathbf{P}^{-1}\mathbf{AP} = \begin{bmatrix} 0 & 1 \\ -1 & 4 \end{bmatrix}\begin{bmatrix} 3 & -4 \\ 0 & 2 \end{bmatrix}\begin{bmatrix} 4 & -1 \\ 1 & 0 \end{bmatrix} = \begin{bmatrix} 2 & 0 \\ 0 & 3 \end{bmatrix}$$

Then

$$e^{\mathbf{A}} = \mathbf{P}e^{\mathbf{\Lambda}}\mathbf{P}^{-1} = \begin{bmatrix} 4 & -1 \\ 1 & 0 \end{bmatrix}\begin{bmatrix} e^2 & 0 \\ 0 & e^3 \end{bmatrix}\begin{bmatrix} 0 & 1 \\ -1 & 4 \end{bmatrix}$$

$$= \begin{bmatrix} e^3 & 4e^2 - 4e^3 \\ 0 & e^2 \end{bmatrix} \doteq \begin{bmatrix} 20.086 & -50.786 \\ 0 & 7.389 \end{bmatrix}$$

7.5.2 Cayley–Hamilton Technique

For the matrix infinite series

$$\mathbf{F}(\mathbf{A}) = k_0 \mathbf{I} + k_1 \mathbf{A} + k_2 \mathbf{A}^2 + \cdots$$

since the matrix $\mathbf{A}$ satisfies its own characteristic equation,

$$\mathbf{A}^n + \alpha_{n-1}\mathbf{A}^{n-1} + \cdots + \alpha_1 \mathbf{A} + \alpha_0 \mathbf{I} = 0$$

all powers of $\mathbf{A}$ higher than $\mathbf{A}^{n-1}$ can be expressed in terms of lower powers of $\mathbf{A}$, giving an alternative finite series expression for the function under consideration:

$$\mathbf{F}(\mathbf{A}) = \mathbf{R}(\mathbf{A}) = d_{n-1}\mathbf{A}^{n-1} + \cdots + d_1 \mathbf{A} + d_0 \mathbf{I} = 0$$

the infinite series $\mathbf{F}(\mathbf{A})$ is said to "collapse" to the finite series $\mathbf{R}(\mathbf{A})$.

The corresponding scalar series

$$\mathbf{F}(\lambda) = k_0 + k_1 \lambda + k_2 \lambda^2 + k_3 \lambda^3 + \cdots$$

does not in general collapse. It is only for the values of λ for which

$$\lambda^n + \alpha_{n-1}\lambda^{n-1} + \cdots + \alpha_1 \lambda + \alpha_0 = 0$$

that is, values of λ that are eigenvalues of $\mathbf{A}$ for which the scalar series collapses. Then it collapses with exactly the same coefficients as the matrix series:

$$\mathbf{F}(\lambda_i) = \mathbf{R}(\lambda_i) = d_{n-1}\lambda_i^{n-1} + d_{n-2}\lambda_i^{n-2} + \cdots + d_1 \lambda_i + d_0$$

If the characteristic values of $\mathbf{A}$ are distinct, the following n equations are linearly independent and can be solved for the n unknowns $d_{n-1}, d_{n-2}, \ldots, d_1, d_0$:

$$F(\lambda_1) = d_{n-1}\lambda_1^{n-1} + d_{n-2}\lambda_1^{n-2} + \cdots + d_1\lambda_1 + d_0$$

$$F(\lambda_2) = d_{n-1}\lambda_2^{n-1} + d_{n-2}\lambda_2^{n-2} + \cdots + d_1\lambda_2 + d_0$$

$$\vdots$$

$$F(\lambda_n) = d_{n-1}\lambda_n^{n-1} + d_{n-2}\lambda_n^{n-2} + \cdots + d_1\lambda_n + d_0$$

Having obtained $d_{n-1}, \ldots, d_0$,

$$\mathbf{F(A)} = \mathbf{R(A)} = d_{n-1}\mathbf{A}^{n-1} + \cdots + d_1\mathbf{A} + d_0\mathbf{I}$$

is then found.

As an example of finding exp $(\mathbf{A})$ using the Cayley–Hamilton technique, consider the matrix

$$\mathbf{A} = \begin{bmatrix} -5 & 1 \\ -6 & 0 \end{bmatrix}$$

The characteristic equation for $\mathbf{A}$ is

$$| \lambda\mathbf{I} - \mathbf{A} | = \begin{vmatrix} \lambda + 5 & -1 \\ 6 & \lambda \end{vmatrix} = \lambda^2 + 5\lambda + 6 = 0$$

so the eigenvalues of $\mathbf{A}$ are $\lambda_1 = -2$ and $\lambda_2 = -3$. Since $\mathbf{A}$ is 2×2,

$$\mathbf{R}(\lambda) = d_1\lambda + d_0$$

and

$$F(\lambda_1) = e^{-2} = -2d_1 + d_0$$

$$F(\lambda_2) = e^{-3} = -3d_1 + d_0$$

which has solution

$$d_1 = e^{-2} - e^{-3} \qquad d_0 = 3e^{-2} - 2e^{-3}$$

Thus

$$\mathbf{R}(\lambda) = (e^{-2} - e^{-3})\lambda + (3e^{-2} - 2e^{-3})$$

and

$$\exp (\mathbf{A}) = \mathbf{R(A)} = (e^{-2} - e^{-3})\mathbf{A} + (3e^{-2} - 2e^{-3})\mathbf{I}$$

$$= \begin{bmatrix} -2e^{-2} + 3e^{-3} & e^{-2} - e^{-3} \\ -6e^{-2} + 6e^{-3} & 3e^{-2} - 2e^{-3} \end{bmatrix} \doteq \begin{bmatrix} -0.121 & 0.086 \\ -0.513 & 0.306 \end{bmatrix}$$

If, in the Cayley–Hamilton technique, the matrix $\mathbf{A}$ has repeated eigenvalues, the equations

$$F(\lambda_i) = \mathbf{R}(\lambda_i)$$

will just be repeated for each root repetition, so there will be an insufficient number of independent equations for a solution. If the matrix $\mathbf{A}$ has a root λ_i repeated p times, additional linearly independent equations of the form

$$\left.\frac{dF}{d\lambda}\right|_{\lambda=\lambda_i} = \left.\frac{dR}{d\lambda}\right|_{\lambda=\lambda_i}$$

$$\left.\frac{d^2F}{d\lambda^2}\right|_{\lambda=\lambda_i} = \left.\frac{d^2R}{d\lambda^2}\right|_{\lambda=\lambda_i}$$

$$\vdots$$

$$\left.\frac{d^pF}{d\lambda^p}\right|_{\lambda=\lambda_i} = \left.\frac{d^pR}{d\lambda^p}\right|_{\lambda=\lambda_i}$$

are used to obtain a solution.

7.5.3 Sylvester's Theorem

An explicit solution for the finite-series expression of a matrix infinite series is known as Sylvester's theorem. The theorem has the simplest form when the matrix $\mathbf{A}$ has distinct eigenvalues. Then, if $\lambda_1, \lambda_2, \ldots, \lambda_n$ are the eigenvalues of $\mathbf{A}$,

$$\mathbf{F}(\mathbf{A}) = \sum_{i=1}^{n} \mathbf{F}(\lambda_i)\mathbf{B}_i$$

where

$$\mathbf{B}_i = \frac{\displaystyle\prod_{j=1 \ \ j\neq i}^{n} (\mathbf{A} - \lambda_j\mathbf{I})}{\displaystyle\prod_{j=1 \ \ j\neq i}^{n} (\lambda_i - \lambda_j)}$$

This result is the matrix equivalent of using evaluation to calculate the coefficients of a partial fraction expansion.

7.6 FADDEEV–LEVERRIER MATRIX INVERSION

Elegant methods for matrix computations are connected with the Faddeev–Leverrier algorithm for matrix inversion. The algorithm, which is recursive, yields the characteristic equation for a square matrix, its adjoint, and (if it exists) its inverse. The determinant, being the negative of the last coefficient in the characteristic equation, is also computed.

7.6.1 Bocher's Formulas

If the characteristic equation of a matrix $\mathbf{A}$ is

$$\lambda^n + \alpha_{n-1}\lambda^{n-1} + \alpha_{n-2}\lambda^{n-2} + \cdots + \alpha_1\lambda + \alpha_0$$

$$= (\lambda - \lambda_1)(\lambda - \lambda_2) \cdots (\lambda - \lambda_n) = 0$$

The coefficient α_{n-1} is the negative of the sum of the eigenvalues of $\mathbf{A}$:

$$\alpha_{n-1} = -(\lambda_1 + \lambda_2 + \cdots + \lambda_n)$$

The coefficient α_{n-2} is the sum of the products of different eigenvalues taken two at a time,

$$\alpha_{n-2} = \lambda_1\lambda_2 + \lambda_1\lambda_3 + \cdots + \lambda_1\lambda_n + \lambda_2\lambda_3 + \cdots + \lambda_{n-2}\lambda_n + \cdots + \lambda_{n-1}\lambda_n$$

the coefficient α_{n-3} is the negative sum of the products of different eigenvalues taken three at a time,

$$\alpha_{n-3} = -(\lambda_1\lambda_2\lambda_3 + \lambda_1\lambda_2\lambda_4 + \cdots + \lambda_{n-2}\lambda_{n-1}\lambda_n)$$

and so on. Finally,

$$\alpha_0 = (-1)^n\lambda_1\lambda_2 \cdots \lambda_n = (-1)^n \, | \, \mathbf{A} \, |$$

The trace of $\mathbf{A}$ is also the sum of $\mathbf{A}$'s eigenvalues, so

$$\alpha_{n-1} = -\,\text{trace}\,[\mathbf{A}] = -(\lambda_1 + \lambda_2 + \cdots + \lambda_n)$$

Since the eigenvalues of $\mathbf{A}^2$ are the squares of the eigenvalues of $\mathbf{A}$,

$$\text{trace}\,(\mathbf{A}^2) = \lambda_1^2 + \lambda_2^2 + \cdots + \lambda_n^2$$

The product of roots taken two at a time is then

$$\alpha_{n-2} = \tfrac{1}{2}\{(\lambda_1 + \lambda_2 + \cdots + \lambda_n)(\lambda_1 + \lambda_2 + \cdots + \lambda_n) - \lambda_1^2 - \lambda_2^2 - \cdots - \lambda_n^2\}$$

$$= -\tfrac{1}{2}\{\alpha_{n-1}\,\text{trace}\,(\mathbf{A}) + \text{trace}\,(\mathbf{A}^2)\}$$

In a similar way, the formulas in Table 7-8 for each of the coefficients of the characteristic equation can be established. This series of computations, known as *Bocher's formulas*,

---

TABLE 7-8 Bocher's Formulas for the Coefficients of the Characteristic Equation of a Matrix

---

If the characteristic equation of the $n \times n$ matrix $\mathbf{A}$ is

$$\lambda^n + \alpha_{n-1}\lambda^{n-1} + \alpha_{n-2}\lambda^{n-2} + \cdots + \alpha_1\lambda + \alpha_0 = 0$$

then

$$\alpha_{n-1} = -\,\text{trace}\,(\mathbf{A})$$

$$\alpha_{n-2} = -\frac{1}{2}\,[\alpha_{n-1}\,\text{trace}\,(\mathbf{A}) + \text{trace}\,(\mathbf{A}^2)]$$

$$\alpha_{n-3} = -\frac{1}{3}\,[\alpha_{n-2}\,\text{trace}\,(\mathbf{A}) + \alpha_{n-1}\,\text{trace}\,(\mathbf{A}^2) + \text{trace}\,(\mathbf{A}^3)]$$

$$\alpha_{n-4} = -\frac{1}{4}\,[\alpha_{n-3}\,\text{trace}\,(\mathbf{A}) + \alpha_{n-2}\,\text{trace}\,(\mathbf{A}^2) + \alpha_{n-1}\,\text{trace}\,(\mathbf{A}^3) + \text{trace}\,(\mathbf{A}^4)]$$

$$\vdots$$

$$\alpha_0 = -\frac{1}{n}\,[\alpha_1\,\text{trace}\,(\mathbf{A}) + \alpha_2\,\text{trace}\,(\mathbf{A}^2) + \cdots + \text{trace}\,(\mathbf{A}^n)]$$

---

show how to determine the coefficients of the characteristic equation of a matrix by algebraically manipulating the traces of powers of the matrix.

7.6.2 Faddeev–Leverrier Algorithm

The formula for a matrix inverse in terms of powers of the matrix and the coefficients of its characteristic equation [equation (7-9)]

$$\mathbf{A}^{-1} = \frac{-1}{\alpha_0} [\mathbf{A}^{n-1} + \alpha_{n-1}\mathbf{A}^{n-2} + \cdots + \alpha_1\mathbf{I}]$$

and Bocher's formulas can be combined to give a powerful computational method for matrix inversion and finding the characteristic equation of a matrix. The result, known as the *Faddeev–Leverrier algorithm*, is given in Table 7-9. A singular matrix is indicated by

$$\alpha_0 = 0$$

and the matrix $\mathbf{D}_{n+1}$, if computed, provides a check on the calculations.

A convenient way of deriving this result is as follows: Let

$$\text{adj } (\lambda\mathbf{I} - \mathbf{A}) = \mathbf{D}_1\lambda^{n-1} + \mathbf{D}_2\lambda^{n-2} + \cdots + \mathbf{D}_{n-1}\lambda + \mathbf{D}_n = \sum_{i=1}^{n} \mathbf{D}_i\lambda^{n-i}$$

and let

$$|\lambda\mathbf{I} - \mathbf{A}| = \lambda^n + \alpha_{n-1}\lambda^{n-1} + \alpha_{n-2}\lambda^{n-2} + \cdots + \alpha_1\lambda + \alpha_0$$

$$= \lambda^n + \sum_{i=1}^{n} \alpha_{n-i}\lambda^{n-i}$$

Since for any square matrix $\mathbf{B}$,

$$\mathbf{B} \text{ adj } (\mathbf{B}) = |\mathbf{B}|\mathbf{I}$$

then

$$(\lambda\mathbf{I} - \mathbf{A}) \text{ adj } (\lambda\mathbf{I} - \mathbf{A}) = |\lambda\mathbf{I} - \mathbf{A}|\mathbf{I}$$

and

$$(\lambda\mathbf{I} - \mathbf{A}) \sum_{i=1}^{n} \mathbf{D}_i\lambda^{n-i} = \left(\lambda^n + \sum_{i=1}^{n} \alpha_{n-i}\lambda^{n-i}\right)\mathbf{I}$$

Equating powers of λ, there results

$$\begin{cases} \mathbf{D}_1\lambda^n & = \mathbf{I}\lambda^n \\ (\mathbf{D}_2 - \mathbf{A}\mathbf{D}_1)\lambda^{n-1} & = \alpha_{n-1}\mathbf{I}\lambda^{n-1} \\ (\mathbf{D}_3 - \mathbf{A}\mathbf{D}_2)\lambda^{n-2} & = \alpha_{n-2}\mathbf{I}\lambda^{n-2} \\ \qquad\qquad \vdots & \\ (\mathbf{D}_n - \mathbf{A}\mathbf{D}_{n-1})\lambda & = \alpha_1\mathbf{I}\lambda \\ -\mathbf{A}\mathbf{D}_n & = \alpha_0\mathbf{I} \end{cases}$$

or

TABLE 7-9   Faddeev–Leverrier Algorithm for Matrix Inversion

For the $n \times n$ matrix $\mathbf{A}$ with characteristic equation

$$|\lambda \mathbf{I} - \mathbf{A}| = \lambda^n + \alpha_{n-1}\lambda^{n-1} + \alpha_{n-2}\lambda^{n-2} + \cdots + \alpha_1\lambda + \alpha_0$$

$$\mathbf{D}_1 = \mathbf{I}$$

$$\alpha_{n-1} = -\text{trace}(\mathbf{AD}_1)$$

$$\mathbf{D}_2 = \mathbf{AD}_1 + \alpha_{n-1}\mathbf{I}$$

$$\alpha_{n-2} = -\frac{1}{2}\text{trace}(\mathbf{AD}_2)$$

$$\mathbf{D}_3 = \mathbf{AD}_2 + \alpha_{n-2}\mathbf{I}$$

$$\alpha_{n-3} = -\frac{1}{3}\text{trace}(\mathbf{AD}_3)$$

$$\vdots$$

$$\mathbf{D}_n = \mathbf{AD}_{n-1} + \alpha_1\mathbf{I}$$

$$\alpha_0 = -\frac{1}{n}\text{trace}(\mathbf{AD}_n)$$

$$\mathbf{D}_{n+1} = \mathbf{AD}_n + \alpha_0\mathbf{I} = \mathbf{0}$$

Then

$$|\lambda \mathbf{I} - \mathbf{A}| = \lambda^n + \alpha_{n-1}\lambda^{n-1} + \alpha_{n-2}\lambda^{n-2} + \cdots + \alpha_1\lambda + \alpha_0$$

$$|\mathbf{A}| = (-1)^n \alpha_0$$

$$\text{adj}(\lambda \mathbf{I} - \mathbf{A}) = \mathbf{D}_1\lambda^{n-1} + \mathbf{D}_2\lambda^{n-2} + \cdots + \mathbf{D}_{n-1}\lambda + \mathbf{D}_n = \sum_{i=1}^{n} \mathbf{D}_i\lambda^{n-i}$$

$$\text{adj}(\mathbf{A}) = (-1)^{n+1}\mathbf{D}_n$$

$$(\lambda \mathbf{I} - \mathbf{A})^{-1} = \frac{\text{adj}(\lambda \mathbf{I} - \mathbf{A})}{|\lambda \mathbf{I} - \mathbf{A}|} \qquad \text{if } (\lambda \mathbf{I} - \mathbf{A}) \text{ is nonsingular}$$

$$\mathbf{A}^{-1} = -\frac{1}{\alpha_0}\mathbf{D}_n \qquad \text{if } \mathbf{A} \text{ is nonsingular}$$

$$\begin{cases} \mathbf{D}_1 = \mathbf{I} \\ \mathbf{D}_2 = \mathbf{AD}_1 + \alpha_{n-1}\mathbf{I} = \mathbf{A} + \alpha_{n-1}\mathbf{I} \\ \mathbf{D}_3 = \mathbf{AD}_2 + \alpha_{n-2}\mathbf{I} = \mathbf{A}^2 + \alpha_{n-1}\mathbf{A} + \alpha_{n-2}\mathbf{I} \\ \vdots \\ \mathbf{D}_n = \mathbf{AD}_{n-1} + \alpha_1\mathbf{I} = \mathbf{A}^{n-1} + \alpha_{n-1}\mathbf{A}^{n-1} + \cdots + \alpha_2\mathbf{A} + \alpha_1\mathbf{I} \end{cases}$$

which is a recursive algorithm for finding the $\mathbf{D}$'s. The relation

$$\mathbf{AD}_n = -\alpha_0\mathbf{I}$$

gives

$$\mathbf{A}^{-1} = -\frac{1}{\alpha_0}\,\mathbf{D}_n$$

if **A** is nonsingular. More generally,

$$\mathbf{D}_n = (-1)^{n+1}\,\text{adj}\,(\mathbf{A})$$

by comparing

$$\mathbf{A}\,\text{adj}\,(\mathbf{A}) = |\,\mathbf{A}\,|\,\mathbf{I}$$

and

$$\mathbf{AD}_n = -\alpha_0\mathbf{I} = -(-1)^n\,|\,\mathbf{A}\,|\,\mathbf{I}$$

These relations also give a simple, general proof of the Cayley–Hamilton theorem, since

$$\mathbf{AD}_n + \alpha_0\mathbf{I} = \mathbf{A}^n + \alpha_{n-1}\mathbf{A}^{n-1} + \cdots + \alpha_1\mathbf{A} + \alpha_0\mathbf{I} = 0$$

7.6.3 Matrix Inversion Program

A computer program in BASIC/FORTRAN that uses the Faddeev–Leverrier algorithm to find the characteristic equation and inverse (if it exists) of a matrix is given in Table 7-10. The characteristic polynomial coefficients are stored in the usual format, and **D**

TABLE 7-10

Computer Program in BASIC for Matrix Inversion Using the Faddeev–Leverrier Algorithm (FADDEEV)

VARIABLES USED

N = dimension of the matrices

$A(20,20)$ = original matrix

$B(20,20)$ = intermediate matrix used to store **AD** products

$D(20,20)$ = succession of **D** matrices

$P(20)$ = characteristic polynomial coefficients; $P(N)$ is the leading coefficient and $P(0)$ is the constant term

I = row index

J = column index

K = column index used for matrix multiplication; column index when summing diagonal elements to form the trace

L = iteration number index

LISTING

```
100     PRINT "FADDEEV MATRIX INVERSION"
110     DIM A(20,20),B(20,20),D(20,20),P(20)
120   REM ENTER DIMENSIONS AND ELEMENTS OF THE A MATRIX
130     PRINT "PLEASE ENTER MATRIX DIMENSION";
140     INPUT N
150     PRINT "ENTER MATRIX A, COLUMN BY COLUMN"
160     FOR J = 1 TO N
170     PRINT "COLUMN ";J
```

(continued)

```
180       FOR I = 1 TO N
190       PRINT "ROW ";I;
200       INPUT A(I,J)
210       NEXT I
220       NEXT J
230     REM SET B=0
240       FOR I = 1 TO N
250       FOR J = 1 TO N
260       B(I,J) = 0
270       NEXT J
280       NEXT I
290     REM SET LEADING POLYNOMIAL COEFFICIENT TO UNITY
300       P(N) = 1
310     REM COMPUTE THE N D MATRICES
320       FOR L = 1 TO N
330     REM SET D=B+P(N-L+1)*I
340       FOR I = 1 TO N
350       FOR J = 1 TO N
360       D(I,J) = B(I,J)
370       NEXT J
380       NEXT I
390       FOR J = 1 TO N
400       D(J,J) = D(J,J) + P(N-L+1)
410       NEXT J
420     REM FORM B=AD
430       FOR I = 1 TO N
440       FOR J = 1 TO N
450       B(I,J) = 0
460       FOR K = 1 TO N
470       B(I,J) = B(I,J) + A(I,K) * D(K,J)
480       NEXT K
490       NEXT J
500       NEXT I
510     REM  FORM P(N-L)=(-1/L)*TRACE(B)
520       P(N-L) = 0
530       FOR K = 1 TO N
540       P(N-L) = P(N-L)-B(K,K)
550       NEXT K
560       P(N-L)= P(N-L) /L
570       NEXT L
580     REM PRINT CHARACTERISTIC EQUATION
590       PRINT "CHARACTERISTIC EQUATION IS"
600       PRINT 1!,"S**";N
610       FOR I = 1 TO N
620       PRINT "+";P(N-I),"S**";N-I
630       NEXT I
640     REM END IF MATRIX IS SINGULAR
650       IF P(0) = 0 THEN PRINT "MATRIX IS SINGULAR"
660       IF P(0) = 0 THEN GOTO 750
670     REM PRINT INVERSE
680       PRINT"INVERSE COLUMNS ARE"
690       FOR J = 1 TO N
700       PRINT "COLUMN ";J
710       FOR I = 1 TO N
720       PRINT -D(I,J) / P(0)
730       NEXT I
740       NEXT J
750       END
```

Computer Program in FORTRAN for Matrix Inversion Using the Faddeev–Leverrier Algorithm (FADDEEV)

```
C ************************************************************************
C PROGRAM FADDEEV -- FADDEEV MATRIX INVERSION
```

```
C *********************************************************************
      REAL A(20,20), B(20,20), D(20,20), P(20)
      DATA ONE /1.0/

      WRITE(*,'(1X,''FADDEEV MATRIX INVERSION'')')

C *** ENTER DIMENSIONS AND ELEMENTS OF THE A MATRIX
      WRITE(*,'(1X,''PLEASE ENTER MATRIX DIMENSION '',\)')
      READ(*,*) N
      WRITE(*,'(1X,''ENTER MATRIX A, COLUMN BY COLUMN '')')
      DO 10 J=1, N
         WRITE(*,'(1X,''COLUMN '',I2)') J
         DO 20 I=1, N
            WRITE(*,'(1X,''ROW '',I2,'' = '',\)') I
            READ(*,*) A(I,J)
 20      CONTINUE
 10   CONTINUE

C *** SET B=0
      DO 30 I=1, N
         DO 40 J=1, N
            B(I,J) = 0.0
 40      CONTINUE
 30   CONTINUE

C *** SET LEADING POLYNOMIAL COEFFICIENT TO UNITY
      P(N+1) = 1.0

C *** COMPUTE THE N D MATRICES
      DO 50 L=1, N

C *** SET D=B+P(N-L+2)*I
         DO 60 I=1, N
            DO 70 J=1, N
               D(I,J) = B(I,J)
 70         CONTINUE
 60      CONTINUE
         DO 80 J=1, N
            D(J,J) = D(J,J) + P(N-L+2)
 80      CONTINUE

C *** FORM B=AD
         DO 90 I=1, N
            DO 100 J=1, N
               B(I,J) = 0.0
               DO 110 K=1, N
                  B(I,J) = B(I,J) + A(I,K)*D(K,J)
 110           CONTINUE
 100        CONTINUE
 90      CONTINUE

C *** FORM P(N-L+1) = (-1/L)*TRACE(B)
         P(N-L+1) = 0.0
         DO 120 K=1, N
            P(N-L+1) = P(N-L+1) - B(K,K)
 120     CONTINUE
         P(N-L+1) = P(N-L+1)/L
 50   CONTINUE

C *** PRINT CHARACTERISTIC EQUATION
      WRITE(*,'(1X,''CHARACTERISTIC EQUATION IS'')')
      WRITE(*,'(1X,''                      S** '',I2)') N
      DO 130 I=1, N
```

(continued)

```
            WRITE(*,'(1X,''+   '',E15.8,'' S** '',I2)') P(N-I+1),N-I
130   CONTINUE

C *** END IF MATRIX IS SINGULAR
      IF (P(1).EQ.0.0) THEN
          WRITE(*,'(1X,''MATRIX IS SINGULAR'')')
          STOP
      ENDIF

C *** PRINT INVERSE
      WRITE(*,'(1X,''INVERSE COLUMNS ARE'')')
      DO 140 J=1, N
          WRITE(*,'(1X,''COLUMN '',I2)') J
          DO 150 I=1, N
              WRITE(*,'(1X,E15.8)') -D(I,J)/P(1)
150       CONTINUE
140   CONTINUE
      STOP
      END
```

matrices are recursively computed, with the latest one being the array D. The B array stores intermediate matrix products.

For the matrix

$$\mathbf{A} = \begin{bmatrix} 2 & 2 & -1 \\ -1 & 0 & 4 \\ 3 & -1 & -3 \end{bmatrix}$$

the FADDEEV program gives

```
            CHARACTERISTIC EQUATION IS
    1                   S**3
    +1                  S**2
    +3                  S**1
    +-25                S**0
    INVERSE COLUMNS ARE
    COLUMN 1
    .16
    .36
    .04
    COLUMN 2
    .28
    -.12
    .32
    COLUMN 3
    .32
    -.28
    .08
```

or

$$\mathbf{A}^{-1} = \begin{bmatrix} 0.16 & 0.28 & 0.32 \\ 0.36 & -0.12 & -0.28 \\ 0.04 & 0.32 & 0.08 \end{bmatrix}$$

The characteristic equation is

$$\lambda^3 + \lambda^2 + 3\lambda - 25 = 0$$

and the matrix inverse is as was found for the same matrix with the INVERSE program. For the singular matrix

$$\mathbf{A} = \begin{bmatrix} 2 & 2 & 0 \\ -1 & 0 & 1 \\ 3 & -1 & -4 \end{bmatrix}$$

the result is

```
CHARACTERISTIC EQUATION IS
1                 S**3
+2                S**2
+-5               S**1
+0                S**0
MATRIX IS SINGULAR
```

7.7 MORE FADDEEV–LEVERRIER METHODS

The Faddeev–Leverrier algorithm also leads to a nice method of eigenvector calculation. To apply it, it is expedient to save each of the $\mathbf{D}$ matrices rather than replacing one by the next. The FADDEEV program is now modified for this purpose; then eigenvector calculation programs, for real and for complex eigenvalues, are developed.

In general, eigenvector calculation requires that the characteristic equation be factored to obtain the eigenvalues. The factoring programs of Chapter 5 are quite suitable for this purpose. By simply combining Faddeev–Leverrier matrix inversion with polynomial factoring and these eigenvector calculation routines, one could enter any square matrix and obtain its determinant, characteristic equation, adjoint, inverse (if it exists), eigenvalues, and eigenvectors.

7.7.1 Eigenvectors

It can be shown that if the eigenvalues of a matrix $\mathbf{A}$ are distinct, the matrices

$$\mathbf{Q}_k = \text{adj} (\lambda_k \mathbf{I} - \mathbf{A})$$

where λ_k is an eigenvalue of $\mathbf{A}$, are nonzero. Using

$$(\lambda_k \mathbf{I} - \mathbf{A}) \, \text{adj} \, (\lambda_k \mathbf{I} - \mathbf{A}) = |\lambda_k \mathbf{I} - \mathbf{A}| \, \mathbf{I} = 0$$

it follows that the columns of $\mathbf{Q}_k$ must be proportional and must be eigenvectors of $\mathbf{A}$ corresponding to λ_k. Since

$$\mathbf{Q}_k = \text{adj} \, (\lambda_k \mathbf{I} - \mathbf{A}) = \sum_{i=1}^{n} \mathbf{D}_i \lambda_k^{n-i}$$

$$= \mathbf{D}_i \lambda_k^{n-1} + \mathbf{D}_2 \lambda_k^{n-2} + \cdots + \mathbf{D}_{n-1} \lambda_k + \mathbf{D}_n \qquad (7\text{-}10)$$

where the matrices $\mathbf{D}_i$ are those obtained by the Faddeev–Leverrier algorithm, it is straightforward to use these to find eigenvectors.

7.7.2 Eigenvector Calculation Program

To compute eigenvectors using (7-10) it is expedient to save the individual **D** matrices produced by the Faddeev–Leverrier algorithm. Table 7-11 lists a modification of the FADDEEV program that saves the succession of N **D** matrices in the array D $(\cdot, \cdot, \cdot)$, where the first index is the matrix number, the second the row, and the third the column. In FADDEEV2, the maximum **A** matrix dimension is changed to seven to reduce the need for memory capacity. When run on systems with adequate memory, this limit can be raised, of course. The statements in FADDEEV involving the array D are changed, and these are evident in the table. Although use of the intermediate array B could be eliminated because the previous D array is now saved, this is not done here so that the FADDEEV program is changed as little as possible.

TABLE 7-11

Modification of the FADDEEV Program in BASIC to Save the D Matrices (FADDEEV2)

This modification to FADDEEV stores each of the **D** matrices. D(20,20) is changed to D(7,7,7) to reduce the computer's memory requirement. The A$(\cdot,\cdot)$, B$(\cdot,\cdot)$ and P$(\cdot,\cdot)$ array dimensions are reduced accordingly.

ADDITIONAL VARIABLES USED

D(7,7,7) = **D** matrices in the Faddeev matrix inversion algorithm, as a function of matrix number, row number, and column number

LISTING

```
100     PRINT "FADDEEV'S ALGORITHM, SAVING THE D MATRICES"
110     DIM A(7,7),B(7,7),D(7,7,7),P(7)
120   REM ENTER DIMENSIONS AND ELEMENTS OF THE A MATRIX
130     PRINT "PLEASE ENTER MATRIX DIMENSION";
140     INPUT N
150     PRINT "ENTER MATRIX A, COLUMN BY COLUMN"
160     FOR J = 1 TO N
170     PRINT "COLUMN ";J
180     FOR I = 1 TO N
190     PRINT "ROW ";I;
200     INPUT A(I,J)
210     NEXT I
220     NEXT J
230   REM SET B=0
240     FOR I = 1 TO N
250     FOR J = 1 TO N
260     B(I,J) = 0
270     NEXT J
280     NEXT I
290   REM SET LEADING POLYNOMIAL COEFFICIENT TO UNITY
300     P(N) = 1
310   REM COMPUTE THE N D MATRICES
320     FOR L = 1 TO N
330   REM SET D=B+P(N-L+1)*I
340     FOR I = 1 TO N
350     FOR J = 1 TO N
360     D(L,I,J) = B(I,J)
370     NEXT J
380     NEXT I
```

```
390        FOR J = 1 TO N
400        D(L,J,J) = D(L,J,J) + P(N-L+1)
410        NEXT J
420   REM FORM B=AD
430        FOR I = 1 TO N
440        FOR J = 1 TO N
450        B(I,J) = 0
460        FOR K = 1 TO N
470        B(I,J)=B(I,J) + A(I,K) *D(L,K,J)
480        NEXT K
490        NEXT J
500        NEXT I
510   REM FORM P(N-L) = (-1/L)*TRACE (B)
520        P(N-L) = 0
530        FOR K = 1 TO N
540        P(N-L) = P(N-L) - B(K,K)
550        NEXT K
560        P(N-L) = P(N-L) / L
570        NEXT L
580   REM PRINT CHARACTERISTIC EQUATION
590        PRINT "CHARACTERISTIC EQUATION IS"
600        PRINT 1!,"S**";N
610        FOR I = 1 TO N
620        PRINT "+";P(N-I),"S**";N-I
630        NEXT I
640   REM END IF MATRIX IS SINGULAR
650        IF P(0) = 0 THEN PRINT "MATRIX IS SINGULAR"
660        IF P(0) =0 THEN GOTO 750
670   REM   PRINT INVERSE
680        PRINT "INVERSE COLUMNS ARE"
690        FOR J = 1 TO N
700        PRINT "COLUMN ";J
710        FOR I = 1 TO N
720        PRINT - D(N,I,J) / P(0)
730        NEXT I
740        NEXT J
750        END
```

Modification of the FADDEEV Program in FORTRAN to Save the D Matrices (FADDEEV2)

```
C ********************************************************************
C PROGRAM FADDEEV2 -- FADDEEV'S ALGORITHM, SAVING THE D MATRICES
C ********************************************************************
      REAL A(7,7), B(7,7), D(7,7,7), P(7)

      WRITE(*,'(1X,''FADDEEVS ALGORITHM, SAVING THE D MATRICES'')')

C *** ENTER DIMENSIONS AND ELEMENTS OF THE A MATRIX
      WRITE(*,'(1X,''PLEASE ENTER MATRIX DIMENSION '',\)')
      READ(*,*) N
      WRITE(*,'(1X,''ENTER MATRIX A, COLUMN BY COLUMN '')')
      DO 10 J=1, N
         WRITE(*,'(1X,''COLUMN '',I2)') J
         DO 20 I=1, N
            WRITE(*,'(1X,''ROW '',I2,'' = '',\)') I
            READ(*,*) A(I,J)
 20      CONTINUE
 10   CONTINUE

C *** SET B=0
      DO 30 I=1, N
         DO 40 J=1, N
```

(continued)

```
              B(I,J) = 0.0
40        CONTINUE
30     CONTINUE

C *** SET LEADING POLYNOMIAL COEFFICIENT TO UNITY
       P(N+1) = 1.0

C *** COMPUTE THE N D MATRICES
       DO 50 L=1, N

C *** SET D=B+P(N-L+2)*I
          DO 60 I=1, N
             DO 70 J=1, N
                D(L,I,J) = B(I,J)
70           CONTINUE
60        CONTINUE
          DO 80 J=1, N
             D(L,J,J) = D(L,J,J) + P(N-L+2)
80        CONTINUE

C *** FORM B=AD
          DO 90 I=1, N
             DO 100 J=1, N
                B(I,J) = 0.0
                DO 110 K=1, N
                   B(I,J) = B(I,J) + A(I,K)*D(L,K,J)
110             CONTINUE
100          CONTINUE
90        CONTINUE

C *** FORM P(N-L+1) = (-1/L)*TRACE(B)
          P(N-L+1) = 0.0
          DO 120 K=1, N
             P(N-L+1) = P(N-L+1) - B(K,K)
120       CONTINUE
          P(N-L+1) = P(N-L+1)/L
50     CONTINUE

C *** PRINT CHARACTERISTIC EQUATION
       WRITE(*,'(1X,''CHARACTERISTIC EQUATION IS'')')
       WRITE(*,'(1X,''                      S** '',I2)') N
       DO 130 I=1, N
          WRITE(*,'(1X,''+  '',E15.8,'' S** '',I2)') P(N-I+1),N-I
130    CONTINUE

C *** END IF MATRIX IS SINGULAR
       IF (P(1).EQ.0.0) THEN
          WRITE(*,'(1X,''MATRIX IS SINGULAR'')')
          STOP
       ENDIF

C *** PRINT INVERSE
       WRITE(*,'(1X,''INVERSE COLUMNS ARE'')')
       DO 140 J=1, N
          WRITE(*,'(1X,''COLUMN '',I2)') J
          DO 150 I=1, N
             WRITE(*,'(1X,E15.8)') -D(N,I,J)/P(1)
150       CONTINUE
140    CONTINUE
       STOP
       END
```

The computer program EIGEN, listed in Table 7-12, uses FADDEEV2 and computes and prints the **Q** matrices given the eigenvalues of **A**. After the characteristic equation and the inverse (if **A** is nonsingular) is found, the user enters an eigenvalue of **A**, the corresponding **Q** matrix is found and printed, and the program loops back for entry of another eigenvalue. An eigenvector could be found, if desired, simply by finding the column of **Q** with the element having the largest magnitude in it. We are assured that at least one column of **Q** will be nonzero. It is also easy to see how one could join a polynomial factoring program such as FACTOR (Table 5-17) with this program to automatically find and substitute the eigenvalues of **A**.

TABLE 7-12

Computer Program in BASIC for Finding Real Eigenvectors Using Faddeev's Method (EIGEN)

ADDITIONAL VARIABLES USED

$Q(7,7)$ = matrix containing the eigenvector

E = real eigenvector of **A**

G = eigenvalue E raised to various powers

LISTING

```
100    PRINT "FADDEEV REAL EIGENVECTOR CALCULATION"
110    DIM A(7,7) ,B(7,7) ,D(7,7,7) ,P(7) ,Q(7,7)
120
```

┌─────────────────────┐
│ FADDEEV2 program;   │
│ TABLE 7-11   │
└─────────────────────┘

```
740
750    REM ENTER EIGENVALUE
760      PRINT "ENTER A EIGENVALUE OF A";
770      INPUT E
780    REM  SET Q=D(N) AND G=1
790      FOR I = 1 TO N
800      FOR J = 1 TO N
810      Q(I,J) = D(N,I,J)
820      NEXT J
830      NEXT I
840      G = 1
850    REM  RECURSIVELY COMPUTE Q
860      FOR K = 1 TO N - 1
870    REM  SET G=E*G AND Q=Q+G*D(N-L)
880      G = E * G
890      FOR I = 1 TO N
900      FOR J = 1 TO N
910      Q(I,J) = Q(I,J) + G * D(N - K,I,J)
920      NEXT J
930      NEXT I
940      NEXT K
950    REM  PRINT THE Q MATRIX
960      PRINT "FOR EIGENVALUE ";E;"Q'S COLUMNS ARE"
970      FOR J = 1 TO N
980      PRINT "COLUMN ";J
```

(continued)

```
990       FOR I = 1 TO N
1000      PRINT Q(I,J)
1010      NEXT I
1020      NEXT J
1030      GOTO 750
1040      END
```

Computer Program in FORTRAN for Finding Real Eigenvectors Using Faddeev's Method (EIGEN)

```
C **********************************************************************
C EIGEN -- FADDEEV REAL EIGENVECTOR CALCULATION
C **********************************************************************
      REAL A(7,7) ,B(7,7) ,D(7,7,7) ,P(7) ,Q(7,7)
      DATA ONE /1.0/

      WRITE(*,'(1X,''FADDEEV REAL EIGENVECTOR CALCULATION'')')
```

FADDEEV2 program; TABLE 7-11

```
C *** ENTER EIGENVALUE AND REPEAT
      DO 230 L=1, 100
         WRITE(*,'(1X,''ENTER A EIGENVALUE OF A'')')
         READ(*,*) E

C ***    SET Q=D(N) AND G=1
         DO 160 I=1, N
            DO 170 J=1, N
               Q(I,J) = D(N,I,J)
170         CONTINUE
160      CONTINUE

         G = 1.0

C ***    RECURSIVELY COMPUTE Q
         DO 180 K=1, N-1

C ***    SET G=E*G AND Q=Q+G*D(N-K)
            G = E * G
            DO 190 I=1, N
               DO 200 J=1, N
                  Q(I,J) = Q(I,J) + G*D(N-K,I,J)
200            CONTINUE
190         CONTINUE
180      CONTINUE

C ***    PRINT THE Q MATRIX
         WRITE(*,'(1X,''FOR EIGENVALUE '',E15.8,'' QS COLUMNS ARE'')') E
         DO 210 J=1, N
            WRITE(*,'(1X,''COLUMN '',I3)') J
            DO 220 I=1, N
               WRITE(*,'(1X,E15.8)') Q(I,J)
220         CONTINUE
210      CONTINUE
230   CONTINUE
      STOP
      END
```

For the matrix

$$\mathbf{A} = \begin{bmatrix} -1 & 2 & 0 \\ -1 & 2 & 0 \\ 1 & -2 & -3 \end{bmatrix}$$

the EIGEN program finds the characteristic equation to be

$$\lambda^3 + 2\lambda^2 - 3\lambda = \lambda(\lambda - 1)(\lambda + 3) = 0$$

and the matrix **A** to be singular. Entering the eigenvalues gives

```
ENTER AN EIGENVALUE OF A?0
FOR EIGENVALUE 0 Q'S COLUMNS ARE
COLUMN 1
-6
-3
0
COLUMN 2
6
3
0
COLUMN 3
0
0
0
ENTER AN EIGENVALUE OF A?1
FOR EIGENVALUE 1 Q'S COLUMNS ARE
COLUMN 1
-4
-4
1
COLUMN 2
8
8
-2
COLUMN 3
0
0
0
ENTER AN EIGENVALUE OF A?-3
FOR EIGENVALUE -3 Q'S COLUMNS ARE
COLUMN 1
0
0
-3
COLUMN 2
0
0
6
COLUMN 3
0
0
12
```

so a set of eigenvectors are

$$\mathbf{p}^1 = \begin{bmatrix} -2 \\ -1 \\ 0 \end{bmatrix} \qquad \mathbf{p}^2 = \begin{bmatrix} -4 \\ -4 \\ 1 \end{bmatrix} \qquad \mathbf{p}^3 = \begin{bmatrix} 0 \\ 0 \\ 1 \end{bmatrix}$$

Entering a number λ that is not an eigenvalue gives

$$\mathbf{Q} = \text{adj}\,(\lambda\mathbf{I} - \mathbf{A})$$

which is not of rank 1. For example, for $\lambda = 2$, there results

```
ENTER AN EIGENVALUE OF A?2
FOR EIGENVALUE 2 Q'S COLUMNS ARE
COLUMN 1
0
-5
2
COLUMN 2
10
15
-4
COLUMN 3
0
0
2
```

or

$$\mathbf{Q} = \begin{bmatrix} 0 & 10 & 0 \\ -5 & 15 & 0 \\ 2 & -4 & 2 \end{bmatrix}$$

EIGEN2, a modification of the EIGEN program to accommodate complex eigenvalues, is given in Table 7-13. The changes are evident in the listing, and the original

---

TABLE 7-13

Computer Program in BASIC for Finding Complex Eigenvectors Using Faddeev's Method (EIGEN2)

This is a modification of the EIGEN program (Table 7-12) to accommodate complex eigenvalues.

VARIABLES USED

N = dimension of the matrices

$A(7,7)$ = original matrix

$B(7,7)$ = intermediate matrix used to store $\mathbf{AD}$ products

$D(7,7,7)$ = $\mathbf{D}$ matrices used in the Faddeev matrix inversion algorithm, as a function of matrix number, row number, and column number

$P(7)$ = characteristic polynomial coefficients; $P(N)$ is the leading coefficient and $P(0)$ is the constant term

$Q(7,7)$ = real part of the matrix containing the eigenvector

$T(7,7)$ = imaginary part of the matrix containing the eigenvector

E = real part of an eigenvalue of $\mathbf{A}$

F = imaginary part of an eigenvalue of $\mathbf{A}$

G = real part of a power of the eigenvalue

H = imaginary part of a power of the eigenvalue

S = temporary storage of G

I = row index

J = column index

K = column index used for matrix multiplication; column index when summing diagonal elements to form the trace

L = iteration number index

LISTING

```
100     PRINT "FADDEEV COMPLEX EIGENVECTOR CALCULATION"
110     DIM A(7,7) ,B(7,7) ,D(7,7,7) ,P(7) ,Q(7,7) ,T(7,7)
120
```

> FADDEEV2 program;
> TABLE 7-11

```
740
750     REM   ENTER EIGENVALUE
760       PRINT "ENTER EIGENVALUE REAL PART";
770       INPUT E
771       PRINT "ENTER EIGENVALUE IMAGINARY PART";
772       INPUT F
780     REM   SET Q=D(N), T=0, G=1 AND H=0
790       FOR I = 1 TO N
800       FOR J = 1 TO N
810       Q(I,J) = D(N,I,J)
811       T(I,J) = 0
820       NEXT J
830       NEXT I
840       G = 1
841       H = 0
850     REM   RECURSIVELY COMPUTE Q
860       FOR K = 1 TO N - 1
870     REM   SET G+IH=(E+IF)*(G+IH) AND (Q+IT)=(G+IH)*D(N-L)
880       S = G
881       G = E * S - F * H
882       H = F * S + E * H
890       FOR I = 1 TO N
900       FOR J = 1 TO N
910       Q(I,J) = Q(I,J) + G * D(N - K,I,J)
911       T(I,J) = T(I,J) + H * D(N - K,I,J)
920       NEXT J
930       NEXT I
940       NEXT K
950     REM   PRINT THE Q MATRIX
960       PRINT " FOR EIGENVALUE ";E;"+I";F;" Q'S COLUMNS ARE"
970       FOR J = 1 TO N
980       PRINT "COLUMN ";J
990       FOR I = 1 TO N
1000      PRINT Q(I,J),"+I ";T(I,J)
1010      NEXT I
1020      NEXT J
1030      GOTO 750
1040      END
```

Computer Program in FORTRAN for Finding Complex Eigenvectors Using Faddeev's Method (EIGEN2)

```
C ******************************************************************
C EIGEN2 -- FADDEEV COMPLEX EIGENVECTOR CALCULATION
C ******************************************************************
      REAL A(7,7) ,B(7,7) ,D(7,7,7) ,P(7) ,Q(7,7), T(7,7)
      DATA ONE /1.0/
```

(continued)

```
        WRITE(*,'(1X,''FADDEEV COMPLEX EIGENVECTOR CALCULATION'')')
```

FADDEEV2 program;
TABLE 7-11

```
C ***   ENTER EIGENVALUE AND REPEAT
        DO 230 L=1, 100
            WRITE(*,'(1X,''ENTER EIGENVALUE REAL PART '',\)')
            READ(*,*) E
            WRITE(*,'(1X,''ENTER EIGENVALUE IMAGINARY PART '',\)')
            READ(*,*) F

C ***   SET Q=D(N), T=0, G=1 AND H=0
            DO 160 I=1, N
                DO 170 J=1, N
                    Q(I,J) = D(N,I,J)
                    T(I,J) = 0.0
170             CONTINUE
160         CONTINUE

            G = 1.0
            H = 0.0

C ***   RECURSIVELY COMPUTE Q
            DO 180 K=1, N-1

C ***   SET G+IH=(E+IF)*(G+IH) AND (Q+IT)=(G+IH)*D(N-K)
            S = G
            G = E*S - F*H
            H = F*S + E*H
            DO 190 I=1, N
                DO 200 J=1, N
                    Q(I,J) = Q(I,J) + G*D(N-K,I,J)
                    T(I,J) = T(I,J) + H*D(N-K,I,J)
200             CONTINUE
190         CONTINUE
180         CONTINUE

C ***   PRINT THE Q MATRIX
            WRITE(*,'(1X,''FOR EIGENVALUE '',E15.8,
     *          '' +I'',E15.8,'' QS COLUMNS ARE'')') E,F
            DO 210 J=1, N
                WRITE(*,'(1X,''COLUMN'',I3,)') J
                DO 220 I=1, N
                    WRITE(*,'(1X,E15.8,'' +I'',E15.8)') Q(I,J),T(I,J)
220             CONTINUE
210         CONTINUE
230     CONTINUE

        STOP
        END
```

line numbers have not been changed, so that it is easy to compare EIGEN and EIGEN2 to see the changes made.

For the matrix

$$\mathbf{A} = \begin{bmatrix} 0 & 1 & 0 \\ 0 & 0 & 1 \\ 13 & -9 & -3 \end{bmatrix}$$

the EIGEN2 program finds the characteristic equation to be

$$|\lambda\mathbf{I} - \mathbf{A}| = \lambda^3 + 3\lambda^2 + 9\lambda - 13 = (\lambda - 1)(\lambda + 2 + \underline{i}3)(\lambda + 2 - \underline{i}3) = 0$$

For the eigenvalue $\lambda_1 = 1 + i0$, EIGEN2 gives

```
FOR EIGENVALUE 1+I0 Q'S COLUMNS ARE
COLUMN I
13                              +I 0
13                              +I 0
13                              +I 0
COLUMN 2
4                               +I 0
4                               +I 0
4                               +I 0
COLUMN 3
1                               +I 0
1                               +I 0
1                               +I 0
```

or

$$\mathbf{p}^1 = \begin{bmatrix} 1 \\ 1 \\ 1 \end{bmatrix}$$

For the eigenvalue $\lambda_2 = -2 + i3$,

```
FOR EIGENVALUE -2+I3 Q'S COLUMNS ARE
COLUMN 1
-2                              +I -3
13                              +I 0
-26                             +I 39
COLUMN 2
1                               +I 3
-11                             +I -3
31                              +I -27
COLUMN 3
1                               +I 0
-2                              +I 3
-5                              +I -12
```

or

$$\mathbf{p}^2 = \begin{bmatrix} 1 \\ -2 + i3 \\ -5 - i12 \end{bmatrix}$$

where the third column of $\mathbf{Q}$ has been used. Columns one and two are each proportional to the third column, since

$$\frac{1 - i3}{10} \begin{bmatrix} 1 + i3 \\ -11 - i3 \\ 31 - i27 \end{bmatrix} = \begin{bmatrix} 1 \\ -2 + i3 \\ -5 - i12 \end{bmatrix}$$

and

$$\frac{-2 + i3}{13} \begin{bmatrix} -2 - i3 \\ 13 \\ -26 + i39 \end{bmatrix} = \begin{bmatrix} 1 \\ -2 + i3 \\ -5 - i12 \end{bmatrix}$$

An eigenvector corresponding to $\lambda_3 = -2 - i3$ is

$$p^3 = p^{2*} = \begin{bmatrix} 1 \\ -2 - i3 \\ -5 + i12 \end{bmatrix}$$

which can also be found using the EIGEN2 program.

7.8 VIBRATION MODES OF A LOSSLESS SYSTEM

Figure 7-1(a) describes a translational mechanical system consisting of three masses and three springs. The positions of each of the masses, $x_1(t)$, $x_2(t)$, and $x_3(t)$, are measured from the rest positions of the masses. There is no energy-loss mechanism in this system, so once it is set in motion, it will continue to oscillate forever. It approximates the behavior of related nearly lossless systems. Using the free-body diagrams of Figure 7-1(b), equations of motion for the system are

$$M_1 \frac{d^2x_1}{dt^2} + (K_1 + K_2)x_1 - K_2x_2 = 0$$

$$M_2 \frac{d^2x_2}{dt^2} - K_2x_1 + (K_2 + K_3)x_2 - K_3x_3 = 0 \qquad (7\text{-}11)$$

$$M_3 \frac{d^2x_3}{dt^2} - K_3x_2 + K_3x_3 = 0$$

We now explore the conditions for which each of the three masses will oscillate sinusoidally with the same frequency f and in phase with one another,

$$x_1(t) = a_1 \cos 2\pi ft \qquad x_2(t) = a_2 \cos 2\pi ft \qquad x_3(t) = a_3 \cos 2\pi ft \qquad (7\text{-}12)$$

where a_1, a_2, and a_3 are the oscillation amplitudes. It will be a very special situation when this happens; otherwise, the motions of each mass will be neither sinusoidal nor in synchronization. Substituting (7-12) into (7-11) and using the numerical values given in Figure 7-1 for the masses and spring constants, there results

$$[-(2\pi f)^2 + 500]a_1 \qquad\qquad - 200a_2 \qquad\qquad\qquad = 0$$
$$- 200a_1 + [-3(2\pi f)^2 + 300]a_2 \qquad\qquad - 100a_3 = 0$$
$$- 100a_2 + [-2(2\pi f)^2 + 100]a_3 = 0$$
$$(7\text{-}13)$$

Satisfaction of this set of three homogeneous linear algebraic equations in three unknowns is the condition for sinusoidal, in-phase oscillations.

Equations (7-13) are functions of the oscillation amplitudes a_1, a_2, and a_3 and of the square of the oscillation frequency f. Letting

$$\lambda = f^2$$

then

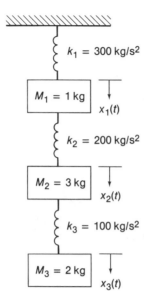

(a) The system

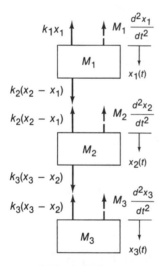

(b) Free-body diagrams

Figure 7-1 Lossless translational mechanical system.

$$\begin{bmatrix} f^2 - 12.68 & 5.07 & 0 \\ 1.69 & f^2 - 2.536 & 0.845 \\ 0 & 1.27 & f^2 - 1.27 \end{bmatrix} \begin{bmatrix} a_1 \\ a_2 \\ a_3 \end{bmatrix} \doteq \begin{bmatrix} 0 \\ 0 \\ 0 \end{bmatrix}$$

or

$$(\lambda \mathbf{I} - \mathbf{M})\mathbf{a} = \mathbf{0}$$

where

$$\mathbf{M} \doteq \begin{bmatrix} 12.68 & -5.07 & 0 \\ -1.69 & 2.536 & -0.845 \\ 0 & -1.27 & 1.27 \end{bmatrix}$$

The characteristic equation of $\mathbf{M}$, found using the FADDEEV program, is

$$\lambda^3 - 16.49\lambda^2 + 41.84\lambda - 16.35 \doteq 0$$

and using the FACTOR program, its characteristic roots are

$$\lambda_1 \doteq 0.478 \qquad \lambda_2 \doteq 2.54 \quad \lambda_3 \doteq 13.47$$

The corresponding oscillation frequencies are

$$f_1 = \sqrt{\lambda_1} \doteq 0.691 \qquad f_2 = \sqrt{\lambda_2} \doteq 1.59 \qquad f_3 = \sqrt{\lambda_3} \doteq 3.67$$

For each oscillation frequency f, there is a solution for the individual mass oscillation amplitudes, which is the eigenvector corresponding to the characteristic root $\lambda = f^2$. These eigenvectors are found using the EIGEN program to be

$$\mathbf{a}^1 \doteq \begin{bmatrix} 4.28 \\ 10.31 \\ 16.54 \end{bmatrix} \qquad \mathbf{a}^2 \doteq \begin{bmatrix} 4.28 \\ 8.61 \\ -8.61 \end{bmatrix} \qquad \mathbf{a}^3 \doteq \begin{bmatrix} 4.28 \\ -0.668 \\ 0.0696 \end{bmatrix}$$

As the scaling of any eigenvector is arbitrary, these serve only to specify the *relative* amplitudes of individual mass oscillations that will result in in-phase sinusoidal oscillation at the characteristic frequency. For example, the positions

$$x_1(t) \doteq 4.28K_1 \cos 4.34t$$

$$x_2(t) \doteq 10.31K_1 \cos 4.34t$$

$$x_3(t) \doteq 16.54K_1 \cos 4.34t$$

where K_1 is any constant, satisfy the equations of motion, as do

$$x_1(t) \doteq 4.28K_2 \cos 9.99t \qquad\qquad x_1(t) \doteq 4.28K_3 \cos 23.06t$$

$$x_2(t) \doteq 8.61K_2 \cos 9.99t \qquad \text{and} \qquad x_2(t) \doteq -0.668K_3 \cos 23.06t$$

$$x_3(t) \doteq -8.61K_2 \cos 9.99t \qquad\qquad x_3(t) \doteq 0.0696K_3 \cos 23.06t$$

where K_2 and K_3 are any constants.

As the equations of motion are linear, any sum of these three characteristic modes are also solutions:

$$x_1(t) \doteq 4.28K_1 \cos 4.34t + 4.28K_2 \cos 9.99t + 4.28K_3 \cos 23.06t$$

$$x_2(t) \doteq 10.31K_1 \cos 4.34t + 8.61K_2 \cos 9.99t - 0.668K_3 \cos 23.06t \qquad (7\text{-}14)$$

$$x_3(t) \doteq 16.54K_1 \cos 4.34t - 8.61K_2 \cos 9.99t + 0.0696K_3 \cos 23.06t$$

The initial mass positions in (7-14) are

$$x_1(0) \doteq 4.28K_1 + 4.28K_2 + 4.28K_3$$

$$x_2(0) \doteq 10.31K_1 + 8.61K_2 - 0.668K_3 \qquad (7\text{-}15)$$

$$x_3(0) \doteq 16.54K_1 - 8.61K_2 + 0.0696K_3$$

and their initial velocities are

$$\dot{x}_1(0) = 0$$

$$\dot{x}_2(0) = 0$$

$$\dot{x}_3(0) = 0$$

If the initial mass positions are known and if their initial velocities are zero, then (7-15) can be solved for K_1, K_2, and K_3, giving the solution (7-14). The solution of (7-15) for the mode amplitudes is the expansion of the initial position vector in terms of the three basis vectors that are eigenvectors in this problem:

$$\mathbf{x}(0) = \mathbf{K}_1\mathbf{a}^1 + \mathbf{K}_2\mathbf{a}^2 + \mathbf{K}_3\mathbf{a}^3$$

For example, for

$$\mathbf{x}(0) = \begin{bmatrix} -1 \\ 3 \\ -2 \end{bmatrix} \qquad \dot{\mathbf{x}}(0) = 0$$

the SOLVE program finds

$$K_1 \doteq 0.0253 \qquad K_2 \doteq 0.2766 \qquad K_3 \doteq -0.5355$$

If the masses have nonzero initial velocities, the individual modes are not each cosine functions, but are of the form

$$x_i(t) = a_i \cos (2\pi ft + \theta_i)$$

or, equivalently,

$$x_i(t) = \alpha_i \cos 2\pi ft + \beta_i \sin 2\pi ft$$

Similar methods apply to problems involving vibrations within solid masses, where there are an infinite number of modes. One method of finding the lowest-frequency (and usually most significant) modes is to approximate the behavior of the continuous mass distribution by a finite number of spring-coupled masses.

PROBLEMS

7-1. Find the characteristic equation, eigenvalues, eigenvectors, and a modal matrix for each of the following matrices. Verify that $\mathbf{P}^{-1}\mathbf{AP} = \mathbf{\Lambda}$.

(a) $\begin{bmatrix} -3 & 1 \\ -2 & 0 \end{bmatrix}$ (b) $\begin{bmatrix} 0 & -2 \\ 2 & 5 \end{bmatrix}$

(c) $\begin{bmatrix} -3 & 1 & 0 \\ -2 & 0 & 1 \\ 0 & 0 & 0 \end{bmatrix}$ (d) $\begin{bmatrix} 2 & 0 & -1 \\ 0 & -3 & 2 \\ 0 & 0 & 1 \end{bmatrix}$

7-2. Show that the characteristic equation for a matrix with the structure

$$
\begin{bmatrix}
-\alpha_{n-1} & 1 & 0 & \cdots & 0 \\
-\alpha_{n-2} & 0 & 1 & \cdots & 0 \\
\vdots & \vdots & \vdots & \vdots & \vdots \\
-\alpha_1 & 0 & 0 & \cdots & 1 \\
-\alpha_0 & 0 & 0 & \cdots & 0
\end{bmatrix}
$$

is

$$
\lambda^n + \alpha_{n-1}\lambda^{n-1} + \alpha_{n-2}\lambda^{n-2} + \cdots + \alpha_1\lambda + \alpha_0 = 0
$$

The structure

$$
\begin{bmatrix}
0 & 1 & 0 & \cdots & 0 & 0 \\
0 & 0 & 1 & \cdots & 0 & 0 \\
\vdots & \vdots & \vdots & \vdots & \vdots & \vdots \\
-\alpha_0 & -\alpha_1 & -\alpha_2 & \cdots & -\alpha_{n-2} & -\alpha_{n-1}
\end{bmatrix}
$$

has the same characteristic equation. These are each convenient ways to construct a matrix that has a desired characteristic equation.

7-3. Show that the characteristic equation of an upper or lower triangular matrix involves only the diagonal elements of the matrix.

7-4. For two nonsingular $n \times n$ square matrices $\mathbf{A}$ and $\mathbf{B}$, show that the eigenvalues of $(\mathbf{AB})$ are the same as the eigenvalues of $(\mathbf{BA})$.

7-5. Find the sum and the product of the eigenvalues of the matrix

$$
\begin{bmatrix}
1 & 0 & -2 & 3 \\
-1 & -3 & 0 & 2 \\
4 & -1 & 0 & 1 \\
-5 & -4 & 3 & -2
\end{bmatrix}
$$

7-6. For the matrix

$$
\mathbf{A} = \begin{bmatrix}
4 & -3 & 2 \\
1 & 0 & -1 \\
3 & 2 & 3
\end{bmatrix}
$$

find the characteristic equation and, from that equation, decide whether or not $\mathbf{A}^{-1}$ exists.

7-7. The matrix

$$
\mathbf{A} = \begin{bmatrix}
-3 & 5 \\
-3 & 4
\end{bmatrix}
$$

has complex eigenvalues. Find a modal matrix $\mathbf{P}$ and verify that $\mathbf{P}^{-1}\mathbf{AP} = \mathbf{\Lambda}$.

7-8. The following matrices have repeated eigenvalues. Determine if a nonsingular modal matrix $\mathbf{P}$ and the corresponding spectral matrix $\mathbf{\Lambda}$ exist for each by examining the rank of $(\lambda\mathbf{I} - \mathbf{A})$. If so, find $\mathbf{P}$ and $\mathbf{\Lambda}$ and verify that $\mathbf{P}^{-1}\mathbf{AP} = \mathbf{\Lambda}$.

(a) $\begin{bmatrix} 0 & 2 \\ -2 & 4 \end{bmatrix}$ (b) $\begin{bmatrix} 2 & 1 \\ 0 & 2 \end{bmatrix}$

(c) $\begin{bmatrix} 2 & 0 \\ 0 & 2 \end{bmatrix}$ (d) $\begin{bmatrix} 2 & 0 & 0 \\ -1 & 3 & 1 \\ 0 & 0 & 2 \end{bmatrix}$

7-9. Are the matrices

$$A = \begin{bmatrix} 2 & -3 & 2 \\ 0 & 1 & 2 \\ 0 & 4 & 3 \end{bmatrix}$$

and

$$B = \begin{bmatrix} 3 & -2 & 2 \\ 0 & 2 & 0 \\ 4 & -1 & 1 \end{bmatrix}$$

similar?

7-10. The two matrices

$$A = \begin{bmatrix} 2 & 1 \\ 0 & 2 \end{bmatrix} \quad \text{and} \quad B = \begin{bmatrix} 2 & 0 \\ 0 & 2 \end{bmatrix}$$

have the same characteristic values. Show that they are *not* similar, that is, that there is no matrix **P** such that

$$P^{-1}BP = A$$

7-11. Show that the trace and the determinant of a matrix are unchanged by a similarity transformation.

7-12. For the matrix

$$A = \begin{bmatrix} 0 & -1 & 2 \\ 0.5 & 1 & -1.5 \\ 3 & -0.5 & -4 \end{bmatrix}$$

use the SERIES program to compute, approximately,

(a) e^{-A} **(b)** $\cos A$

Ensure that each element in the resulting matrix is accurate to at least three significant digits.

7-13. Use the diagonalization method to find exp (**A**) for the following matrices:

(a) $A = \begin{bmatrix} 3 & -1 \\ -2 & 2 \end{bmatrix}$ **(b)** $A = \begin{bmatrix} -6 & 1 & 0 \\ -11 & 0 & 1 \\ -6 & 0 & 0 \end{bmatrix}$

7-14. Use the Cayley–Hamilton method to find cos **A** for the following matrices.

(a) $A = \begin{bmatrix} -4 & 1 \\ -2 & -1 \end{bmatrix}$ **(b)** $A = \begin{bmatrix} -7 & 0 & 0 \\ 0 & 2 & 0 \\ 0 & 0 & 1 \end{bmatrix}$

7-15. Use Sylvester's theorem to find exp (**A**) for the following matrices.

(a) $A = \begin{bmatrix} -3 & 1 \\ 0 & -1 \end{bmatrix}$ **(b)** $A = \begin{bmatrix} 3 & 1 & 0 \\ -2 & 0 & 1 \\ 0 & 0 & 0 \end{bmatrix}$

7-16. The matrix exponential function

$$e^{At} = \exp (At) = I + (At) + \frac{(At)^2}{2!} + \frac{(At)^3}{3!} + \cdots$$

where t is a variable, plays much the same role in the solution of higher-order linear differential equations as the scalar exponential function,

$$e^{at} = 1 + (at) + \frac{(at)^2}{2!} + \frac{(at)^3}{3!} + \cdots$$

does in the solution of first-order linear differential equations. For the matrix

$$\mathbf{A} = \begin{bmatrix} 0 & 1 \\ -3 & -4 \end{bmatrix}$$

find $\exp(\mathbf{A}t)$. The elements of the matrix $\exp(\mathbf{A}t)$ will generally be functions of the variable t.

7-17. For the matrix

$$\mathbf{A} = \begin{bmatrix} 0 & 1 & 0 \\ 0 & 0 & 1 \\ 2 & -1 & 4 \end{bmatrix}$$

(a) Express the matrix polynomial

$$\mathbf{A}^4 + 2\mathbf{A}^3 + \mathbf{A}^2 - \mathbf{A} + 3\mathbf{I}$$

in terms of $\mathbf{A}^2$, $\mathbf{A}$, and $\mathbf{I}$.

(b) Find $\mathbf{A}^{-1}$ by first expressing it in terms of $\mathbf{A}^2$, $\mathbf{A}$, and $\mathbf{I}$.

7-18. Although every matrix satisfies its own characteristic equation, some matrices satisfy polynomial equations of lower degree than the characteristic equation. The polynomial equation of lowest degree that is satisfied by a matrix is called its *minimal polynomial*. Verify that the 3×3 matrix

$$\mathbf{A} = \begin{bmatrix} 3 & 0 & 0 \\ 0 & 1 & 0 \\ 0 & 0 & 3 \end{bmatrix}$$

satisfies the second-degree polynomial equation

$$\mathbf{A}^2 - 4\mathbf{A} + 3\mathbf{I} = 0$$

7-19. Find the characteristic equations of the following matrices by applying Bocher's formulas.

(a) $\begin{bmatrix} 1 & -2 \\ 3 & 4 \end{bmatrix}$ (b) $\begin{bmatrix} 1 & 0 & -1 \\ 4 & 0 & 3 \\ -3 & -2 & 0 \end{bmatrix}$

7-20. Use the FADDEEV program to find the characteristic equation, the determinant, and (if it exists) the inverse of each of the following matrices. Check the results by forming the product of the matrix and its inverse.

(a) $\begin{bmatrix} -1 & 3 \\ 2 & -4 \end{bmatrix}$ (b) $\begin{bmatrix} 2 & 1 & -2 \\ 4 & 3 & 0 \\ -1 & -1 & 0 \end{bmatrix}$

(c) $\begin{bmatrix} 3.7 & 0 & 3.3 \\ -2.4 & -0.6 & -7.0 \\ 1.5 & 5.4 & -0.8 \end{bmatrix}$

7-21. Modify the FADDEEV program so that it prints the adjoint matrix, in addition to the characteristic equation and (if it exists) the inverse. Find the adjoints of the following matrices with the program and check the results by forming the product of the matrix and its adjoint.

(a) $\begin{bmatrix} 3 & -2 \\ -6 & 4 \end{bmatrix}$

(b) $\begin{bmatrix} 1 & 3 \\ 2 & -4 \end{bmatrix}$

(c) $\begin{bmatrix} 1 & 4 & 7 \\ -2 & -5 & -8 \\ 3 & 6 & 9 \end{bmatrix}$

7-22. Use the FADDEEV program to find the characteristic polynomials of each of the following matrices. For each, factor the characteristic polynomial, perhaps using the FACTOR program of Chapter 5, then use the EIGEN program to find the eigenvectors. Check the results by forming a modal matrix and calculating the spectral matrix.

(a) $\begin{bmatrix} 0 & 1 & 0 \\ 0 & 0 & 1 \\ 0 & -5 & -5 \end{bmatrix}$

(b) $\begin{bmatrix} -4.7 & 0 & 0.6 \\ 0 & -0.4 & 0 \\ 3.0 & -2.2 & 1.5 \end{bmatrix}$

7-23. For the three-mass lossless system of Section 7.8, let $M_2 = 2$ kg and $K_2 = 400$ kg/s^2 instead of the values given in Figure 7-1. Find the oscillation frequencies, the relative amplitudes of the three characteristic modes, and the solution for $x_2(t)$ when $x_2(0) = 10$ and all other initial conditions are zero.

7-24. Find a lossless translational mechanical system of the type described in Section 7.8 but having just two masses and two springs for which the oscillation frequencies are $f_1 = 2$ and $f_2 = 3$.

8

QUADRATIC FORMS
AND LEAST SQUARES

8.1 PREVIEW

Some sophisticated and powerful applications of matrix algebra and the characteristic value problem solution are now discussed. Quadratic forms occur in a variety of practical physical problems, including the description of mechanical stress, dynamics, wave motion, and other phenomena. We begin, in Sections 8.2 and 8.3, by discussing quadratic forms, then showing that by an orthogonal change of variables they can always be reduced to a sum of squares of each of the new variables. In this endeavor, the characteristic value problem plays a key role.

Discussion of numerical determination of the rank of a matrix has largely been delayed until now, when we have discussed concepts needed for a good introduction to the subject. For most matrices of relatively low order, Gauss–Jordan pivoting will usually suffice to determine matrix rank. Numerical inaccuracies can destroy or create linear independence of matrix rows or columns, but making a judgment as to whether or not this probably has occurred is seldom really difficult. In many practical problems, "almost" linearly dependent equations are as unacceptable as are truly linearly dependent ones. When the matrix dimensions are large, however, a judgment as to matrix rank can be very difficult when there are small numerical inaccuracies in the calculations. Singular-value decomposition is then an important tool. This is the subject of Section 8.5.

When a system is described by more than the minimum number of linear algebraic equations needed for a unique solution and when these equations, because of numerical inaccuracies, are inconsistent, a solution of the equations does not exist. The system from which the equations were derived certainly exists and has a solution; it is just that the equations for it that we have are not entirely accurate. A popular way of estimating what the solution would be if the equations had been accurate is called least squares estimation. We derive the least squares estimate and develop a computer program for its calculation in Section 8.6. This subject leads naturally, in Section 8.7, to the subject of ''best fits'' (in the least squares sense) of curves to data. An example of curve fitting to economic data is discussed and criticized.

The chapter concludes with a recursive formulation of the least squares algorithm. Instead of processing all of the data in one batch to obtain the least squares estimate, new data can be processed as it becomes available, updating the previous estimate without the need for complete recalculation. A computer program is developed for this purpose. Incorporating navigational data for estimation of a ship's position provides a nice illustrative example of using recursive least squares.

8.2 SYMMETRIC MATRICES AND QUADRATIC FORMS

A quadratic form is a special kind of nonlinear function of several variables consisting of a linear combination of the squares of the variables and products of one variable with another. Quadratic forms are most conveniently expressed in terms of symmetric matrices, so our study begins with these.

8.2.1 Symmetric Matrix Properties

A *symmetric* matrix is a square matrix $\mathbf{S}$ with the property that

$$\mathbf{S}^\dagger = \mathbf{S}$$

Elements off the diagonal are symmetric about the diagonal,

$$\mathbf{S}_{ji} = \mathbf{S}_{ij}$$

as in the symmetric matrix

$$\mathbf{S} = \begin{bmatrix} 3 & 1 & -3 & 7 & 8 \\ 1 & -2 & 0 & 6 & -1 \\ -3 & 0 & 4 & 5 & 9 \\ 7 & 6 & 5 & 0 & -7 \\ 8 & -1 & 9 & -7 & 10 \end{bmatrix}$$

As will be demonstrated, any two different eigenvectors corresponding to different eigenvalues of a symmetric matrix are orthogonal to one another. Further, the eigenvalues of a symmetric matrix composed of real elements are always real numbers. As a consequence, the corresponding eigenvectors are always composed of real numbers. Table 8-1 lists symmetric matrix properties.

Consider two different eigenvalues, λ_1 and λ_2, and the corresponding eigenvectors, $\mathbf{p}^1$ and $\mathbf{p}^2$, of a symmetric matrix $\mathbf{S}$:

$$(\lambda_1 \mathbf{I} - \mathbf{S})\mathbf{p}^1 = 0 \qquad \mathbf{S}\mathbf{p}^1 = \lambda_1 \mathbf{p}^1$$

$$(\lambda_2 \mathbf{I} - \mathbf{S})\mathbf{p}^2 = 0 \qquad \mathbf{S}\mathbf{p}^2 = \lambda_2 \mathbf{p}^2$$

TABLE 8-1   Properties of Symmetric Matrices

DEFINING PROPERTY

$$\mathbf{S}^\dagger = \mathbf{S}$$

FURTHER PROPERTIES

(1) The eigenvalues of a symmetric matrix with real elements are always real numbers. In consequence, the corresponding eigenvectors are always real.

(2) Any two eigenvectors corresponding to different eigenvalues are orthogonal to one another.

(3) If, for a real symmetric matrix $\mathbf{S}$, an eigenvalue λ_i is repeated, the rank of $\lambda_i\mathbf{I} - \mathbf{S}$ is always reduced by the number of occurrences of the root λ_i.

(4) Property (3) means that in the characteristic value problem,

$$\mathbf{Sp} = \lambda\mathbf{p}$$

where

$$\mathbf{p} = [\mathbf{p}^1 \mid \mathbf{p}^2 \mid \cdots \mid \mathbf{p}^n]$$

a transformation can always be found such that

$$\mathbf{P}^{-1}\mathbf{SP} = \Lambda$$

even if $\mathbf{S}$ has repeated eigenvalues.

(5) Property (4) means that in the characteristic value problem, a transformation $\mathbf{P}$ can always be found that is orthogonal:

$$\mathbf{P}^{-1} = \mathbf{P}\dagger$$

This is done by choosing the eigenvectors corresponding to repeated eigenvalues to be orthogonal to one another and by normalizing each eigenvector:

$$\|\mathbf{p}^i\| = 1 \qquad i = 1, 2, \ldots, n$$

From these relations,

$$\mathbf{p}^{1\dagger}\mathbf{S}^\dagger\mathbf{p}^2 = \lambda_1\mathbf{p}^{1\dagger}\mathbf{p}^2 \tag{8-1}$$

Using $\mathbf{S} = \mathbf{S}^\dagger$, then,

$$\mathbf{p}^{1\dagger}\mathbf{S}^\dagger\mathbf{p}^2 = \lambda_1\mathbf{p}^{1\dagger}\mathbf{p}^2 \qquad \mathbf{p}^{1\dagger}\lambda_2\mathbf{p}^2 = \lambda_1\mathbf{p}^{1\dagger}\mathbf{p}^2$$

so that, using (8-1),

$$(\lambda_2 - \lambda_1)\mathbf{p}^{1\dagger}\mathbf{p}^2 = 0$$

For λ_1 and λ_2 different from one another,

$$\mathbf{p}^{1\dagger}\mathbf{p}^2 = 0 \tag{8-2}$$

which is to say that the eigenvectors of a symmetric matrix are pairwise orthogonal to one another, provided that their elements are real. As n different vectors can be pairwise orthogonal only if they span n dimensions, distinct eigenvectors of a symmetric matrix are mutually orthogonal.

Because the eigenvalues of a matrix composed of real elements occur in complex conjugate pairs, the corresponding eigenvectors (when chosen to be of equal magnitude) are complex conjugates of one another. Suppose that a symmetric matrix has conjugate eigenvectors:

$$\mathbf{p}^1 = \mathbf{u} + i\mathbf{v}$$

$$\mathbf{p}^2 = \mathbf{p}^{1*} = \mathbf{u} - i\mathbf{v}$$

Using (8-2), we obtain

$$\mathbf{p}^{1\dagger}\mathbf{p}^2 = \mathbf{u}^\dagger\mathbf{u} + \mathbf{v}^\dagger\mathbf{v} = \mathbf{0}$$

which can hold only if

$$\mathbf{u} = \mathbf{v} = \mathbf{0}$$

a contradiction.

The eigenvectors of a real, symmetric matrix with distinct eigenvalues are orthogonal and can be scaled as desired, so they may be chosen to be orthonormal. In that event, the modal matrix $\mathbf{P}_0$, which is then composed of orthonormal eigenvectors as columns, is an orthogonal matrix:

$$\mathbf{P}_0^{-1} = \mathbf{P}_0^\dagger$$

As a numerical example, consider the symmetric matrix

$$\mathbf{S} = \begin{bmatrix} 2 & -36 & 0 \\ -36 & 23 & 0 \\ 0 & 0 & -75 \end{bmatrix} \tag{8-3}$$

which has characteristic equation

$$\begin{vmatrix} \lambda - 2 & 36 & 0 \\ 36 & \lambda - 23 & 0 \\ 0 & 0 & \lambda + 75 \end{vmatrix} = (\lambda + 75)(\lambda^2 - 25\lambda - 1250)$$

$$= (\lambda + 75)(\lambda - 50)(\lambda + 25) = 0$$

The eigenvalues of $\mathbf{S}$ are $\lambda_1 = -25$, $\lambda_2 = 50$, $\lambda_3 = -75$, which are real and distinct, as they should be. The eigenvector $\mathbf{p}^1$ corresponding to λ_1 satisfies

$$-27p_{11} + 36p_{21} = 0$$

$$36p_{11} - 48p_{21} = 0$$

$$50p_{31} = 0$$

Deleting the second equation, which is the same as the first equation, and choosing $p_{21} = 3$,

$$\mathbf{p}^1 = \begin{bmatrix} 4 \\ 3 \\ 0 \end{bmatrix}$$

The eigenvector $\mathbf{p}^2$ corresponding to λ_2 satisfies

$$48p_{12} + 36p_{22} = 0$$

$$36p_{12} + 27p_{22} = 0$$

$$125p_{32} = 0$$

Deleting the second equation, which is redundant, and choosing $p_{22} = -4$ for convenience,

$$\mathbf{p}^2 = \begin{bmatrix} 3 \\ -4 \\ 0 \end{bmatrix}$$

Similarly, the eigenvector $\mathbf{p}^3$ satisfies

$$-77p_{31} + 36p_{32} = 0$$

$$36p_{31} - 98p_{32} = 0$$

and has a solution

$$\mathbf{p}^3 = \begin{bmatrix} 0 \\ 0 \\ 1 \end{bmatrix}$$

The eigenvectors are mutually orthogonal, as can easily be demonstrated. Normalizing them and forming a modal matrix, there results

$$\mathbf{P}_0 = \begin{bmatrix} \frac{4}{5} & \frac{3}{5} & 0 \\ \frac{3}{5} & -\frac{4}{5} & 0 \\ 0 & 0 & 1 \end{bmatrix} \tag{8-4}$$

which is orthogonal,

$$\mathbf{P}_0^{-1} = \mathbf{P}_0^{\dagger}$$

and for which

$$\mathbf{P}_0^{-1}\mathbf{S}\mathbf{P}_0 = \mathbf{P}_0^{\dagger}\mathbf{S}\mathbf{P}_0 = \begin{bmatrix} -25 & 0 & 0 \\ 0 & 50 & 0 \\ 0 & 0 & -75 \end{bmatrix}$$

If, for a real, symmetric matrix, a root λ_i is repeated, the rank of $\lambda_i\mathbf{I} - \mathbf{S}$ is always reduced by the number of occurrences of the root λ_i. Thus it is always possible to find n linearly independent eigenvectors and hence a nonsingular modal matrix. In fact, even with repeated roots, the eigenvectors can always be chosen to be orthonormal, so that the modal matrix is orthogonal.

8.2.2 Quadratic Forms and Properties

A quadratic form is a multiple-variable nonlinear function of the form

$$f(x_1, x_2, \ldots, x_n) = f(\mathbf{x}) = \mathbf{x}^\dagger \mathbf{S} \mathbf{x}$$

$$= [x_1 \quad x_2 \quad \cdots \quad x_n] \begin{bmatrix} s_{11} & s_{12} & \cdots & s_{1n} \\ s_{21} & s_{22} & \cdots & s_{2n} \\ \vdots & \vdots & \vdots & \vdots \\ s_{n1} & s_{n2} & \cdots & s_{nn} \end{bmatrix} \begin{bmatrix} x_1 \\ x_2 \\ \vdots \\ x_n \end{bmatrix}$$

$$= s_{11}x_1^2 + s_{12}x_1x_2 + s_{13}x_1x_3 + \cdots + s_{1n}x_1x_n$$

$$+ s_{21}x_2x_1 + s_{22}x_2^2$$

$$+ s_{23}x_2x_3 + \cdots + s_{2n}x_2x_n + \cdots + s_{n1}x_nx_1$$

$$+ s_{n2}x_nx_2 + \cdots + s_{nn}x_n^2$$

Quadratic forms are linear combinations of the two-at-a-time products of the variables $x_i x_j$, including the squares of each variable. For example,

$$f(\mathbf{x}) = 4x_1^2 - 2x_1x_2 + 6x_1x_3 + 5x_2^2 - 4x_2x_3$$

$$= [x_1 \quad x_2 \quad x_3] \begin{bmatrix} 4 & -1 & 3 \\ -1 & 5 & -2 \\ 3 & -2 & 0 \end{bmatrix} \begin{bmatrix} x_1 \\ x_2 \\ x_3 \end{bmatrix}$$

is a three-variable quadratic form. A quadratic form can be expressed in terms of many different matrices $\mathbf{S}$, but it will always be possible to choose $\mathbf{S}$ to be symmetric, as was done above and as will be assumed hereafter.

8.3 TRANSFORMATION OF QUADRATIC FORM TO A SUM OF SQUARES

For every quadratic form, there is a linear transformation of the variables for which the function is just a linear combination of the squares of the new variables. The character of the function depends on the algebraic signs of the coefficients of these new variables.

8.3.1 Quadratic Form Transformations

For a quadratic form

$$f(\mathbf{x}) = \mathbf{x}^\dagger \mathbf{S} \mathbf{x}$$

expressed in terms of a symmetric matrix $\mathbf{S}$, it is always possible to find an orthogonal change of variables

$$\mathbf{x} = \mathbf{P}_0 \mathbf{x}' \qquad \mathbf{x}' = \mathbf{P}_0^\dagger \mathbf{x}$$

so that

$$f(\mathbf{x}) = \mathbf{x}^\dagger \mathbf{S} \mathbf{x} = \mathbf{x}'^\dagger (\mathbf{P}_0^\dagger \mathbf{S} \mathbf{P}_0) \mathbf{x}'$$

$$= \mathbf{x}'^\dagger (\mathbf{P}_0^{-1} \mathbf{S} \mathbf{P}_0) \mathbf{x}' = \mathbf{x}'^\dagger \Lambda \mathbf{x}'$$

where Λ is diagonal. The orthogonal matrix $\mathbf{P}_0$ is one that diagonalizes $\mathbf{S}$.

For the quadratic form defined by the symmetric matrix (8-3), for example,

$$f(\mathbf{x}) = [x_1 \quad x_2 \quad x_3] \begin{bmatrix} 2 & -36 & 0 \\ -36 & 23 & 0 \\ 0 & 0 & -75 \end{bmatrix} \begin{bmatrix} x_1 \\ x_2 \\ x_3 \end{bmatrix}$$

$$= 2x_1^2 - 72x_1 x_2 + 23x_2^2 - 75x_3^2$$

the orthogonal change of variables,

$$\mathbf{x} = \mathbf{P}_0 \mathbf{x}' \qquad \mathbf{x}' = \mathbf{P}_0^{-1} \mathbf{x} = \mathbf{P}_0^\dagger \mathbf{x}$$

using

$$\mathbf{P}_0 = \begin{bmatrix} \frac{4}{5} & \frac{3}{5} & 0 \\ \frac{3}{5} & -\frac{4}{5} & 0 \\ 0 & 0 & 1 \end{bmatrix}$$

which was calculated in (8-4) to diagonalize $\mathbf{S}$, gives

$$f(\mathbf{x}') = (\mathbf{P}_0 \mathbf{x}')^\dagger \mathbf{S}(\mathbf{P}_0 \mathbf{x}') = \mathbf{x}'^\dagger (\mathbf{P}_0^\dagger \mathbf{S} \mathbf{P}_0) \mathbf{x}' = \mathbf{x}'^\dagger \Lambda \mathbf{x}'$$

$$= [x_1' \quad x_2' \quad x_3'] \begin{bmatrix} -25 & 0 & 0 \\ 0 & 50 & 0 \\ 0 & 0 & -70 \end{bmatrix} \begin{bmatrix} x_1' \\ x_2' \\ x_3' \end{bmatrix} = -25x_1'^2 + 50x_2'^2 - 70x_3'^2$$

which is a pure sum of squares.

8.3.2 Sign Definiteness

A quadratic form $f(\mathbf{x})$ is called *positive definite* if

$$f(\mathbf{x}) = \mathbf{x}^\dagger \mathbf{S} \mathbf{x} > 0$$

for all nonzero real vectors $\mathbf{x}$. That is, the value of $\mathbf{x}^\dagger \mathbf{S} \mathbf{x}$ is positive for every $\mathbf{x}$ except $\mathbf{x} = \mathbf{0}$, for which $f(\mathbf{x} = \mathbf{0}) = \mathbf{0}$. A quadratic forms is said to be *positive semidefinite* if

$$f(\mathbf{x}) = \mathbf{x}^\dagger \mathbf{S} \mathbf{x} \geq 0$$

for all real vectors $\mathbf{x}$. If

$$f(\mathbf{x}) = \mathbf{x}^\dagger \mathbf{S} \mathbf{x} < 0$$

for all nonzero real vectors $\mathbf{x}$, it is termed *negative definite*, and it is *negative semidefinite* if

$$f(\mathbf{x}) = \mathbf{x}^\dagger \mathbf{S} \mathbf{x} \leq 0$$

for all real $\mathbf{x}$. The symmetric matrix $\mathbf{S}$ that defines a quadratic form is also termed *positive definite*, *positive semidefinite*, and so on, in accordance with the property of the associated quadratic form. These definitions are summarized in Table 8-2.

Sign definiteness of a quadratic form is unchanged by a nonsingular transformation of variables

$$\mathbf{x} = \mathbf{P}\mathbf{x}' \qquad \mathbf{x}' = \mathbf{P}^{-1}\mathbf{x}$$

since $\mathbf{x}' = 0$ only for $\mathbf{x} = 0$ and since for every $\mathbf{x}$ there is a corresponding $\mathbf{x}'$, and vice versa. Hence a quadratic form's sign definiteness, if any, is the same as when it is transformed by an orthogonal similarity transformation to a sum of squares:

$$f(\mathbf{x}') = (\mathbf{P}_0\mathbf{x}')^\dagger \mathbf{S}(\mathbf{P}_0\mathbf{x}') = \mathbf{x}'^\dagger(\mathbf{P}_0^\dagger\mathbf{S}\mathbf{P}_0)\mathbf{x}' = \mathbf{x}'^\dagger\boldsymbol{\Lambda}\mathbf{x}'$$

$$= \lambda_1 x_1'^2 + \lambda_2 x_2'^2 + \cdots + \lambda_n x_n'^2$$

For $\mathbf{S}$ to be positive definite, it is necessary and sufficient for all of the eigenvalues λ_1, $\lambda_2, \ldots, \lambda_n$ of $\mathbf{S}$ (which are real since $\mathbf{S}$ is symmetric) to be positive.

If all of the eigenvalues of $\mathbf{S}$ are nonnegative, with one or more roots equal to zero, the quadratic form is positive semidefinite. If all of the characteristic roots of $\mathbf{S}$ are

TABLE 8-2   Sign Definiteness of Quadratic Forms

A quadratic form

$$f(\mathbf{x}) = \mathbf{x}^\dagger \mathbf{S} \mathbf{x}$$

and its associated symmetric matrix $\mathbf{S}$ is classified as sign definite as follows:

(A) If for all nonzero real vectors $\mathbf{x}$

$$\mathbf{x}^\dagger \mathbf{S} \mathbf{x} > 0$$

the quadratic form is *positive definite*.

(B) If for all nonzero real vectors $\mathbf{x}$

$$\mathbf{x}^\dagger \mathbf{S} \mathbf{x} \geq 0$$

the quadratic form is *positive semidefinite*.

(C) If for all nonzero real vectors $\mathbf{x}$

$$\mathbf{x}^\dagger \mathbf{S} \mathbf{x} < 0$$

the quadratic form is *negative definite*.

(D) If for all nonzero real vectors $\mathbf{x}$

$$\mathbf{x}^\dagger \mathbf{S} \mathbf{x} \leq 0$$

the quadratic form is *negative semidefinite*.

Otherwise, the quadratic form is not sign definite.

negative, f is negative definite, and so on. For example, for the quadratic form $\mathbf{x}^t\mathbf{S}\mathbf{x}$, where

$$\mathbf{S} = \begin{bmatrix} 0 & 0 & 0 \\ 0 & -3 & 1 \\ 0 & 1 & 1 \end{bmatrix}$$

the characteristic equation of $\mathbf{S}$ is

$$|\lambda\mathbf{I} - \mathbf{S}| = \begin{vmatrix} \lambda & 0 & 0 \\ 0 & \lambda + 3 & -1 \\ 0 & -1 & \lambda - 1 \end{vmatrix} = \lambda(\lambda^2 + 2\lambda - 4)$$

$$= \lambda(\lambda + 1 + \sqrt{5})(\lambda + 1 - \sqrt{5}) = 0$$

As one of the eigenvalues of $\mathbf{S}$ is positive and the other is negative, this quadratic form is not sign definite. A Routh–Hurwitz test of the characteristic equation of $\mathbf{S}$ will suffice to determine sign definiteness without the need for factoring.

8.3.3 Multidimensional Extrema

When a continuous function of several variables, $f(x_1, x_2, x_3, \ldots, x_n) = f(\mathbf{x})$ having the needed partial derivatives is expanded in a Taylor series about a point

$$\mathbf{x}_0 = \begin{bmatrix} a \\ b \\ c \end{bmatrix}$$

there results

$$f(\mathbf{x}) = f(\mathbf{x}_0) + \frac{\partial f(\mathbf{x}_0)}{\partial \mathbf{x}}(\mathbf{x} - \mathbf{x}_0) + (\mathbf{x} - \mathbf{x}_0)^\dagger J(\mathbf{x} - \mathbf{x}_0) + \cdots$$

where J is the square, symmetric matrix of second partial derivatives, evaluated at $\mathbf{x} = \mathbf{x}_0$:

$$J = \begin{bmatrix} \dfrac{\partial^2 f(\mathbf{x}_0)}{\partial x_1^2} & \dfrac{\partial^2 f(\mathbf{x}_0)}{\partial x_1\,\partial x_2} & \cdots & \dfrac{\partial^2 f(\mathbf{x}_0)}{\partial x_1\,\partial x_n} \\[2ex] \dfrac{\partial^2 f(\mathbf{x}_0)}{\partial x_2\,\partial x_1} & \dfrac{\partial^2 f(\mathbf{x}_0)}{\partial x_2^2} & \cdots & \dfrac{\partial^2 f(\mathbf{x}_0)}{\partial x_2\,\partial x_n} \\[2ex] \vdots & \vdots & \vdots & \vdots \\[2ex] \dfrac{\partial^2 f(\mathbf{x}_0)}{\partial x_n\,\partial x_1} & \dfrac{\partial^2 f(\mathbf{x}_0)}{\partial x_n\,\partial x_2} & \cdots & \dfrac{\partial^2 f(\mathbf{x}_0)}{\partial x_n^2} \end{bmatrix}$$

For example, for a function of three variables,

$$f(x_1, x_2, x_3) = f(a, b, c) + \frac{\partial f(a, b, c)}{\partial x_1}(x_1 - a) + \frac{\partial f(a, b, c)}{\partial x_2}(x_2 - b)$$

$$+ \frac{\partial f(a, b, c)}{\partial x_3}(x_3 - c) + \frac{\partial^2 f(a, b, c)}{\partial x_1^2}(x_1 - a)^2$$

$$+ 2 \frac{\partial^2 f(a, b, c)}{\partial x_1 \, \partial x_2} (x_1 - a)(x_2 - b)$$

$$+ 2 \frac{\partial^2 f(a, b, c)}{\partial x_1 \, \partial x_3} (x_1 - a)(x_3 - b) + \frac{\partial^2 f(a, b, c)}{\partial x_2^2} (x_2 - b)^2$$

$$+ 2 \frac{\partial^2 f(a, b, c)}{\partial x_2 \, \partial x_3} (x_2 - b)(x_3 - c)$$

$$+ \frac{\partial^2 f(a, b, c)}{\partial x_3^2} (x_3 - c)^2 + \cdots$$

The function f has a critical point at

$$\mathbf{x}_0 = \begin{bmatrix} a \\ b \\ c \end{bmatrix}$$

if

$$\frac{\partial f(\mathbf{x}_0)}{\partial \mathbf{x}} = \begin{bmatrix} \dfrac{\partial f(\mathbf{x}_0)}{\partial x_1} & \dfrac{\partial f(\mathbf{x}_0)}{\partial x_2} & \cdots & \dfrac{\partial f(\mathbf{x}_0)}{\partial x_n} \end{bmatrix} = 0$$

Expanding f about a critical point x_0 yields

$$f(\mathbf{x}) = f(\mathbf{x}_0) + (\mathbf{x} - \mathbf{x}_0)^\dagger J(\mathbf{x} - \mathbf{x}_0) + \cdots$$

The function has a relative minimum at $\mathbf{x} = \mathbf{x}_0$ if J is positive definite, in which event the quadratic form term

$$(\mathbf{x} - \mathbf{x}_0)^\dagger J(\mathbf{x} - \mathbf{x}_0)$$

is positive for all $\mathbf{x}$. If J is negative definite, the function has a relative maximum at $\mathbf{x} = \mathbf{x}_0$. If J is not sign definite, a value of $\mathbf{x}$ very close to $\mathbf{x}_0$ can be found for which the quadratic form term is postive, and for another value of $\mathbf{x}$ very close to $\mathbf{x}_0$, the quadratic form term will be negative. The critical point is then neither a relative minimum nor a relative maximum. If $J = 0$ at a critical point $\mathbf{x}_0$, higher-order terms in the Taylor series expansion of f about $\mathbf{x}_0$ determine whether $\mathbf{x}_0$ is a relative minimum, a relative maximum, or neither.

For example, consider the function

$$f(x_1, x_2) = x_1^3 + 4x_1 x_2 + x_2^2 + 4x_1$$

This function has critical points where

$$\frac{\partial f}{\partial x_1} = 3x_1^2 + 4x_2 + 4 = 0$$

$$\frac{\partial f}{\partial x_2} = 4x_1 + 2x_2 = 0$$

which are at $(x_1 = 2, x_2 = -4)$ and $(x_1 = \frac{2}{3}, x_2 = -\frac{4}{3})$. The matrix of second partial derivatives is

$$
\mathbf{J}(x_1, x_2) = \begin{bmatrix} \dfrac{\partial^2 f}{\partial x_1^2} & \dfrac{\partial^2 f}{\partial x_1 x_2} \\[2ex] \dfrac{\partial^2 f}{\partial x_2 x_1} & \dfrac{\partial^2 f}{\partial x_2^2} \end{bmatrix} = \begin{bmatrix} 6x_1 & 4 \\ 4 & 2 \end{bmatrix}
$$

When evaluated at the first of these two critical points,

$$
\mathbf{J}_1 = \mathbf{J}(x_1 = 2, x_2 = -4) = \begin{bmatrix} 12 & 4 \\ 4 & 2 \end{bmatrix}
$$

The characteristic equation of J_1 is

$$
|\lambda \mathbf{I} - \mathbf{J}_1| = \begin{vmatrix} \lambda - 12 & -4 \\ -4 & \lambda - 2 \end{vmatrix} = \lambda^2 - 14\lambda + 8 = 0
$$

so its eigenvalues are

$$
\lambda_1, \lambda_2 = \frac{14 \pm \sqrt{196 - 32}}{2}
$$

which are both positive. $\mathbf{J}_1$ is thus positive definite and the point $(x_1 = 2, x_2 = -4)$ is a relative minimum.

When evaluated at the second critical point

$$
J_2 = J\left(x_1 = \frac{2}{3}, x_2 = -\frac{4}{3}\right) = \begin{bmatrix} 4 & 4 \\ 4 & 2 \end{bmatrix}
$$

The characteristic equation of J_2 is

$$
|\lambda \mathbf{I} - J_2| = \begin{vmatrix} \lambda - 4 & -4 \\ -4 & \lambda - 2 \end{vmatrix} = \lambda^2 - 6\lambda - 8 = 0
$$

and its eigenvalues are

$$
\lambda_1, \lambda_2 = \frac{6 \pm \sqrt{36 + 32}}{2}
$$

One of these eigenvalues is positive and the other is negative, so $(x_1 = \frac{2}{3}, x_2 = -\frac{4}{3})$ is neither a relative minimum nor a relative maximum.

8.3.4 Square Roots of Symmetric Matrices

Every symmetric matrix S can be factored in a form

$$
\mathbf{S} = \mathbf{\Psi}^\dagger \mathbf{\Psi}
$$

where $\mathbf{\Psi}$ is termed a square root of $\mathbf{S}$. Square roots of symmetric matrices are generally not unique, and then can involve elements that are complex numbers.

Positive definite and positive semidefinite matrices have square roots that are especially easy to construct and visualize. Let the matrix be diagonalized by the orthogonal tranformation $\mathbf{P}_0$:

$$\mathbf{\Lambda} = \begin{bmatrix} \lambda_1 & 0 & \cdots & 0 & 0 \\ 0 & \lambda_2 & \cdots & 0 & 0 \\ \vdots & \vdots & \vdots & \vdots & \vdots \\ 0 & 0 & \cdots & 0 & \lambda_n \end{bmatrix} = \mathbf{P}_0^\dagger \mathbf{S} \mathbf{P}_0 = \mathbf{P}_0^{-1} \mathbf{S} \mathbf{P}_0$$

Then

$$\mathbf{S} = \mathbf{P}_0 \mathbf{\Lambda} \mathbf{P}_0^{-1} = \mathbf{P}_0 \mathbf{\Lambda} \mathbf{P}_0^\dagger$$

Because the eigenvalues $\lambda_1, \lambda_2, \ldots, \lambda_n$ of $\mathbf{S}$ and $\mathbf{\Lambda}$ are real and nonnegative, their square roots are also real numbers. Defining

$$\mathbf{\Lambda}^{1/2} = \begin{bmatrix} \sqrt{\lambda_1} & 0 & \cdots & 0 & 0 \\ 0 & \sqrt{\lambda_2} & \cdots & 0 & 0 \\ \vdots & \vdots & \vdots & \vdots & \vdots \\ 0 & 0 & \cdots & 0 & \sqrt{\lambda_n} \end{bmatrix} = (\mathbf{\Lambda}^{1/2})^\dagger$$

so that

$$\mathbf{S} = \mathbf{P}_0 (\mathbf{\Lambda}^{1/2})^\dagger \mathbf{\Lambda}^{1/2} \mathbf{P}_0^{-1} = [\mathbf{P}_0 (\mathbf{\Lambda}^{1/2})^\dagger \mathbf{P}_0^{-1}] [\mathbf{P}_0 \mathbf{\Lambda}^{1/2} \mathbf{P}_0^{-1}]$$

$$= [\mathbf{P}_0 (\mathbf{\Lambda}^{1/2})^\dagger \mathbf{P}_0^\dagger] [\mathbf{P}_0 \mathbf{\Lambda}^{1/2} \mathbf{P}_0^\dagger] = (\mathbf{P}_0 \mathbf{\Lambda}^{1/2} \mathbf{P}_0^\dagger)^\dagger (\mathbf{P}_0 \mathbf{\Lambda}^{1/2} \mathbf{P}_0^\dagger)$$

and

$$\mathbf{\Psi} = \mathbf{P}_0 \mathbf{\Lambda}^{1/2} \mathbf{P}_0^\dagger$$

is a square root of $\mathbf{S}$. Negative algebraic signs attached to various of the diagonal elements of $\mathbf{\Lambda}^{1/2}$ are also square roots of $\mathbf{S}$.

Another way of determining the square root of a symmetric positive definite matrix that does not require diagonalization is to use the LU decomposition method discussed in Chapter 6. If the matrix $\mathbf{\Psi}$ is chosen to be upper triangular of the form

$$\mathbf{\Psi} = \begin{bmatrix} \mathbf{\Psi}_{11} & \mathbf{\Psi}_{12} & \mathbf{\Psi}_{13} & \cdots & \mathbf{\Psi}_{1n} \\ 0 & \mathbf{\Psi}_{22} & \mathbf{\Psi}_{23} & \cdots & \mathbf{\Psi}_{2n} \\ 0 & 0 & \mathbf{\Psi}_{33} & \cdots & \mathbf{\Psi}_{3n} \\ \vdots & \vdots & \vdots & \vdots & \vdots \\ 0 & 0 & 0 & \cdots & \mathbf{\Psi}_{nn} \end{bmatrix}$$

its elements can be determined by direct multiplication of $\mathbf{\Psi}^\dagger$ by $\mathbf{\Psi}$ and equating the results to the corresponding elements of the matrix $\mathbf{S}$. Hence it can easily be verified that the elements of the $\mathbf{\Psi}$ matrix satisfy the following recursive formula:

$$\mathbf{\Psi}_{11} = \sqrt{s_{11}}$$

$$\mathbf{\Psi}_{1j} = \frac{s_{1j}}{\mathbf{\Psi}_{11}} \qquad\qquad (j = 2, 3, \ldots, n)$$

$$\Psi_{ii} = \sqrt{s_{ii} - \sum_{k=1}^{i-1} \Psi_{ki}^2} \qquad\qquad (i = 2, 3, \ldots, n) \qquad\qquad (8\text{-}5)$$

$$\Psi_{ij} = \frac{1}{\Psi_{ii}} \left[s_{ij} - \sum_{k=1}^{i-1} \Psi_{ki} \Psi_{kj} \right] \qquad \begin{cases} (j = i + 1, i + 2, \ldots, n) \\ \text{incrementing } j \text{ for every } i \end{cases}$$

$$\Psi_{ij} = 0 \qquad\qquad\qquad\qquad i > j$$

As an example, consider the symmetric positive definite matrix

$$\mathbf{S} = \begin{bmatrix} 4 & 2 & -4 \\ 2 & 2 & 0 \\ -4 & 0 & 9 \end{bmatrix}$$

using the set of equations above,

$$\Psi_{11} = \sqrt{4} = 2$$

$$\Psi_{12} = \frac{s_{12}}{2} = 1 \qquad \Psi_{13} = \frac{s_{13}}{2} = -2$$

$$\Psi_{22} = \sqrt{2 - \Psi_{12}^2} = \sqrt{2 - 1} = 1$$

$$\Psi_{23} = \frac{1}{1} [s_{23} - \Psi_{12} \Psi_{13}] = 2$$

and

$$\Psi_{33} = \sqrt{s_{33} - (\Psi_{13}^2 + \Psi_{23}^2)} = \sqrt{9 - (4 + 4)} = 1$$

Thus

$$\mathbf{\Psi} = \begin{bmatrix} 2 & 1 & -2 \\ 0 & 1 & 2 \\ 0 & 0 & 1 \end{bmatrix}$$

This method of determining the square root of a matrix is also termed Cholesky's decomposition method.

8.4 PRINCIPAL AXES

An important application of the methods of quadratic forms is to the determination of the directions of coordinate axes for which the description of a quadratic function is most simple. When an orthogonal change of variables

$$\mathbf{x} = \mathbf{P}_0 \mathbf{x}' \qquad \mathbf{x}' = \mathbf{P}_0^{-1} \mathbf{x} = \mathbf{P}_0^{\dagger} \mathbf{x}$$

is made, reducing a quadratic form

$$f(\mathbf{x}) = \mathbf{x}^{\dagger} \mathbf{S} \mathbf{x}$$

to a sum of squares, the result is

$$f(\mathbf{x'}) = \lambda_1 x_1'^2 + \lambda_2 x_2'^2 + \cdots + \lambda_n x_n'^2 \tag{8-6}$$

where the λ's are the (real) eigenvalues of the symmetric matrix $\mathbf{S}$. A quadratic equation

$$f(\mathbf{x}) = c \qquad c \text{ a constant}$$

becomes

$$f(\mathbf{x'}) = \lambda_1 x_1'^2 + \lambda_2 x_2'^2 + \cdots + \lambda_n x_n'^2 = c \tag{8-7}$$

in terms of the new variables.

Functions of the form (8-7), involving the squares of the variables, are ellipsoids or hyperbolids or functions with a combination of elliptic and hyperbolic properties in various hyperplanes. The functions (8-6) must also be of this type, because the transformation $\mathbf{P}_0$ is orthogonal, representing a rotation (and possibly reversals) of axes. For two variables (8-7) is

$$f(\mathbf{x'}) = \lambda_1 x_1'^2 + \lambda_2 x_2'^2 = c \tag{8-8}$$

If (8-8) can be written in the form

$$\left(\frac{x_1'}{a}\right)^2 + \left(\frac{x_2'}{a}\right)^2 = 1$$

it is a circle about the origin of radius a. If it can be written as

$$\left(\frac{x_1'}{a}\right)^2 + \left(\frac{x_2'}{b}\right)^2 = 1$$

it is an ellipse, aligned with the x_1' and x_2' axes, with major and minor axes of length $2a$ and $2b$. If (8-8) can be expressed in the form

$$\left(\frac{x_1'}{a}\right)^2 - \left(\frac{x_2'}{b}\right)^2 = 1$$

or

$$-\left(\frac{x_1'}{a}\right)^2 + \left(\frac{x_2'}{b}\right)^2 = 1$$

it is a hyperbola.

In three and more dimensions, a quadratic function can be a sphere (or spheroid), an ellipsoid, a hyperbolic, or having a combination of elliptic and hyperbolic properties. In the sum-of-squares form (8-7) the quadratic function is aligned with the coordinates axes. In the general form (8-6), it is rotated with respect to these *principal axes* through the orthogonal transformation $\mathbf{P}_0$.

8.4.1 Polarization Axes of Light

The electric field vector of a plane light wave has components in two orthogonal directions in a plane perpendicular to the direction of propagation

$$E_x(t) = A \cos \omega t \tag{8-9}$$

$$E_y(t) = B \cos (\omega t + \phi)$$

where ω is the radian frequency of the wave and A, B, and ϕ are constants. The origin of the time scale has been chosen to make E_x a pure cosine, to simplify the algebra. The description of other plane waves, for example acoustic waves, is similar.

Equations (8-9) are parametric equations in terms of time, t, for the electric field vector

$$\mathbf{E}(t) = \mathbf{E}_x(t)x + \mathbf{E}_y(t)y$$

the tip of which, as we will see, generally traces an ellipse. When $B = 0$, the wave is polarized in the x direction. When $A = 0$, the wave is polarized in the y direction. A general plane wave can be considered to be the sum of two orthogonal linearly polarized waves, as in (8-9).

Using the trigonometric identities

$$\cos (\alpha + \beta) = \cos \alpha \cos \beta - \sin \alpha \sin \beta \qquad \sin \omega t = \sqrt{1 - \cos^2 \omega t}$$

the parameter t can be eliminated from (8-9) giving the equation of the curve traced by the tip of the E vector in the x-y plane:

$$y = B \cos (\omega t + \phi) = B \cos \omega t \cos \phi - B \sin \phi \sin \omega t$$

$$= B \left(\frac{x}{A}\right) \cos \phi - B \sin \phi \sqrt{1 - \left(\frac{x}{A}\right)^2}$$

Then

$$\left(y - \frac{Bx \cos \phi}{A}\right)^2 = y^2 - \frac{2B \cos \phi}{A} xy + \frac{B^2 x^2 \cos^2 \phi}{A^2}$$

$$= B^2 \sin^2 \phi \left(1 - \frac{x^2}{A^2}\right)$$

or, using $\sin^2 \phi + \cos^2 \phi = 1$,

$$y^2 - \left(\frac{2B}{A} \cos \phi\right) xy + \left(\frac{B^2}{A^2}\right) x^2 = B^2 \sin^2 \phi \tag{8-10}$$

Generally, the elliptical curve defined by (8-10) is not aligned with the x and y axes. As it is a quadratic function of x and y, it is straightforward to find the directions of the major and minor axes of an elliptical polarization. For example, if $A = 2$, $B = 1$, and $\phi = -7\pi/8$, then

$$E_x(t) = 2 \cos \omega t$$

$$E_y(t) = \cos \left(\omega t - \frac{7\pi}{8}\right)$$

and the electric field vector traces the curve

$$y^2 + 0.924xy + 0.25x^2 \approx 0.146 \tag{8-11}$$

or

$$[x \quad y]\begin{bmatrix} 0.25 & 0.462 \\ 0.462 & 1 \end{bmatrix}\begin{bmatrix} x \\ y \end{bmatrix} = [x \quad y]\, \mathbf{S}\begin{bmatrix} x \\ y \end{bmatrix} \approx 0.146$$

The eigenvalues of the symmetric matrix $\mathbf{S}$ are given by

$$|\, \boldsymbol{\lambda}\mathbf{I} - \mathbf{S}\,| = \begin{vmatrix} \lambda - 0.25 & -0.462 \\ -0.462 & \lambda - 1 \end{vmatrix} \approx \lambda^2 - 1.25\lambda + 0.037 = 0$$

so that $\lambda_1 \approx 1.22$ and $\lambda_2 \approx 0.031$. The eigenvector $\mathbf{P}^1$, corresponding to λ_1, satisfies

$$0.97p_1^1 - 0.462p_1^2 = 0$$

$$-0.462p_1^1 + 0.22p_1^2 = 0$$

when normalized,

$$\mathbf{p}^1 = \begin{bmatrix} 0.430 \\ 0.903 \end{bmatrix}$$

Similarly, the eigenvector $\mathbf{p}^2$, corresponding to λ_2, satisfies

$$-0.219p_1^2 - 0.462p_2^2 = 0$$

$$-0.462p_1^2 - 0.969p_2^2 = 0$$

giving a normalized

$$\mathbf{p}^2 = \begin{bmatrix} -0.903 \\ 0.430 \end{bmatrix}$$

This choice of $\mathbf{p}^2$ rather than its negative is used so that the resulting orthogonal transformation has no axis direction reversals.

The change of variables

$$\begin{bmatrix} x \\ y \end{bmatrix} = \mathbf{P}_0\begin{bmatrix} x' \\ y' \end{bmatrix} \qquad \begin{bmatrix} x' \\ y' \end{bmatrix} = \mathbf{P}_0^{-1}\begin{bmatrix} x \\ y \end{bmatrix} = \mathbf{P}_0^{\dagger}\begin{bmatrix} x \\ y \end{bmatrix}$$

where

$$\mathbf{P}_0 = [\mathbf{p}^1 \,|\, \mathbf{p}^2] = \begin{bmatrix} 0.430 & -0.903 \\ 0.903 & 0.430 \end{bmatrix}$$

transforms (8-11) to a pure sum of squares:

$$[x' \quad y']\,(\mathbf{P}_0^{\dagger}\mathbf{S}\mathbf{P}_0)\begin{bmatrix} x' \\ y' \end{bmatrix} = [x' \quad y']\begin{bmatrix} 1.22 & 0 \\ 0 & 0.031 \end{bmatrix}\begin{bmatrix} x' \\ y' \end{bmatrix}$$

or

$$1.22x'^2 + 0.031y'^2 = 0.146$$

or

$$\left(\frac{x'}{0.346}\right)^2 + \left(\frac{y'}{2.17}\right)^2 = 1$$

This is the equation of an ellipse, centered on the origin, with major axis of 4.34 units along y' and minor axis of 0.692 unit along x' as illustrated in Figure 8-1. The orthogonal transformation $\mathbf{P}_0$ is of the form of a rotation of axes through the angle ϕ,

$$\mathbf{P}_0 = \begin{bmatrix} \cos\phi & -\sin\phi \\ \sin\phi & \cos\phi \end{bmatrix} = \begin{bmatrix} 0.430 & -0.903 \\ 0.903 & 0.430 \end{bmatrix}$$

where

$$\cos\phi = 0.430 \qquad \sin\phi = 0.903$$

or

$$\phi = 1.126 \text{ radians} = 64.5°$$

8.4.2 Linear–Quadratic Functions

A function that contains both quadratic and linear terms has principal axes that generally involve both rotation and translation of the original axes. It is easiest to rotate the axes first, then translate them, because the quadratic terms that are not squares of the variables are then eliminated, leaving just squares and linear terms equated to a constant. By completing the square for each variable, the additional required translation of coordinates is then easily found.

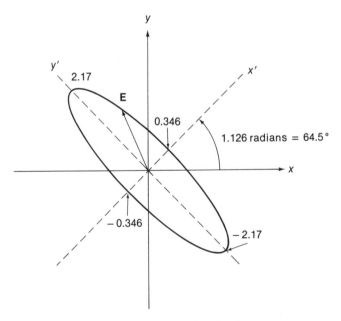

Figure 8-1   Elliptical polarization of a plane wave.

A linear–quadratic function has the form

$$\mathbf{x}^\dagger \mathbf{S}\mathbf{x} + \mathbf{r}^\dagger \mathbf{x} = c, \text{ a constant}$$

An orthogonal transformation

$$\mathbf{x} = \mathbf{P}_0 \mathbf{x}' \qquad \mathbf{x}' = \mathbf{P}_0^{-1}\mathbf{x} = \mathbf{P}_0^\dagger \mathbf{x}$$

(a rotation of axes) that diagonalizes the real symmetric matrix $\mathbf{S}$ converts the function to the form

$$\mathbf{x}'^\dagger(\mathbf{P}_0^\dagger \mathbf{S}\mathbf{P}_0)\mathbf{x}' + (\mathbf{r}^\dagger \, \mathbf{P}_0)\mathbf{x}' = c$$

or

$$\lambda_1 x_1'^2 + \lambda_2 x_2'^2 + \cdots + \lambda_n x_n'^2 + \delta_1 x_1' + \delta_2 x_2' + \cdots + \delta_n x_n' = c$$

in terms of the new variables $\mathbf{x}'$, where the eigenvalues λ are real numbers. For nonzero λ's, completing the squares for each variable gives

$$\lambda_1\left[\left(x_1' + \frac{\delta_1}{2\lambda_1}\right)^2 - \left(\frac{\delta_1}{2\lambda_1}\right)^2\right] + \lambda_2\left[\left(x_2' + \frac{\delta_2}{2\lambda_2}\right)^2 - \left(\frac{\delta_2}{2\lambda_2}\right)^2\right]$$

$$+ \cdots + \lambda_n\left[\left(x_n' + \frac{\delta_n}{2\lambda_n}\right)^2 - \left(\frac{\delta_n}{2\lambda_n}\right)^2\right] = c$$

or

$$\lambda_1\left(x_1' + \frac{\delta_1}{2\lambda_1}\right)^2 + \lambda_2\left(x_2' + \frac{\delta_2}{2\lambda_2}\right)^2 + \cdots + \lambda_n\left(x_n' + \frac{\delta_n}{2\lambda_n}\right)^2$$

$$= c + \frac{\delta_1^2}{4\lambda_1} + \frac{\delta_2^2}{4\lambda_2} + \cdots + \frac{\delta_n^2}{4\lambda_n} = d$$

The additional translation of variables

$$x_1'' = x_1' + \frac{\delta_1}{2\lambda_1}$$

$$x_2'' = x_2' + \frac{\delta_2}{2\lambda_2}$$

$$\vdots$$

$$x_n'' = x_n' + \frac{\delta_n}{2\lambda_n}$$

then result in the pure sum of squares

$$\lambda_1 x_1''^2 + \lambda_2 x_2''^2 + \cdots + \lambda_n x_n''^2 = d$$

As a numerical example, consider the linear-quadratic function of two variables

$$2x^2 + 4xy - y^2 + 3x - 5y = 6$$

The quadratic terms in the function are the quadratic form

$$f(x, y) = 2x^2 + 4xy - y^2 = [x \quad y] \begin{bmatrix} 2 & 2 \\ 2 & -1 \end{bmatrix} \begin{bmatrix} x \\ y \end{bmatrix} = [x \quad y] \, S \begin{bmatrix} x \\ y \end{bmatrix}$$

An orthogonal change of variables,

$$\begin{bmatrix} x \\ y \end{bmatrix} = P_0 \begin{bmatrix} x' \\ y' \end{bmatrix} \qquad \begin{bmatrix} x' \\ y' \end{bmatrix} = P_0^{-1} \begin{bmatrix} x \\ y \end{bmatrix} = P_0^{\dagger} \begin{bmatrix} x \\ y \end{bmatrix}$$

that transforms f into a pure sum of squares is

$$P_0 = \begin{bmatrix} 0.447 & 0.894 \\ -0.894 & 0.447 \end{bmatrix}$$

for which

$$f(x', y') = [x' \quad y'] \, P_0^{\dagger} S P_0 \begin{bmatrix} x' \\ y' \end{bmatrix} = [x' \quad y'] \begin{bmatrix} -2 & 0 \\ 0 & 3 \end{bmatrix} \begin{bmatrix} x' \\ y' \end{bmatrix} = -2x'^2 + 3y'^2$$

This transformation is a rotation of axes by the angle ϕ given by

$$P_0 = \begin{bmatrix} \cos \phi & -\sin \phi \\ \sin \phi & \cos \phi \end{bmatrix} = \begin{bmatrix} 0.447 & 0.894 \\ -0.894 & 0.447 \end{bmatrix}$$

so $\phi \approx -1.11$ radians or $-63.4°$.

     Changing variables in the original function gives

$$-2x'^2 + 3y'^2 + 3(0.447x' + 0.894y') - 5(-0.894x' + 0.447y')$$

$$= -2x'^2 + 5.811x' + 3y'^2 + 0.447y' = 6$$

Completing the squares in x' and y', the linear quadratic function becomes

$$-2[(x' - 1.45)^2 - (1.45)^2] + 3[(y' + 0.0745)^2 - (0.0745)^2] = 6$$

or

$$-2(x' - 1.45)^2 + 3(y' + 0.0745)^2 = 1.81$$

The translation of variables

$$x'' = x' - 1.45$$

$$y'' = y' + 0.0745$$

then gives

$$-2x''^2 + 3y''^2 = 1.81$$

or

$$-\left(\frac{x''}{0.95}\right)^2 + \left(\frac{y''}{0.777}\right)^2 = 1$$

which is the equation of a hyperbola. Figure 8-2 shows the relations between the coordinates for this rotation followed by a translation.

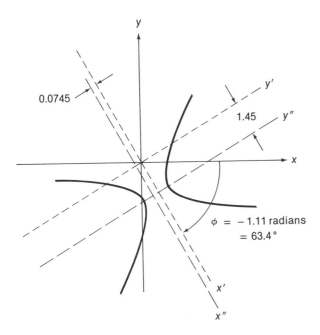

Figure 8-2 Finding principal axes of a linear–quadratic function.

8.5 MATRIX RANK DETERMINATION

Many applications of matrix methods require determination of the ranks of matrices. A simple example is in finding whether or not a set of linear algebraic equations is linearly independent and consistent. Numerical calculation of the rank of a matrix is difficult because, in a calculation of rank, the presence of small numerical errors due to roundoff can easily change linearly dependent matrix rows or columns to independent ones.

8.5.1 Pivoting to Determine Rank

A straightforward way of finding the rank of an $m \times n$ **A** is to perform Gauss–Jordan pivoting on its rows or columns. For example, for the matrix

$$\mathbf{A} = \begin{bmatrix} 1 & 2 \\ 2 & 4 \\ -1 & -2 \end{bmatrix}$$

adding the third row to the top row and adding twice the third row to the second row gives

$$\mathbf{A}' = \mathbf{PA} = \begin{bmatrix} 1 & 2 \\ 0 & 0 \\ 0 & 0 \end{bmatrix}$$

where **P** is a nonsingular product of elementary row operation matrices. The rank of **A** and **A**′ are the same, and **A**′ has rank equal to the number of nonzero rows after full Gauss–Jordan pivoting, which is rank 1.

When numerical Gauss–Jordan pivoting is done, however, small numerical errors such as those from roundoff can easily convert a linearly dependent row (or column) to one that is linearly independent. For example, for the matrix

$$\mathbf{A} = \begin{bmatrix} 1 & 2 & 3 \\ 4 & 5 & 6 \\ 7 & 8 & 9 \\ 10 & 11 & 12 \end{bmatrix}$$

the PIVOT program (Table 2-4), ignoring the knowns, gives

```
FULLY PIVOTED EQUATIONS ARE
COEFFICIENTS OF EQUATION 1
0
.5
1
KNOWN IS 0
COEFFICIENTS OF EQUATION 2
1
.5
0
KNOWN IS 0
COEFFICIENTS OF EQUATION 3
0
4.65661287E-10
0
KNOWN IS 0
COEFFICIENTS OF EQUATION 4
0
0
0
KNOWN IS 0
```

or

$$\mathbf{A}' = \mathbf{PA} = \begin{bmatrix} 0 & 0.5 & 1 \\ 1 & 0.5 & 0 \\ 0 & 4.65 \times 10^{-10} & 0 \\ 0 & 0 & 0 \end{bmatrix}$$

where $\mathbf{P}$ is some nonsingular product of elementary row operation matrices. It seems obvious that the 4.65×10^{-10} element of $\mathbf{A}'$ should really be zero (and it should be), but some judgment is generally needed in this decision. If $\mathbf{A}$ had, instead, been

$$\mathbf{A} = \begin{bmatrix} 10^{-10} & 2 \times 10^{-10} & 3 \times 10^{-10} \\ 4 & 5 & 6 \\ 7 & 8 & 9 \\ 10 & 11 & 12 \end{bmatrix}$$

and the result for $\mathbf{A}'$ were similar, the decision as to rank might be much more difficult.

Gauss–Jordan pivoting is a useful tool in matrix rank determination, but it should always be used with judgment and care. It is significant, too, that different orders of pivoting can give quite different numerical results.

8.5.2 Singular Values of a Matrix

For an $m \times n$ matrix $\mathbf{A}$, with $m \geq n$, the rank of the related symmetric $n \times n$ matrix

$$\mathbf{S} = \mathbf{A}^{\dagger}\mathbf{A}$$

is the same as the rank of $\mathbf{A}$. This can be proved as follows. Any $m \times n$ matrix $\mathbf{A}$ with $m \geq n$ can be transformed with a sequence of elementary column operations $\mathbf{Q}$ to a form consisting of rows containing all zeros or a single one and all other elements in the row zero. Each one that occurs in a row is in a different column. For example, a 4×3 matrix $\mathbf{A}$ might transform to

$$\mathbf{AQ} = \begin{bmatrix} 0 & 0 & 0 \\ 0 & 1 & 0 \\ 0 & 0 & 0 \\ 0 & 0 & 1 \end{bmatrix} = \mathbf{G} \tag{8-12}$$

In Section 6.3.4 additional elementary row operations were imagined to interchange the rows of $\mathbf{AQ}$ to transform $\mathbf{A}$ to normal form,

$$\mathbf{PAQ} = \begin{bmatrix} \mathbf{I}_r & | & \mathbf{0} \\ - & - & - \\ \mathbf{0} & | & \mathbf{0} \end{bmatrix}$$

Or, one could operate on the rows and then interchange columns, or do some other combination of column and row operations to get to normal form. This will not be done here.

The rank r of $\mathbf{AQ} = \mathbf{G}$ is the number of rows (or columns) of $\mathbf{G}$ containing ones, and the rank of $\mathbf{A}$ is the same as the rank of $\mathbf{G}$ because the two matrices are related by a nonsingular transformation $\mathbf{Q}$. The symmetric $n \times n$ matrix

$$\mathbf{G}^{\dagger}\mathbf{G} = (\mathbf{AQ})^{\dagger}\mathbf{AQ} = \mathbf{Q}^{\dagger}\mathbf{A}^{\dagger}\mathbf{AQ}$$

has r ones along its diagonal and thus is of rank r. For the example (8-12),

$$\mathbf{G}^{\dagger}\mathbf{G} = \begin{bmatrix} 0 & 0 & 0 \\ 0 & 1 & 0 \\ 0 & 0 & 1 \end{bmatrix}$$

Then

$$\mathbf{A}^{\dagger}\mathbf{A} = (\mathbf{Q}^{\dagger})^{-1}\mathbf{G}^{\dagger}\mathbf{G}\mathbf{Q}^{-1}$$

must also be of rank r because it is the product of $\mathbf{G}^{\dagger}\mathbf{G}$ with nonsingular matrices.

As $\mathbf{S} = \mathbf{A}^{\dagger}\mathbf{A}$ is real and symmetric, its eigenvalues are real numbers and can always be diagonalized with a real, orthogonal similarity transformation $\mathbf{P}$:

$$\mathbf{P}_0^{-1}(\mathbf{A}^{\dagger}\mathbf{A})\mathbf{P}_0 = \mathbf{P}_0^{-1}\mathbf{S}\mathbf{P}_0 = \mathbf{P}_0^{\dagger}\mathbf{S}\mathbf{P}_0 = \mathbf{\Lambda} = \begin{bmatrix} \lambda_1 & 0 & 0 & \cdots & 0 \\ 0 & \lambda_2 & 0 & \cdots & 0 \\ \vdots & \vdots & \vdots & \vdots & \vdots \\ 0 & 0 & 0 & \cdots & \lambda_n \end{bmatrix}$$

Since $\mathbf{P}$ is nonsingular, the rank of $\mathbf{S}$, and thus of $\mathbf{A}$, is the rank of Λ, which is the number of nonzero eigenvalues $\lambda_1, \lambda_2, \ldots, \lambda_n$. The positive square roots of the eigenvalues of $\mathbf{S} = \mathbf{A}^\dagger\mathbf{A}$ are termed the *singular values* of $\mathbf{A}$.

Unlike the elements of a fully pivoted matrix, the singular values of a matrix do not depend on the order or type of operations that are performed to get the result. Hopefully, we can simply judge the closeness to zero or near-zero singular values of a matrix to decide on its rank.

Since the rank of $\mathbf{A}$ and $\mathbf{A}^\dagger$ are the same, if $m < n$ (we have so far assumed that $m \geq n$), $\mathbf{A}^\dagger$ has more rows than columns, and

$$\mathbf{S} = \mathbf{A}\mathbf{A}^\dagger$$

instead of $\mathbf{S} = \mathbf{A}^\dagger\mathbf{A}$. That is, the symmetric matrix $\mathbf{S}$, with eigenvalues that are the squares of the singular values of a $m \times n$ matrix $\mathbf{A}$ is the $n \times n$ matrix

$$\mathbf{S} = \mathbf{A}^\dagger\mathbf{A}$$

when $m \geq n$ and is the $m \times n$ matrix

$$\mathbf{S} = \mathbf{A}\mathbf{A}^\dagger$$

when $n > m$. When m and n differ, it is the smaller-dimensioned of the products $\mathbf{A}^\dagger\mathbf{A}$ and $\mathbf{A}\mathbf{A}^\dagger$. When $m = n$,

$$\mathbf{A}^\dagger\mathbf{A} = \mathbf{A}\mathbf{A}^\dagger$$

so either product is the matrix $\mathbf{S}$. The eigenvalues of the symmetric matrix $\mathbf{S}$ and their square roots, the singular values of $\mathbf{A}$, are always real numbers. In addition, the eigenvalues of $\mathbf{S}$ and the singular values of $\mathbf{A}$ (of a matrix $\mathbf{A}$ with real elements) are nonnegative. This follows from the fact that

$$\Psi = \mathbf{P}_0\mathbf{A}$$

is a square root of the diagonal matrix Λ, and there is no way to obtain negative diagonal elements of Λ from

$$\Psi^\dagger\Psi = (\mathbf{P}_0\mathbf{A})^\dagger(\mathbf{P}_0\mathbf{A}) = \mathbf{P}_0^\dagger(\mathbf{A}^\dagger\mathbf{A})\mathbf{P}_0 = \Lambda$$

unless Ψ has one or more complex elements. But $\mathbf{A}$ and $\mathbf{P}_0$ are both real, so their product is real.

As a numerical example, consider the rank 2 matrix

$$\mathbf{A} = \begin{bmatrix} 3.01 & 4.01 & 5.01 \\ 3.99 & 4.99 & 5.99 \\ 1.07 & 1.08 & 1.09 \\ 2.00 & 2.10 & 2.20 \end{bmatrix} \qquad \mathbf{S} = \begin{bmatrix} 30.1251 & 37.3358 & 44.5465 \\ 37.3358 & 46.5566 & 55.7774 \\ 44.5465 & 55.7774 & 67.0083 \end{bmatrix}$$

Using the FADDEEV program (Table 7-10), the characteristic equation of $\mathbf{S}$ is found to be

$$s^3 - 143.69s^2 + 51.3616153s + 3.01264226 \times 10^{-5} = 0$$

Using the REALFACT program (Table 5-9), the roots of this polynomial are

$$\lambda_1 \approx -2.149 \times 10^{-8} \qquad \lambda_2 \approx 0.358 \qquad \lambda_3 \approx 143.332$$

The squares of the singular values λ_2 and λ_3 being a factor of 10^8 or greater in magnitude than λ_1 shows that it is likely that the rank of **A** is 2. Numerical imprecision, probably in the characteristic polynomial calculation, has made the λ_1 root negative. This is not a very good way to find the singular values because it deals with their squares. It is much better from a numerical standpoint to operate on the matrix **A** directly with a technique called UV decomposition.

When the elements of a matrix have widely varied values, it is usually a good idea to *scale* it before attempting a rank determination. A popular scaling method is to multiply each of the columns by factors that make the largest element in the new column unity, then do the same to the rows. The rank of the resulting matrix is unchanged by these operations, although numerical inaccuracy may, as usual, destroy or create linearly dependence of a row or a column. The philosophy behind such scaling is that it is better to do so first, where numerical inaccuracy effects on the rank are relatively easy to understand, than to risk much worse numerical difficulties later.

8.6 LEAST SQUARES ESTIMATION

There are many situations in practice that are described by an overdetermined set of linear algebraic equations that are, because of errors in measurements, inconsistent. One way of selecting an approximate solution is to find the combination of the variables for which the sum of the squares of equation errors is minimum, a process called least squares estimation. In this section the methods of least squares estimation are developed and a computer program for least squares calculation is developed. Weighted least squares estimation is also discussed.

8.6.1 Matrix Derivatives

The derivative of a matrix **A** with respect to a scalar is the matrix composed of elements that are the derivatives of the corresponding elements of **A**. Higher derivatives, partial derivatives, and integrals are defined similarly. Table 8-3 lists derivative and partial derivative relations for matrices. These can be verified easily by expansion. Using the relation for the derivative of a product of two matrices,

$$\frac{d}{dt}(\mathbf{AB}) = \frac{d\mathbf{A}}{dt}\mathbf{B} + \mathbf{A}\frac{d\mathbf{B}}{dt}$$

the derivative of the inverse of a matrix is derived as follows:

$$\frac{d}{dt}(\mathbf{A}^{-1}\mathbf{A}) = \frac{d\mathbf{A}^{-1}}{dt}\mathbf{A} + \mathbf{A}^{-1}\frac{d\mathbf{A}}{dt} = \mathbf{0}$$

$$\frac{d\mathbf{A}^{-1}}{dt} = -\mathbf{A}^{-1}\frac{d\mathbf{A}}{dt}\mathbf{A}^{-1}$$

The partial derivative of a scalar function f with respect to a vector **x** is a row vector that is composed of the derivatives of f with respect to each of the components of **x**:

TABLE 8-3   Properties of Matrix Derivatives

DERIVATIVES OF A MATRIX WITH RESPECT TO A SCALAR

(1) $\dfrac{d}{dt}(k\mathbf{A}) = k\dfrac{d\mathbf{A}}{dt} + \dfrac{dk}{dt}\mathbf{A}$ if k is a scalar

$\quad\ \dfrac{\partial}{\partial t}(k\mathbf{A}) = k\dfrac{\partial\mathbf{A}}{\partial t} + \dfrac{\partial k}{\partial t}\mathbf{A}$ if k is a scalar

(2) $\dfrac{d}{dt}(\mathbf{A} + \mathbf{B}) = \dfrac{d\mathbf{A}}{dt} + \dfrac{d\mathbf{B}}{dt}$

$\quad\ \dfrac{\partial}{\partial t}(\mathbf{A} + \mathbf{B}) = \dfrac{\partial\mathbf{A}}{\partial t} + \dfrac{\partial\mathbf{B}}{\partial t}$

(3) $\dfrac{d}{dt}(\mathbf{AB}) = \dfrac{d\mathbf{A}}{dt}\mathbf{B} + \mathbf{A}\dfrac{d\mathbf{B}}{dt}$

$\quad\ \dfrac{\partial}{\partial t}(\mathbf{AB}) = \dfrac{\partial\mathbf{A}}{\partial t}\mathbf{B} + \mathbf{A}\dfrac{\partial\mathbf{B}}{\partial t}$

DERIVATIVE OF A MATRIX INVERSE

(4) $\dfrac{d}{dt}(\mathbf{A}^{-1}) = -\mathbf{A}\dfrac{d\mathbf{A}}{dt}\mathbf{A}^{-1}$

DERIVATIVE OF A SCALAR WITH RESPECT TO A VECTOR

(5) $\dfrac{\partial}{\partial\mathbf{x}}(\mathbf{x}^\dagger\mathbf{y}) = \dfrac{\partial}{\partial\mathbf{x}}(\mathbf{y}^\dagger\mathbf{x}) = \mathbf{y}^\dagger$

(6) $\dfrac{\partial}{\partial\mathbf{x}}(\mathbf{x}^\dagger\mathbf{A}\mathbf{x}) = \mathbf{x}^\dagger\mathbf{A}^\dagger + \mathbf{x}^\dagger\mathbf{A}$

DERIVATIVE OF A VECTOR WITH RESPECT TO A VECTOR

(7) $\dfrac{\partial}{\partial\mathbf{x}}(\mathbf{A}\mathbf{x}) = \mathbf{A}$

$$\frac{\partial f}{d\mathbf{x}} = \left[\begin{array}{cccc} \dfrac{\partial f}{dx_1} & \dfrac{\partial f}{dx_2} & \cdots & \dfrac{\partial f}{dx_n} \end{array}\right]$$

This is also called the *gradient operation*. Properties of the gradient are also listed in Table 8-3. The partial derivative of a vector $\mathbf{y}$ with respect to a vector $\mathbf{x}$ is then defined as

$$\frac{\partial\mathbf{y}}{\partial\mathbf{x}} = \begin{bmatrix} \dfrac{\partial y_1}{d\mathbf{x}} \\[2mm] \dfrac{\partial y_2}{d\mathbf{x}} \\[2mm] \vdots \\[2mm] \dfrac{\partial y_m}{d\mathbf{x}} \end{bmatrix} = \begin{bmatrix} \dfrac{\partial y_1}{\partial x_1} & \dfrac{\partial y_1}{\partial x_2} & \cdots & \dfrac{\partial y_1}{\partial x_n} \\[2mm] \dfrac{\partial y_2}{\partial x_1} & \dfrac{\partial y_2}{\partial x_2} & \cdots & \dfrac{\partial y_2}{\partial x_n} \\[2mm] \vdots & \vdots & \vdots & \vdots \\[2mm] \dfrac{\partial y_m}{\partial x_1} & \dfrac{\partial y_m}{\partial x_2} & \cdots & \dfrac{\partial y_m}{\partial x_n} \end{bmatrix}$$

8.6.2 Least Squares Problem

The fundamental least squares problem involves an overdetermined set of linear algebraic equations

$$\mathbf{Ax} = \mathbf{b}$$

where the matrix $\mathbf{A}$ is $m \times n$ and $m > n$. Because the equations have no solution, since they are inconsistent, we seek an approximate solution vector $\hat{\mathbf{x}}$ such that the sum of squares of the errors between the actual knowns $\mathbf{b}$ and the knowns $\mathbf{A}\hat{\mathbf{x}}$ that would be necessary for the equations to be consistent,

$$J(\hat{\mathbf{x}}) = (\mathbf{b} - \mathbf{A}\hat{\mathbf{x}})^{\dagger}(\mathbf{b} - \mathbf{A}\hat{\mathbf{x}}) = \mathbf{v}^{\dagger}\mathbf{v} \qquad (8\text{-}13)$$

is minimized. That is, an estimated solution $\hat{\mathbf{x}}$ is to be found such that in

$$\mathbf{A}\hat{\mathbf{x}} + \mathbf{v} = \mathbf{b}$$

the sum of the squares of the elements of the error vector $\mathbf{v}$ is minimum.

To find the minimum, the partial derivatives of J in (8-13) with respect to each of the elements of $\hat{\mathbf{x}}$ are equated to zero. Using properties (5) and (6) of Table 8-3, then,

$$\left(\frac{\partial J}{\partial \hat{\mathbf{x}}}\right)^{\dagger} = \begin{bmatrix} \dfrac{\partial J}{\partial \hat{\mathbf{x}}_1} \\[2mm] \dfrac{\partial J}{\partial \hat{\mathbf{x}}_2} \\[1mm] \vdots \\[1mm] \dfrac{\partial J}{\partial \hat{\mathbf{x}}_n} \end{bmatrix} = \mathbf{A}^{\dagger}(\mathbf{b} - \mathbf{A}\hat{\mathbf{x}}) = \mathbf{0}$$

This gives

$$\mathbf{A}^{\dagger}\mathbf{A}\hat{\mathbf{x}} = \mathbf{A}^{\dagger}\mathbf{b}$$

or, if the inverse of $\mathbf{A}^{\dagger}\mathbf{A}$ exists,

$$\hat{\mathbf{x}} = (\mathbf{A}^{\dagger}\mathbf{A})^{-1}\mathbf{A}^{\dagger}\mathbf{b} \qquad (8\text{-}14)$$

If $\mathbf{A}$ is of full rank, then $\mathbf{A}^{\dagger}\mathbf{A}$, which is an $m \times m$ matrix, is of full rank, that is, nonsingular. Thus $(\mathbf{A}^{\dagger}\mathbf{A})^{-1}$ exists and the least squares estimate $\hat{\mathbf{x}}$, given by (8-14), is unique. It is a minimum because of the matrix of second derivatives of J,

$$\frac{\partial^2 J}{\partial \hat{\mathbf{x}}^2} = \frac{\partial}{\partial \hat{\mathbf{x}}}\left(\frac{\partial J}{\partial \hat{\mathbf{x}}}\right) = \mathbf{A}^{\dagger}\mathbf{A}$$

which is symmetric, is positive definite if $\mathbf{A}$ is of full rank.

As a single-variable example of least-squares estimation, consider the three linear algebraic equations

$$2x = 2.2$$

$$3x = 2.9$$

$$-x = -1.3$$

which are of the form

$$\begin{bmatrix} 2 \\ 3 \\ -1 \end{bmatrix} x = \begin{bmatrix} 2.2 \\ 2.9 \\ -1.3 \end{bmatrix} \quad \text{or} \quad \mathbf{a}x = \mathbf{b}$$

These might represent three different measurements of the bearing of a ship, each slightly in error. The least squares estimate of x is

$$\hat{x} = (\mathbf{a}^\dagger \mathbf{a})^{-1} \mathbf{a}^\dagger \mathbf{b} = \left(\begin{bmatrix} 2 & 3 & -1 \end{bmatrix} \begin{bmatrix} 2 \\ 3 \\ -1 \end{bmatrix} \right)^{-1}$$

$$\times \begin{bmatrix} 2 & 3 & -1 \end{bmatrix} \begin{bmatrix} 2.2 \\ 2.9 \\ -1.3 \end{bmatrix} = \frac{1}{14}(14.4) \approx 1.029$$

For the overdetermined equations

$$2x_1 - x_2 = 4.8$$

$$3x_1 + 4x_2 = 2.1$$

$$x_1 - 3x_2 = 5.2$$

$$6x_1 + 5x_2 = 8.0$$

which are of the form

$$\begin{bmatrix} 2 & -1 \\ 3 & 4 \\ 1 & -3 \\ 6 & 5 \end{bmatrix} \begin{bmatrix} x_1 \\ x_2 \end{bmatrix} = \begin{bmatrix} 4.8 \\ 2.1 \\ 5.2 \\ 8.0 \end{bmatrix} \quad \text{or} \quad \mathbf{A}\mathbf{x} = \mathbf{b}$$

the least squares estimate of the vector $\mathbf{x}$ is

$$\hat{\mathbf{x}} = (\mathbf{A}^\dagger \mathbf{A})^{-1} \mathbf{A}^\dagger \mathbf{b}$$

$$= \left(\begin{bmatrix} 2 & 3 & 1 & 6 \\ -1 & 4 & -3 & 5 \end{bmatrix} \begin{bmatrix} 2 & -1 \\ 3 & 4 \\ 1 & -3 \\ 6 & 5 \end{bmatrix} \right)^{-1}$$

$$\times \begin{bmatrix} 2 & 3 & 1 & 6 \\ -1 & 4 & -3 & 5 \end{bmatrix} \begin{bmatrix} 4.8 \\ 2.1 \\ 5.2 \\ 8.0 \end{bmatrix} \approx \begin{bmatrix} 2.107 \\ -0.979 \end{bmatrix}$$

The least squares estimate $\hat{\mathbf{x}}$ might be interpreted as the "most likely" value of the vector $\mathbf{x}$ to have produced the set of measurements $\mathbf{b}$.

Karl Frederick Gauss (1777–1855) invented the method of least squares estimation in 1795 for application to the calculation of planetary and comet orbits from telescopic measurements. Six precise measurements would determine the six parameters of each orbit, but the individual measurements available were likely to be quite inaccurate. Many more than the minimum number of measurements were used, and a "best fit" to an orbit was found by minimizing the sum of squares of the orbital parameter measurement errors. Adrien Marie Legendre (1707–1783) independently developed least squares estimation and, in 1806, was first to publish the method.

8.6.3 Least Squares Program

The least squares estimate

$$\hat{\mathbf{x}} = (\mathbf{A}^{\dagger}\mathbf{A})^{-1}\mathbf{A}^{\dagger}\mathbf{b}$$

for a set of overdetermined linear algebraic equations is performed by the LEAST computer program listed in Table 8-4. The needed matrix inversion is done with the Faddeev–Leverrier method in lines 380–820. These steps are the same as in the earlier FADDEEV program (Table 7-10), but with different variables. Since there is no need to store any of the characteristic equation coefficients, a single variable C is used as temporary storage of each coefficient, eventually being equal to the determinant of $\mathbf{R} = \mathbf{A}^{\dagger}\mathbf{A}$. Previously, these coefficients were stored in a one-dimensional array $P(\cdot)$. The END instruction is at line 1900 because additional instructions will be inserted in this program in the next section.

TABLE 8-4

Computer Program in BASIC for Least Squares Estimation (LEAST)

VARIABLES USED

M	=	number of rows of $\mathbf{A}$
N	=	number of columns of $\mathbf{A}$
A(20,20)	=	$\mathbf{A}$ matrix
B(20)	=	$\mathbf{b}$ vector
R(20,20)	=	intermediate matrices such as $\mathbf{A}^{\dagger}\mathbf{A}$
P(20,20)	=	intermediate matrices such as $(\mathbf{A}^{\dagger}\mathbf{A})^{-1}$
Q(20,20)	=	intermediate matrices such as those used in computing a matrix inverse with Faddeev's method
C	=	characteristic equation coefficients calculated during matrix inversion
G(20)	=	intermediate vector such as $\mathbf{A}^{\dagger}\mathbf{b}$
X(20)	=	$\mathbf{x}$ vector
I	=	row index
J	=	column index

LISTING

```
100     PRINT "BATCH LEAST SQUARES ESTIMATION"
110     DIM A(20,20),B(20),P(20,20),R(20,20),Q(20,20)
120     DIM G(20),X(20)
130     PRINT "PLEASE ENTER NO. OF EQUATIONS";
140     INPUT M
150     PRINT "PLEASE ENTER NO. OF VARIABLES";
160     INPUT N
170     IF N > M THEN PRINT "NOT ENOUGH EQUATIONS"
180     IF N >M THEN GOTO 130
190     REM   ENTER EQUATIONS
200     FOR I = 1 TO M
210     PRINT "ENTER COEFFICIENTS OF EQUATION NO.";I
220     FOR J = 1 TO N
230     PRINT "COEFFICIENTS OF VARIABLE NO.";J;
240     INPUT A(I,J)
250     NEXT J
260     PRINT "KNOWN FOR EQUATION";I;
270     INPUT B(I)
280     NEXT I
290     REM   COMPUTE R=A TRANSPOSE TIMES A
300     FOR I = 1 TO N
310     FOR J = 1 TO N
320     R(I,J) = 0
330     FOR K = 1 TO M
340     R(I,J) = R(I,J) + A(K,I) * A(K,J)
350     NEXT K
360     NEXT J
370     NEXT I
380     REM   COMPUTE P=R INVERSE USING FADDEEV'S METHOD
390     REM   SET Q=0
400     FOR I = 1 TO N
410     FOR J = 1 TO N
420     Q(I,J) = 0
430     NEXT J
440     NEXT I
450     REM   SET LEADING POLYNOMIAL COEFFICIENT TO UNITY
460     C = 1
470     REM   FIND THE N P MATRICES
480     FOR L = 1 TO N
490     REM   FORM P=Q PLUS C TIMES I
500     FOR I = 1 TO N
510     FOR J = 1 TO N
520     P(I,J) = Q(I,J)
530     NEXT J
540     NEXT I
550     FOR J = 1 TO N
560     P(J,J) = P(J,J) + C
570     NEXT J
580     REM   Q=R TIMES P
590     FOR I = 1 TO N
600     FOR J = 1 TO N
610     Q(I,J) = 0
620     FOR K = 1 TO N
630     Q(I,J) = Q(I,J) + R(I,K) * P(K,J)
640     NEXT K
650     NEXT J
660     NEXT I
670     REM   FORM C=(-1/L)*TRACE(Q)
680     C = 0
690     FOR K = 1 TO N
700     C = C - Q(K,K)
```

(continued)

```
710      NEXT K
720      C = C / L
730      NEXT L
740    REM PRINT MESSAGE AND END IF R IS SINGULAR
750      IF C = 0 THEN PRINT "EQUATIONS ARE SINGULAR"
760      IF C = 0 THEN  GOTO 1900
770    REM  FORM P=R INVERSE
780      FOR I = 1 TO N
790      FOR J = 1 TO N
800      P(I,J) = - P(I,J) / C
810      NEXT J
820      NEXT I
830    REM  COMPUTE AND PRINT ESTIMATE
840    REM  FORM G=A TRANSPOSE TIMES B
850      FOR I = 1 TO N
860      G(I) = 0
870      FOR J = 1 TO M
880      G(I) = G(I) + A(J,I) * B(J)
890      NEXT J
900      NEXT I
910    REM  FORM X=P TIMES G
920      FOR I = 1 TO N
930      X(I) = 0
940      FOR J = 1 TO N
950      X(I) = X(I) + P(I,J) * G(J)
960      NEXT J
970      NEXT I
980    REM  PRINT RESULT
990      PRINT "LEAST SQUARES ESTIMATE VECTOR IS"
1000     FOR I = 1 TO N
1010     PRINT X(I)
1020     NEXT I
1900     END
```

Computer Program in FORTRAN for Least Squares Estimation (LEAST)

```
C ********************************************************************
C PROGRAM LEAST -- BATCH LEAST SQUARES ESTIMATION
C ********************************************************************
      REAL A(20,20),B(20),P(20,20),R(20,20),Q(20,20)
      REAL G(20),X(20)

      WRITE(*,'(1X,''BATCH LEAST SQUARES ESTIMATION'')')

  5   WRITE(*,'(1X,''PLEASE ENTER NO. OF EQUATIONS '',\)')
      READ(*,*) M
      WRITE(*,'(1X,''PLEASE ENTER NO. OF VARIABLES '',\)')
      READ(*,*) N
      IF (N.GT.M) THEN
         WRITE(*,'(1X,''NOT ENOUGH EQUATIONS'')')
         GOTO 5
      ENDIF

C ***  ENTER EQUATIONS
      DO 10 I=1, M
         WRITE(*,'(1X,''ENTER COEFFICIENTS OF EQUATION NO. '',I2)') I
            DO 20 J=1, N
               WRITE(*,'(1X,''COEFFICIENTS OF VARIABLE NO. ''
     *          I2,'' = '',\)') J
               READ(*,*) A(I,J)
 20         CONTINUE
         WRITE(*,'(1X,''KNOWN FOR EQUATION '',I2,'' = '',\)') I
         READ(*,*) B(I)
 10   CONTINUE
```

(continued)

```
C ***  COMPUTE R=A TRANSPOSE TIMES A
      DO 30 I=1, N
         DO 40 J=1, N
            R(I,J) = 0.0
            DO 50 K=1, M
               R(I,J) = R(I,J) + A(K,I) * A(K,J)
50          CONTINUE
40       CONTINUE
30    CONTINUE

C ***  COMPUTE P=R INVERSE USING FADDEEV'S METHOD
C ***  SET Q=0
      DO 60 I=1, N
         DO 70 J=1, N
            Q(I,J) = 0.0
70       CONTINUE
60    CONTINUE

C ***  SET LEADING POLYNOMIAL COEFFICIENT TO UNITY
      C = 1.0

C ***  FIND THE N P MATRICES
      DO 80 L=1, N

C ***  FORM P=Q PLUS C TIMES I
         DO 90 I=1, N
            DO 100 J=1, N
               P(I,J) = Q(I,J)
100         CONTINUE
90       CONTINUE
         DO 110 J=1, N
            P(J,J) = P(J,J) + C
110      CONTINUE

C ***  Q=R TIMES P
         DO 120 I=1, N
            DO 130 J=1, N
               Q(I,J) = 0.0
               DO 140 K=1, N
                  Q(I,J) = Q(I,J) + R(I,K)*P(K,J)
140            CONTINUE
130         CONTINUE
120      CONTINUE

C ***  FORM C=(-1/L)*TRACE(Q)
         C = 0.0
         DO 150 K=1, N
            C = C - Q(K,K)
150      CONTINUE
         C = C / L
80    CONTINUE

C ***  PRINT MESSAGE AND END IF R IS SINGULAR
      IF (C.EQ.0.0) THEN
         WRITE(*,'(1X,''EQUATIONS ARE SINGULAR'')')
         STOP
      ENDIF

C ***  FORM P=R INVERSE
      DO 160 I=1, N
         DO 170 J=1, N
            P(I,J) = - P(I,J)/C
170      CONTINUE
160   CONTINUE
```

(continued)

```
C ***  COMPUTE AND PRINT ESTIMATE
C ***  FORM G=A TRANSPOSE TIMES B
       DO 180 I=1, N
          G(I) = 0.0
          DO 190 J=1, M
             G(I) = G(I) + A(J,I)*B(J)
 190      CONTINUE
 180   CONTINUE

C ***  FORM X=P TIMES G
       DO 200 I=1, N
          X(I) = 0.0
          DO 210 J=1, N
             X(I) = X(I) + P(I,J)*G(J)
 210      CONTINUE
 200   CONTINUE

C ***  PRINT RESULT
       WRITE(*,'(1X,''LEAST SQUARES ESTIMATE VECTOR IS'')')
       DO 220 I=1, N
          WRITE(*,'(1X,E15.8)') X(I)
 220   CONTINUE
       STOP
       END
```

8.6.4 Weighted Least Squares

A more general least squares problem minimizes

$$J(\hat{\mathbf{x}}) = (\mathbf{b} - \mathbf{A}\hat{\mathbf{x}})^\dagger \mathbf{W}(\mathbf{b} - \mathbf{A}\hat{\mathbf{x}}) = \mathbf{v}^\dagger \mathbf{W}\mathbf{v}$$

where $\mathbf{W}$ is a symmetric, positive definite weighting matrix. J is then a quadratic form in the measurement errors $\mathbf{v}$. When one is more confident in the accuracy of some of the measurements than of others, the elements of $\mathbf{W}$ can be chosen to weigh those measurements more heavily than the others. From a statistical standpoint, if the errors $\mathbf{v}$ are zero-mean (E denotes expectation)

$$E[\mathbf{v}] = \mathbf{0}$$

with a known positive covariance matrix

$$E[\mathbf{v}\mathbf{v}^\dagger] = \mathbf{R}$$

then it is natural to choose $\mathbf{W} = \mathbf{R}^{-1}$.

For a set of overdetermined equations

$$\mathbf{A}\mathbf{x} = \mathbf{b}$$

the least squares estimate $\hat{\mathbf{x}}$ of $\mathbf{x}$ in

$$\mathbf{A}\hat{\mathbf{x}} + \mathbf{v} = \mathbf{b}$$

that minimizes

$$J(\hat{\mathbf{x}}) = (\mathbf{b} - \mathbf{A}\hat{\mathbf{x}})^\dagger \mathbf{W}(\mathbf{b} - \mathbf{A}\hat{\mathbf{x}}) = \mathbf{v}^\dagger \mathbf{W}\mathbf{v} \qquad (8\text{-}15)$$

is

$$\hat{\mathbf{x}} = (\mathbf{A}^\dagger \mathbf{W}\mathbf{A})^{-1}\mathbf{A}^\dagger \mathbf{W}\mathbf{b}$$

This result is easily derived by expressing the positive definite weighting matrix $\mathbf{W}$ in terms of a square root as

$$\mathbf{W} = \mathbf{\Psi}^{\dagger}\,\mathbf{\Psi}$$

Then

$$J(\hat{\mathbf{x}}) = [\mathbf{\Psi b} - (\mathbf{\Psi A})\hat{\mathbf{x}}]^{\dagger}[\mathbf{\Psi b} - (\mathbf{\Psi A})\hat{\mathbf{x}}]$$

which is of the form of (8-15) with $\mathbf{b}$ replaced by $\mathbf{\Psi b}$ and $\mathbf{A}$ replaced by $\mathbf{\Psi A}$. Making these substitutions into the fundamental least squares solution (8-14) gives the stated result:

$$\hat{\mathbf{x}} = [(\mathbf{A}^{\dagger}\,\mathbf{\Psi}^{\dagger})(\mathbf{\Psi A})]^{-1}(\mathbf{A}^{\dagger}\,\mathbf{\Psi}^{\dagger})(\mathbf{\Psi b}) = (\mathbf{A}^{\dagger}\,\mathbf{WA})^{-1}\mathbf{A}^{\dagger}\mathbf{WB}$$

For the overdetermined equations

$$\begin{bmatrix} 3 & 1 \\ 0 & 2 \\ -1 & 2 \\ 2 & -4 \end{bmatrix} \begin{bmatrix} x_1 \\ x_2 \end{bmatrix} = \begin{bmatrix} -5.1 \\ 1.8 \\ 4.4 \\ -8.0 \end{bmatrix} \qquad \text{or} \qquad \mathbf{Ax} = \mathbf{b}$$

suppose it is desired to find the least squares estimate with weighing

$$\mathbf{W} = \begin{bmatrix} 3 & 0 & 0 & 0 \\ 0 & 3 & 0 & 0 \\ 0 & 0 & 2 & 0 \\ 0 & 0 & 0 & 1 \end{bmatrix}$$

That is, we have the most confidence in the accuracy of the first two equations, and decreasing confidence in the other two equations. The weighted least squares estimate is

$$\hat{\mathbf{x}} = (\mathbf{A}^{\dagger}\,\mathbf{WA})^{-1}\mathbf{A}^{\dagger}\,\mathbf{Wb}$$

$$= \left(\begin{bmatrix} 3 & 0 & -1 & 2 \\ 1 & 2 & 2 & -4 \end{bmatrix} \begin{bmatrix} 3 & 0 & 0 & 0 \\ 0 & 3 & 0 & 0 \\ 0 & 0 & 2 & 0 \\ 0 & 0 & 0 & 1 \end{bmatrix} \begin{bmatrix} 3 & 1 \\ 0 & 2 \\ -1 & 2 \\ 2 & -4 \end{bmatrix} \right)^{-1}$$

$$\times \begin{bmatrix} 3 & 0 & -1 & 2 \\ 1 & 2 & 2 & -4 \end{bmatrix} \begin{bmatrix} 3 & 0 & 0 & 0 \\ 0 & 3 & 0 & 0 \\ 0 & 0 & 2 & 0 \\ 0 & 0 & 0 & 1 \end{bmatrix} \begin{bmatrix} -5.1 \\ 1.8 \\ 4.4 \\ -8.0 \end{bmatrix}$$

$$\approx \begin{bmatrix} -2.052 \\ 0.999 \end{bmatrix}$$

It is easy to interpret the meaning of $\mathbf{W}$ when $\mathbf{W}$ is diagonal. In the example above, we expect the square errors, v_1^2 and v_2^2, in the first two equations to be equal. We expect the square error in the third equation, v_3^2, to be $\frac{3}{2}$ as large as in either the first or second equations. And we expect v_4^2 to be three times as large as v_1^2 or v_2^2.

8.7 CURVE FITTING

Least squares estimation is now applied to the problem of fitting curves to data. First, we consider *interpolating curves* that pass through all data points. Then, least squares approximations of curves to data are developed. An economic prediction example illustrates some of the pitfalls of interpreting curve fits.

8.7.1 Interpolating Curves

If a curve is described by n parameters, then, except in special cases, it can be made to pass exactly through n given data points. For example, consider the straight line

$$y = f(x) = \alpha_1 + \alpha_2 x$$

where α_1 and α_2 are parameters to be determined. If the curve is to pass through the two points $(1, 1)$ and $(2, -1)$, the parameters must satisfy the two simultaneous linear algebraic equations

$$\alpha_1 + \alpha_2 = 1$$

$$\alpha_1 + 2\alpha_2 = -1$$

giving

$$y = f(x) = 3 - 2x$$

as illustrated in Figure 8-3. A constant-plus sinusoid curve of the form

$$y = \alpha_1 + \alpha_2 \cos x + \alpha_3 \sin x$$

passes through the three points $(0, -2)$, $(1, 0)$, and $(3, 4)$ when the α's satisfy

$$\alpha_1 + \cos(0)\,\alpha_2 + \sin(0)\,\alpha_3 = -2 \qquad \alpha_1 + \qquad \alpha_2 \qquad \approx -2$$

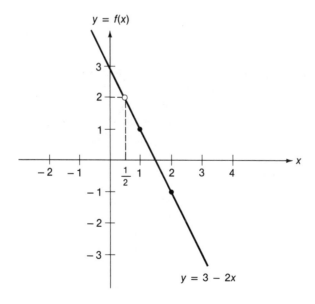

$$y = 3 - 2x$$

Figure 8-3 Straight line that passes through two given points.

$$\alpha_1 + \cos(1)\, \alpha_2 + \sin(1)\, \alpha_3 = 0 \quad \text{or} \quad \alpha_1 + 0.54\alpha_2 + 0.84\alpha_3 \approx 0$$

$$\alpha_1 + \cos(3)\, \alpha_2 + \sin(3)\, \alpha_3 = 4 \qquad\qquad \alpha_1 - 0.99\alpha_2 + 0.14\alpha_3 \approx 4$$

In general, when a curve is described by a linear function of its parameters, as in the examples above, the equations for curve fitting are linear algebraic ones. Such curves are of the form

$$y = f(x) = \alpha_1 b_1(x) + \alpha_2 b_2(x) + \cdots + \alpha_n b_n(x)$$

where the known functions $b_i(x)$ are termed the *basis functions* of the curve and the n constants α_i are the parameters of the curve. If $f(x)$ must pass through n given points (x_1, y_1), (x_2, y_2), . . ., (x_n, y_n), then the n parameters must satisfy the n linear algebraic equations

$$\alpha_1 b_1(x_1) + \alpha_2 b_2(x_1) + \cdots + \alpha_n b_n(x_1) = y_1$$

$$\alpha_1 b_1(x_2) + \alpha_2 b_2(x_2) + \cdots + \alpha_n b_n(x_2) = y_2$$

$$\vdots$$

$$\alpha_1 b_1(x_n) + \alpha_2 b_2(x_n) + \cdots + \alpha_n b_n(x_n) = y_n$$

If these equations are linearly independent (the basis functions are then said to be linearly independent of one another), there is a unique solution for the parameters and thus for the curve.

If the curve is a nonlinear function of parameters, the resulting equations are nonlinear. For example, the damped sinusoidal curve with exponential constant and frequency to be determined,

$$y = e^{-\alpha_1 x} \cos \alpha_2 x$$

passes through the points $(1, 2)$ and $(4, -3)$ if

$$e^{-\alpha_1} \cos \alpha_2 = 2$$

$$e^{-4\alpha_1} \cos 4\alpha_2 = -3$$

We will emphasize curves with linear parameters in the discussion to follow.

Curves that exactly pass through all the data points are called *interpolating curves*. They are commonly used to estimate values of an underlying function between the data points, a process called *interpolation*. They can also be used to *extrapolate* the data, predicting values of a function outside the range of the data points.

8.7.2 Least Squares Curve Fitting

If there are more data points than curve parameters and if the resulting equations for the parameters are inconsistent because the data does not all exactly fit a curve, a "best fit" of a curve to the data is sought. A least squares fit that selects the parameters so that the sum of squares of the errors between the data points and the curve is minimum is popular and widely used.

For an approximating curve consisting of a linear combination of n basis functions,

$$y = f(x) = \alpha_1 b_1(x) + \alpha_2 b_2(x) + \cdots + \alpha_n b_n(x)$$

and $m > n$ data points (x_1, y_1), (x_2, y_2), . . ., (x_n, y_n), the equations for the curve's parameters are

$$\alpha_1 b_1(x_1) + \alpha_2 b_2(x_1) + \cdots + \alpha_n b_n(x_1) = y_1$$

$$\alpha_1 b_1(x_2) + \alpha_2 b_2(x_2) + \cdots + \alpha_n b_n(x_2) = y_2$$

$$\vdots$$

$$\alpha_1 b_1(x_m) + \alpha_2 b_2(x_m) + \cdots + \alpha_n b_n(x_m) = y_m$$

The least squares estimate of the unknown curve parameters in these equations gives the minimum sum of squares of errors between the curve and the data points, as illustrated in Figure 8-4. These errors are known as the *residuals* of the approximation.

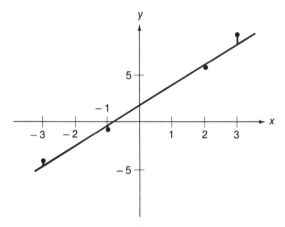

(a) Least squares fit of a straight line

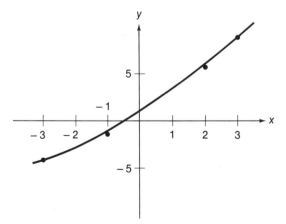

(b) Least squares fit of a parabola

Figure 8-4 Least squares curve fits.

For example, the least squares fit of a straight line to the points $(-3, -4)$, $(-1, -1)$, $(2, 6)$, $(3, 9)$ is given by the curve

$$y = f(x) = \hat{\alpha}_1 + \hat{\alpha}_2 x$$

where $\hat{\alpha}_1$ and $\hat{\alpha}_2$ are the least squares solution to the overdetermined equations

$$\alpha_1 - 3\alpha_2 = -4$$
$$\alpha_1 - \alpha_2 = -1$$
$$\alpha_1 + 2\alpha_2 = 6$$
$$\alpha_1 + 3\alpha_2 = 9$$

Using the LEAST program, the least squares solution is $\hat{\alpha}_1 \approx 1.956$, $\hat{\alpha}_2 \approx 2.176$, so the line

$$y = 1.956 + 2.176x$$

is the best fit to this data in the least squares sense, as shown in Figure 8-4(a).
The least squares fit of a parabola,

$$y = f(x) = \hat{\alpha}_1 + \hat{\alpha}_2 x + \hat{\alpha}_3 x^2$$

to these same data points is the least squares solution to the equations

$$\alpha_1 - 3\alpha_2 + 9\alpha_3 = -4$$
$$\alpha_1 - \alpha_2 + \alpha_3 = -1$$
$$\alpha_1 + 2\alpha_2 + 4\alpha_3 = 6$$
$$\alpha_1 + 3\alpha_2 + 9\alpha_3 = 9$$

which is $\hat{\alpha}_1 \approx 1.000$, $\hat{\alpha}_2 \approx 2.167$, $\hat{\alpha}_3 \approx 0.167$, as shown in Figure 8-4(b).
Least squares estimates and curve fits are affected much more by large errors than small ones. This means that if occasional large data errors occur, these errors can have large, undesirable effects on the result. Of course, if one knows that a particular data point has a high probability of error, it can be incorporated accordingly in a weighted least squares estimate. But if there is no a priori knowledge of the relative accuracy of the data, it is often useful to use the collection of data, itself, to identify occasional data points that are likely to be grossly in error.
These data points, if they occur, are called *outliers*. For example, in collecting and processing temperature data at sea level at the equator, it is a good bet that a temperature measurement of below freezing is an outlier. Outliers are characterized by unexpectedly large errors from the least squares (or other) curve fit, and it is common simply to remove them and perform a revised least squares calculation. Doing so is, to various degrees in some circumstances a way of manipulating results for the unwary.

8.7.3 Predicting Economic Trends

Stock market analysts and others commonly attempt to predict such things as the Dow Jones average or the future values of stocks by fitting curves to past data. Often the predictions are simply straight-line projections from the past two data points, the low dependability of which is easily demonstrated. There is a tendency for many to assume that curves based on many data points are bound to be more reliable, but this is not necessarily true. Similar analysis has been applied from time to time to such phenomena as glacier growth and earthquake activity with predictably poor results.

Table 8-5 lists five-year averages of the gross national product (GNP) of the United States for the years 1950–1980. Letting x denote the year number after 1950, a straight-line least squares curve fit to the data points is governed by

$$y = f(x) = \hat{\alpha}_1 + \hat{\alpha}_2 x$$

where

$$\alpha_1 = 0.530$$
$$\alpha_1 + \alpha_2 = 0.656$$
$$\alpha_1 + 2\alpha_2 = 0.733$$
$$\alpha_1 + 3\alpha_2 = 0.934$$

TABLE 8-5 U.S. Gross National Product for the Years 1950–1980

Year	Real GNP (trillions of dollars)[a]
1950	0.530
1955	0.656
1960	0.733
1965	0.934
1970	1.091
1975	1.230
1980	1.484

Source: U.S. Department of Commerce. [a]In terms of 1972 U.S. dollars.

Using the LEAST program, $\hat{\alpha}_1 \approx 0.5199$ and $\hat{\alpha}_2 \approx 0.1289$, giving the fit shown in Figure 8-5(a). With this straight line, the predicted 1970 GNP is

$$\hat{\alpha}_1 + 4\hat{\alpha}_2 \approx 1.0355$$

an error of
$$\frac{1.091 - 1.0355}{1.0355} = +5.087\%$$

from the actual 1970 GNP.

Instead, using a least squares quadratic curve fit,

$$y = f(x) = \hat{\alpha}_1 + \hat{\alpha}_2 x + \hat{\alpha}_3 x^2$$

the same four data points result in the overdetermined equations

$$\alpha_1 = 0.530$$

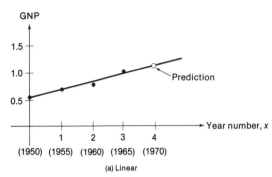

(a) Linear

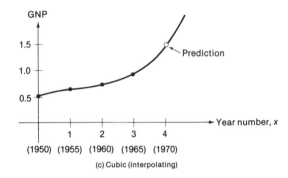

(c) Cubic (interpolating)

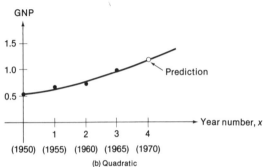

(b) Quadratic

Figure 8-5 Least squares curve fits to economic data.

$$\alpha_1 + \alpha_2 + \alpha_3 = 0.656$$

$$\alpha_1 + 2\alpha_2 + 4\alpha_3 = 0.733$$

$$\alpha_1 + 3\alpha_2 + 9\alpha_3 = 0.934$$

and $\hat{\alpha}_1 \approx 0.5387$, $\hat{\alpha}_2 \approx 0.07265$, $\hat{\alpha}_3 \approx 0.01874$ is the least squares estimate of their solution. Using this curve, the predicted 1970 GNP is

$$\hat{\alpha}_1 + 4\,\hat{\alpha}_2 + 16\,\hat{\alpha}_3 \approx 1.12924$$

an error from the actual value of

$$\frac{1.091 - 1.12924}{1.091} = -3.5\%$$

This result is shown in Figure 8-5(b).

A cubic interpolation of the same four data points

$$y = f(x) = \hat{\alpha}_1 + \hat{\alpha}_2 x + \hat{\alpha}_3 x^2 + \hat{\alpha}_4 x^3$$

has solution $\hat{\alpha}_1 \approx 0.5300$, $\hat{\alpha}_2 \approx 0.2082$, $\hat{\alpha}_3 \approx -0.1110$, $\hat{\alpha}_4 \approx 0.02883$. As shown in Figure 8-5(c), this curve predicts a GNP of

$$\hat{\alpha}_1 + 4\,\hat{\alpha}_2 + 16\,\hat{\alpha}_3 + 64\,\hat{\alpha}_4 \approx 1.4319$$

for 1970, an error of

$$\frac{1.091 - 1.4319}{1.091} = -31.25\%$$

from the actual value.

Quadratic and higher-degree polynomial basis functions are usually poor for making long-term predictions from data because they "blow up" with large values of their arguments. Very nice for predicting huge success or disaster, but seldom accurate.

8.8 RECURSIVE LEAST SQUARES

Recursive least squares is an arrangement of the least squares solution in which each new equation is used to update the previous least squares estimate that was based on previous equations. Instead of processing all of the equations at once, they are processed one at a time, with each new equation causing a modification in the current estimate. Least squares estimates are linear transformations of the knowns. The least squares estimate based on the first $k + 1$ equations can therefore be expressed as a linear transformation of the least squares estimate based on the first k equations, plus a linear correction term based on the $(k + 1)$th equation alone.

8.8.1 Derivation of the Recursive Algorithm

Denoting the number of equations used by arguments, the least squares estimate based on the first k equations is

$$\hat{\mathbf{x}}(k) = [\mathbf{A}^\dagger(k)\mathbf{A}(k)]^{-1}\mathbf{A}^\dagger(k)\mathbf{b}(k)$$

The least squares estimate based on $k + 1$ equations is

$$\hat{\mathbf{x}}(k + 1) = [\mathbf{A}^\dagger(k + 1)\mathbf{A}(k + 1)]^{-1}\mathbf{A}^\dagger(k + 1)\mathbf{b}(k + 1)$$

where $\mathbf{A}(k + 1)$ is $\mathbf{A}(k)$ with an additional row, $\mathbf{a}^\dagger(k + 1)$,

$$\mathbf{A}(k + 1) = \begin{bmatrix} \mathbf{A}(k) \\ \mathbf{a}^\dagger(k + 1) \end{bmatrix}$$

and the vector of the knowns $\mathbf{b}(k + 1)$ is the vector $\mathbf{b}$ with one additional element, b_{k+1}:

$$\mathbf{b}(k + 1) = \begin{bmatrix} \mathbf{b}(k) \\ b_{k+1} \end{bmatrix}$$

In terms of these quantities,

$$\mathbf{A}^\dagger(k + 1)\,\mathbf{A}(k + 1) = [\mathbf{A}^\dagger(k) : \mathbf{a}(k + 1)]\begin{bmatrix} \mathbf{A}(k) \\ \mathbf{a}^\dagger(k + 1) \end{bmatrix}$$

$$= \mathbf{A}^\dagger(k)\mathbf{A}(k) + \mathbf{a}(k + 1)\mathbf{a}^\dagger(k + 1)$$

Defining

$$P(k) = [A^\dagger(k)A(k)]^{-1} \qquad P(k + 1) = [A^\dagger(k + 1)A(k + 1)]^{-1}$$

then

$$
\begin{aligned}
P(k + 1) &= [A^\dagger(k + 1)A(k + 1)]^{-1} \\
&= [A^\dagger(k)A(k) + a(k + 1)a^\dagger(k + 1)]^{-1} \\
&= [P^{-1}(k) + a(k + 1)a^\dagger(k + 1)]^{-1}
\end{aligned}
\tag{8-16}
$$

Relation (8-16) is in the form for which the *matrix inversion lemma*,

$$(\Gamma + uv^\dagger)^{-1} = \Gamma^{-1} - \frac{\Gamma^{-1}uv^\dagger\Gamma^{-1}}{1 + v^\dagger\Gamma^{-1}u} \tag{8-17}$$

applies. The lemma is proved by multiplying both sides of (8-17) by $\Gamma + uv^\dagger$, obtaining the identity matrix:

$$
\begin{aligned}
(\Gamma + uv^\dagger)^{-1}(\Gamma + uv^\dagger) &= \Gamma^{-1}(\Gamma + uv^\dagger) - \frac{\Gamma^{-1}uv^\dagger\Gamma^{-1}(\Gamma + uv^\dagger)}{1 + v^\dagger\Gamma^{-1}u} \\
&= I + \Gamma^{-1}uv^\dagger - \frac{\Gamma^{-1}u(v^\dagger + v^\dagger\Gamma^{-1}uv^\dagger)}{1 + v^\dagger\Gamma^{-1}u} \\
&= I + \Gamma^{-1}uv^\dagger - \frac{\Gamma^{-1}u(1 + v^\dagger\Gamma^{-1}u)v^\dagger}{1 + v^\dagger\Gamma^{-1}u}
\end{aligned}
$$

Applying the matrix inversion lemma to (8-16) gives

$$P(k + 1) = P(k) - \frac{P(k)a(k + 1)a^\dagger(k + 1)P(k)}{1 + a^\dagger(k + 1)P(k)a(k + 1)} \tag{8-18}$$

which shows how $\qquad P(k) = [A^\dagger(k)A(k)]^{-1}$

changes when a new equation is incorporated. It is helpful now to define the quantities

$$\delta(k + 1) = 1 + a^\dagger(k + 1)P(k)a(k + 1)$$

$$g(k + 1) = P(k)a(k + 1)\delta^{-1}(k + 1)$$

to simplify the expression (8-18) to

$$
\begin{aligned}
P(k + 1) &= P(k) - P(k)a(k + 1)\delta^{-1}(k + 1)a^\dagger(k + 1)P(k) \\
&= [I - g(k + 1)a^\dagger(k + 1)]P(k)
\end{aligned}
$$

The least squares estimate with $k + 1$ equations, in terms of the estimate with k equations is then

$$\hat{x}(k + 1) = P(k + 1)A^\dagger(k + 1)b(k + 1)$$

$$= P(k + 1)[A^\dagger(k) : a(k + 1)]\begin{bmatrix} b(k) \\ \text{--------} \\ b_{k+1} \end{bmatrix}$$

$$= \mathbf{P}(k + 1)[\mathbf{A}^{\dagger}(k)\mathbf{b}(k) + \mathbf{a}(k + 1)b_{k+1}]$$

$$= [\mathbf{P}(k) - \mathbf{g}(k + 1)\mathbf{a}^{\dagger}(k + 1)\mathbf{P}(k)] \cdot \qquad (8\text{-}19)$$

$$[\mathbf{A}^{\dagger}(k)\mathbf{b}(k) + \mathbf{a}(k + 1)b_{k+1}]$$

$$= \mathbf{P}(k)\mathbf{A}^{\dagger}(k)\mathbf{b}(k) + \mathbf{P}(k)\mathbf{a}(k + 1)b_{k+1} - \mathbf{g}(k + 1)$$

$$\times [\mathbf{a}^{\dagger}(k + 1)\mathbf{P}(k)\mathbf{A}^{\dagger}(k)\mathbf{A}^{\dagger}9k)\mathbf{b}(k) + \mathbf{a}^{\dagger}(k + 1)\mathbf{P}(k)\mathbf{a}(k + 1)b_{k+1}]$$

using

$$\hat{\mathbf{x}}(k) = \mathbf{P}(k)\mathbf{A}^{\dagger}(k)\mathbf{b}(k)$$

and

$$\mathbf{g}(k + 1) = \mathbf{P}(k)\mathbf{a}(k + 1) - \mathbf{g}(k + 1)\mathbf{a}^{\dagger}(k + 1)\mathbf{P}(k)\mathbf{a}(k + 1)$$

(8-19) becomes

$$\hat{\mathbf{x}}(k + 1) = \hat{\mathbf{x}}(k) + \mathbf{g}(k + 1)[b_{k+1} - \mathbf{a}^{\dagger}(k + 1)\hat{\mathbf{x}}(k)]$$

This elegant result is that the least squares estimate based on $k + 1$ equations is the least squares estimate based on the first k equations, plus a correction term that is a vector $\mathbf{g}$ times the new known b_{k+1} minus the new predicted known,

$$\mathbf{a}^{\dagger}(k + 1)\hat{\mathbf{x}}(k)$$

based on the estimate from the first k equations. The vector "gain" $\mathbf{g}$ depends only on the equation coefficients, not on the knowns (or "data").

The equations for recursive least squares estimation are summarized in Table 8-6. To apply them, one begins with the first n equations in n variables. For these,

$$\mathbf{P}(n) = [\mathbf{A}^{\dagger}(n)\mathbf{A}(n)]^{-1}$$

and

$$\hat{\mathbf{x}}(n) = \mathbf{P}(n)\mathbf{A}^{\dagger}(n)\mathbf{b}(n) = \mathbf{A}^{-1}(n)\mathbf{b}(n)$$

are calculated. Then new equations are added to the set one at a time, and for each new equation, the least squares estimate is corrected, giving the estimate based on the previous equations and one more equation.

TABLE 8-6   Recursive Least Squares Estimation

PROBLEM STATEMENT

For a sequence of overdetermined linear algebraic equations,

$$a_{11}x_1 + a_{12}x_2 + \cdots + a_{1n}x_n = b_1$$

$$a_{21}x_1 + a_{22}x_2 + \cdots + a_{2n}x_n = b_2$$

$$\vdots$$

$$a_{k1}x_1 + a_{k2}x_2 + \cdots + a_{kn}x_n = b_k$$

$$\vdots$$

it is desired to find the sequence of least squares estimate $\hat{\mathbf{x}}(k)$, based on the first k of these equations, for $k = n, n + 1, \ldots$. The quantities used are

$$\mathbf{A}(k) = \begin{bmatrix} a_{11} & a_{12} & \cdots & a_{1n} \\ a_{21} & a_{22} & \cdots & a_{2n} \\ \vdots & \vdots & \vdots & \vdots \\ a_{k1} & a_{k2} & \cdots & a_{kn} \end{bmatrix} \qquad k = n, n + 1, \ldots$$

$$\mathbf{a}^\dagger(k) = [a_{k1} \quad a_{k2} \quad \cdots \quad a_{kn}] \qquad k = n + 1, n + 2, \ldots$$

$$\mathbf{b}(k) = \begin{bmatrix} b_1 \\ b_2 \\ \vdots \\ b_k \end{bmatrix} \qquad k = n, n + 1, \ldots$$

INITIAL ESTIMATE

$$\mathbf{P}(n) = [\mathbf{A}^\dagger(n)\mathbf{A}(n)]^{-1}$$
$$\hat{\mathbf{x}}(n) = \mathbf{P}(n)\mathbf{A}^\dagger(n)\mathbf{b}(n) = \mathbf{A}^{-1}(n)\mathbf{b}(n)$$

CORRECTOR GAIN

$$\delta(k + 1) = 1 + \mathbf{a}^\dagger(k + 1)\mathbf{P}(k)\mathbf{a}(k + 1)$$
$$\mathbf{g}(k + 1) = \mathbf{P}(k)\mathbf{a}(k + 1)\delta^{-1}(k + 1)$$
$$\mathbf{P}(k + 1) = [\mathbf{I} - \mathbf{g}(k + 1)\mathbf{a}^\dagger(k + 1)]\mathbf{P}(k)$$

PREDICTOR–CORRECTOR FOR UPDATED ESTIMATE

$$\hat{\mathbf{x}}(k + 1) = \hat{\mathbf{x}}(k) + \mathbf{g}(k + 1)[b_{k+1} - \mathbf{a}^\dagger(k + 1)\hat{\mathbf{x}}(k + 1)]$$

As a numerical example, consider the overdetermined equations

$$x_1 - 3x_2 = -5$$
$$-2x_1 + 2x_2 = 8$$
$$-2x_1 + x_2 = 8$$
$$4x_1 - x_2 = -15$$

The initialization step uses the first two of these equations to obtain the initial $\mathbf{P}(2)$ and $\hat{\mathbf{x}}(2)$:

$$\mathbf{P}(2) = [\mathbf{A}^\dagger(2)\mathbf{A}(2)]^{-1}$$

$$= \left\{ \begin{bmatrix} 1 & -2 \\ -3 & 2 \end{bmatrix} \begin{bmatrix} 1 & -3 \\ -2 & 2 \end{bmatrix} \right\}^{-1} = \begin{bmatrix} 0.8125 & 0.4375 \\ 0.4375 & 0.3125 \end{bmatrix}$$

$$\hat{\mathbf{x}}(2) = \mathbf{P}(2)\mathbf{A}^\dagger(2)\mathbf{b}(2)$$

$$= \begin{bmatrix} 0.8125 & 0.4375 \\ 0.4375 & 0.3125 \end{bmatrix} \begin{bmatrix} 1 & -2 \\ -3 & 2 \end{bmatrix} \begin{bmatrix} -5 \\ 8 \end{bmatrix} = \begin{bmatrix} -3.5 \\ 0.5 \end{bmatrix}$$

Hereafter, the recursive relations are used. The least squares estimate of the first three equations is given by

$$\delta(3) = 1 + \mathbf{a}^\dagger(3)\mathbf{P}(2)\mathbf{a}(3)$$

$$\approx 1 + [-2 \quad 1] \begin{bmatrix} 0.8125 & 0.4375 \\ 0.4375 & 0.3125 \end{bmatrix} \begin{bmatrix} -2 \\ 1 \end{bmatrix} = 2.8125$$

$$\mathbf{g}(3) = \mathbf{P}(2)\mathbf{a}(3)\delta^{-1}(3)$$

$$\approx \begin{bmatrix} 0.8125 & 0.4375 \\ 0.4375 & 0.3125 \end{bmatrix} \begin{bmatrix} -2 \\ 1 \end{bmatrix} \left(\frac{1}{2.8125}\right) = \begin{bmatrix} -0.422 \\ -0.2 \end{bmatrix}$$

$$\mathbf{P}(3) = [\mathbf{I} - \mathbf{g}(3)\mathbf{a}^\dagger(3)]\mathbf{P}(2)]$$

$$\approx \left(\begin{bmatrix} 1 & 0 \\ 0 & 1 \end{bmatrix} - \begin{bmatrix} -0.422 \\ -0.2 \end{bmatrix} [-2 \quad 1]\right) \begin{bmatrix} 0.8125 & 0.4375 \\ 0.4375 & 0.3125 \end{bmatrix}$$

$$= \begin{bmatrix} 0.311 & 0.2 \\ 0.2 & 0.2 \end{bmatrix}$$

$$\hat{\mathbf{x}}(3) = \hat{\mathbf{x}}(2) + \mathbf{g}(3)[b_3 - \mathbf{a}^\dagger(3)\hat{\mathbf{x}}(2)]$$

$$\approx \begin{bmatrix} -3.5 \\ 0.5 \end{bmatrix} + \begin{bmatrix} -0.422 \\ -0.2 \end{bmatrix} \left(8 - [-2 \quad 1] \begin{bmatrix} -3.5 \\ 0.5 \end{bmatrix}\right)$$

$$= \begin{bmatrix} -3.711 \\ 0.4 \end{bmatrix}$$

The least squares estimate based on all four equations is given by

$$\delta(4) = 1 + \mathbf{a}^\dagger(4)\mathbf{P}(3)\mathbf{a}(4)$$

$$\approx 1 + [4 \quad -1] \begin{bmatrix} 0.311 & 0.2 \\ 0.2 & 0.2 \end{bmatrix} \begin{bmatrix} 4 \\ -1 \end{bmatrix} = 4.578$$

$$\mathbf{g}(4) = \mathbf{P}(3)\mathbf{a}(4)\delta^{-1}(4)$$

$$\approx \begin{bmatrix} 0.311 & 0.2 \\ 0.2 & 0.2 \end{bmatrix} \begin{bmatrix} 4 \\ -1 \end{bmatrix} \left(\frac{1}{4.578}\right) = \begin{bmatrix} 0.228 \\ 0.131 \end{bmatrix}$$

$$\mathbf{P}(4) = [\mathbf{I} - \mathbf{g}(4)\mathbf{a}^\dagger(4)]\mathbf{P}(3)$$

$$\approx \left(\begin{bmatrix} 1 & 0 \\ 0 & 1 \end{bmatrix} - \begin{bmatrix} 0.228 \\ 0.131 \end{bmatrix} [4 \quad -1]\right) \begin{bmatrix} 0.311 & 0.2 \\ 0.2 & 0.2 \end{bmatrix}$$

$$= \begin{bmatrix} 0.085 & 0.075 \\ 0.063 & 0.121 \end{bmatrix}$$

$$\hat{\mathbf{x}}(4) = \hat{\mathbf{x}}(3) + \mathbf{g}(4)[b_4 - \mathbf{a}^\dagger(4)\hat{\mathbf{x}}(3)]$$

$$\approx \begin{bmatrix} -3.711 \\ 0.4 \end{bmatrix} + \begin{bmatrix} 0.228 \\ 0.131 \end{bmatrix} \left(-15 - \begin{bmatrix} 4 & -1 \end{bmatrix} \begin{bmatrix} -3.711 \\ 0.4 \end{bmatrix} \right)$$

$$\hat{\mathbf{x}}(4) = \begin{bmatrix} -3.65 \\ 0.43 \end{bmatrix}$$

and so on.

For Gauss, using recursive least squares would allow the incorporation of new planetary orbit data as it becomes available, instead of having to "start over" whenever there is new information. Many modern problems are of a nature that is well suited to the recursive formulation. For example, in navigation one typically has an estimated position and desires to update that estimate periodically by using additional data as it becomes available.

8.8.2 Recursive Least Squares Program

The recursive least squares algorithm of Table 8-6 is arranged as a computer program, called RECURSIV, in Table 8-7. The previous batch least squares program LEAST is used as the first part of this routine, to obtain an initial least squares solution, after which the new equations are entered one at a time, resulting in updated estimates.

For writing programs such as this one, it could be helpful to use a language for which matrix operations are an integral part. A complete BASIC compiler does include

TABLE 8-7

Computer Program in BASIC for Recursive Least Squares Estimation (RECURSIV)

VARIABLES USED

The same variables are used as in the LEAST program (Table 8-4). After the initial solution, only an N-vector $\mathbf{a}$ is needed, instead of an $M \times N$ matrix $\mathbf{A}$, so

$$A(20,1) = \mathbf{a} \text{ vector}$$

Similarly, only a scalar b is needed, so

$$B(1) = b$$

In addition, the following variables are used:

$$H(20) = \text{vector of intermediate results}$$
$$D = \delta$$
$$E = \text{prediction error, } b - \mathbf{a}^\dagger \hat{\mathbf{x}}$$

LISTING

```
100     PRINT "RECURSIVE LEAST SQUARES ESTIMATION"
110     DIM A(20,20),B(20),P(20,20),R(20,20),Q(20,20)
120     DIM G(20),X(20),H(20)
130
```

LEAST program;
TABLE 8-4

```
1020
1030    REM ENTER NEXT EQUATION
1040     PRINT "ENTER NEXT EQUATION"
```

```
1050      FOR J = 1 TO N
1060      PRINT "COEFFICIENT OF VARIABLE ";J;
1070      INPUT A(J,1)
1080      NEXT J
1090      PRINT "KNOWN";
1100      INPUT B(1)
1110   REM   COMPUTE D
1120   REM   FIND H=P TIMES A
1130      FOR I = 1 TO N
1140      H(I) = 0
1150      FOR J = 1 TO N
1160      H(I) = H(I) + P(I,J) * A(J,1)
1170      NEXT J
1180      NEXT I
1190   REM   FIND D=A TRANSPOSE TIMES H
1200      D = 1
1210      FOR J = 1 TO N
1220      D = D + A(J,1) * H(J)
1230      NEXT J
1240   REM   COMPUTE G
1250      FOR I = 1 TO N
1260      G(I) = H(I) / D
1270      NEXT I
1280   REM   COMPUTE P
1290   REM   FIND Q= MINUS G TIMES A TRANSPOSE
1300      FOR I = 1 TO N
1310      FOR J = 1 TO N
1320      Q(I,J) = - G(I) - A(J,1)
1330      NEXT J
1340      NEXT I
1350   REM   FIND Q=I PLUS Q
1360      FOR I = 1 TO N
1370      Q(I,I) = 1 + Q(I,I)
1380      NEXT I
1390   REM   FIND R=Q TIMES P
1400      FOR I = 1 TO N
1410      FOR J = 1 TO N
1420      R(I,J) = 0
1430      FOR K = 1 TO N
1440      R(I,J) = R(I,J) + Q(I,K) * P(K,J)
1450      NEXT K
1460      NEXT J
1470      NEXT I
1480   REM   SET P=R
1490      FOR I = 1 TO N
1500      FOR J = 1 TO N
1510      P(I,J) = R(I,J)
1520      NEXT J
1530      NEXT I
1540   REM   COMPUTE AND PRINT UPDATED ESTIMATE
1550   REM   FORM E=B MINUS A TRANSPOSE TIMES X
1560      E = B(1)
1570      FOR I = 1 TO N
1580      E = E - A(I,1) *  X(I)
1590      NEXT I
1600   REM   FORM X=X PLUS G TIMES E
1610      FOR I = 1 TO N
1620      X(I) = X(I) + G(I) * E
1630      NEXT I
1640   REM   PRINT RESULT AND LOOP BACK
1650      FOR I = 1 TO N
1660      PRINT X(I)
1670      NEXT I
1680      GOTO 1030
```

Computer Program in FORTRAN for Recursive Least Squares Estimation (RECURSIV)

```
C  ******************************************************************
C  PROGRAM RECURSIV -- RECURSIVE LEAST SQUARES ESTIMATION
C  ******************************************************************
       REAL A(20,20), B(20), P(20,20), R(20,20), Q(20,20)
       REAL G(20),    X(20), H(20)

       WRITE(*,'(1X,''RECURSIVE LEAST SQUARES ESTIMATION'')')
```

◄——— LEAST program;
                                                    TABLE 8-4

```
C *** ENTER NEXT EQUATION, MAXIMUM OF 100 LOOPS
       DO 235 II=1, 100
           WRITE(*,'(1X,''ENTER NEXT EQUATION'')')
           DO 240 J=1, N
               WRITE(*,'(1X,''COEFFICIENT OF VARIABLE '',I2,'' = '',\)') J
               READ(*,*) A(J,1)
 240       CONTINUE
           WRITE(*,'(1X,''KNOWN = '',\)')
           READ(*,*) B(1)

C *** COMPUTE D
C *** FIND H=P TIMES A
           DO 250 I=1, N
               H(I) = 0.0
               DO 260 J=1, N
                   H(I) = H(I) + P(I,J)*A(J,1)
 260           CONTINUE
 250       CONTINUE

C *** FIND D=A TRANSPOSE TIMES H
           D = 1.0
           DO 270 J=1, N
               D = D + A(J,1)*H(J)
 270       CONTINUE

C *** COMPUTE G
           DO 280 I=1, N
               G(I) = H(I)/D
 280       CONTINUE

C *** COMPUTE P
C *** FIND Q= MINUS G TIMES A TRANSPOSE
           DO 290 I=1, N
               DO 300 J=1, N
                   Q(I,J) = -G(I) - A(J,1)
 300           CONTINUE
 290       CONTINUE

C *** FIND Q=I PLUS Q
           DO 310 I=1, N
               Q(I,I) = 1.0 + Q(I,I)
 310       CONTINUE

C *** FIND R=Q TIMES P
           DO 320 I=1, N
               DO 330 J=1, N
                   R(I,J) = 0.0
                   DO 340 K=1, N
                       R(I,J) = R(I,J) + Q(I,K)*P(K,J)
```

```
340                CONTINUE
330                CONTINUE
320            CONTINUE

C *** SET P=R
          DO 350 I=1, N
              DO 360 J=1, N
                  P(I,J) = R(I,J)
360           CONTINUE
350        CONTINUE

C *** COMPUTE AND PRINT UPDATED ESTIMATE
C *** FORM E=B MINUS A TRANSPOSE TIMES X
          E = B(1)
          DO 370 I=1, N
              E = E - A(I,1)*X(I)
370        CONTINUE

C *** FORM X=X PLUS G TIMES E
          DO 380 I=1, N
              X(I) = X(I) + G(I)*E
380        CONTINUE

C *** PRINT RESULT AND LOOP BACK
          DO 390 I=1, N
              WRITE(*,'(1X,E15.8)') X(I)
390        CONTINUE
235     CONTINUE

        STOP
        END
```

matrix operations, but these are omitted from many common versions. In BASIC, and even more so in other languages, there is an overhead in learning and remembering time and in dealing with idiosyncrasies that can make the stringing together of blocks of instructions more desirable for the occasional user.

For example, when matrix operations are supported in BASIC, the array dimensions used in the dimension instructions are taken to be the matrix dimensions. If redimensioning arrays or even setting their dimensions with an input variable is not allowed, the writing of interactive programs and programs where the same array is to be used for matrices of various dimensions (such as in LEAST and RECURSIV) can be awkward. There are many additional questions that occur. How are the rows and columns of larger matrices conveniently manipulated? Does the inverse function halt execution of the program when the matrix involved is singular? This would mean that one would often want to use a string of instructions involving calculation of the determinant before calculating the inverse.

Other languages often share these kinds of difficulties and, in addition, require an involved definition of the "data types" to be used for matrices. All in all, stringing together blocks of instructions is not too bad an approach to doing matrix operations given the state of today's most generally available programming languages. After all, that is exactly what the compiler does in generating machine-language instructions from higher-level ones. One day, more user-friendly tools will be commonplace.

8.8.3 Recursive Estimation of a Ship's Course

The navigation officer aboard a cargo ship in the vicinity of 24°S latitude, 155° longitude (near the Society Islands in the South Pacific) wishes to compute the ship's position on the basis of several measurements. For simplicity, there will not be a correction for the curvature of the earth, and it will be assumed that the ship travels at a constant velocity during the period of time the measurements are taken. The ship's longitude x and the latitude y are thus modeled by the parametric equations

$$x = x_0 + v_x t$$

$$y = y_0 + v_y t$$

where the position at time $t = 0$ is (x_0, y_0) is unknown and the x and y velocity components v_x and v_y are unknown.

Figure 8-6 summarizes the measurement data to be used. The position of the ship was previously computed to be (154.85°, 23.82°S) at time $t = 0$, giving the equations

$$x_0 = 154.85° \tag{8-20a}$$

$$y_0 = 23.82°S \tag{8-20b}$$

At time $t = 3.10$ hours, the latitude of the ship is measured from the angle to the horizon of the North Star to be 24.08°S, giving the equation

$$y_0 + 3.10v_y = 24.08°S \tag{8-21}$$

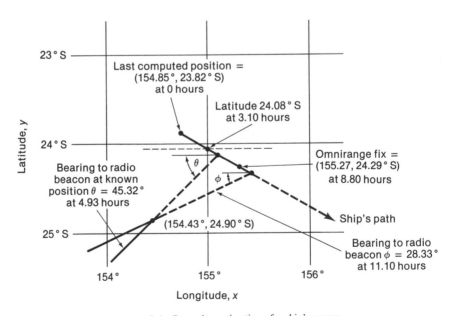

Figure 8-6   Recursive estimation of a ship's course.

The bearing from the ship to a known radio beacon at $(154.43°, 24.90°S)$ is measured to be $35.32°$ at time 4.93 hours. The straight line passing through the coordinates $(154.43°, 24.90°S)$ with angle $\phi = 45.32°$ has equation

$$y = -1.0112x + 181.060$$

so that $\qquad y_0 + v_y(4.93) = -1.0112[x_0 + v_x(4.93)] + 181.060$

or $\qquad 1.0112x_0 + y_0 + 4.9852v_x + 4.93v_y = 181.060 \qquad (8\text{-}22)$

When the four equations (8-20a), (8-20b), (8-21), and (8-22) are solved for the four unknowns, x_0, y_0, v_x, and v_y the result is $x_0 = 154.85°$, $y_0 = 23.82°S$, $v_x = 0.0486°$ per hour, and $v_y = 0.0839°$ per hour. The RECURSIV program is used for this calculation because additional data will be incorporated later. The ship's estimated position as a function of time, based on these minimal measurements is

$$x = 154.85 + 0.0486t$$

$$y = 23.82 + 0.0839t$$

At time $t = 8.80$ hours, the omnirange position computer on the ship is finally repaired and produces an approximate position of $(155.27°, 24.29°S)$. Entering the additional equations

$$x_0 + 8.80v_x = 155.27°$$

$$y_0 + 8.80v_y = 24.29°S$$

gives $\hat{x}_0 = 154.88°$, $\hat{y}_0 = 23.88°S$, $\hat{v}_x = 0.0487°$ per hour, and $\hat{v}_y = 0.0530°$ per hour. As these values are not greatly different from the previous estimates, it is assumed that the omnirange system is operating correctly.

Finally, at time $t = 11.10$ hours, the bearing from the ship to the radio beacon at $(154.43°, 24.90°S)$ is measured to be $\phi = 28.33°$. The straight line passing through these coordinates with angle ϕ has the equation

$$y = -0.539x + 108.16$$

so $\qquad y_0 + v_y(11.10) = -0.539[x_0 + v_x(11.10)] = 108.16$

or $\qquad 0.539x_0 + y_0 + 5.98v_x + 11.10v_y = 108.16$

Incorporating this equation into the least squares estimate gives $\hat{x}_0 = 154.89°$, $\hat{y}_0 = 23.89°S$, $\hat{v}_x = 0.0463°$ per hour, and $\hat{v}_y = 0.0484°$ per hour. The ship's estimated position at time 11.10 is thus

$$x = \hat{x}_0 + \hat{v}_x t = 155.40° \text{ longitude}$$

$$y = \hat{y}_0 + \hat{v}_y t = 24.43°S \text{ latitude}$$

Here, each of the measurements was weighted equally in the least squares estimate. In practice, the weighting would probably not be equal.

PROBLEMS

8-1. Show that the following matrices are symmetric.
    (a) $A + A^\dagger$ for A square   **(b)** $A^\dagger A$ **(c)** $AA^\dagger$

8-2. Find orthogonal similarity transformations that diagonalize each of the following symmetric matrices.

(a) $\begin{bmatrix} 1 & 2 \\ 2 & 3 \end{bmatrix}$ **(b)** $\begin{bmatrix} 3 & 1 & 0 \\ 1 & -4 & -2 \\ 0 & -2 & -5 \end{bmatrix}$ **(c)** $\begin{bmatrix} 0 & -1 & 0 & -3 \\ -1 & 1 & 6 & 2 \\ 0 & 6 & 4 & 1 \\ -3 & 2 & 1 & 2 \end{bmatrix}$

8-3. Write the following quadratic forms in terms of symmetric matrices.

    (a) $f_1(\mathbf{x}) = -3x_1^2 + x_1 x_2 - x_1 x_3 + x_4^2$

    (b) $f_2(\mathbf{x}) = x_1^2 + 2x_1 x_2 - 3x_1 x_3 + 4x_2^2 + 6x_2 x_3 - 7x_3^2$

8-4. Find orthogonal changes of variables that convert the following quadratic forms to pure sums of squares.

    (a) $f_1(\mathbf{x}) = 2x_1^2 - 2x_1 x_2 + 7x_2^2$ **(b)** $f_2(\mathbf{x}) = \mathbf{x}^\dagger \begin{bmatrix} 2 & 0 & -1 \\ 0 & 3 & -2 \\ -1 & -2 & 4 \end{bmatrix} \mathbf{x}$

    (c) $f_3(\mathbf{x}) = \mathbf{x}^\dagger \begin{bmatrix} 5 & 2 & 1 \\ 2 & 0 & -1 \\ 1 & -1 & 3 \end{bmatrix} \mathbf{x}$

8-5. Determine the sign definiteness, if any, of each of the following quadratic forms.

    (a) $f_1(\mathbf{x}) = 4x_1^2 + 2x_1 x_2 + x_2^2$ **(b)** $f_2(\mathbf{x}) = \mathbf{x}^\dagger \begin{bmatrix} 4 & -1 & 3 \\ -1 & 1 & 2 \\ 3 & 2 & 2 \end{bmatrix} \mathbf{x}$

    (c) $f_3(\mathbf{x}) = \mathbf{x}^\dagger \begin{bmatrix} 3 & 1 & 4 \\ 1 & 3 & 2 \\ 4 & 2 & 3 \end{bmatrix} \mathbf{x}$ **(d)** $f_4(\mathbf{x}) = -2x_1 - x_1 x_2 - 3x_2^2 + x_2 x_3 - x_3^2$

8-6. The function

$$f(x, y, z) = -x^3 + 3xyz + \tfrac{3}{2}y^2 - 3z$$

has a critical point at $(-1, -1, -1)$. Use the matrix of second partial derivatives to determine if this point is a relative maximum, a relative minimum, or neither.

8-7. **(a)** Find four different square roots of the symmetric matrix

$$S = \begin{bmatrix} 3 & 1 \\ 1 & 2 \end{bmatrix}$$

    (b) Find a single square root of the symmetric matrix

$$S = \begin{bmatrix} 2 & 3 \\ 3 & -1 \end{bmatrix}$$

    (c) Using Cholesky's decomposition method, find the square root of the matrix

$$\mathbf{s} = \begin{bmatrix} 9 & 6 & 3 & 15 \\ 6 & 5 & 3 & 11 \\ 3 & 3 & 6 & 14 \\ 15 & 11 & 14 & 43 \end{bmatrix}$$

Verify by direct multiplication that it is a square root.

8-8. Suppose that the electric field vector **E** of a plane light wave has components

$$Ex(t) = \cos \omega t$$

$$Ey(t) = 3 \cos \left(\omega t + \frac{\pi}{8} \right)$$

instead of those used in the example of Section 8.4. Find the principal directions of the elliptical polarization and the lengths of the axes of the ellipse traced by the tip of the **E** vector.

8-9. For the linear-quadratic function of two variables,

$$x^2 + 3xy + 2y^2 - 6x + 4y = 5$$

determine the function's type and find the directions and coordinates of its principal axes.

8-10. Modify the PIVOT program (Table 2-4) to eliminate the entry and pivoting of the knowns, giving a program for matrix rank determination by pivoting. Use the modified program to find the ranks of the following matrices.

(a) $\begin{bmatrix} 2 & 10 & -3 \\ 4 & 20 & -6 \\ 6 & -30 & 9 \end{bmatrix}$
(b) $\begin{bmatrix} 5 & 6 & 6 & 8 \\ 1 & 2 & 3 & 4 \\ 13 & 14 & 15 & 16 \\ 9 & 10 & 11 & 12 \end{bmatrix}$

8-11. Find the singular values of each of the following matrices.

(a) $\begin{bmatrix} 2 & 3 \\ 1 & -1 \end{bmatrix}$
(b) $\begin{bmatrix} 1 & 4 \\ 2 & 5 \\ 3 & 6 \end{bmatrix}$
(c) $\begin{bmatrix} 1 & 1 & 0 & 1 \\ 3 & 0 & 3 & 3 \\ 2 & 0 & 0 & 2 \end{bmatrix}$

8-12. For an $m \times n$ matrix **A**, $\mathbf{A}^\dagger\mathbf{A}$ is symmetric and $n \times n$ and $\mathbf{A}\mathbf{A}^\dagger$ is symmetric and $m \times m$. The matrices $\mathbf{A}^\dagger\mathbf{A}$ and $\mathbf{A}\mathbf{A}^\dagger$ share the same eigenvalues with the larger-dimensioned matrix having its other eigenvalues $\lambda = 0$. Verify this property for

$$\mathbf{A} = \begin{bmatrix} 1 & 2 & -3 & 1 \\ -1 & 0 & 4 & -1 \end{bmatrix}$$

8-13. Using

$$\frac{d}{dt} (\mathbf{AB}) = \frac{d\mathbf{A}}{dt} \mathbf{B} + \mathbf{A} \frac{d\mathbf{B}}{dt}$$

find

$$\frac{d}{dt} (\mathbf{A}^3)$$

8-14. Prove that, if **A** does not depend on **x**,

(a) $\dfrac{\partial}{\partial \mathbf{x}} (\mathbf{Ax}) = \mathbf{A}$
(b) $\dfrac{\partial}{\partial \mathbf{x}} (\mathbf{x}^\dagger \mathbf{Ax}) = \mathbf{x}^\dagger \mathbf{A}^\dagger + \mathbf{x}^\dagger \mathbf{A}$

8-15. Use the LEAST program to find least squares estimates for the following sets of equations.

(a) $3x = 5.2$

$4x = 7.3$

$-2x = -4.0$

$2x = 3.8$

(b) $4.1x_1 - 0.9x_2 = -10.6$

$2.2x_1 - 2.2x_2 = -8.5$

$-1.1x_1 - 3.1x_2 = -5.8$

$-2.0x_1 + 4.4x_2 = 16.2$

(c)
$$\begin{bmatrix} 1 & -1 & 0 & 2 \\ 3 & 2 & -1 & 0 \\ 1 & 1 & -1 & 1 \\ 2 & 1 & 2 & -1 \\ -1 & 2 & 1 & 0 \\ 0 & 0 & -2 & 3 \end{bmatrix} X = \begin{bmatrix} 3 \\ 2.5 \\ 4 \\ -4 \\ 0.5 \\ 7.6 \end{bmatrix}$$

8-16. If the matrix **A** in a least squares problem,

$$\mathbf{AX} = \mathbf{b}$$

is not of full rank, the least squares estimate cannot be found because $\mathbf{A}^\dagger\mathbf{A}$ is then singular. When **A** is not of full rank, one or more of its columns are linearly dependent on the other columns. This means that there are one or more combinations of variables that can have any value; they are not determined (or overdetermined) by the equations. Carefully interpret the situation for

$$\begin{bmatrix} 2 & 1 & 1 \\ 3 & 2 & 1 \\ 1 & -1 & 2 \\ -4 & 4 & -8 \\ -2 & 1 & -3 \end{bmatrix} \mathbf{x} = \mathbf{Ax} = \begin{bmatrix} 3 \\ 5 \\ 7 \\ -22 \\ -8 \end{bmatrix} = \mathbf{b}$$

and find least squares estimates for the combinations of variables that are affected by the equations.

8-17. Modify the LEAST program so that it will perform weighted least squares estimates. Then find the weighted least squares estimate of the solution to the following equations with the weighting function given.

$$\begin{bmatrix} 3 & -1 & 0 \\ 2 & 1 & 2 \\ 1 & 2 & -1 \\ 0 & 1 & 1 \\ -1 & -1 & -1 \end{bmatrix} \mathbf{x} = \begin{bmatrix} -6 \\ -3 \\ 0 \\ 1 \\ 2 \end{bmatrix} \quad \mathbf{W} = \begin{bmatrix} 2 & 1 & -1 & 1 & 0 \\ 1 & 3 & 0 & 0 & 0 \\ -1 & 0 & 4 & 0 & 0 \\ 1 & 0 & 0 & 1 & 0 \\ 0 & 0 & 0 & 0 & 1 \end{bmatrix}$$

8-18. Find interpolating curves of the forms indicated that pass through the following sets of points.

(a) $y = a_1 + a_2x$; $(-4, 0), (-3, 2)$

(b) $y = a_1 + a_2x + a_3x^2$; $(1, 1), (2, 2), (3, 0)$

(c) $x = a_1e^{-y} + a_2e^{-2y}$; $(1.5, 0), (0, 0.5)$

(d) $z = a_1 + a_2x + a_3y$; $(0, 0, 0), (1, 0 -1), (2, 1, 1)$

8-19. Find least squares curve fits for curves of the forms indicated, with the data points given.

(a) $y = a_1 + a_2x$; $(0, -4)$, $(1, -2)$, $(-1, -5)$, $(2, 0)$

(b) $y = a_1 + a_2x + a_3x^2$;

   $(1.7, -0.1)$, $(2.3, 0.2)$, $(6.2, 6.6)$, $(0.1, 0.0)$, $(8.5, 13.8)$, $(-1.0, 1.2)$

(c) $y = a_1 + a_2e^x + a_3e^{1.5x}$; $(-1, -1)$, $(0, -1)$, $(1, -1.5)$, $(2, -7)$, $(3, -30)$

(d) $y = a_1 + a_2 \sin 2x + a_3 \cos 2x$;

   $(0, -4)$, $(1, -3)$, $(2, -3)$, $(-1, -5)$, $(-2, -3)$, $(3, -5)$

8-20. For a least squares curve fit of $y = a_1 + a_2x$

to the data points $(-2, -4)$, $(-1, -7)$, $(0, -3)$, $(1, -3)$, and $(2, -2)$, find the one most likely outlier, delete it, and find the revised curve.

8-21. Use the GNP data in Table 8-5 to find least squares curve fits and predictions as follows:
   (a) A straight line fit to the five data points from 1950 through 1970, the predicted 1975 and 1980 GNP, and their errors from the actual 1975 and 1980 GNP.
   (b) A quadratic curve fit to the five data points from 1950 through 1970 and the predicted 1975 and 1980 GNP.
   (c) A cubic interpolating curve for the four data points from 1955 through 1970, the predicted 1975 and 1980 GNP, and their errors from the actual 1975 and 1980 GNP.

8-22. Use the RECURSIV program to find a sequence of least squares estimates for the following equations in the order given.

(a) $3x = -2$ (b) $7.3x_1 - 1.1x_2 = -12.9$ (c) $2x_1 + 3x_2 + x_3 = 1$

$-2x = 2$ $2.1x_1 + 3.7x_2 = -8.0$ $x_1 - 2x_2 - 4x_3 = 5$

$4x = -3$ $-3.4x_1 - 4.2x_2 = 11.2$ $-3x_1 - x_2 + 2x_3 = -5$

$7x = -8$ $6.6x_1 - 0.8x_2 = -13.1$ $-2x_1 + 2x_2 - 3x_3 = 0$

$-4x = 5$ $5.8x_1 + 2.6x_2 = -15.3$ $4x_1 + 3x_2 - 3x_3 = 9$

                                                            $3x_1 - 5x_2 + 6x_3 = 8$

8-23. For the ship navigation problem of Section 8.8.3, suppose that the initial position computation of $(154.85°, 23.82°S)$ at $t = 0$ is not available. Using the other data, recursively estimate the ship's position and velocity. With all the data (except the $t = 0$ data) estimate what the ship's position was at $t = 0$.

8-24. For the ship navigation problem of Section 8.8.3, let the ship's longitude x and latitude y be modeled, instead, by the parametric equations

$$x = x_0 + v_xt + 0.001t + 2.5 \times 10^{-5}t^2$$

$$y = y_0 + v_yt - 0.002t + 3.7 \times 10^{-5}t^2$$

which account for known velocity components due to ocean currents. Use the data in that section to recursively estimate the position and the relative velocities v_x and v_y at the four times of the measurements.

9

DIFFERENTIAL EQUATIONS

9.1 PREVIEW

In this final chapter, we study differential equations and their solutions. The symbol t, rather than x, is used as the variable for solutions $y(t)$ because x is commonly used in this field to denote state variables. Often in physical problems, the variable t represents time, but that is not always the case.

Our study begins with linear, constant-coefficient differential equations because the analytical theory of their solution is well developed. Both classical and Laplace transform solution methods are discussed, in Sections 9.2 and 9.3, respectively. Then, because analytical solutions are not known for most other differential equations, our attention turns to numerical methods of solution. Numerical integration is fundamental to this endeavor, and in developing this topic in Section 9.4, we discuss the trade-off between calculation complexity and computational error.

Numerical methods for generating approximate numerical solutions to differential equations with known initial conditions progresses from the general, nonlinear case in Section 9.5, to the case of linear equations in Section 9.6, then to linear equations with constant coefficients in Section 9.8. Computer solution programs are given for each class of problems. When simple recursive approximations result in excessive numerical error, more sophisticated approximations are used. These are discussed in Section 9.7.

Finally, boundary value problems, where the boundary conditions are at more than a single value of the variable, are solved. It is shown that the solution of boundary value problems with linear differential equations can be determined from the solution of several initial value problems. For nonlinear differential equations, an interactive zero-crossing search process is generally necessary.

9.2 LINEAR, CONSTANT-COEFFICIENT DIFFERENTIAL EQUATIONS

Our study of differential equations begins with the simplest kind, linear equations with constant coefficients. For these, there is a general solution procedure for commonly encountered driving functions.

9.2.1 Properties and Homogeneous Solutions

An equation of the form

$$\alpha_n \frac{d^n y}{dt^n} + \alpha_{n-1} \frac{d^{n-1} y}{dt^{n-1}} + \cdots + \alpha_1 \frac{dy}{dt} + \alpha_0 y = g(t)$$

where $\alpha_n, \alpha_{n-1}, \ldots, \alpha_1, \alpha_0$ are constants, is a linear, constant-coefficient differential equation. It is termed *linear* because of the function $y(t)$ and derivatives of $y(t)$ appear linearly in the equation. The order of the equation is the degree, n, of the highest derivative term in the equation. The function $g(t)$ is the *driving function* of the equation. A function $y(t)$ that satisfies a differential equation is a *solution* to that equation. Linear differential equations have many solutions, however. The *general solution* of a differential equation is the collection of all possible solutions. The general solution of any nth-order linear differential equation involves n independent arbitrary constants, so its general solution is characterized by an n-parameter family of functions. For example, any function of the form

$$y(t) = Ke^{-3t} + 4$$

for any value of the constant K satisfies the linear, constant-coefficient differential equation

$$\frac{dy}{dt} + 3y = 12$$

It involves one arbitrary constant, K, and is the general solution of this first-order equation.

A homogeneous differential equation has zero driving function. The first-order homogeneous, linear, constant-coefficient equation is

$$\alpha_1 \frac{dy}{dt} + \alpha_0 y = 0$$

or

$$\frac{dy}{dt} = -\frac{\alpha_0}{\alpha_1} y$$

Its solutions are functions that are proportional to their own derivatives, namely

$$y(t) = Ke^{-(\alpha_0/\alpha_1)t}$$

An easy way to determine this solution is to substitute a possible solution of the form

$$y(t) = Ke^{st}$$

into the equation and to find, if possible, the value of s that works. For the equation

$$3\frac{dy}{dt} - 4y = 0$$

this substitution gives

$$(3s - 4) = 0 \qquad s = \tfrac{4}{3}$$

so any function of the form

$$y(t) = Ke^{(4/3)t}$$

satisfies the equation. As the equation is linear and first-order, this must be its general solution because it involves one arbitrary constant.

Exponential functions also comprise the solution to higher-order homogeneous linear, constant-coefficient differential equations. For a general constant-coefficient homogeneous equation of nth order,

$$\alpha_n \frac{d^n y}{dt^n} + \alpha_{n-1} \frac{d^{n-1} y}{dt^{n-1}} + \cdots + \alpha_1 \frac{dy}{dt} + \alpha_0 y = 0 \tag{9-1}$$

Substituting a possible solution of the form

$$y(t) = Ke^{st}$$

gives the nth-degree polynomial *characteristic equation*

$$\alpha_n s^n + \alpha_{n-1} s^{n-1} + \cdots + \alpha_1 s + \alpha_0 = 0$$

The *characteristic roots* of the differential equation are the n roots of this polynomial, $s_1, s_2, \ldots, s_n$.

Any function of the form

$$y(t) = K_1 e^{s_1 t} + K_2 e^{s_2 t} + \cdots + K_n e^{s_n t} \tag{9-2}$$

satisfies the differential equation for any values of the n arbitrary constants $K_1, K_2, \ldots, K_n$.

If the characteristic roots are distinct, the arbitrary constants are independent of one another, and (9-2) is the general solution to (9-1). For example, for the third-order differential equation

$$\frac{d^3 y}{dt^3} - 5\frac{d^2 y}{dt^2} + 4\frac{dy}{dt} + 2y = 0$$

the characteristic equation is

$$s^3 - 5s^2 + 4s + 2 = 0$$

Using the FACTOR program (Table 5-17), the characteristic roots are $s_1 = 3.814$, $s_2 = 1.530$, $s_3 = -0.343$, so the general solution of the differential equation is, approximately,

$$y(t) = K_1 e^{3.814t} + K_2 e^{1.530t} + K_3 e^{-0.343t}$$

If there are repeated characteristic roots, the terms in the general solution corresponding to root repetitions are successive powers of the variable t times the exponential. For the differential equation

$$\frac{d^4y}{dt^4} + 6\frac{d^3y}{dt^3} + 12\frac{d^2y}{dt^2} + 8\frac{dy}{dt} = 0$$

for example, the characteristic equation is

$$s^4 + 6s^3 + 12s^2 + 8s = s(s + 2)^3 = 0$$

The characteristic root $s_1 = 0$ is not repeated, but the root $s_2 = s_3 = s_4 = -2$ is repeated. The general solution is

$$y(t) = K_1 e^{0t} + K_2 e^{-2t} + K_3 t e^{-2t} + K_3 t^2 e^{-2t}$$

Since $\exp(0) = 1$, this can be written as

$$y(t) = K_1 + K_2 e^{-2t} + K_3 t e^{-2t} + K_3 t^2 e^{-2t}$$

For differential equations with real coefficients, complex roots to the characteristic equation occur in complex conjugate pairs. For example, for the equation

$$\frac{d^3y}{dt^3} + 3\frac{d^2y}{dt^2} - 2\frac{dy}{dt} + 4y = 0$$

the characteristic equation is

$$s^3 + 3s^2 - 2s + 4 = 0$$

and using the FACTOR program, the characteristic roots are $s_1 = 0.401 + i0.944$, $s_2 = s_1^* = 0.401 - i0.944$, $s_3 = -3.802$. The general solution to the equation is thus

$$y(t) = K_1 e^{(0.401 + i0.944)t} + K_2 e^{(0.0401 - i0.944)t} + K_3 e^{-3.802t} \tag{9-3}$$

Because there are complex characteristic roots, the arbitrary constants, K_1 and K_2, associated with those roots are likely to be complex.

Combining pairs of complex conjugate terms gives the form

$$K_1 e^{(a + ib)t} + K_2 e^{(a - ib)t} = e^{at}(K_1 e^{ibt} + K_2 e^{-ibt})$$

$$= e^{at}[(K_1(\cos bt + i \sin bt) + K_2(\cos bt - i \sin bt)]$$

$$= e^{at}[(K_1 + K_2) \cos bt + (iK_1 - iK_2) \sin bt]$$

$$= e^{at}(A \cos bt + B \sin bt)$$

Hence an alternative expression for a pair of conjugate exponentials is a real exponential function times a sinusoid. The general solution (9-3) is alternatively of the form

$$y(t) = e^{0.401t}(A \cos 0.944t + B \sin 0.944t) + K_3 e^{-3.802t}$$

where A, B, and K_3 are the arbitrary constants.

9.2.2 General Solutions to Driven Equations

For a linear differential equation with a nonzero driving function, the corresponding homogeneous equation is the original equation with the driving function replaced by zero. The general solution to the entire equation is the *general* solution to the homogeneous equation, plus *one* solution to the entire equation. The solution to the homogeneous equation is termed the *homogeneous* (or *natural* or *transient*) component of the solution. The single solution to the whole equation is the *particular* (or *forced* or *steady-state*) component of the solution. The sum of the homogeneous and particular solutions satisfies the entire linear differential equation. Containing, as it does, the general solution to the homogeneous equation, it involves the correct numbers of independent arbitrary constants, so it is the general solution to the differential equation with nonzero driving function.

Analytical solutions to linear differential equation with coefficients that are not constant are usually difficult to obtain. Except perhaps for those first- and second-order linear variable-coefficient equations that have been studied extensively in the past, one usually is forced to seek numerical solutions.

For linear, constant-coefficient equations,

$$\alpha_n \frac{d^n y}{dt^n} + \alpha_{n-1} \frac{d^{n-1} y}{dt^{n-1}} + \cdots + \alpha_1 \frac{dy}{dt} + \alpha_0 y = g(t)$$

finding the homogeneous component of the solution, the general solution to

$$\alpha_n \frac{d^n y_h}{dt^n} + \alpha_{n-1} \frac{d^{n-1} y_h}{dt^{n-1}} + \cdots + \alpha_1 \frac{dy_h}{dt} + \alpha_0 y_h = 0$$

involves factoring the characteristic equation,

$$\alpha_n s^n + \alpha_{n-1} s^{n-1} + \cdots + \alpha_1 s + \alpha_0 = 0$$

which, again except in low-order and special cases, will be a numerical solution. Particular solutions are easy to find analytically in some special cases but are not so in general.

The easy special cases of finding particular solutions to linear, constant-coefficient linear differential equations are listed in Table 9-1. Fortunately, these kinds of driving functions are of considerable practical interest. Except in a special case, if the driving

TABLE 9-1 Particular Solution to Some Linear, Constant-Coefficient
Differential Equations

Driving Function	Particular Solution
(1) Constant, A	Another constant, B
(2) Real exponential, Ae^{at}	Another real exponential with the same exponential constant, Be^{at}
(3) Sinusoid, $A_1 \cos bt$	Another sinusoid with the same radian frequency, $B_1 \cos bt + B_2 \sin bt$
(4) Polynomial of the form, $A_k t^k + A_{k-1} t^{k-1} + \cdots + A_1 t + A_0$	Another polynomial of the same degree, $B_k t^k + B_{k-1} t^{k-1} + \cdots + B_1 t + B_0$

functions of a linear, constant-coefficient differential equation is a constant, a particular solution to that equation is also a constant.

For example, for the equation

$$\frac{d^2y}{dt^2} + 2\frac{dy}{dt} + 10y = 12$$

substituting

$$y_p = A,$$

a constant to be determined, gives

$$0 + 2(0) + 10A = 12$$

A particular solution is thus

$$y_p = A = \tfrac{6}{5}$$

The homogeneous equation is

$$\frac{d^2y_h}{dt^2} + 2\frac{dy_h}{dt} + 10y_h = 0$$

which has characteristic equation

$$s^2 + 2s + 10 = (s + 1 + \underline{i}3)(s + 1 - \underline{i}3) = 0$$

so the homogeneous component of the solution is

$$y_h(t) = K_1e^{-(1+\underline{i}3)t} + K_2e^{-(1-\underline{i}3)t}$$

$$= e^{-t}(K_1' \cos 3t + K_2' \sin 3t)$$

The general solution is

$$y(t) = y_h(t) + y_p = e^{-t}(K_1' \cos 3t + K_2' \sin 3t) + \tfrac{6}{5}$$

It satisfies the original second-order differential equation and has two independent arbitrary constants.

The special case occurs when the homogeneous solution includes a constant term, as it does for the equation

$$\frac{d^3y}{dt^3} + \frac{d^2y}{dt^2} - 2\frac{dy}{dt} = 10 \qquad (9\text{-}4)$$

The homogeneous equation is

$$\frac{d^3y_h}{dt^3} + \frac{d^2y_h}{dt^2} - 2\frac{dy_h}{dt} = 0$$

and the characteristic equation is

$$s^3 + s^2 - 2s = s(s - 1)(s + 2) = 0$$

The homogeneous component of the general solution of (9-4) is then

$$y_h(t) = K_1 + K_2 e^t + K_3 e^{-2t}$$

A constant particular solution

$$y_p = A$$

when substituted into (9-4) gives

$$0 = 10$$

and so does not work. The particular solution in this special case is of the form

$$y_p(t) = At$$

Substituting into (9-4) yields

$$-2A = 10 \qquad A = -5 \qquad y_p(t) = -5t$$

so the general solution of this equation is

$$y(t) = y_h(t) + y_p(t) = K_1 + K_2 e^t + K_3 e^{-2t} - 5t$$

Except in the special case where the homogeneous solution has a term of the same form as the driving function, if the driving function of a linear, constant-coefficient differential equation is an exponential function, the particular solution is another exponential function with the same exponential constant. The equation

$$\frac{d^2 y}{dt^2} + 5 \frac{dy}{dt} + 3y = 4e^{-2t} \tag{9-5}$$

has homogeneous solution

$$y_h(t) = K_1 e^{-0.697t} + K_2 e^{-4.303t}$$

Substituting a particular solution of the form

$$y_p(t) = Ae^{-2t}$$

gives

$$4A - 10A + 3A = 4 \qquad A = -\tfrac{4}{3} \qquad y_p = -\tfrac{4}{3} e^{-2t}$$

so the general solution to (9-5) is

$$y(t) = y_h(t) + y_p(t) = K_1 e^{-0.697t} + K_2 e^{-4.303t} - \tfrac{4}{3} e^{-2t}$$

Similar considerations apply to linear, constant-coefficient differential equations with sinusoidal and with polynomial driving functions. Because of the equation's linearity, sums of driving function terms result in particular solutions that are sums of the particular solutions for each term considered separately.

9.2.3 Boundary Conditions and Specific Solutions

The general solutions to the differential equations describing systems indicate that there is a whole family of possible solutions, different possibilities for each different set of arbitrary constants. Of course, only one of all the possible solutions applies to any specific situation. Which of the possible solutions is the one that applies is described with *boundary conditions*. For example, the solution of the differential equation

$$\frac{dy}{dt} + 2y = 8 \qquad y(t) = Ke^{-2t} + 4$$

with the boundary condition

$$y(0) = 3$$

is the specific function

$$y(t) = -e^{-2t} + 4$$

For the differential equation

$$\frac{d^2y}{dt^2} + 6\frac{dy}{dt} - 5y = 3e^{-2t}$$

which has general solution

$$y(t) = K_1 e^{0.742t} + K_2 e^{-6.742t} - \frac{3}{13} e^{-2t} \tag{9-6}$$

If the boundary conditions are

$$y(0) = 4 \qquad \frac{dy}{dt}\bigg|_{t=0} = \dot{y}(0) = -1$$

the specific solution that applies has

$$
\begin{aligned}
K_1 + K_2 - \tfrac{3}{13} &= 4 \\
0.742K_1 - 6.742K_2 + \tfrac{6}{13} &= -1
\end{aligned}
\qquad K_1 = 3.616 \text{ and } K_2 = 0.615
$$

so that

$$y(t) = 3.616e^{0.742t} + 0.615e^{-6.742t} - \frac{3}{13} e^{-2t}$$

If, instead, the boundary conditions on (9-6) are not both *initial conditions* but are

$$y(0) = 4 \qquad y(0.1) = 0$$

Then the arbitrary constants must satisfy

$$
\begin{aligned}
K_1 + K_2 - \tfrac{3}{13} &= 4; & K_1 + K_2 &= 4.231 \\
e^{0.0742}K_1 + e^{-0.6742}K_2 - \tfrac{3}{13}e^{-0.2} &= 0; & 1.077K_1 + 0.510K_2 &= 0.189
\end{aligned}
$$

so

$$K_1 = -3.47 \qquad K_2 = 7.70$$

and

$$y(t) = -3.47e^{0.742t} + 7.70e^{-6.742t} - \frac{3}{13}e^{-2t}$$

9.2.4 Solution of a Rotational Mechanical System

The rotational mechanical system of Figure 9-1(a) is described by the equations

$$4\frac{d^2\theta_1}{dt^2} + \frac{d\theta_1}{dt} + \tau_1 = \tau$$

$$6\theta_2 + \tau_2 = 0$$

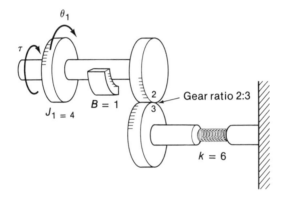

(a) Rotational system

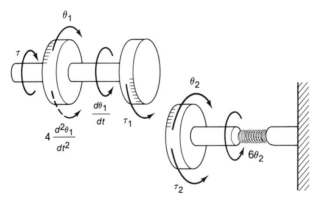

(b) Free-body diagrams

Figure 9-1 Rotational mechanical system.

which result from equating torques in the free-body diagrams of Figure 9-1(b). The gear torques and positions are related by

$$\tau_2 = \tfrac{3}{2}\tau_1 \qquad \theta_2 = -\tfrac{2}{3}\theta_1$$

giving

$$4\frac{d^2\theta_1}{dt^2} + \frac{d\theta_1}{dt} + \tfrac{8}{3}\theta_1 = \tau \tag{9-7}$$

where θ_1 is the angle of the inertia J from its rest position when $\tau = 0$ and τ is the applied torque, as shown.

The homogeneous equation for (9-7) is

$$4\frac{d^2\theta_{1h}}{dt^2} + \frac{d\theta_{1h}}{dt} + \tfrac{8}{3}\theta_{1h} = 0$$

the characteristic equation is

$$4s^2 + s + \tfrac{8}{3} = 0$$

and the characteristic roots are $s_1, s_2 = -0.125 \pm i0.807$. The homogeneous component of the solution for $\theta_1(t)$ is thus

$$\theta_{1h}(t) = e^{-0.125t}(k_1 \cos 0.807t + k_2 \sin 0.807t)$$

If there is no applied torque, $\tau = 0$, and if there is an initial nonzero position and zero initial velocity,

$$\theta_1(0) = 10 \qquad \frac{d\theta_1}{dt}\bigg|_{t=0} = 0$$

then

$$\theta_1(t) = \theta_{1h}(t)$$

and

$$\theta_1(0) = k_1 = 10 \qquad\qquad k_1 = 10$$

$$\frac{d\theta_1}{dt}\bigg|_{t=0} = 0 = -0.125k_1 + 0.807k_2 = 0 \qquad k_2 = 1.55$$

giving

$$\theta_1(t) = e^{-0.125t}(10 \cos 0.807t + 1.55 \sin 0.807t)$$

This solution is plotted in Figure 9-2(a).

If, instead, there is no applied torque, a zero initial position and a nonzero initial velocity,

$$\theta_1(0) = 0 \qquad \frac{d\theta_1}{dt}\bigg|_{t=0} = 1$$

then

$$\theta_1(0) = k_1 = 0 \qquad\qquad\qquad k_1 = 0$$

$$\frac{d\theta_1}{dt}\bigg|_{t=0} = -0.125k_1 + 0.807k_2 = 1 \qquad k_2 = 1.24$$

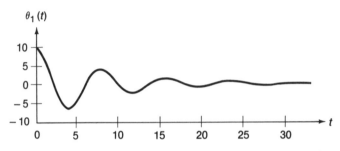

(a) Zero applied torque, nonzero initial position, and zero initial velocity

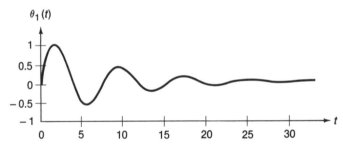

(b) Zero applied torque, zero initial position, and nonzero initial velocity

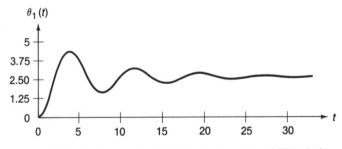

(c) Constant applied torque, zero initial position, and zero initial velocity

Figure 9-2   Response of the rotational mechanical system for various initial conditions and driving functions.

so

$$\theta_1(t) = 1.24 e^{-0.125t} \sin 0.807t$$

as plotted in Figure 9-2(b).

For a constant applied torque, $\tau = 7$, the particular solution to (9-7) is $\theta_{1p} = \frac{21}{8}$, so the general solution is

$$\theta_1(t) = e^{-0.125t}(k_1 \cos 0.807t + k_2 \sin 0.807t) + \frac{21}{8}$$

If the initial position and velocity are each zero, then

$$\theta_1(0) = k_1 + \frac{21}{8} = 0 \qquad\qquad\qquad k_1 = -\frac{21}{8}$$

$$\left.\frac{d\theta_1}{dt}\right|_{t=0} = -0.125k_1 + 0.807k_2 = 0 \qquad k_2 = -0.407$$

so that the specific solution is

$$\theta_1(t) = e^{-0.125t}\left(-\frac{21}{8} \cos 0.807t - 0.407 \sin 0.807t\right) + \frac{21}{8}$$

This solution is plotted in Figure 9-2(c).

9.3 LAPLACE TRANSFORMATION

The Laplace transformation is a tool for manipulating and solving single and simultaneous constant-coefficient linear differential equations. It was first used by Pierre Simon Laplace (1749–1827) in connection with his study of the stability of the solar system. It was discovered independently and popularized by Oliver Heaviside (1850–1925) with the name "operational calculus." Laplace transformation converts constant-coefficient linear differential equations to equivalent linear algebraic equations and so, today, is used nearly universally in connection with this class of problems.

9.3.1 Laplace Transforms and Properties

The Laplace transform of a function $f(t)$ is the function of the variable s

$$\mathscr{L}[y(t)] = Y(s) = \int_0^\infty y(t)e^{-st}\, dt$$

As the integration begins at $t = 0$, only values of $y(t)$ for $t \geq 0$ contribute to $Y(s)$. For all but the most pathological of functions, the function for $t \geq 0$ can be recovered via the inverse Laplace transform,

$$y(t)u(t) = \frac{1}{2\pi} \int_{-\infty}^{\infty} Y(s = \sigma + \underline{i}\omega)e^{(\sigma + \underline{i}\omega)t} \, d\omega$$

$$= \frac{1}{2\pi\underline{i}} \int_{\sigma - \underline{i}\infty}^{\sigma + \underline{i}\infty} Y(s)e^{st} \, ds$$

(9-8)

In (9-8), $u(t)$ is the unit step function,

$$u(t) = \begin{cases} 0 & t < 0 \\ 1 & t \geq 0 \end{cases}$$

and the *convergence constant* σ is any real number for which

$$\int_{-\infty}^{\infty} |y(t)| \, e^{-\sigma t} \, dt$$

is finite. As an example of transform calculation, consider the Laplace transform of

$$y(t) = e^{-at}$$

The transform is

$$Y(s) = \int_0^{\infty} e^{-at} e^{-st} \, dt = \int_0^{\infty} e^{-(a+s)t} \, dt$$

$$= \left. \frac{e^{-(a+s)t}}{-(a + s)} \right|_0^{\infty} = \frac{1}{a + s}$$

(9-9)

In (9-9), the convergence constant σ in $s = \sigma + \underline{i}\omega$ is chosen to be sufficiently large so that the value of the exponential at the top limit is zero.

The Laplace transformation has a number of important properties that are listed in Table 9-2. We will be most interested in the linearity and derivative properties here, but the others listed are highly useful in deriving transformations and in relating applications. The derivative property (7) is shown as follows:

$$\mathscr{L}\left[\frac{dy}{dt}\right] = \int_0^{\infty} \frac{dy}{dt} e^{-st} \, dt$$

Integrating by parts, with

$$u = e^{-st} \qquad du = -se^{-st} \, dt \qquad dv = dy \qquad v = y(t)$$

then

$$\mathscr{L}\left[\frac{dy}{dt}\right] = e^{-st} y(t)|_0^{\infty} - \int_0^{\infty} y(t) \, (-se^{-st} \, dt)$$

$$= -y(0) + s \int_0^{\infty} y(t)e^{-st} \, dt = sY(s) - y(0)$$

The Laplace transform of the derivative of a function is s times the transform of the function itself, minus the initial value of the function. Applying this result twice yields

TABLE 9-2   Laplace Transform Properties

DEFINITION

$$Y(s) = \int_0^\infty y(t)e^{-st}\, dt$$

$$y(t)u(t) = \frac{1}{2\pi}\int_{-\infty}^\infty Y(s = \sigma + \underline{i}\omega)e^{(\sigma + \underline{i}\omega)t}\, d\omega = \frac{1}{2\pi\underline{i}}\int_{\sigma - \underline{i}\infty}^{\sigma + \underline{i}\infty} Y(s)e^{st}\, ds$$

PROPERTIES

(1) $\mathcal{L}[ky(t)] = kY(s)$, k a constant

(2) $\mathcal{L}[y_1(t) + y_2(t)] = Y_1(s) + Y_2(s)$

(3) $\mathcal{L}[y_1(t)y_2(t)]$ does not equal $Y_1(s)Y_2(s)$

(4) $\mathcal{L}[y(t - T)] = e^{-sT}Y(s)$, T a constant, provided that $y(t)$ and $y(t - T)$ are both zero prior to $t = 0$

(5) $\mathcal{L}[y(at)] = \frac{1}{a}Y(\frac{s}{a})$, a a positive constant

(6) $\mathcal{L}[e^{-at}y(t)] = Y(s + a)$

(7) $\mathcal{L}\left[\dfrac{dy}{dt}\right] = sY(s) - y(0^-)$

(8) $\mathcal{L}\left[\dfrac{d^2y}{dt^2}\right] = s^2Y(s) - sy(0^-) - \dot{y}(0^-)$

(9) $\mathcal{L}\left[\dfrac{d^ny}{dt^n}\right] = s^nY(s) - s^{n-1}y(0^-) - s^{n-2}\dot{y}(0^-) - \cdots - sy^{(n-2)}(0^-) - y^{(n-1)}(0^-)$

(10) $\mathcal{L}\left[\displaystyle\int_{0^-}^t y(t)\, dt\right] = \dfrac{Y(s)}{s}$

(11) $\mathcal{L}\left[\displaystyle\int_{-\infty}^t y(t)\, dt\right] = \dfrac{Y(s)}{s} + \dfrac{1}{s}\displaystyle\int_{-\infty}^{0^-} y(t)\, dt$

(12) $\mathcal{L}[t\, y(t)] = -\dfrac{dY(s)}{ds}$

(13) $\mathcal{L}[t^2 y(t)] = \dfrac{d^2Y(s)}{ds^2}$

(14) $\mathcal{L}[t^n y(t)] = (-1)^n\dfrac{d^nY(s)}{ds^n}$

$$\mathcal{L}\left[\frac{d^2y}{dt^2}\right] = \mathcal{L}\left[\frac{d}{dt}\left(\frac{dy}{dt}\right)\right] = s\mathcal{L}\left[\frac{dy}{dt}\right] - \frac{dy}{dt}\bigg|_{t=0}$$

$$= s[sY(s) - y(0)] - \frac{dy}{dt}\bigg|_{t=0} = s^2Y(s) - sy(0) - \frac{dy}{dt}\bigg|_{t=0}$$

and so on.

The inverse Laplace transform can be computed with the real integration in terms of ω or with the complex integration involving the complex variable s. Inverse transforms are seldom calculated in practice because, since the result is unique, one can compile transform pairs, as in Table 9-3. If a transform of the form of one of the entries in the

TABLE 9-3 Laplace Transform Pairs

$y(t)$	$Y(s)$
$u(t)$, unit step	$\dfrac{1}{s}$
$tu(t)$	$\dfrac{1}{s^2}$
$t^n u(t)$	$\dfrac{n!}{s^{n+1}}$
$e^{-at}u(t)$, a a constant	$\dfrac{1}{s+a}$
$te^{-at}u(t)$	$\dfrac{1}{(s+a)^2}$
$t^n e^{-at}u(t)$	$\dfrac{n!}{(s+a)^{n+1}}$
$(\sin bt)u(t)$, b a constant	$\dfrac{b}{s^2+b^2}$
$(\cos bt)u(t)$	$\dfrac{s}{s^2+b^2}$
$(t\sin bt)u(t)$	$\dfrac{2bs}{(s^2+b^2)^2}$
$(t\cos bt)u(t)$	$\dfrac{s^2-b^2}{(s^2+b^2)^2}$
$(e^{-at}\sin bt)u(t)$	$\dfrac{b}{(s+a)^2+b^2}$
$(e^{-at}\cos bt)u(t)$	$\dfrac{(s+a)}{(s+a)^2+b^2}$

table is encountered, the inverse transform must be the corresponding $y(t)$ entry for $t \geq 0$.

9.3.2 Partial Fraction Expansion

Partial fraction expansion is used to express the high-degree polynomial ratios that typically result from Laplace transformation of differential equations to sums of lower-degree terms to which transform table entries apply. For example, the Laplace transform

$$Y(s) = \frac{6s^2 + 10s + 2}{s(s+1)(s+2)} = \frac{K_1}{s} + \frac{K_2}{s+1} + \frac{K_3}{s+2}$$

expands in partial fractions as shown, for some values of the constants K_1, K_2, and K_3.

Placing the expression to the right over a common denominator and equating numerators, there results

$$Y(s) = \frac{K_1(s + 1)(s + 2) + K_2 s(s + 2) + K_3 s(s + 1)}{s(s + 1)(s + 2)}$$

$$= \frac{(K_1 + K_2 + K_3)s^2 + (3K_1 + 2K_2 + K_3)s + 2K_1}{s(s + 1)(s + 2)}$$

$$= \frac{6s^2 + 10s + 2}{s(s + 1)(s + 2)}$$

$$\begin{aligned} K_1 + K_2 + K_3 &= 6 \\ 3K_1 + 2K_2 + K_3 &= 10 \\ 2K_1 &= 2 \end{aligned}$$

Using the SOLVE program (Table 6-7) gives

$$K_1 = 1 \qquad K_2 = 2 \qquad K_3 = 3$$

so that $Y(s)$ expands in partial fractions as

$$Y(s) = \frac{1}{s} + \frac{2}{s + 1} + \frac{3}{s + 2}$$

Using the Laplace transform table, we have

$$y(t) = 1 + 2e^{-t} + 3e^{-2t} \qquad t \geq 0$$

When a Laplace transform that is a ratio of polynomials in s has a denominator polynomial with a repeated root, the corresponding partial fraction expansion terms involve powers of the denominator factor. For example, the following rational function has a partial fraction expansion of the given form:

$$Y(s) = \frac{7s^3 + 8}{(s + 3)(s - 2)^3} = \frac{K_1}{s + 3} + \frac{K_2}{s - 2} + \frac{K_3}{(s - 2)^2} + \frac{K_4}{(s - 2)^3}$$

Equating powers of s in the numerator,

$$K_1(s - 2)^3 + K_2(s + 3)(s - 2)^2 + K_3(s + 3)(s - 2) + K_4(s + 3)$$

$$= (K_1 + K_2)s^3 + (-6K_1 - K_2 + K_3)s^2$$

$$+ (12K_1 - 8K_2 + K_3 + K_4)s + (-8K_1 + 12K_2 - 6K_3 + 3K_4)$$

$$= 7s^3 + 8$$

$$\begin{aligned} K_1 + K_2 &= 7 \\ -6K_1 - K_2 + K_3 &= 0 \\ 12K_1 - 8K_2 + K_3 + K_4 &= 0 \\ -8K_1 + 12K_2 - 6K_3 + 3K_4 &= 8 \end{aligned}$$

Then

$$K_1 = 1.448 \qquad K_2 = 5.552 \qquad K_3 = 14.24 \qquad K_4 = 12.8$$

The function $Y(s)$ thus expands in partial fractions as

$$Y(s) = \frac{1.448}{s + 3} + \frac{5.552}{s - 2} + \frac{14.24}{(s - 2)^2} + \frac{12.8}{(s - 2)^3}$$

so

$$y(t) = 1.448e^{-3t} + 5.552e^{2t} + 14.24te^{2t} + 6.4t^2e^{2t}, \ t \geq 0$$

9.3.3 Laplace Transform Differential Equation Solutions

When a linear, constant-coefficient differential equation is Laplace transformed, the result is a linear algebraic equation in terms of initial conditions. When the initial conditions are known, the transform of the differential equation solution can be found and the transform inverted to find the differential equation solution. For example, for the equation

$$\frac{d^2y}{dt^2} + 5\frac{dy}{dt} + 6y = 10$$

with

$$y(0) = 4$$

$$\left.\frac{dy}{dt}\right|_{t=0} = -2$$

Laplace transforming gives

$$s^2Y(s) - sy(0) - \left.\frac{dy}{dt}\right|_{t=0} + 5[sY(s) - y(0)] + 6Y(s) = \frac{10}{s}$$

$$Y(s) = \frac{4s^2 + 18s + 10}{s(s^2 + 5s + 6)} = \frac{\frac{5}{3}}{s} + \frac{5}{s + 2} + \frac{-\frac{8}{3}}{s + 3}$$

and

$$y(t) = \tfrac{5}{3} + 5e^{-2t} - \tfrac{8}{3}e^{-3t} \qquad t \geq 0$$

9.3.4 Solution of a Translational Mechanical System

The translational mechanical system of Figure 9-3(a) is described by two coupled simultaneous linear, constant-coefficient differential equations. From the free-body diagrams of Figure 9-3(b), they are

$$4\frac{d^2x_1}{dt^2} + 2\frac{dx_1}{dt} - 2\frac{dx_2}{dt} + 4.5x_1 - 1.5x_2 = 0$$

$$\frac{d^2x_2}{dt^2} + 2\frac{dx_2}{dt} - 2\frac{dx_1}{dt} + 2x_2 - 1.5x_1 = f(t)$$

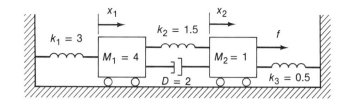

(a) Translational system

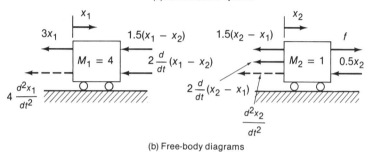

(b) Free-body diagrams

Figure 9-3   A translational mechanical system.

If the initial velocities of the two masses are zero, and their initial displacements from their static equilibrium positions are as given,

$$\frac{dx_1}{dt}\bigg|_{t=0} = \frac{dx_2}{dt}\bigg|_{t=0} = 0 \qquad x_1(0) = 3 \qquad x_2(0) = -1$$

and if the applied force is

$$f(t) = 6 \sin 10t$$

then the Laplace transformed equations are

$$(4s^2 + 2s + 4.5)X_1(s) - (2s + 1.5)X_2(s) = 12s + 8 \tag{9-10}$$

$$-(2s + 1.5)X_1(s) + (s^2 + 2s + 2)X_2(s) = -s - 8 + \frac{60}{s^2 + 100}$$

Solving (9-10) for $X_1(s)$ and $X_2(s)$ gives

$$X_1(s) = \frac{3s^5 + 7.5s^4 + 305.625s^3 + 751s^2 + 592.5s + 122.5}{(s^2 + 100)(s^4 + 2.5s^3 + 3.125s^2 + 1.75s + 1.6875)}$$

$$X_2(s) = \frac{-s^5 - 2.5s^4 + 3.375s^3 - 196s^2 + 267.5s - 532.5}{(s^2 + 100)(s^4 + 2.5s^3 + 3.125s^2 + 1.75s + 1.6875)}$$

Using the FACTOR program, the characteristic polynomial of the system factors is

$$(s^2 + 100)(s^4 + 2.5s^3 + 3.125s^2 + 1.75s + 1.6875)$$
$$= (s + 10\underline{i})(s - 10\underline{i})(s + 0.00171 + \underline{i}0.833)(s + 0.00171 \tag{9-11}$$
$$- \underline{i}0.833)(s + 1.248 + \underline{i}0.928)(s + 1.248 - \underline{i}0.928)$$

So $X_1(s)$ and $X_2(s)$ expand in partial fractions as

$$X_1(s) = \frac{K_1}{s + 10i} + \frac{K_2}{s - 10i} + \frac{K_3}{s + 0.00171 + i0.833} + \frac{K_4}{s + 0.00171 - i0.833}$$

$$+ \frac{K_5}{s + 1.248 + i0.928} + \frac{K_6}{s + 1.248 - i0.928}$$

$$X_2(s) = \frac{K_7}{s + 10i} + \frac{K_8}{s - 10i} + \frac{K_9}{s + 0.00171 + i0.833} + \frac{K_{10}}{s + 0.00171 - i0.833}$$

$$+ \frac{K_{11}}{s + 1.248 + i0.928} + \frac{K_{12}}{s + 1.248 - i0.928}$$

The coefficients will generally be complex, and those of conjugate terms will be complex conjugates of one another:

$$K_2 = K_1^* \qquad K_4 = K_3^* \qquad K_6 = K_5^* \qquad K_8 = K_7^* \qquad K_{10} = K_9^* \qquad K_{12} = K_{11}^*$$

It is helpful, then, to combine the pairs of complex conjugate terms into the equivalent forms

$$X_1(s) = \frac{K_1's + K_2'}{s_2 + 100} + \frac{K_3's + K_4'}{s^2 + 0.00342s + 0.694} + \frac{K_5's + K_6'}{s^2 + 2.50s + 2.419}$$

$$X_2(s) = \frac{K_7's + K_8'}{s^2 + 100} + \frac{K_9's + K_{10}'}{s^2 + 0.00342s + 0.694} + \frac{K_{11}'s + K_{12}'}{s^2 + 2.50s + 2.419}$$

where all the K's are real numbers.

    Finding these constants involves placing each expansion over the common denominator and equating coefficients, resulting in two sets of six simultaneous linear algebraic equations (or some equivalent problem) in six variables each. Once the expansions of $X_1(s)$ and $X_2(s)$ are found, the Laplace transform pair entries in Table 9-2 are used to find $x_1(t)$ and $x_2(t)$ for $t \geq 0$. These are plotted in Figure 9-4.

    Although, from a physical standpoint, this is a relatively simple mechanical system, it is very complicated to solve analytically. Without the help of the Laplace transform, obtaining the solution would be considerably more involved because we would have to deal with the coupled differential equations directly, instead of linear algebraic equations, and incorporation of the initial conditions would be more difficult. Along the way, numerical procedures are needed for factoring the characteristic equation and for the solution of the two sets of six simultaneous linear algebraic equations in six unknowns for the partial fraction expansion coefficients.

    It should not be surprising, then, that for systems such as this one and for the much more complicated systems that routinely occur in practice, analytical solutions are seldom sought. From the characteristic equation (9-11) it is known that $x_1(t)$ and $x_2(t)$ will each consist of the particular sinusoidal response plus a sum of terms (called *modes*) that are a damped sinusoid

$$e^{-0.00171t}(A_1 \cos 0.833t + B_1 \sin 0.833t)$$

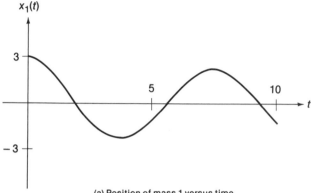

(a) Position of mass 1 versus time

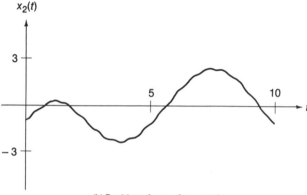

(b) Position of mass 2 versus time

Figure 9-4 Mass positions as functions of time for the translational mechanical system problem.

and another damped sinusoid

$$e^{-1.248t}(A_2 \cos 0.928t + B_2 \sin 0.928t)$$

The individual amplitudes of modes will be such as to, with the particular components, meet the initial conditions. Perhaps this is enough information for the purpose at hand. If response plots are needed, they are alternatively (and much more easily) obtained by numerically solving the differential equations with the methods of the next sections.

9.4 NUMERICAL INTEGRATION

The simplest differential equation is

$$\frac{dy}{dt} = f(t)$$

which is just the problem of integration,

$$y(t) = \int f(t)\, dt + k$$

Approximate numerical calculation of integrals is an important computation problem in its own right and is fundamental to numerical differential equation solution. For convenience, most of the integrals involved here will begin at $t = 0$. It is a simple matter to shift the variable and cast a problem in this simpler form.

9.4.1 Polynomial Approximations

The most common way of performing numerical integration is to divide the interval of integration into many subintervals Δt and approximate the contribution to the integral in each Δt strip by the integral of a polynomial approximation to the integrand in that strip. The first several polynomial integral approximations of increasing order are listed in Table 9-4. The zeroth-order one simply approximates the integrand by a constant equal to the value of the integrand at the left endpoint of each Δt subinterval, as shown in Figure 9-5(a). The first-order approximation approximates the integrand with a straight line, giving the trapezoidal result illustrated in Figure 9-5(b). The second-order approximation, Figure 9-5(c) uses a parabola approximation for each two Δt subintervals, and so on.

There is very little difference in the result of a zero-order (Euler) and a first-order (trapezoidal) integral approximation when the overall interval of integration is divided into many Δt subintervals. As shown in Figure 9-6,

$$\int_{t_0}^{t_0 + m\Delta t} f(t)\, dt = \{f(t_0) + f(t_0 + \Delta t) + f(t_0 + 2\Delta t)$$

$$+ \cdots + f[t_0 + (m - 1)\,\Delta t]\}\,\Delta t$$

with the zero-order approximation. For the first-order approximation,

$$\int_{t_0}^{t_0 + m\Delta t} f(t)\, dt = \tfrac{1}{2}\,[f(t_0) + f(t_0 + \Delta t)]\,\Delta t + \tfrac{1}{2}\,[f(t_0 + \Delta t) + f[t_0 + 2\Delta t]\,\Delta t$$

$$+ \cdots + \tfrac{1}{2}\{f[t_0 + (m - 1)\,\Delta t] + f(t_0 + m\,\Delta t)\}\,\Delta t$$

$$= \{\tfrac{1}{2}f(t_0) + f(t_0 + \Delta t) + f(t_0 + 2\Delta t)$$

$$+ \cdots + f([t_0 + (m - 1)\,\Delta t] + \tfrac{1}{2}f(t_0 + m\,\Delta t)\}\,\Delta t$$

TABLE 9-4   Polynomial Integral Approximations

ZEROTH-ORDER (EULER)

$$\int_{t_0}^{t_0 + \Delta t} f(t)\, dt = f(t_0)\,\Delta t$$

FIRST-ORDER (TRAPEZOIDAL)

$$\int_{t_0}^{t_0 + \Delta t} f(t)\, dt = \left[\tfrac{1}{2}f(t_0) + \tfrac{1}{2}f(t_0 + \Delta t)\right]\Delta t$$

SECOND-ORDER (SIMPSON'S ONE-THIRD RULE)

$$\int_{t_0}^{t_0 + 2\Delta t} f(t)\, dt = \left[\tfrac{1}{3}f(t_0) + \tfrac{4}{3}f(t_0 + \Delta t) + \tfrac{1}{3}f(t_0 + 2\Delta t)\right]\Delta t$$

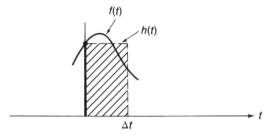

(a) Zeroth-order (Euler) approximation

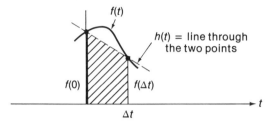

(b) First-order (trapezoidal) approximation

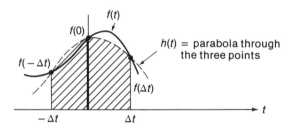

(c) Second-order (**Simpson's one-third rule**) approximation

Figure 9-5   Polynomial integration approximations.

The two approximations differ only in that the zero-order one uses $f(0)$ and not $f(m\,\Delta t)$ while the first-order approximation uses $\frac{1}{2}f(0)$ and $\frac{1}{2}f(m\,\Delta t)$ instead. These *end effects* generally become less and less significant as the number of subintervals used is increased.

In the second-order polynomial approximation, the function $f(t)$ is approximated by a quadratic polynomial

$$h(t) = at^2 + bt + c$$

where a, b, and c are constants to be determined. Requiring the polynomial to pass through the values of f at $t = -\Delta t$, $t = 0$, and $t = \Delta t$ gives

$$a\,\Delta t^2 - b\,\Delta t + c = f(-\Delta t)$$
$$c = f(0)$$
$$a\,\Delta t^2 + b\,\Delta t + c = f(\Delta t)$$

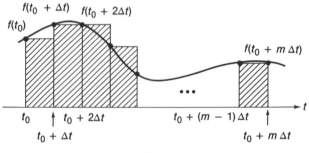

(a) Zeroth-order

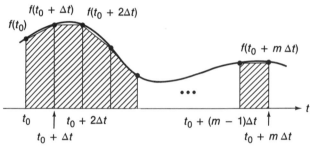

(b) First-order

Figure 9-6   Comparing zeroth-order and first-order integration approximations.

or

$$a = \frac{f(-\Delta t) + f(\Delta t) - 2f(0)}{2\Delta t^2}$$

$$b = \frac{f(\Delta t) - f(-\Delta t)}{2\Delta t}$$

$$c = f(0)$$

The area under the parabola $h(t)$ between $t = -\Delta t$ and $t = \Delta t$ is

$$\int_{-\Delta t}^{\Delta t} h(t)\, dt = \int_{-\Delta t}^{\Delta t} (at^2 + bt + c)\, dt = \left[\frac{at^3}{3} + \frac{bt^2}{2} + ct \right]_{-\Delta t}^{\Delta t}$$

$$= \left[\tfrac{1}{3} f(-\Delta t) + \tfrac{4}{3} f(0) + \tfrac{1}{3} f(\Delta t) \right] \Delta t$$

Translating this result, known as *Simpson's one-third rule*, to other values of the variable t gives the result in Table 9-4.

These derivations can be continued, giving a third-order approximation to

$$\int_{t}^{t+3\Delta t} f(t)\, dt$$

by finding the integral of the cubic polynomial that passes through the four points $f(\Delta t)$, $f(t + \Delta)$, $f(t + 2\Delta t)$, and $f(t + 3\Delta t)$, and so on. The order of the polynomial approximation is the degree of the polynomial that is integrated perfectly by the method. For example, if f is a zero-degree polynomial, a constant, in the integration interval, the zeroth-order Euler approximation is not an approximation: It gives the correct value of the integral. If $f(t)$ is any straight line

$$f(t) = at + b$$

in the integration interval, the first-order, trapezoidal, approximation gives the value of the integral without error, and so on.

One must be careful in comparing one polynomial approximation with another because all but the zeroth-order and first-order ones involve different integration subintervals. The integration interval for the second-order polynomial integral approximation is $2\Delta t$, for example, and should properly be compared with the result of two steps of first-order approximation, each using Δt. As illustrated in Figure 9-7, two steps of trapezoidal integral approximation give

$$\int_{t}^{t+2\Delta t} f(t)\, dt \simeq \{\tfrac{1}{2}\,[f(t) + f(t + \Delta t)] + \tfrac{1}{2}\,[f(t + \Delta t) + f(t + 2\Delta t)]\}\Delta t$$
$$= \{\tfrac{1}{2} f(t) + f(t + \Delta t) + \tfrac{1}{2} f(t + 2\Delta t)\}\Delta t \tag{9-12}$$

which compares with

$$\int_{t}^{t+2\Delta t} f(t)\, dt \simeq \{\tfrac{1}{3} f(t) + \tfrac{4}{3} f(t + \Delta t) + \tfrac{1}{3} f(t + 2\Delta t)\}\Delta t \tag{9-13}$$

for the second-order approximation.

A higher-order approximation is not inherently more accurate than one of lower order. If $f(t)$ more closely follows a parabolic curve in the $2\Delta t$ interval, the second-order approximation is more accurate. If $f(t)$ is closer to the two-segment trapezoidal approximation, that approximation will be more accurate. For an overall interval of integration that is divided into the same number of Δt subintervals, there is no inherent difference in computational complexity. The $\tfrac{1}{2}$, 1, $\tfrac{1}{2}$ coefficients in (9-12) are just different numbers than the $\tfrac{1}{3}$, $\tfrac{4}{3}$, $\tfrac{1}{3}$ in (9-13), for instance, and the same sort of thing occurs for higher-order approximations.

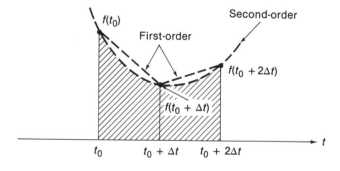

Figure 9-7 Comparing second-order and first-order integration approximations.

9.4.2 Polynomial Approximation Errors

The advantage of using a higher-order polynomial integral approximation is that, for integrands with continuous derivatives of sufficiently high degree, the user can bound the approximation error, and the bound is lower the higher the approximation order. Normally, the actual error of the approximation is not known; if it were known, there would be no need to approximate at all.

Consider the numerical approximation of the integral of a function over one sub-interval Δt,

$$I = \int_{t_0}^{t_0 + \Delta t} f(t) \, dt = y(t_0 + \Delta t) - y(t_0)$$

where

$$y(t) = \int f(t) \, dt$$

is the indefinite integral (or antiderivative). Expanding $y(t_0 + \Delta t)$ in a Taylor series about $t = t_0$,

$$y(t_0 + \Delta t) = y(t_0) + \frac{\dot{y}(t_0) \, \Delta t}{1!} + \frac{\ddot{y}(t_0) \, \Delta t^2}{2!} + \frac{\dddot{y}(t_0) \, \Delta t^3}{3!} + \cdots$$

$$= y(t_0) + \frac{f(t_0) \, \Delta t}{1!} + \frac{\dot{f}(t_0) \, \Delta t^2}{2!} + \frac{\ddot{f}(t_0) \, \Delta t^3}{3!} + \cdots$$

so that

$$I = y(t_0 + \Delta t) - y(t_0) = \frac{f(t_0) \, \Delta t}{1!} + \frac{\dot{f}(t_0) \, \Delta t^2}{2!} + \frac{\ddot{f}(t_0) \, \Delta t^3}{3!} + \cdots \tag{9-14}$$

Similarly expanding $f(t_0 + \Delta t)$ in a Taylor series about $t = t_0$,

$$f(t_0 + \Delta t) = f(t_0) + \frac{\dot{f}(t_0) \, \Delta t}{1!} + \frac{\ddot{f}(t_0) \, \Delta t^2}{2!} + \cdots$$

or

$$\dot{f}(t_0) \, \Delta t = f(t_0 + \Delta t) - f(t_0) - \frac{\ddot{f}(t_0) \, \Delta t^2}{2!} - \cdots \tag{9-15}$$

Substituting for $\dot{f}(t_0) \, \Delta t$ in (9-15) into (9-14) gives

$$I = \underbrace{\tfrac{1}{2} [f(t_0 + \Delta t) + f(t_0)] \, \Delta t}_{\text{first-order approximation}} \underbrace{- \tfrac{1}{12} \ddot{f}(t_0) \, \Delta t^3 - \cdots}_{\text{error of the approximation}} \tag{9-16}$$

which explicitly shows the error of the first-order (trapezoidal) approximation. For sufficiently small Δt, the $\ddot{f}(t_0)$ term in (9-14) dominates the error. The per-step error is then proportional, with some unknown constant of proportionality, C depending on the function $f(t)$, to Δt^3:

$$E_{\text{per step}} = C \, \Delta t^3$$

The entire integral, over $m \, \Delta t$ steps, has approximation error

$$E_{\text{total}} = mC \, \Delta t^3 = (m \, \Delta t)C \, \Delta t^2 = C' \, \Delta t^2$$

which is proportional to Δt^2 since $m \, \Delta t$ is the fixed overall integration interval.

In a similar way, for sufficiently small Δt it can be shown that the total approximation error of second-order polynomial integral approximation is proportional to Δt^4, and so on. The trade-off between using a low-order integral approximation and a smaller step size versus a high-order approximation and larger step size is this: If arranged in the best possible way numerical computational errors will be approximately proportional to the number of Δt steps. As we have seen, comparing (9-12) and (9-13), for instance, the required computations are of the same character, no matter what order of approximation is used. The integral approximation errors can be made increasingly smaller, for sufficiently small Δt, by increasing the approximation order.

When a low-order approximation with small Δt results in too many computational errors, a high-order approximation with larger Δt will give the same approximation error but less computational error. It is silly, though, to invest energy in implementing high-order integral approximations unless they are needed. To get some feeling for the situation, consider the integral

$$\int_0^1 t \log (1 + t) \, dt = \tfrac{1}{4}$$

Its exact value is known from analytical considerations. Numerical integration using the zeroth-order approximation gives the results of Table 9-5 on a typical inexpensive personal computer. One should have to subdivide the overall integration interval into more than a million Δt's before a dominance of the numerical error would be apparent. It is true that this could be a situation where the various errors, including those in evaluating the log function, tend to cancel one other, but there is really nothing special about this integral.

9.4.3 Precision Integration

When an integration extends over a very long interval, as in long-term inertial navigation where velocity is integrated to obtain position, there can be a special need for extremely accurate numerical integration. Over the years, especially in connection with navigation and similar problems, many high-accuracy integration approximation techniques have

TABLE 9-5   Result of Numerical Integration

Δt	Computed Integral
10^{-1}	0.216336584
10^{-2}	0.246544207
10^{-3}	0.249653523
10^{-4}	0.249965356
10^{-5}	0.249997078
10^{-6}	0.249997704

been developed. In a typical precision integration problem, samples of the integrand $y(0)$, $y(t_1)$, $y(t_2)$, . . . are available sequentially in time, and it is desired to compute

$$y(t) = \int_0^t f(t) \, dt$$

from these data, where t, the upper limit on the integral, is the present time or some time (such as that of an earth satellite orbital rendezvous) of special importance. The computaton is thus *ongoing*. As new samples are received, the integral can be found with greater precision. And one might wish periodically to change the upper limit t.

One approach to the problem is to approximate the integrand by a polynomial that passes through all of the integrand samples,

$$f(t) = h(t) = \beta_0 + \beta_1 t + \beta_2 t^2 + \cdots + \beta_{m-1} t^{m-1}$$

where the constants β are chosen so that

$$h(0) = f(0)$$

$$h(t_1) = f(t_1)$$

$$h(t_2) = f(t_2)$$

$$\vdots$$

$$h(t_m) = f(t_m)$$

The integral is then approximated by the area under the approximating polynomial curve:

$$y(t) = \int_0^t f(t) \, dt = \int_0^t h(t) \, dt = \beta_0 t + \frac{\beta_1 t^2}{2} + \frac{\beta_2 t^3}{3} + \cdots + \frac{\beta_{m-1} t^m}{m}$$

As new additional data is to be incorporated, the polynomial degree is increased. Much of the sophistication of the resulting algorithms is connected with correcting previous integral calculations with new data, instead of starting over with each new data point $[t_i, f(t_i)]$, and in arranging the calculations so that the effects of numerical errors are minimum. To this end, predictor–correctors are used in much the same manner as for recursive least squares. The new data point is predicted on the basis of the past estimate, then the difference is used to correct the previous calculations. When the prediction and correction calculations tend to have errors with opposite algebraic signs, the overall error is likely to be reduced. Usually, linear combinations of other functions that are composed of polynomials, such as the Legendre polynomials (Table 9-6), are used rather than the power series polynomial form. Another common approach is to approximate the integrand in an interval by a fixed-degree least squares polynomial fit.

Yet another popular technique is to approximate the integrand by a *spline*. Splines are piecewise fixed-degree polynomial functions that join the data points and have the maximum number of continuous derivatives where they join. Splines are also commonly used to give smooth interpolation between computed points on lines and surfaces in computer-aided design (CAD) and computer-aided manufacturing (CAM) systems.

TABLE 9-6   Legendre Polynomial

The Legendre polynomials are functions $p_n(t)$ that are specific polynomials that are given by Rodrigues's formula,

$$p_n(t) = \frac{1}{2^n \, n!} \frac{d^n}{dt^n} (t^2 - 1)^n$$

The first several Legendre polynomials are

$$p_0(t) = 1$$

$$p_1(t) = t$$

$$p_2(t) = \frac{3}{2} t^2 - \frac{1}{2}$$

$$p_3(t) = \frac{5}{2} t^3 - \frac{3}{2} t$$

$$p_4(t) = \frac{35}{8} t^4 - \frac{30}{8} t^2 + \frac{3}{8}$$

$$\vdots$$

They satisfy the recurrence relation

$$(n + 1)p_{n+1}(t) = (2n + 1)tP_n(t) + np_{n-1}(t)$$

and have the orthogonality properties

$$\int_{-1}^{1} p_m(t)p_n(t) \, dt = 0 \qquad m \neq n$$

$$\int_{-1}^{1} p_n^2(t) \, dt = \frac{2}{2n + 1}$$

9.5 NUMERICAL DIFFERENTIAL EQUATION APPROXIMATION

Considering the number of different differential equations of practical interest, analytical solutions of very few are known. In fact, just the subject of solutions of first-order and second-order *linear* differential equations is very involved. This should not be surprising since integral tables are the solutions of first-order differential equations of the simplest type, and even these are inadequate for the integration of all functions of interest and importance.

We now examine methods for the numerical computation of differential equation solutions. To do so, it is expedient to express an nth-order differential equation as an equivalent set of n coupled first-order differential equations, in what is called *state variable* form. The emphasis here is on those solutions that are specified by initial conditions; boundary conditions that involve different values of the variable are considered later. For convenience, we take $t = 0$ to be the value of the variable at which the boundary conditions apply.

Except in very special situations (usually when a change of variables can be found that converts a nonlinear differential equation to a linear, constant-coefficient one), the only option for nonlinear differential equations and many linear differential equations is to obtain a numerical approximation to the solution. It is only for constant-coefficient linear differential equations that general analytical solution procedures exist and, even for these, numerical polynomial factoring is generally required and numerical linear algebraic equation solution is usually desirable.

We begin by considering the general problem. Then we will specialize to the cases of linear and linear, constant-coefficient equations.

9.5.1 Equations in State Variable Form

In general, an nth-order differential equation can be expressed in the form

$$\frac{d^n y}{dt^n} = f\left(\frac{d^{n-1}y}{dt^{n-1}}, \frac{d^{n-2}y}{dt^{n-2}}, \ldots, \frac{dy}{dt}, y, t\right) \tag{9-17}$$

where f denotes some function of the indicated quantities. That is, the highest derivative equals a function of the lower derivatives, the function y and the variable t. For a linear differential equation the function f is linear, and

$$\frac{d^n y}{dt^n} = -\alpha_{n-1}(t)\frac{d^{n-1}y}{dt^{n-1}} - \alpha_{n-2}(t)\frac{d^{n-2}y}{dt^{n-2}} - \cdots - \alpha_1(t)\frac{dy}{dt} - \alpha_0(t)y + g(t)$$

If a differential equation is linear and constant-coefficient, then

$$\frac{d^n y}{dt^n} = -\alpha_{n-1}\frac{d^{n-1}y}{dt^{n-1}} - \alpha_{n-2}\frac{d^{n-2}y}{dt^{n-2}} - \cdots - \alpha_1\frac{dy}{dt} - \alpha_0 y + g(t)$$

where the α's are constants.

Rather than dealing with single differential equations of various orders, it is very helpful to transform an nth-order differential equation to an equivalent set of n coupled first-order differential equations. Defining the n *state variables*

$$x_1(t) = y(t)$$

$$x_2(t) = \frac{dx_1}{dt} = \frac{dy}{dt}$$

$$x_3(t) = \frac{dx_2}{dt} = \frac{d^2 y}{dt^2}$$

$$\vdots$$

$$x_{n-1}(t) = \frac{dx_{n-2}}{dt} = \frac{d^{n-2}y}{dt^{n-2}}$$

$$x_n(t) = \frac{dx_{n-1}}{dt} = \frac{d^{n-1}y}{dt^{n-1}}$$

an equivalent state variable form for equation (9-17) is

$$\frac{dx_n}{dt} = \frac{d^n y}{dt^n} = f(x_n, x_{n-1}, \ldots, x_2, x_1, t)$$

and the relations

$$\frac{dx_1}{dt} = x_2$$

$$\frac{dx_2}{dt} = x_3$$

$$\vdots$$

$$\frac{dx_{n-1}}{dt} = x_n$$

For example, the nonlinear differential equation

$$(3t + 1)\frac{d^3y}{dt^3} + \left(\frac{d^2y}{dt^2}\right)\left(\frac{dy}{dt}\right) - \sqrt{\frac{dy}{dt}} \sin ty = \ln t$$

or

$$\frac{d^3y}{dt^3} = -\frac{1}{3t + 1}\left(\frac{d^2y}{dt^2}\right)\left(\frac{dy}{dt}\right) + \frac{1}{3t + 1}\sqrt{\frac{dy}{dt}} \sin ty + \frac{\ln t}{3t + 1}$$

Defining the state variables

$$x_1(t) = y(t)$$

$$x_2(t) = \frac{dx_1}{dt} = \frac{dy}{dt}$$

$$x_3(t) = \frac{dx_2}{dt} = \frac{d^2y}{dt^2}$$

an equivalent state variable form is

$$\frac{dx_1}{dt} = x_2$$

$$\frac{dx_2}{dt} = x_3$$

$$\frac{dx_3}{dt} = \frac{d^3y}{dt^3} = -\frac{1}{3t + 1} x_2 x_3 + \frac{1}{3t + 1}\sqrt{x_2 \sin tx_1} + \frac{\ln t}{3t + 1}$$

with

$$y(t) = x_1(t)$$

The initial conditions are, in terms of the state variables,

$$x_1(0) = y(0) \qquad x_2(0) = \dot{y}(0) \qquad x_3(0) = \ddot{y}(0)$$

In matrix form the state variable equations are

$$\frac{d\mathbf{x}}{dt} = \mathbf{f}(\mathbf{x}, t)$$

$$y(t) = [1 \; 0 \; \cdots \; 0]\mathbf{x} = \mathbf{c}^\dagger \mathbf{x}(t)$$

where f is now a vector function. The initial conditions of y and its derivatives are the initial values of the state variables:

$$\mathbf{x}(0) = \begin{bmatrix} x_1(0) \\ x_2(0) \\ \vdots \\ x_n(0) \end{bmatrix} = \begin{bmatrix} y(0) \\ \dot{y}(0) \\ \vdots \\ y^{[n-1]}(0) \end{bmatrix}$$

If the differential equation is linear, the state variable equations are linear and are, in matrix form,

$$\frac{d\mathbf{x}}{dt} = \begin{bmatrix} 0 & 1 & 0 & \cdots & 0 & 0 \\ 0 & 0 & 1 & \cdots & 0 & 0 \\ \vdots & \vdots & \vdots & \vdots & \vdots & \vdots \\ 0 & 0 & 0 & \cdots & 0 & 1 \\ -\alpha_0(t) & -\alpha_1(t) & -\alpha_2(t) & \cdots & -\alpha_{n-2}(t) & -\alpha_{n-1}(t) \end{bmatrix} \mathbf{x} + \begin{bmatrix} 0 \\ 0 \\ \vdots \\ 0 \\ 1 \end{bmatrix} g(t)$$

$$= \mathbf{A}(t)\mathbf{x}(t) + \mathbf{b}g(t)$$

$$y(t) = [1 \quad 0 \quad \cdots \quad 0 \quad 0]\mathbf{x}(t) = \mathbf{c}^\dagger \mathbf{x}$$

For example, a state variable equivalent of

$$3\frac{d^4 y}{dt^4} + 2t\frac{d^2 y}{dt^2} + 4\sin t\frac{dy}{dt} + 6y = \cos 5t$$

is

$$\begin{bmatrix} \frac{dx_1}{dt} \\ \frac{dx_2}{dt} \\ \frac{dx_3}{dt} \\ \frac{dx_4}{dt} \end{bmatrix} = \begin{bmatrix} 0 & 1 & 0 & 0 \\ 0 & 0 & 1 & 0 \\ 0 & 0 & 0 & 1 \\ -2 & -\frac{4}{3}\sin t & -\frac{2}{3}t & 0 \end{bmatrix} \begin{bmatrix} x_1(t) \\ x_2(t) \\ x_3(t) \\ x_4(t) \end{bmatrix} + \begin{bmatrix} 0 \\ 0 \\ 0 \\ \frac{1}{3} \end{bmatrix} \cos 5t$$

$$= \mathbf{A}(t)\mathbf{x}(t) + \mathbf{b}g(t)$$

$$y(t) = [1 \quad 0 \quad 0 \quad 0] \begin{bmatrix} x_1(t) \\ x_2(t) \\ x_3(t) \\ x_4(t) \end{bmatrix} = \mathbf{c}^\dagger \mathbf{x}(t)$$

The initial conditions are

$$\mathbf{x}(0) = \begin{bmatrix} x_1(0) \\ x_2(0) \\ x_3(0) \\ x_4(0) \end{bmatrix} = \begin{bmatrix} y(0) \\ \dot{y}(0) \\ \ddot{y}(0) \\ \dddot{y}(0) \end{bmatrix}$$

For a constant-coefficient linear differential equation, the *state coupling matrix* **A** does not vary with t.

A linear transformation of these state variables yields new equations in terms of the new variables that are of the same state variable form. Usually, there is little benefit from a change of state variables except for constant-coefficient linear equations.

9.5.2 Recursive Approximations

There are two basic methods of numerical approximation of state variable equations to obtain an approximation of the solution of an nth-order differential equation,

$$\frac{d\mathbf{x}}{dt} = \mathbf{f}(\mathbf{x}, t)$$

In the first method, the derivatives of the n state variables, $d\mathbf{x}/dt$, are approximated and the resulting equations solved. For t in the vicinity of $t = k \, \Delta t$, $k = 0, 1, 2, \ldots$, using

$$\frac{d\mathbf{x}}{dt} = \frac{\mathbf{x}[(k + 1) \, \Delta t] - \mathbf{x}(k \, \Delta t)}{\Delta t}$$

and approximating $\mathbf{f}$ by its value at $t = k \, \Delta t$ gives

$$\mathbf{x}[(k + 1) \, \Delta t] = \mathbf{x}(k \, \Delta t) + \mathbf{f}[\mathbf{x}(k \, \Delta t), t = k \, \Delta t] \, \Delta t \qquad (9\text{-}18)$$

This is a recursive recipe for finding, approximately, $\mathbf{x}$ at the next Δt step from $\mathbf{x}$ at the present step. Beginning with known initial conditions $\mathbf{x}(0)$, then

$$\mathbf{x}(\Delta t) = \mathbf{x}(0) + \mathbf{f}[\mathbf{x}(0), 0] \, \Delta t$$

$$\mathbf{x}(2\Delta t) = \mathbf{x}(\Delta t) + \mathbf{f}[\mathbf{x}(\Delta t), \Delta t] \, \Delta t$$

$$\mathbf{x}(3\Delta t) = \mathbf{x}(2\Delta t) + \mathbf{f}[\mathbf{x}(2\Delta t), 2\Delta t] \, \Delta t$$

$$\vdots$$

and so on. In using this algorithm, we generate evenly spaced *samples* of an approximation of the state $\mathbf{x}$.

If the differential equation is linear so that its state variable representation is

$$\frac{d\mathbf{x}}{dt} = \mathbf{A}(t)\mathbf{x}(t) + \mathbf{b}g(t)$$

$$y(t) = \mathbf{c}^\dagger \mathbf{x}(t)$$

then this derivative approximation is

$$\mathbf{x}[(k + 1) \, \Delta t] = [\mathbf{I} + \mathbf{A}(k \, \Delta t) \, \Delta t]\mathbf{x}(k \, \Delta t) + \mathbf{b} \, \Delta t \, g(k \, \Delta t)$$

$$y(k \, \Delta t) = \mathbf{c}^\dagger \mathbf{x}(k \, \Delta t)$$

If the differential equation is linear and constant-coefficient, the state coupling matrix $\mathbf{A}$ is constant.

To improve the approximation in (9-18), Δt is made smaller. Even so, there are occasional situations in practice where, with limited numerical precision, the systematic errors introduced by the application of (9-18) are not acceptable. Table 9-7 lists a higher-order, potentially more accurate, first derivative approximation for t in the vicinity of $t = k \, \Delta t$. These are *backward-difference* approximations, involving values of t before $t = (k + 1) \, \Delta t$ because we seek a *recursive* approximation of samples of the state. For sufficiently small Δt, the first-order approximation has per-step approximation error that is proportional to Δt^2. The second-order approximation has error proportional to Δt^3. One can go on with a third-order approximation with approximation error proportional to Δt^4 for small Δt, and so on. Applying the second-order approximation

$$\frac{d\mathbf{x}}{dt} = \frac{3\mathbf{x}[(k + 1) \, \Delta t] - 4\mathbf{x}(k \, \Delta t) + \mathbf{x}[(k - 1) \, \Delta t]}{2\Delta t}$$

gives

$$\mathbf{x}[(k + 1) \, \Delta t] = -\tfrac{1}{3}\mathbf{x}[(k - 1) \, \Delta t] + \tfrac{4}{3}\mathbf{x}(k \, \Delta t) + \tfrac{2}{3}\mathbf{f}[x(k \, \Delta t), t = k \, \Delta t] \, \Delta t$$

The other main approach to approximate numerical differential equation solution is approximate integration. Since

$$\frac{d\mathbf{x}}{dt} = \mathbf{f}[\mathbf{x}(t), t]$$

then

$$\mathbf{x}(t) = \mathbf{x}(t_0) + \int_{t_0}^{t} \mathbf{f}(\mathbf{x}(t), t) \, dt \qquad t \geq t_0$$

For evenly spaced sample times $t = k \, \Delta t$, $k = 0, 1, 2, \ldots$,

$$\mathbf{x}[(k + 1) \, \Delta t] = \mathbf{x}(k \, \Delta t) + \int_{k\Delta t}^{(k + 1)\Delta t} \mathbf{f}(\mathbf{x}(t), t) \, dt$$

TABLE 9-7 Some Backward-Difference Derivative Approximations

FIRST-ORDER

$$\frac{d\mathbf{x}}{dt} = \frac{\mathbf{x}[(k + 1) \, \Delta t] - \mathbf{x}(k \, \Delta t)}{\Delta t} \qquad k \, \Delta t \leq t \leq (k + 1) \, \Delta t$$

SECOND-ORDER

$$\frac{d\mathbf{x}}{dt} = \frac{3\mathbf{x}[(k + 1) \, \Delta t] - 4\mathbf{x}(k \, \Delta t) + \mathbf{x}[(k - 1) \, \Delta t]}{2\Delta t} \qquad k \, \Delta t \leq t \leq (k + 1) \, \Delta t$$

Approximating an integral by the Euler approximation,

$$\int_{k\Delta t}^{(k+1)\Delta t} \mathbf{f}(\mathbf{x}(t), t)\, dt = \mathbf{f}[\mathbf{x}(k\,\Delta t), t = k\,\Delta t]\,\Delta t$$

gives the recursive approximation solution

$$\mathbf{x}(\Delta t) = \mathbf{x}(0) + \mathbf{f}[\mathbf{x}(0), 0]\,\Delta t$$

$$\mathbf{x}(2\Delta t) = \mathbf{x}(\Delta t) + \mathbf{f}[\mathbf{x}(\Delta t), \Delta t]\,\Delta t$$

$$\mathbf{x}(3\Delta t) = \mathbf{x}(2\Delta t) + \mathbf{f}[\mathbf{x}(2\Delta t), 2\Delta t]\,\Delta t$$

$$\vdots$$

and so on. This is the same result as was obtained with the first-order derivative approximation. The advantage of integration for higher-order approximations is twofold: (1) all we know about numerical integration particularly precision methods when they are needed can be brought to bear upon this problem, and (2) and perhaps most important, the approximations are of $\mathbf{x}$, not $d\mathbf{x}/dt$, so the nature of the approximation errors is more apparent.

9.5.3 General Nonlinear Differential Equation Solution Program

The program listed in Table 9-8, called NONLINEA, is for general numerical solution of nonlinear differential equations

$$\frac{d^n y}{dt^n} = f\left(\frac{d^{n-1}y}{dt^{n-1}}, \frac{d^{n-2}y}{dt^{n-2}}, \ldots, \frac{dy}{dt}, y, t\right)$$

TABLE 9-8

Computer Program in BASIC for Solution of a Nonlinear Differential Equation (NONLINEA)

VARIABLES USED

N = equation order

D = step size

$X(20)$ = present state vector

$Z(20)$ = new state vector

T = value of the variable

F = Nth-state variable in terms of the state variables and the driving function

LISTING

```
100     PRINT "NONLINEAR DIFFERENTIAL EQUATION SOLUTION"
110     DIM X(2), F(2)
120   REM ENTER STEP SIZE AND INITIAL CONDITIONS
130     PRINT "PLEASE ENTER STEP SIZE"
140     INPUT D
150     PRINT "ENTER INITIAL CONDITIONS"
160     PRINT "INITIAL VALUE OF FUNCTION = ";
170     INPUT X(1)
```

(continued)

```
180      PRINT "INITIAL VALUE OF DERIVATIVE = ";
190      INPUT X(2)
200   REM PRINT HEADING AND INITIALIZE VARIABLE
210      PRINT "VARIABLE","SOLUTION"
220      T = 0
230   REM PRINT VARIABLE AND SOLUTION
240      FOR I=1 TO 1000
250      PRINT T,X(1)
260   REM  UPDATE STATE AND TIME
270      GOSUB 1000
280      X(1) = X(1) + D * F(1)
290      X(2) = X(2) + D * F(2)
300      T = T + D
310      NEXT I
320      END
1000  REM SUBROUTINE TO CALCULATE STATE FUNCTION
1010     F(1) = X(2)
1020     F(2) = -X(1) + X(2) - X(1) * X(1) * X(2)
1030     RETURN
```

Computer Program in FORTRAN for Solution of a Nonlinear Differential Equation (NONLINEA)

```
C ********************************************************************
C PROGRAM NONLINEA -- NONLINEAR DIFFERENTIAL EQUATION SOLUTION
C ********************************************************************
      REAL X(2), F(2)

      WRITE(*,'(1X,''NONLINEAR DIFFERENTIAL EQUATION SOLUTION'')')

C *** ENTER STEP SIZE AND INITIAL CONDITIONS
      WRITE(*,'(1X,''PLEASE ENTER STEP SIZE '',\)')
      READ(*,*) D
      WRITE(*,'(1X,''ENTER INITIAL CONDITIONS'')')
      WRITE(*,'(1X,''INITIAL VALUE OF FUNCTION = '',\)')
      READ(*,*) X(1)
      WRITE(*,'(1X,''INITIAL VALUE OF DERIVATIVE = '',\)')
      READ(*,*) X(2)

C *** PRINT HEADING AND INITIALIZE VARIABLE
      WRITE(*,'(1X,''    VARIABLE        SOLUTION'')')
      T = 0.0

C *** PRINT VARIABLE AND SOLUTION, MAXIMUM OF 1000 ITERATIONS
      DO 10 I=1, 1000
         WRITE(*,'(1X,E15.8,1X,E15.8)') T, X(1)

C ***  UPDATE STATE AND TIME
         CALL FUNCT(X,F)
         X(1) = X(1) + D*F(1)
         X(2) = X(2) + D*F(2)
         T = T + D
 10   CONTINUE
      STOP
      END

C ********************************************************************
      SUBROUTINE FUNCT(X,F)
C ********************************************************************
      REAL X(2), F(2)

C *** SUBROUTINE TO CALCULATE STATE FUNCTION
      F(1) = X(2)
```

(continued)

```
F(2) = -X(1) + X(2) - X(1) * X(1) * X(2)
RETURN
END
```

in state variable form

$$x_1(t) = y(t)$$

$$x_2(t) = \frac{dx_1}{dt} = \frac{dy}{dt}$$

$$x_3(t) = \frac{dx_2}{dt} = \frac{d^2y}{dt^2}$$

$$\vdots$$

$$x_n(t) = \frac{dx_{n-1}}{dt} = \frac{d^{n-1}y}{dt^{n-1}}$$

$$\frac{dx_n}{dt} = \frac{d^ny}{dt^n} = f(x_n, x_{n-1}, \ldots, x_2, x_1, t)$$

or

$$\frac{dx_1}{dt} = x_2(t)$$

$$\frac{dx_2}{dt} = x_3(t)$$

$$\vdots$$

$$\frac{dx_{n-1}}{dt} = x_n(t)$$

$$\frac{dx_n}{dt} = f(x_n, x_{n-1}, \ldots, x_2, x_1, t)$$

where

$$y(t) = x_1(t)$$

The Euler approximations

$$x_1[(k + 1)\,\Delta t] = x_1(k\,\Delta t) + x_2(k\,\Delta t)\,\Delta t$$

$$x_2[(k + 1)\,\Delta t] = x_2(k\,\Delta t) + x_3(k\,\Delta t)\,\Delta t$$

$$\vdots$$

$$x_{n-1}[(k + 1)\,\Delta t] = x_{n-1}(k\,\Delta t) + x_n(k\,\Delta t)\,\Delta t$$

$$x_n[(k + 1)\,\Delta t] = x_n(k\,\Delta t) + f[x_n(k\,\Delta t), x_{n-1}(k\,\Delta t), \ldots ,$$

$$x_2(k\,\Delta t), x_1(k\,\Delta t), k\,\Delta t]\,\Delta t$$

with

$$y(k\,\Delta t) = x_1(k\,\Delta t)$$

are used. The function f is computed in subroutine 1000 and is supplied by the user. The listed function is for the nonlinear Van der Pol differential equation,

$$\frac{d^2y}{dt^2} + (y^2 - 1)\frac{dy}{dt} + y = 0 \tag{9-19}$$

This equation describes a mechanical spring–mass–damper system with position-dependent damping and an important kind of electronic oscillator.

Defining

$$x_1(t) = y(t)$$

$$x_2(t) = \frac{dx_1}{dt} = \frac{dy}{dt}$$

a state equation representation of (9-19) is

$$\begin{bmatrix} \dfrac{dx_1}{dt} \\[2mm] \dfrac{dx_2}{dt} \end{bmatrix} = \begin{bmatrix} 0 & 1 \\ -1 & 1 - x_1^2 \end{bmatrix} \begin{bmatrix} x_1(t) \\ x_2(t) \end{bmatrix}$$

$$y(t) = \begin{bmatrix} 1 & 0 \end{bmatrix} \begin{bmatrix} x_1(t) \\ x_2(t) \end{bmatrix}$$

Figure 9-8 shows the approximate solutions obtained for the initial conditions

$$y(0) = x_1(0) = 0.5 \qquad \text{and} \qquad \dot{y}(0) = x_2(0) = 0$$

and several step sizes.

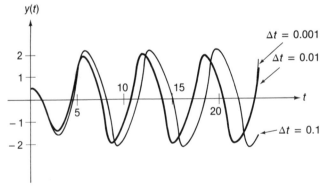

Figure 9-8   Numerical solutions to the Van der Pol equation.

9.6 NUMERICAL APPROXIMATION OF LINEAR EQUATIONS

The special properties of linear differential equations are now considered. A general solution approximation program using the Euler integral approximation is developed and applied. For a linear differential equation, state equations are of the form

$$\frac{d\mathbf{x}}{dt} = \mathbf{A}(t)\mathbf{x}(t) + \mathbf{b}g(t)$$

$$y(t) = \mathbf{c}^\dagger \mathbf{x}(t)$$

and

$$\mathbf{x}(t) = \mathbf{x}(t_0) + \int_{t_0}^{t} [\mathbf{A}(t)\mathbf{x}(t) + \mathbf{b}g(t)]\, dt \qquad t \geq t_0$$

For evenly spaced samples at $t = k\,\Delta t$, $k = 0, 1, 2, \ldots$,

$$\mathbf{x}[(k + 1)\,\Delta t] = \mathbf{x}(k\,\Delta t) + \int_{k\Delta t}^{(k+1)\Delta t} [\mathbf{A}(t)\mathbf{x}(t) + \mathbf{b}g(t)]\, dt$$

An Euler approximation of the integral gives the approximate recursive relations

$$\mathbf{x}[(k + 1)\,\Delta t] = [\mathbf{I} + \mathbf{A}(k\,\Delta t)\,\Delta t]\mathbf{x}(k\,\Delta t) + \mathbf{b}\,\Delta t\, g(k\,\Delta t)$$

$$k = 0, 1, 2, \ldots \qquad (9\text{-}20)$$

9.6.1 General Linear Equation Approximation Program

A general program for numerical approximation of linear differential equations, called LINEARDE, is given in Table 9-9. It performs the approximation (9-20). The equation order N is set by line 130, the equation coefficients are computed by subroutine 1000,

TABLE 9-9

Computer Program in BASIC for Linear Differential Equation Approximation (LINEARDE)

VARIABLES USED

N	=	order of equation
D	=	step size
X(20)	=	present state vector
Z(20)	=	new state vector
T	=	the variable
P(20)	=	coefficients of the differential equation; P(N − 1) is the coefficient of the (N − 1)th derivative term and P(0) is the coefficient of the function
G	=	value of the driving function
I	=	index

(continued)

LISTING

```
100     PRINT "LINEAR DIFFERENTIAL EQUATION SOLUTION"
110     DIM X (20) ,Z (20) , P (20)
120   REM SET ORDER
130     N = 2
140   REM  ENTER STEP SIZE AND INITIAL CONDITIONS
150     PRINT "PLEASE ENTER STEP SIZE";
160     INPUT D
170     PRINT "ENTER INITIAL CONDITIONS"
180     PRINT "INITIAL VALUE = ";
190     INPUT X(1)
200     IF N = 1 THEN GOTO 270
210     FOR I = 2 TO N
220     PRINT "INITIAL ";I - 1;" DERIVATIVE =";
230     INPUT X(I)
240     NEXT I
250   REM INITIALIZE VARIABLE
260     T = 0
270   REM PRINT HEADING
280     PRINT "VARIABLE", "SOLUTION"
290   REM  PRINT VARIABLE AND SOLUTION
300     FOR K=1 TO 1000
310     PRINT T,X(1)
320   REM UPDATE STATE AND TIME
330     IF N = 1 THEN GOTO 370
340     FOR I = 1 TO N - 1
350     Z(I) = X(I) + D * X(I + 1)
360     NEXT I
370     GOSUB 1000
380     GOSUB 2000
390     Z(N) = X(N) + D * G
400     FOR I = 1 TO N
410     Z(N) = Z(N) - D * P(I-1) * X(I)
420     NEXT I
430     FOR I = 1 TO N
440     X(I) = Z(I)
450     NEXT I
460     T = T+ D
470     NEXT K
480     END
1000  REM  SUBROUTINE TO CALCULATE COEFFICIENTS
1010    P(1) = 3
1020    P(0) = 2
1030    RETURN
2000  REM  SUBROUTINE TO CALCULATE DRIVING FUNCTION
2010    G = 3
2020    RETURN
```

Computer Program in FORTRAN for Linear Differential Equation Approximation (LINEARDE)

```
C *****************************************************************
C PROGRAM LINEARDE -- LINEAR DIFFERENTIAL EQUATION SOLUTION
C *****************************************************************
      REAL X(20), Z(20), P(20)

      WRITE(*,'(1X,''LINEAR DIFFERENTIAL EQUATION SOLUTION'')')

C *** SET ORDER
      N = 2

C *** ENTER STEP SIZE AND INITIAL CONDITIONS
      WRITE(*,'(1X,''PLEASE ENTER STEP SIZE '',\)')
      READ(*,*) D
```

```
          WRITE(*,'(1X,''ENTER INITIAL CONDITIONS'')')
          WRITE(*,'(1X,''INITIAL VALUE = '',\)')
          READ(*,*) X(1)
          IF (N.NE.1) THEN
              DO 10 I=2, N
                  WRITE(*,'(1X,''INITIAL '',I2,'' DERIVATIVE = '',\)') I-1
                  READ(*,*) X(I)
   10         CONTINUE

C *** INITIALIZE VARIABLE
              T = 0.0
          ENDIF

C *** PRINT HEADING
          WRITE(*,'(1X,''   VARIABLE         SOLUTION'')')

C *** PRINT VARIABLE AND SOLUTION FOR MAXIMUM OF 1000 INTERATIONS
          DO 15 K=1, 1000

              WRITE(*,'(1X,E15.8,1X,E15.8)') T,X(1)

C *** UPDATE STATE AND TIME
              IF (N.NE.1) THEN
                  DO 20 I=1, N-1
                      Z(I) = X(I) + D*X(I+1)
   20             CONTINUE
              ENDIF

              CALL FUNCT1(P)
              CALL FUNCT2(G)
              Z(N) = X(N) + D*G

              DO 30 I=1, N
                  Z(N) = Z(N) - D*P(I)*X(I)
   30         CONTINUE

              DO 40 I=1, N
                  X(I) = Z(I)
   40         CONTINUE

              T = T+D
   15     CONTINUE

          STOP
          END

C ****************************************************************
          SUBROUTINE FUNCT1(P)
C ****************************************************************

C *** SUBROUTINE TO CALCULATE COEFFICIENTS
          REAL P(20)

          P(2) = 3
          P(1) = 2
          RETURN
          END

C ****************************************************************
          SUBROUTINE FUNCT2(G)
C ****************************************************************
```

(continued)

```
C *** SUBROUTINE TO CALCULATE DRIVING FUNCTION
      REAL G

      G = 3
      RETURN
      END
```

and the driving function is computed by subroutine 2000. The user enters the step size and the initial conditions.

The listed coefficients and driving function are for the differential equation

$$\frac{d^2y}{dt^2} + 3\frac{dy}{dt} + 2y = 3$$

These and the order N are to be changed by the user as needed. For the listed equation, and the initial conditions

$$y(0) = 10 \qquad \dot{y}(0) = 0$$

Figure 9-9 shows the calculation results for various step sizes D. For these plots the calculated points are simply connected by straight-line segments. As the step size is reduced, the approximate numerical solution approaches the actual solution of the differential equation.

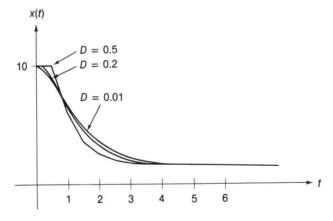

Figure 9-9   Various step sizes in the approximation of a differential equation.

9.6.2 Approximation Solutions of Bessel's Equation

Bessel's equation,

$$t^2\frac{dy^2}{dt^2} + t\frac{dy}{dx} + (t^2 - m^2)y = 0$$

provides another good example of linear differential equation solution. Bessel's equation occurs in many practical problems, including the vibration of a circular membrane and the distribution of frequencies in an FM signal. In many applications, the variable represents distance or some other quantity, not time.

The mth-degree Bessel functions are defined by the series

$$J_m(t) = \frac{t^m}{2^m \, m!} \left[1 - \frac{t^2}{2^2(m+1)} + \frac{t^4}{2^4 \cdot 2! \, (m+1)(m+2)} \right.$$

$$\left. - \frac{t^6}{2^6 \cdot 3! \, (m+1)(m+2)(m+3)} + \cdots \right]$$

in much the same way as sine, cosine, exponential, and other functions can be defined in terms of series. The zeroth-degree Bessel function

$$J_0(t) = 1 - \frac{t^2}{2^2} + \frac{t^4}{2^2 \cdot 4^2} - \frac{t^6}{2^2 \cdot 4^2 \cdot 6^2} + \cdots$$

is one of the solutions of Bessel's equation with $m = 0$, $J_1(t)$ is one of the solutions of Bessel's equation with $m = 1$, and so on. For each integer m, there is a second solution also, the Neumann function (or Bessel function of the second kind), $N_m(t)$, so that the general solution to Bessel's equation has the requisite two independent arbitrary constants, K_1 and K_2:

$$y(t) = K_1 J_m(t) + K_2 N_m(t)$$

The Neumann functions are infinite at $t = 0$, so very often in physical problems, $K_2 = 0$.

In using the LINEARDE program to solve Bessel's equation, the equation is first divided to make the coefficient of the highest-derivative term unity:

$$\frac{d^2 y}{dt^2} + \frac{1}{t} \frac{dy}{dt} + \left(1 - \frac{m^2}{t^2} \right) y = 0$$

There is then a possible problem in beginning the solution at $t = 0$ because the divisions by t in the equation coefficients. This is easily resolved by beginning the solution at $t = \Delta t$ instead of $t = 0$.

For the zero-degree case and

$$\frac{d^2 y}{dt^2} + \frac{1}{t} \frac{dy}{dt} + y = 0$$

the zero-degree Bessel function is the solution for which

$$J_0(0) = y(0) = 1 \qquad \text{and} \qquad \dot{J}_0(0) = \dot{y}(0) = 0$$

For a sufficiently small increment Δt, $y(\Delta t)$ will still be very nearly unity and $\dot{y}(\Delta t)$ will still be nearly zero. Modifying the LINEARDE program with

```
260 T = D
1010 P(1) = 1/T
1020 P(0) = 1
2010 G = 0
```

and entering the initial values of $y(\Delta t) = 1$ and $\dot{y}(\Delta t) = 0$ gives the result plotted in Figure 9-10(a). A step size $D = 0.001$ was used for this plot.

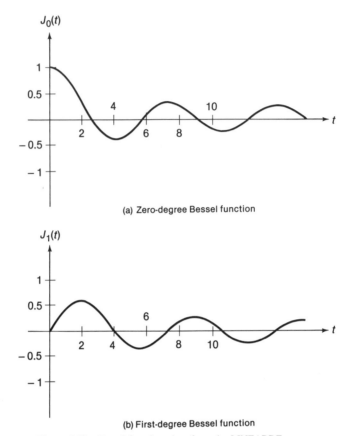

(a) Zero-degree Bessel function

(b) First-degree Bessel function

Figure 9-10   Bessel function plots from the LINEARDE program.

For the first-degree Bessel function,

$$J_1(0) = y(0) = 0$$

and

$$\dot{J}_1(0) = \dot{y}(0) = \tfrac{1}{2}$$

If we again begin the solution at $t = \Delta t$, then $y(\Delta t) = \tfrac{1}{2}$, but, using this slope, $y(\Delta t) = \Delta t/2$. Modifying the LINEARDE program with

```
260 T = D
1010 P(1) = 1/T
1020 P(0) = 1-1/(T*T)
2010 G = 0
```

using the step size $D = 0.001$ and entering the initial values $y(\Delta t) = 0.0005$, $\dot{y}(\Delta t) = \tfrac{1}{2}$ gives the result plotted in Figure 9-10(b).

9.7 HIGHER-ACCURACY DIFFERENTIAL EQUATION INTEGRATION

When adequate accuracy in numerical differential equation solution cannot be obtained with the zeroth-order Euler integral approximation, even for extremely small step size Δt, then a higher-order approximation must be used. Among the best of these are the higher-order polynomial approximations. In this section the use of first-order (trapezoidal) integral approximations is examined in some detail and the nature of higher-order polynomial integral approximations is outlined.

The trapezoidal approximation to the integral in

$$\mathbf{x}[(k + 1) \Delta t] = \mathbf{x}(k \Delta t) + \int_{k\Delta t}^{(k+1\Delta t)} \mathbf{f}(\mathbf{x}(t), t) \, dt \tag{9-21}$$

gives

$$\mathbf{x}[(k + 1) \Delta t] = \mathbf{x}(k \Delta t) + \tfrac{1}{2}\mathbf{f}[\mathbf{x}(k \Delta t), k \Delta t] \Delta t$$

$$+ \tfrac{1}{2}\mathbf{f}\{\mathbf{x}(k + 1) \Delta t, (k + 1) \Delta t\} \Delta t \tag{9-22}$$

which, unfortunately, requires $\mathbf{x}[(k + 1) \Delta t]$ in order to compute $\mathbf{f}\{\mathbf{x}(k + 1) \Delta t, (k + 1) \Delta t\}$ which is needed to compute $\mathbf{x}[(k + 1) \Delta t]$. An iterative method of solving (9-22) is by modified Euler-algorithm, which is of a predictor–corrector type.

The numerical solution of the nonlinear equation

$$\mathbf{x}[(k + 1) \Delta t] = \mathbf{x}(k \Delta t) + \tfrac{1}{2}\mathbf{f}[\mathbf{x}(k \Delta t), k \Delta t] \Delta t$$

$$+ \tfrac{1}{2}\mathbf{f}\{\mathbf{x}(k + 1) \Delta t, (k + 1) \Delta t\} \Delta t \tag{9-23}$$

for $\mathbf{x}[(k + 1) \Delta t]$ given $\mathbf{x}(k \Delta t)$ can be viewed as a straightforward problem of the type discussed in Chapters 3 and 4. In the form of state equations where the state variables are successive derivatives of the function of interest, the first $n - 1$ algebraic equations of (9-23) are linear, whereas the nth equation is nonlinear. These can be solved in many ways, but for sufficiently small Δt, $\mathbf{x}[(k + 1) \Delta t]$ is close to the known vector $\mathbf{x}(\Delta t)$, so a gradient-like algorithm can be used.

Denoting a sequence of approximate numerical solutions to (9-23) by superscripts,

$$\mathbf{x}^0[(k + 1) \Delta t], \mathbf{x}^1[(k + 1) \Delta t], x^2[(k + 1) \Delta t], \ldots$$

and so on, the algorithm in Table 9-10 will give increasingly accurate results for $\mathbf{x}[(k + 1) \Delta t]$, given $\mathbf{x}(k \Delta t)$, for a sufficiently smooth function $\mathbf{f}$ and small step size Δt. The first trial value x^0, the prediction, uses the zeroth-order integral approximation. Then this prediction is corrected with a trapezoidal integral approximation that uses the predicted value of $\mathbf{x}[(k + 1) \Delta t]$. If necessary, the correction equation is repeated for improved accuracy, until the amount of the correction is less than some selected value ε.

For a *linear* differential equation, for which

$$\mathbf{x}[(k + 1) \Delta t] = \mathbf{x}(k \Delta t) + \int_{k\Delta t}^{(k+1)\Delta t} [A(t)\mathbf{x}(t) + \mathbf{b}g(t)] \, dt$$

TABLE 9-10 Trapezoidal Predictor–Corrector (or Modified Euler)
Algorithm Predictor

PREDICTOR

$$\mathbf{x}^0[(k + 1)\,\Delta t] = \mathbf{x}(k\,\Delta t) + \mathbf{f}[\mathbf{x}(k\,\Delta t), k\,\Delta t]\,\Delta t$$

CORRECTOR

$$\mathbf{x}^1[(k + 1)\,\Delta t] = \mathbf{x}(k\,\Delta t) + \tfrac{1}{2}\mathbf{f}[\mathbf{x}(k\,\Delta t), k\,\Delta t]\,\Delta t$$

$$+ \tfrac{1}{2}\mathbf{f}\{\mathbf{x}^0[(k + 1)\,\Delta t], (k + 1)\,\Delta t\}\,\Delta t$$

CORRECTOR ITERATION

$$\mathbf{x}^i[(k + 1)\,\Delta t] = \mathbf{x}(k\,\Delta t) + \tfrac{1}{2}\mathbf{f}[\mathbf{x}(k\,\Delta t), k\,\Delta t]\,\Delta t$$

$$+ \tfrac{1}{2}\mathbf{f}\{\mathbf{x}^{i-1}[(k + 1)\,\Delta t, (k + 1)\,\Delta t\}\,\Delta t \qquad i = 2, 3, \ldots$$

until

$$\|\mathbf{x}^i - \mathbf{x}^{i-1}\| < \varepsilon$$

---

the trapezoidal integral approximation is

$$\mathbf{x}[(k + 1)\,\Delta t] \simeq \mathbf{x}(k\,\Delta t) + \tfrac{1}{2}\mathbf{A}(k\,\Delta t)\mathbf{x}(k\,\Delta t)\,\Delta t + \tfrac{1}{2}\mathbf{A}[(k + 1)\,\Delta t]$$

$$\cdot\,\mathbf{x}[(k + 1)\,\Delta t]\,\Delta t + \tfrac{1}{2}\mathbf{b}g(k\,\Delta t)\,\Delta t + \tfrac{1}{2}\mathbf{b}g[(k + 1)\,\Delta t]\,\Delta t$$

or

$$\{\mathbf{I} - \tfrac{1}{2}\mathbf{A}[(k + 1)\,\Delta t]\,\Delta t\}\mathbf{x}[(k + 1)\,\Delta t] = [\mathbf{I} + \tfrac{1}{2}\mathbf{A}(k\,\Delta t)\,\Delta t]\mathbf{x}(k\,\Delta t)$$

$$+ \tfrac{1}{2}\mathbf{b}g(k\,\Delta t)\,\Delta t + \tfrac{1}{2}\mathbf{b}g[(k + 1)\,\Delta t]\,\Delta t$$

or

$$\mathbf{x}[(k + 1)\,\Delta t] = \{\mathbf{I} - \tfrac{1}{2}\mathbf{A}[(k + 1)\,\Delta t]\,\Delta t\}^{-1}\{[\mathbf{I} + \tfrac{1}{2}\mathbf{A}(k\,\Delta t)\,\Delta t]\mathbf{x}(k\,\Delta t)$$

$$+ \tfrac{1}{2}\mathbf{b}g(k\,\Delta t)\,\Delta t + \tfrac{1}{2}\mathbf{b}g[(k + 1)\,\Delta t]\,\Delta t\} \qquad (9\text{-}24)$$

The inverse in (9-24) exists for Δt sufficiently small. Because of the linearity of the differential equation, there is no need to iterate to find the trapezoidal integral approximation as it is when the differential equation is nonlinear.

Higher-order polynomial integral approximations can be done in much the same way, but for a recursive numerical differential equation solution, an integral approximation can only involve present and past values of the integrand, not future ones. Table 9-11 lists the first several such approximations. The result of a higher-order integral approximation is a vector nonlinear equation similar to (9-23) but involving more than a single past sample of $\mathbf{x}$, so that additional equations must be used to start the algorithm. The

TABLE 9-11   Some Integral Approximations Using Present and Past
Integral Samples

ZEROTH-ORDER (EULER)

$$\int_{k\Delta t}^{(k+1)\Delta t} f(t)\, dt \simeq f(k\, \Delta t)\, \Delta t$$

FIRST-ORDER (TRAPEZOIDAL)

$$\int_{k\Delta t}^{(k+1)\Delta t} f(t)\, dt \simeq \tfrac{1}{2} f[(k+1)\, \Delta t]\, \Delta t + \tfrac{1}{2} f(k\, \Delta t)\, \Delta t$$

SECOND-ORDER

$$\int_{k\Delta t}^{(k+1)\Delta t} f(t)\, dt \simeq \tfrac{5}{12} f[(k+1)\, \Delta t]\, \Delta t + \tfrac{8}{12} f(k\, \Delta t)\, \Delta t - \tfrac{1}{12} f[(k-1)\, \Delta t]\, \Delta t$$

solution of this nonlinear equation is found by an analogous predictor–corrector of comparable complexity.

Runge–Kutta methods are another kind of integral approximation for differential equation solution. They do not use iterative methods to solve the nonlinear equation (9-21) and they are self-starting, not requiring more than the immediate past value of **x**. However, several function evaluations per integration step are needed and it is difficult to estimate or control errors.

One should not be too hasty in abandoning the simple Euler approximation for a higher-order approximation. In modeling physical systems, poor accuracy of the Euler approximation with very small step size is often indicative of an underlying lack of physical robustness that probably ought to be carefully investigated. The differential equation solution routines in system analysis program packages usually use higher-order approximations simply because the program's uses are not known in advance. By doing so, an important check on the sensitivity of the solution to small errors can easily be lost.

9.8 NUMERICAL SOLUTION OF CONSTANT-COEFFICIENT EQUATIONS

The special propereties of *constant-coefficient* linear differential equations are now examined. The general integral solution of the first-order equation is developed, then the higher-order equation solution is found in terms of state equations and the matrix exponential function. It is shown that for distinct eigenvalues of the state coupling matrix **A**, a change of state variables can result in a similarity transformation of **A** that diagonalizes the new state coupling matrix. In terms of the new state variables, the first-order state equations are then decoupled from one another.

9.8.1 General Solution of the First-Order Equation

For any first-order linear, constant-coefficient differential equation

$$\frac{dy}{dt} + \alpha_0 y = g(t)$$

there is a general solution in terms of an integral. Multiplying each side of the equation by the *integrating factor* exp $(\alpha_0 t)$,

$$e^{\alpha_0 t} \frac{dy}{dt} + \alpha_0 e^{\alpha_0 t} y = e^{\alpha_0 t} g(t)$$

gives

$$\frac{d}{dt} [e^{\alpha_0 t} y(t)] = e^{\alpha_0 t} g(t)$$

so that

$$e^{\alpha_0 t} y(t) = \int e^{\alpha_0 t} g(t) \, dt + k$$

where k is an arbitrary constant, or

$$y(t) = e^{-\alpha_0 t} \left[\int e^{\alpha_0 t} g(t) \, dt + k \right]$$

In terms of the initial value of y,

$$y(t) = e^{-\alpha_0 t} [y(0) + \int_0^t e^{\alpha_0 t} g(t) \, dt] \qquad t \geq 0 \qquad (9\text{-}25)$$

using a dummy variable of integration τ and bringing the exponential in t under the integral sign gives

$$y(t) = e^{-\alpha_0 t} y(0) + \int_0^t e^{-\alpha_0(t-\tau)} g(\tau) \, d\tau$$

The integral involved is the *convolution* of the function exp $(-\alpha_0 t)$ with the driving function $g(t)$. For evenly spaced values of the variable $t = k \, \Delta t$, $k = 0, 1, 2, \ldots$, starting initially at $t = 0$,

$$y[(k+1) \, \Delta t] = e^{-\alpha_0[(k+1)\Delta t]} [y(0) + \int_0^{(k+1)\Delta t} e^{\alpha_0 \tau} g(\tau) \, d\tau]$$

$$= e^{-\alpha_0 \Delta t} \{ e^{-\alpha_0 k \Delta t} [y(0) + \int_0^{k\Delta t} e^{\alpha_0 \tau} g(\tau) \, d\tau]$$

$$+ \int_{k\Delta t}^{(k+1)\Delta t} e^{-\alpha_0(k\Delta t - \tau)} g(\tau) \, d\tau \} \qquad (9\text{-}26)$$

$$= e^{-\alpha_0 \Delta t} [y(k \, \Delta t) + \int_{k\Delta t}^{(k+1)\Delta t} e^{-\alpha_0(k\Delta t - \tau)} g(\tau) \, d\tau]$$

Equation (9-26) is a recursive formula for obtaining evenly spaced values of the *exact* solution of any first-order constant-coefficient equation. To use it, one must either use analytical methods to find the values of the integrals needed. Or, an approximate solution can be obtained by numerically approximating the integrals. Letting

$$\tau' = \tau - k \, \Delta t \qquad \tau = \tau' + k \, \Delta t$$

(9-26) becomes

$$y[(k + 1) \Delta t] = e^{-\alpha_0 \Delta t} [y(k \Delta t) + \int_0^{\Delta t} e^{\alpha_0 \tau'} g(k \Delta t + \tau') \, d\tau']$$

which has easier limits on the integral. Beginning with $y(0)$, successive evenly spaced values of y are found recursively from

$$y(\Delta t) = e^{-\alpha_0 \Delta t} [y(0) + \int_0^{\Delta t} e^{\alpha_0 \tau'} g(\tau') \, d\tau']$$

$$y(2\Delta t) = e^{-\alpha_0 \Delta t} [y(\Delta t) + \int_0^{\Delta t} e^{\alpha_0 \tau'} g(\Delta t + \tau') \, d\tau']$$

$$y(3\Delta t) = e^{-\alpha_0 \Delta t} [y(2\Delta t) + \int_0^{\Delta t} e^{\alpha_0 \tau'} g(2\Delta t + \tau') \, d\tau']$$

$$\vdots$$

and so on. If the integrals are replaced by zeroth-order Euler approximations, then approximate evenly spaced values of y are found recursively from

$$y(\Delta t) = e^{-\alpha_0 \Delta t} [y(0) + g(0) \, \Delta t]$$

$$y(2\Delta t) = e^{-\alpha_0 \Delta t} [y(\Delta t) + g(\Delta t) \, \Delta t]$$

$$y(3\Delta t) = e^{-\alpha_0 \Delta t} [y(2\Delta t) + g(2\Delta t) \, \Delta t]$$

$$\vdots$$

9.8.2 Higher-Order General Solutions

Finding general solutions to higher-order constant-coefficient linear differential equations is most conveniently done in terms of state variables. As we will see, the matrix exponential function $\exp(\mathbf{A} \Delta t)$ plays much the same role in the solution of higher-order equations as the scalar exponential $\exp(-\alpha_0 t)$ does in first-order equation solutions. For a constant-coefficient linear differential equation, the state variable representation,

$$\dot{\mathbf{x}}(t) = \mathbf{A}\mathbf{x}(t) + \mathbf{b}g(t)$$

$$y(t) = \mathbf{c}^\dagger \mathbf{x}(t)$$

has a constant state coupling matrix $\mathbf{A}$. Any constant nonsingular change in the state variables,

$$\mathbf{x}(t) = \mathbf{P}\mathbf{x}'(t) \qquad \mathbf{x}'(t) = \mathbf{P}^{-1}\mathbf{x}(t)$$

gives new equations in terms of $\mathbf{x}'(t)$,

$$\mathbf{P}\dot{\mathbf{x}}'(t) = \mathbf{A}\mathbf{P}\mathbf{x}'(t) + \mathbf{b}g(t) \qquad \dot{\mathbf{x}}'(t) = (\mathbf{P}^{-1}\mathbf{A}\mathbf{P})\mathbf{x}'(t) + (\mathbf{P}^{-1}\mathbf{b})g(t)$$

or

$$\dot{\mathbf{x}}'(t) = \mathbf{A}'\mathbf{x}'(t) + \mathbf{b}'g(t)$$

$$y(t) = \mathbf{c}'^\dagger \mathbf{x}'(t)$$

where

$$\mathbf{A}' = \mathbf{P}^{-1}\mathbf{A}\mathbf{P}$$

$$\mathbf{b}'(t) = \mathbf{P}^{-1}\mathbf{b}$$

$$\mathbf{c}'^{\dagger} = \mathbf{c}^{\dagger}\mathbf{P}$$

which are also of the state variable form. The new state coupling matrix $\mathbf{A}'$ is a similarity transformation of the old one.

If the original state coupling matrix $\mathbf{A}$ has distinct eigenvalues, a transformation of state variables $\mathbf{P}$ can be found such that $\mathbf{A}'$ is diagonal:

$$\mathbf{A}' = \mathbf{P}^{-1}\mathbf{A}\mathbf{P} = \Lambda$$

In terms of the new state variables, the state equations,

$$
\begin{bmatrix} \dot{\mathbf{x}}_1'(t) \\ \dot{\mathbf{x}}_2'(t) \\ \vdots \\ \dot{\mathbf{x}}_n'(t) \end{bmatrix}
=
\begin{bmatrix} \lambda_1 & 0 & 0 & \cdots & 0 \\ 0 & \lambda_2 & 0 & \cdots & 0 \\ \vdots & \vdots & & \vdots & \\ 0 & 0 & 0 & \cdots & \lambda_n \end{bmatrix}
\begin{bmatrix} x_1'(t) \\ x_2'(t) \\ \vdots \\ x_n'(t) \end{bmatrix}
+
\begin{bmatrix} b_1' \\ b_2' \\ \vdots \\ b_n' \end{bmatrix} g(t)
$$

are decoupled from one another. They consist of a set of n first-order linear constant-coefficient differential equations,

$$\dot{x}_1'(t) - \lambda_1 x_1'(t) = b_1' g(t)$$
$$\dot{x}_2'(t) - \lambda_2 x_2'(t) = b_2' g(t)$$
$$\vdots$$
$$\dot{x}_n'(t) - \lambda_n x_n'(t) = b_n' g(t)$$

The solution of each of these first-order equations is (9-25)

$$x_i'(t) = e^{\lambda_i t}[x_i'(0) + \int_0^t e^{-\lambda_i t} g(t)\, dt]$$

after $t = 0$. In matrix form,

$$\mathbf{x}'(t) = e^{\Lambda t}[\mathbf{x}'(0) + \int_0^t e^{-\Lambda t}\mathbf{b}' g(t)\, dt] \tag{9-27}$$

$$= e^{\Lambda t}\mathbf{x}'(0) + \int_0^t e^{\Lambda(t-\tau)}\mathbf{b}' g(\tau)\, d\tau \qquad t \geq 0$$

In general, the solution for the original state vector is

$$\mathbf{x}(t) = e^{\mathbf{A}t}[\mathbf{x}(0) + \int_0^t e^{-\mathbf{A}t}\mathbf{b} g(t)\, dt] \tag{9-28}$$

$$= e^{\mathbf{A}t}\mathbf{x}(0) + \int_0^t e^{\mathbf{A}(t-\tau)}\mathbf{b} g(\tau)\, d\tau \qquad t \geq 0$$

as can be verified by transformation of (9-27) back in terms of $\mathbf{x}$ in the case of distinct eigenvalues or, in general, by substitution:

$$\frac{d\mathbf{x}}{dt} = \mathbf{A}e^{\mathbf{A}t}[\mathbf{x}(0) + \int_0^t e^{-\mathbf{A}t}\mathbf{b}g(t)\,dt] + e^{\mathbf{A}t}e^{-\mathbf{A}t}\mathbf{b}g(t)$$

$$= \mathbf{A}\mathbf{x}(t) + \mathbf{b}g(t)$$

Recursive solutions of (9-28) are obtained by using

$$\mathbf{x}[(k+1)\,\Delta t] = e^{\mathbf{A}(k+1)\Delta t}\mathbf{x}(0) + \int_0^{(k+1)\Delta t} e^{\mathbf{A}[(k+1)\Delta t - \tau]}\mathbf{b}g(\tau)\,d\tau$$

$$= e^{\mathbf{A}\Delta t}[e^{\mathbf{A}k\Delta t}\mathbf{x}(0) + \int_0^{k\Delta t} e^{\mathbf{A}(k\Delta t - \tau)}\mathbf{b}g(\tau)\,d\tau]$$

$$+ \int_{k\Delta t}^{(k+1)\Delta t} e^{\mathbf{A}[(k+1)\Delta t - \tau]}\mathbf{b}g(\tau)\,d\tau$$

$$= e^{\mathbf{A}\Delta t}\mathbf{x}(k\,\Delta t) + \int_{k\Delta t}^{(k+1)\Delta t} e^{\mathbf{A}[(k+1)\Delta t - \tau]}\mathbf{b}g(\tau)\,d\tau$$

Letting $\tau' = \tau - k\,\Delta t$ in the integral,

$$\mathbf{x}[(k+1)\,\Delta t] = e^{\mathbf{A}\Delta t}\mathbf{x}(k\,\Delta t) + \int_0^{\Delta t} e^{\mathbf{A}(\Delta t - \tau')}\mathbf{b}g(k\,\Delta t + \tau')\,d\tau \tag{9-29}$$

$$= e^{\mathbf{A}\Delta t}[\mathbf{x}(k\,\Delta t) + \int_0^{\Delta t} e^{-\mathbf{A}\tau'}\mathbf{b}g(k\,\Delta t + \tau')\,d\tau']$$

When the matrix exponential exp $(\mathbf{A}\,\Delta t)$ and the integral are calculated exactly instead of approximated values of the exact solution, not an approximation to them, is obtained, although probably with some computational error. Expressing the matrix exponential in series form,

$$e^{\mathbf{A}\Delta t} = \mathbf{I} + \frac{\mathbf{A}\,\Delta t}{1!} + \frac{\mathbf{A}^2\,\Delta t^2}{2!} + \frac{\mathbf{A}^3\,\Delta t^3}{3!} + \cdots$$

it is seen how the exact solution,

$$\mathbf{x}[(k+1)\,\Delta t] = (\mathbf{I} + \mathbf{A}\,\Delta t + \tfrac{1}{2}\mathbf{A}^2\,\Delta t^2 + \tfrac{1}{6}\mathbf{A}^3\,\Delta t^3 + \cdots)\mathbf{x}(k\,\Delta t)$$

$$+ \int_0^{\Delta t} [\mathbf{I} + \mathbf{A}(\Delta t - \tau') + \tfrac{1}{2}\mathbf{A}^2(\Delta t - \tau')^2$$

$$+ \tfrac{1}{6}\mathbf{A}^3(\Delta t - \tau')^3 + \cdots]\mathbf{b}g(k\,\Delta t + \tau')\,d\tau'$$

compares with, say, the zeroth-degree Euler approximation:

$$\mathbf{x}[(k+1)\,\Delta t] \simeq (\mathbf{I} + \mathbf{A}\,\Delta t)\mathbf{x}(k\,\Delta t) + \mathbf{b}g(k\,\Delta t)\,\Delta t$$

The key point about the required integral in (9-29) is that it is an integration of a known function. It does not require knowledge of the state for its evaluation. Its approximation can involve values of t after $t = (k+1)\,\Delta t$ and can use a step size smaller than Δt.

9.8.3 Solution Program

A general program for the numerical solution of constant-coefficient differential equations
is listed in Table 9-12. The user enters the order of the equation, the step size to be used,
and the equation coefficients. The driving function is defined through subroutine 1000.
The first 10 terms of

$$e^{\mathbf{A}\Delta t} = \mathbf{I} + \frac{\mathbf{A}}{1!} + \frac{\mathbf{A}^2}{2!} + \frac{\mathbf{A}^3}{3!} + \cdots$$

are then computed, and

$$\mathbf{x}[(k + 1)\,\Delta t] = e^{\mathbf{A}\Delta t}\mathbf{x}(k\,\Delta t) + \int_0^{\Delta t} e^{\mathbf{A}(\Delta t - \tau')}\mathbf{b}g(k\,\Delta t + \tau')\,\Delta\tau'$$

$$y(k\,\Delta t) = \mathbf{c}^\dagger\mathbf{x}(k\,\Delta t)$$

are used to produce samples of the solution $y(t)$ spaced at Δt intervals, beginning with
the user-entered initial conditions.

Subroutine 1000 is to produce values of or an approximation to

$$h = \int_0^{\Delta t} e^{\mathbf{A}(\Delta t - \tau')}\mathbf{b}g(\tau' + k\,\Delta t)\,d\tau'$$

as a function of the current value of the variable $t = k\,\Delta t$. A simple approximation to
$\mathbf{h}$ is

$$\mathbf{h} \simeq \mathbf{b}g(k\,\Delta t)\,\Delta t$$

but the user may wish to program a better integral approximation or the correct value of
the integral obtained from analytical considerations.

The previous plots of the solution of the translation mechanical system in Figure
9-4 were produced by this program. Closely spaced samples of each of the state variables
representing the mass positions are connected by straight-line segments to create the
illusion of continuous plots. In subroutine 1000,

$$\mathbf{h} = 6 \int_{k\,\Delta t}^{(k+1)\,\Delta t} e^{-\mathbf{A}\tau}\mathbf{b} \sin \tau\, d\tau$$

$$= 6 \int_{k\,\Delta t}^{(k+1)\,\Delta t} (\mathbf{I} - \mathbf{A}\tau + \tfrac{1}{2}\mathbf{A}^2\tau^2 - \tfrac{1}{6}\mathbf{A}^3\tau^3 + \cdots) \begin{bmatrix} 0 \\ 0 \\ \vdots \\ 0 \\ 1 \end{bmatrix} \sin \tau\, d\tau$$

is approximated as

$$\mathbf{h} = \begin{bmatrix} 0 \\ 0 \\ \vdots \\ 6\,\Delta t \sin (k\,\Delta t) \end{bmatrix}$$

because the sampling interval Δt is very small.

TABLE 9-12

Computer Program in BASIC for Constant-Coefficient Differential Equation Solution (CONSTANT)

VARIABLES USED

$$N = \text{order of equation}$$
$$A(20,20) = \text{state coupling matrix } \mathbf{A}$$
$$D = \text{step size}$$
$$E(20,20) = \text{10-term approximation to exp } (\mathbf{A} \, \Delta t)$$
$$Q(20,20) = \text{intermediate matrix, used to store matrix products in the exp } (\mathbf{A} \, \Delta t) \text{ calculation}$$
$$R(20,20) = \text{intermediate matrix, used to accumulate powers of } \mathbf{A} \text{ in the exp } (\mathbf{A} \, \Delta t) \text{ calculation}$$
$$T = \text{the variable}$$
$$H(20) = \text{driving function integral vector}$$
$$X(20) = \text{present state vector}$$
$$Z(20) = \text{new state vector}$$
$$I = \text{vector element index; matrix row index}$$
$$J = \text{matrix column index}$$
$$K = \text{matrix element summation index}$$
$$L = \text{exponential series term index}$$

LISTING

```
100       PRINT "CONSTANT-COEFFICIENT D.E. SOLUTION"
110       DIM A(20,20),E(20,20),Q(20,20),R(20,20)
120       DIM H(20),X(20),Z(20)
130    REM  ENTER ORDER, COEFFICIENTS, AND STEP SIZE
140       PRINT "PLEASE ENTER EQUATION ORDER";
150       INPUT N
160       PRINT "ENTER EQUATION COEFFICIENTS"
170       FOR I=1 TO N
180       PRINT "COEFFICIENT OF ";N-I;"DERIVATIVE = ";
190       INPUT A(N,N - I + 1)
200       A(N,N - I + 1)=-A(N,N - I + 1)
210       NEXT I
220       PRINT "ENTER STEP SIZE"
230       INPUT D
240    REM   COMPUTE E=EXP(AD)
250    REM   FORM THE A MATRIX
260       FOR I = 1 TO N - 1
270       FOR J= 1 TO N
280       A(I,J) = 0
290       NEXT J
300       NEXT I
310       FOR I = 1 TO N - 1
320       A(I,I + 1) = 1
330       NEXT I
340    REM   SET E=I AND Q=I
350       FOR I = 1 TO N
360       FOR J = 1 TO N
370       E(I,J) = 0
380       Q(I,J) = 0
390       NEXT J
400       NEXT I
```

(continued)

```
410        FOR I = 1 TO N
420        E(I,I) = 1
430        Q(I,I) = 1
440        NEXT I
450   REM    LOOP TO ADD SERIES TERMS
460        FOR L = 1 TO 9
470   REM    SET R=AD TIMES Q
480        FOR I = 1 TO N
490        FOR J = 1 TO N
500        R(I,J) = 0
510        FOR K = 1 TO N
520        R(I,J) = R(I,J) + D * A(I,K) * Q(K,J)
530        NEXT K
540        NEXT J
550        NEXT I
560   REM    SET Q=(1/L) TIMES R AND ADD Q TO E
570        FOR I = 1 TO N
580        FOR J = 1 TO N
590        Q(I,J) = (1/L) * R(I,J)
600        E(I,J) = E(I,J) + Q(I,J)
610        NEXT J
620        NEXT I
630        NEXT L
640   REM    ENTER INITIAL CONDITIONS
650        PRINT "ENTER INITIAL CONDITIONS"
660        FOR I = 1 TO N
670        PRINT "INITIAL ";I - 1;" DERIVATIVE = ";
680        INPUT X(I)
690        NEXT I
700   REM    INITIALIZE VARIABLE AND PRINT HEADING
710        T = 0
720        PRINT "VARIABLE","SOLUTION"
730   REM    PRINT VARIABLE AND SOLUTION
740        FOR K = 1 TO 1000
750        PRINT T,X(1)
760   REM    UPDATE STATE AND TIME
770        GOSUB 1000
780        FOR I = 1 TO N
790        Z(I) =   H(J)
800        FOR J = 1 TO N
810        Z(I) = Z(I) + E(I,J) * X(J)
820        NEXT J
830        NEXT I
840        Z(N) = Z(N) + H
850        FOR I = 1 TO N
860        X(I) = Z(I)
870        NEXT I
880        T = T + D
890        NEXT K
900        END
1000  REM    SUBROUTINE TO COMPUTE DRIVING FUNCTION
1010       FOR I = 1 TO N - 1
1020       H(I) = 0
1030       H(N) = 6 * D * SIN (T)
1035       NEXT I
1040       RETURN
```

Computer Program in FORTRAN for Constant-Coefficient Differential Equation Solution (CONSTANT)

```
C ****************************************************************
C PROGRAM CONSTANT -- CONSTANT-COEFFICIENT D.E. SOLUTION
C ****************************************************************
      REAL A(20,20), E(20,20), Q(20,20), R(20,20),
     *     H(20),    X(20),     Z(20)
```

```
            WRITE(*,'(1X,''CONSTANT-COEFFICIENT D.E. SOLUTION'')')

C *** ENTER ORDER, COEFFICIENTS, AND STEP SIZE
            WRITE(*,'(1X,''PLEASE ENTER EQUATION ORDER '',\)')
            READ(*,*) N
            WRITE(*,'(1X,''ENTER EQUATION COEFFICIENTS'')')
            DO 10 I=1, N
                WRITE(*,'(1X,''COEFFICIENT OF'',I3,'' DERIVATIVE = '',\)') N-I
                READ(*,*) A(N, N-I+1)
                A(N, N-I+1) = - A(N, N-I+1)
   10       CONTINUE
            WRITE(*,'(1X,''ENTER STEP SIZE '',\)')
            READ(*,*) D

C *** COMPUTE E=EXP(AD)
C *** FORM THE A MATRIX
            DO 20 I=1, N-1
                DO 30 J=1, N
                    A(I,J) = 0.0
   30           CONTINUE
   20       CONTINUE

            DO 40 I=1, N-1
                A(I,I+1) = 1.0
   40       CONTINUE

C *** SET E=I AND Q=I
            DO 50 I=1, N
                DO 60 J=1, N
                    E(I,J) = 0.0
                    Q(I,J) = 0.0
   60           CONTINUE
   50       CONTINUE

            DO 70 I=1, N
                E(I,I) = 1.0
                Q(I,I) = 1.0
   70       CONTINUE

C *** LOOP TO ADD SERIES TERMS
            DO 80 L=1, 9

C *** SET R=AD TIMES Q
            DO 90 I=1, N
                DO 100 J=1, N
                    R(I,J) = 0.0
                    DO 110 K=1, N
                        R(I,J) = R(I,J) + D * A(I,K) * Q(K,J)
  110               CONTINUE
  100           CONTINUE
   90       CONTINUE

C *** SET Q=(1/L) TIMES R AND ADD Q TO E
            DO 120 I=1, N
                DO 130 J=1, N
                    Q(I,J) = (1/L) * R(I,J)
                    E(I,J) = E(I,J) + Q(I,J)
  130           CONTINUE
  120       CONTINUE
   80       CONTINUE

C *** ENTER INITIAL CONDITIONS
            WRITE(*,'(1X,''ENTER INITIAL CONDITIONS'')')
```

(continued)

```
      DO 140 I=1, N
         WRITE(*,'(1X,''INITIAL '',I3,'' DERIVATIVE = '',\)') I-1
         READ(*,*) X(I)
  140 CONTINUE

C *** INITIALIZE VARIABLE AND PRINT HEADING
      T = 0.0
      WRITE(*,'(1X,''  VARIABLE       SOLUTION'')')

C *** PRINT VARIABLE AND SOLUTION
      DO 180 K=1, 1000
         WRITE(*,'(1X,2E15.8)') T, X(1)

C *** UPDATE STATE AND TIME
         CALL FUNCT(N,T,H)

         DO 150 I=1, N
            Z(I) = H(I)
            DO 160 J=1, N
               Z(I) = Z(I) + E(I,J) * X(J)
  160       CONTINUE
  150    CONTINUE

         Z(N) = Z(N) + H(N)

         DO 170 I=1, N
            X(I) = Z(I)
  170    CONTINUE

         T = T + D

  180 CONTINUE

      STOP
      END

C ********************************************************************
      SUBROUTINE FUNCT(N,T,H)
C ********************************************************************
      REAL H(20), T
      INTEGER I

C *** SUBROUTINE TO COMPUTE DRIVING FUNCTION TERM
      DO 10 I=1, N-1
         H(I) = 0.0
         H(N) = 6.0 * D *  SIN(T)
   10 CONTINUE
      RETURN
      END
```

9.9 BOUNDARY VALUE PROBLEMS

If all the boundary conditions on a differential equation involve the same value of the variable, one has an initial value problem such as the problems discussed previously. When the boundary conditions involve more than one value of the variable, the situation is a *boundary value problem*. We will here emphasize *two-point* boundary value problems, where only two different values of the variable are involved in the boundary conditions because of their practical importance and to best convey the basic ideas involved.

9.9.1 Equivalent Simultaneous Algebraic Equations

In principle, it is straightforward to solve boundary value problems when the differential equation is linear and constant-coefficient because the general solution to the differential equation can be obtained for most driving functions of interest. The boundary conditions then give simultaneous linear algebraic equations where the arbitrary constants are the variables. For example, the general solution of

$$\frac{d^2y}{dt^2} + 4\frac{dy}{dt} + 3y = 18$$

is

$$y(t) = K_1 e^{-t} + K_2 e^{-3t} + 6$$

If the boundary conditions are

$$y(0) = 0 \qquad y(2) = -3$$

then the arbitrary constants K_1 and K_2 must satisfy

$$K_1 + \quad K_2 + 6 = \quad 0$$
$$e^{-2}K_1 + e^{-6}K_2 + 6 = -3$$

which has solution $K_1 = -67.63$, $K_2 = 61.63$.

For linear differential equations with coefficients that are not constant and for nonlinear differential equations, the general solution of the equation is often not known. Even for constant-coefficient equations, it may be altogether too complicated to find the general solution. Then a numerical procedure for finding the solution that meets the boundary conditions is needed.

Recall, it is knowing the initial conditions that allow recursive numerical differential equation solution. We begin with the initial state vector and integrate forward in time to find the solution at later times. If desired, we could also integrate backward in time to find the solution before the initial time. What we are really doing, however, is solving a huge set of simultaneous equations. Those equations are all linear if the differential equation is linear; some of them are nonlinear if the differential equation is nonlinear. When this can be done recursively, it means that the first several equations can be solved, by themselves. Then this solution can be used to solve the next several equations, that solution to solve the next several equations, and so on.

For example, consider the approximate numerical solution of

$$\frac{d^2y}{dt^2} + 2\frac{dy}{dt} + 5y = 7$$

with the initial conditions

$$y(0) = 3 \qquad \dot{y}(0) = 4$$

Letting

$$x_1(t) = y(t)$$
$$x_2(t) = \dot{x}_1(t) = \dot{y}(t)$$

state equations for this problem are

$$\begin{bmatrix} \dot{x}_1(t) \\ \dot{x}_2(t) \end{bmatrix} = \begin{bmatrix} 0 & 1 \\ -5 & -2 \end{bmatrix} \begin{bmatrix} x_1(t) \\ x_2(t) \end{bmatrix} + \begin{bmatrix} 0 \\ 1 \end{bmatrix} \quad (7) = \mathbf{A}\mathbf{x}(t) + \mathbf{b}(7)$$

$$y(t) = \begin{bmatrix} 1 & 0 \end{bmatrix} \begin{bmatrix} x_1(t) \\ x_2(t) \end{bmatrix} = \mathbf{c}^\dagger \mathbf{x}(t)$$

with

$$x_1(0) = 3 \quad \text{and} \quad x_2(0) = 4$$

If these equations are approximated numerically with an Euler integral approximation with step size 0.01 then, approximately,

$$\mathbf{x}[(k + 1)\,\Delta t] = (\mathbf{I} + \mathbf{A}\,\Delta t)\mathbf{x}(k\,\Delta t) + 7\Delta t\,\mathbf{b}$$

or

$$\begin{bmatrix} x_1[0.01(k + 1)] \\ x_2[0.01(k + 1)] \end{bmatrix} = \begin{bmatrix} 1.0 & 0.01 \\ -0.05 & 0.98 \end{bmatrix} \begin{bmatrix} x_1(0.01k) \\ x_2(0.01k) \end{bmatrix} + \begin{bmatrix} 0 \\ 0.07 \end{bmatrix}$$

where for $k = 0, 1, 2, \ldots,$

$$y(0.01k) = x_1(0.01k)$$

The equations to be solved are then

$$
\begin{aligned}
x_1(0) &= 3 \\
x_2(0) &= 4 \\
-x_1(0) - 0.01x_2(0) + x_1(0.01) &= 0 \\
0.05x_1(0) - 0.98x_2(0) + x_2(0.01) &= 0.07 \\
-x_1(0.01) - 0.01x_2(0.01) + x_1(0.02) &= 0 \\
0.05x_1(0.01) - 0.98x_2(0.01) + x_2(0.02) &= 0.07
\end{aligned}
\quad (9\text{-}30)
$$

and so on. These can be solved recursively, two at a time. Knowing the values of $x_1(0)$ and $x_2(0)$ from the first two equations, $x_1(0.01)$ and $x_2(0.01)$ can be found from the third and fourth equations. Knowing $x_1(0.01)$ and $x_2(0.01)$, then $x_1(0.02)$ and $x_2(0.02)$ can be found from the fifth and sixth equations, and so on.

But now, suppose that the boundary conditions on this equation are not all at the same value of t but are

$$y(0) = x_1(0) = 3 \quad \text{and} \quad y(1) = x_1(1) = -2$$

instead. The equations would be those of (9-30) without the second equation. If the unknowns are ordered in time, eventually there will be equations

$$\vdots$$

$$
\begin{aligned}
-x_1(0.98) - 0.01x_2(0.98) + x_1(0.99) &= 0 \\
0.05x_1(0.98) - 0.98x_2(0.98) + x_2(0.99) &= 0.07 \\
-x_1(0.99) - 0.01x_2(0.99) + x_1(1) &= 0 \\
x_1(1) &= -2
\end{aligned}
\quad (9\text{-}31)
$$

It is only now, after 201 equations in the unknowns $x_1(0)$, $x_1(0.01)$, $x_1(0.02)$, . . ., $x_1(1)$ and $x_2(0)$, $x_2(0.01)$, $x_2(0.02)$, . . ., $x_2(0.99)$, that there are as many equations as unknowns. These equations can be solved for all the variables and thus for the approximate differential equation solution between $t = 0$ and $t = 1$; it is just that they are not in a form where the solution is *recursive* until after $t = 1$.

A higher-degree integration approximation makes the corresponding linear algebraic equations more complicated, with fewer zero coefficients, but they have the same character. If the differential equation is linear but not constant-coefficient, the same simultaneous equation coefficients do not repeat as they do in this example. For nonlinear differential equations, the corresponding set of simultaneous equations contains nonlinear equations.

9.9.2 Solution for the Case of a Linear Differential Equation

When the differential equation for a two-point boundary value problem is linear, the underlying larger set of simultaneous equations such as (9-30) through (9-31) are linear. Perhaps the easiest solution method uses the fact that the solution of a set of simultaneous linear algebraic equations is a linear combination of the knowns. The solution of the simultaneous equations for the unknown arbitrary constants can thus be expressed as a linear combination of the initial conditions and the driving function.

Solving the equations recursively to the second boundary with the known initial conditions and driving function terms, but zero for the unknown initial conditions calculates the effects of these knowns on the conditions at the second boundary. Then, solving the equations recursively again with these terms all replaced by zero and a unit value for one of the unknown initial conditions will calculate the effect of that initial condition on the solution at the second boundary. Repeating, with only a unit value for each of the unknown initial conditions in turn shows how each affects the conditions at the second boundary. It is then a simple matter to solve for the unknown initial conditions that result in the correct boundary conditions.

9.9.3 Nonlinear Differential Equations

When the differential equation in a two-point boundary value problem is nonlinear, the conditions at the second boundary can be considered to be nonlinear algebraic functions of those of the initial conditions that are unknown. The fundamental problem is then one of finding roots of nonlinear algebraic equations, although there is the additional complication that each nonlinear function evaluation requires a numerical differential equation solution.

For example, if the boundary conditions on the Van der Pol equation, with a nonzero driving function,

$$\frac{d^2y}{dt^2} + (y^2 - 1)\frac{dy}{dt} + y = 4$$

which has state variable equations

$$\begin{bmatrix} \dfrac{dx_1}{dt} \\ \dfrac{dx_2}{dt} \end{bmatrix} = \begin{bmatrix} 0 & 1 \\ -1 & 1-x_1^2 \end{bmatrix} \begin{bmatrix} x_1(t) \\ x_2(t) \end{bmatrix} + \begin{bmatrix} 0 \\ 1 \end{bmatrix} \quad (4)$$

$$y(t) = \begin{bmatrix} 1 & 0 \end{bmatrix} \begin{bmatrix} x_1(t) \\ x_2(t) \end{bmatrix}$$

are

$$y(0) = x_1(0) = 3 \quad \text{and} \quad y(1) = x_1(1) = 1$$

Running the NONLINEA program for this equation with the nonzero driving function, the known initial condition, and various initial values of $\dot{y}(0) = x_2(0)$ gives the $y(1)$ versus $\dot{y}(0)$ curve of Figure 9-11. A bisection zero crossing search of $[y(1) - 1]$ as a function of $\dot{y}(0)$ shows that the solution for which

$$\dot{y}(0) = x_2(0) = -9.04$$

very nearly meets the boundary condition at $t = 1$. There might, of course, be multiple solutions.

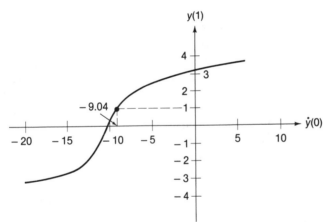

Figure 9-11   Conditions at $t = 1$ for a nonlinear differential equation with various initial conditions.

There you have it, dear reader—Nine chapters of analytical, numerical, and computational methods beginning with a review and overview of the BASIC and FORTRAN77 Languages and ending with differential equations and their solutions. We sincerely hope that you have enjoyed the tour as much as we have and that you will find this material highly useful now and in the future.

PROBLEMS

9-1. Use the classical methods of Section 9.2 to find the solutions of the following differential equations with the boundary conditions indicated.

(a) $\dfrac{d^2y}{dt^2} + 3\dfrac{dy}{dt} + y = 10; \quad y(0) = 2; \quad \dot{y}(0) = -4$

(b) $3\dfrac{d^3y}{dt^3} + 4\dfrac{d^2y}{dt^2} + \dfrac{dy}{dt} = 2t; \quad y(0) = 0; \quad \dot{y}(0) = 0; \quad \ddot{y}(0) = 0$

(c) $\dfrac{d^2y}{dt^2} + 5\dfrac{dy}{dt} + 6y = 3e^{-t}; \quad y(0) = 4; \quad \dot{y}(1) = -2$

(d) $\dfrac{dy}{dy} + 3y = 4\sin 2t; \quad y(0) = 6$

9-2. For the rotational mechanical system of Section 9.2.4, let the gear ratio be 1:3 instead of 2:3. With all other parameters and conditions the same, find $\theta_1(t)$.

9-3. Find a differential equation describing the angle θ in the rotational mechanical system of Figure 9-12, then solve it for $\theta(t)$, $t \geq 0$ when $\theta(0) = 10$ and $\dot{\theta}(0) = -2$. The spring exerts no force when $\theta = 0$. Sketch the solution, noting that an interval of 2π radians in θ is one complete revolution.

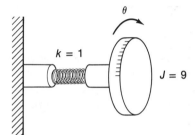

Figure 9-12

9-4. Use Laplace transform methods to find the solutions of the following differential equations with the initial conditions indicated. Use computer aid where convenient.

(a) $\dfrac{d^2y}{dt^2} + 3\dfrac{dy}{dt} + 2y = 4e^{-5t}; \quad y(0) = -1; \quad \dot{y}(0) = 0$

(b) $2\dfrac{d^3y}{dt^3} + \dfrac{d^2y}{dt^2} - 3\dfrac{dy}{dt} = 5; \quad y(0) = 0; \quad \dot{y}(0) = 0; \quad \ddot{y}(0) = 0$

(c) $\dfrac{dx_1}{dt} = -4x_1 + x_2$

$\dfrac{dx_2}{dt} = -x_1 + 5t$

$y(t) = 3x_1(t) - x_2(t)$

$x_1(0) = -4; \quad x_2(0) = -5.$

9-5. In using Laplace transform methods to solve boundary value problems with linear, constant-coefficient differential equations, those initial conditions that the Laplace transform requires but are not known (boundary conditions at some other value of the variable are known instead) are unknown arbitrary constants and can be treated as such. Use Laplace transform methods to solve

$$\dfrac{d^2y}{dt^2} + 7\dfrac{dy}{dt} + 12y = 4$$

with the boundary conditions

$$\dot{y}(0) = 0 \qquad \text{and} \qquad y(1) = -1$$

9-6. One way to find the coefficients of a partial fraction expansion is the method of evaluation. For an expansion of the form

$$F(s) = \frac{4s^2 - 2s + 5}{(s + 1)(s + 2)(s - 3)} = \frac{K_1}{s + 1} + \frac{K_2}{s + 2} + \frac{K_3}{s - 3}$$

then

$$K_1 = [F(s)(s + 1)]_{s = -1} = \frac{4s^2 - 2s + 5}{(s + 2)(s - 3)}\bigg|_{s = -1} = -\frac{11}{4}$$

$$K_2 = [F(s)(s + 2)]_{s = -2} = \frac{4s^2 - 2s + 5}{(s + 1)(s - 3)}\bigg|_{s = -2} = 5$$

$$K_3 = [F(s)(s - 3)]_{s = 3} = \frac{4s^2 - 2s + 5}{(s + 2)(s + 1)}\bigg|_{s = 3} = \frac{35}{20} = \frac{7}{4}$$

(a) Use evaluation to find K_1 and K_2 in

$$F(s) = \frac{-4s + 5}{s^2 + 3s + 2} = \frac{K_1}{s + 1} + \frac{K_2}{s + 2}$$

(b) Use evaluation to show that the coefficients of complex conjugate partial fraction terms are always complex conjugates.

9-7. In the translational mechanical system of Section 9.3.4, let the spring constant K_2 be, instead, $K_2 = 6$. Find the Laplace transforms $X_1(s)$ and $X_2(s)$.

9-8. Find a differential equation describing the distance x in the translational mechanical system of Figure 9-13. Then use Laplace transform methods to solve for $x(t)$, $t \geq 0$ when $x(0) = 10$ and $\dot{x}(0) = 0$

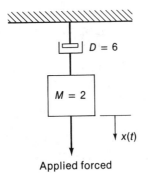

Applied forced
$f(t) = 3 \sin 2t$ **Figure 9-13**

9-9. Write, test, and debug a computer program to compute accurately the following integral using zeroth-order (Euler) approximation:

$$I = \int_2^5 \frac{1}{2 + \sin 5t} \, dt$$

9-10. Write, test, and debug a computer program to compute accurately the following integral using first-order (trapezoidal) approximation:

$$I = \int_{-0.9}^{0.9} \frac{\sin 3t}{\ln |t|} \, dt$$

9-11. Write, test, debug, and run a computer program to compute accurately the following integral using second-order (Simpson's one-third rule) polynomial approximation:

$$I = \int_0^\tau e^{-0.2t} \cos t\ 3t\ dt$$

for $\tau = 0.1, 0.2, 0.3, 0.4$, and 0.5.

9-12. By varying the step size as appropriate, use a computer program to compute

$$I = \int_1^{10} \frac{1}{t}\ dt$$

Carefully explain how you know that the computed value is accurate.

9-13. The flower-shaped curve of Figure 9-14 is described by the parametric equations

$$r(t) = 1 + \cos 3t$$
$$\theta(t) = t$$

in polar coordinates for $0 \le t < 2\pi$. An element of arc length along the curve is given by

$$dl = \sqrt{dr^2 + (r\ d\theta)^2} = \sqrt{9 \sin^2 3t + (1 + \cos 3t)^2}$$

so that the total arc length is given by

$$L = \int_0^{2\pi} \sqrt{9 \sin^2 3t + (1 + \cos 3t)^2}\ dt$$

Use a computer program to find L.

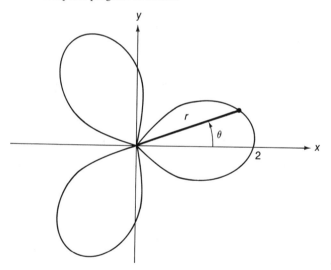

Figure 9-14

9-14. Find state variable equivalents of the following differential equations. For the boundary conditions indicated, what is the initial state vector?

(a) $\dfrac{d^2y}{dt^2} + y\dfrac{dy}{dt} + \left(\dfrac{dy}{dt}\right)^2 = 9e^{-3t}; \quad y(0) = 2; \quad \dot{y}(0) = -5$

(b) $\dfrac{d^3y}{dt^3} + \sin t\dfrac{d^2y}{dt^2} - 2\dfrac{d^2y}{dt^2} = 3t^2; \quad y(0) = 1; \quad \dot{y}(0) = 0; \quad \ddot{y}(0) = 4$

(c) $\dfrac{d^4y}{dt^4} + 5\dfrac{d^3y}{dt^3} - 4\dfrac{dy}{dt} + 2y = 10; \quad y(0) = 0; \quad \dot{y}(0) = -3;$

$$\ddot{y}(0) = 10; \quad \dddot{y}(0) = -6$$

9-15. Find recursive approximation equations using the zeroth-order Euler integral approximation, beginning with a known $\mathbf{x}(0)$ and in terms of Δt, for the following state variable equations:

(a) $\dfrac{dx_1}{dt} = x_2$

$\dfrac{dx_2}{dt} = x_2 \sin x_1 - x_2^2 + 10$

$y(t) = x_1(t)$

(b)
$$
\begin{bmatrix} \dfrac{dx_1}{dt} \\[2mm] \dfrac{dx_2}{dt} \\[2mm] \dfrac{dx_3}{dt} \end{bmatrix}
=
\begin{bmatrix} 0 & 1 & 0 \\ 0 & 0 & 1 \\ \cos t & \dfrac{1}{t^2+1} & -2 \end{bmatrix}
\begin{bmatrix} x_1(t) \\ x_2(t) \\ x_3(t) \end{bmatrix}
+
\begin{bmatrix} 0 \\ 0 \\ 1 \end{bmatrix} \sin t
$$

$$
y(t) = \begin{bmatrix} 1 & 0 & 0 \end{bmatrix} \begin{bmatrix} x_1(t) \\ x_2(t) \\ x_3(t) \end{bmatrix}
$$

(c) $\dfrac{d\mathbf{x}}{dt} = \begin{bmatrix} 0 & 1 & 0 \\ 0 & 0 & 1 \\ 3 & -2 & 1 \end{bmatrix} \mathbf{x} + \begin{bmatrix} 0 \\ 0 \\ 1 \end{bmatrix} 6e^{-4t}$

$y(t) = \begin{bmatrix} 1 & 0 & 0 \end{bmatrix} \mathbf{x}(t)$

9-16. Use the NONLINEA program to obtain numerical solutions to the following differential equations with the initial conditions indicated. Sketch or plot the solutions obtained.

(a) $\dfrac{d^3y}{dt^3} + y\dfrac{dy}{dt} = e^{-t}\cos 4t; \quad y(0) = 2; \quad \dot{y}(0) = -3; \quad \ddot{y}(0) = 0$

(b) $\dfrac{d^2y}{dt^2} + 3\dfrac{dy}{dt} + 2y^2 = 4\sin 5t; \quad y(0) = 6; \quad \dot{y}(0) = -7$

9-17. When the mass of the rod is insignificant, the motion of the pendulum of Figure 9-15 is governed by the differential equation

$$\dfrac{d^2\theta}{dt^2} + \dfrac{g}{L}\sin\theta = 0$$

where $g = 9.8$ m/s^2 is the acceleration of gravity and L is the distance from the pivot to the center of mass M. Use a computer program to find the period of oscillation of the pendulum for the following conditions.

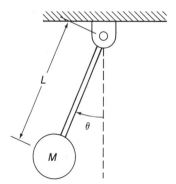

Figure 9-15

(a) $L = 0.3$ m, $\theta(0) = \pi/2$, $\dot{\theta}(0) = 0$

(b) $L = 0.2$ m, $\theta(0) = \pi/3$, $\dot{\theta}(0) = 0.1$ rad/s

9-18. Use the LINEARDE program to obtain numerical solutions to the following differential equations with the indicated initial conditions. Sketch or plot the solutions obtained.

(a) $\dfrac{d^2y}{dt^2} + t\dfrac{dy}{dt} + \dfrac{1}{1+t}y = \cos t$; $y(0) = 5$; $\dot{y}(0) = 0$

(b) $\dfrac{d^3y}{dt^3} + 2\dfrac{d^2y}{dt^2} + \sin t\dfrac{dy}{dt} = 0$; $y(0) = 0$; $\dot{y}(0) = 0$; $\ddot{y}(0) = 1$

9-19. Use the LINEARDE program to find an accurate numerical solution to

$$\frac{d^2y}{dt^2} + e^{-t}\frac{dy}{dt} + 4y = 5$$

for $t \geq 2$ when

$$y(2) = 3 \qquad \text{and} \qquad \dot{y}(2) = -2$$

9-20. Use the LINEARDE program to locate accurately the first three zeros of the following Bessel functions, for $m = 1, 2, \ldots, J_m(0) = 0$ and $\dot{J}_m(0) > 0$.

(a) $J_0(t)$; these are approximately 2.405, 5.520, and 8.654

(b) $J_1(t)$; these are zero and approximately 3.832 and 7.016

(c) $J_2(t)$

9-21. Use the integrating factor method to find the general solution to

$$\frac{dy}{dt} + \frac{1}{t}y = 3$$

9-22. Use the CONSTANT program to obtain numerical solutions to the following differential equations with zero initial conditions. Sketch or plot the solutions obtained.

(a) $\dfrac{d^3y}{dt^3} + \dfrac{d^2y}{dt^2} + 3.7\dfrac{dy}{dt} + 2.3y = 6.4t\sin 5.1t$

(b) $\dfrac{d^2z_1}{dt^2} + 2z_1 - 3z_2 = 10$

$\dfrac{d^2z_2}{dt^2} - 2z_1 + \dfrac{dz_2}{dt} + 4z_2 = 0$

$y(t) = z_1(t) - z_2(t)$

9-23. Use the CONSTANT program to find an accurate numerical approximation to the integral

$$I = \int_0^{10} \frac{\ln(1 + \cos t)}{t^2 + 3t + 4}\,dt$$

9-24. Replace the zeroth-order integral approximation in the CONSTANT in Table 9.12 with a first order trapezoidal one, as discussed in section 9.4. Then compare results of the two approximations for different step sizes with the differential equation

$$\frac{d^2y}{dt^2} + 4\frac{dy}{dt} = 5e^{-t}$$

9-25. Use the trapezoidal integral approximation described in Section 9.7 to obtain a numerical solution of the linear differential equation

$$\frac{d^2y}{dt^2} + 2\frac{dy}{dt} + 10e^{-0.2t}y = 0$$

with $y(0) = 10$ and $\dot{y}(0) = 0$.

9-26. Solve the following linear two-point boundary value problems by the method of Section 9.9.1:

(a) $\dfrac{d^2y}{dt^2} + 3\dfrac{dy}{dt} + 2y = 5; \quad \dot{y}(0) = 4; \quad y(4) = 0$

(b) $\dfrac{d^3y}{dt^3} + 5\dfrac{d^2y}{dt^2} + 6\dfrac{dy}{dt} = e^{4t}; \quad y(0) = 0; \quad \dot{y}(0) = 0; \quad \dot{y}(3) = 7$

9-27. Find numerical solutions to the following differential equations with the indicated initial conditions:

(a) $y\dfrac{d^2y}{dt^2} + y^2\dfrac{dy}{dt} + 4y = 0; \quad y(0) = 3; \quad \dot{y}(0) = 2$

(b) $3\dfrac{d^3y}{dt^3} - 2\dfrac{d^2y}{dt^2} + \dfrac{dy}{dt} + 4y = 10; \quad y(1) = 0; \quad \dot{y}(1) = 3; \quad \ddot{y}(1) = 0$

(c) $\dfrac{dy}{dt} + (\cos t)y = \sin 3t; \quad \dot{y}(-1) = 2$

9-28. Solve the following nonlinear two-point boundary-value problems.

(a) $\dfrac{d^2y}{dt^2} + (y^2 - 1)\dfrac{dy}{dt} + y = \sin t; \quad y(0) = -4; \quad y(2) = 3$

(b) $\left(\dfrac{dy}{dt}\right)^2 + y\dfrac{dy}{dt} = 2; \quad \dot{y}(0) = 0; \quad y(3) = 4$

ANSWERS TO SELECTED PROBLEMS

Chapter 2

2-1 (a) $5.83e^{i1.03}$;   (c) $6.55e^{i1.90}$

2-2 (a) $-1.25 + i2.73$;   (d) $1.70 - i2.64$

2-4 (b) $m = -1.168 + \underline{h}0.331$

2-8 One possibility is

$$\begin{cases} 2x_1 + 3x_2 + 4x_3 = 5 \\ 4x_1 + 6x_2 + 8x_3 = 12 \end{cases}$$

2-9 One possibility is

$$\begin{cases} 2x_1 + 3x_2 = -1 \\ 4x_1 - x_2 = 5 \\ 6x_1 + 2x_2 = 4 \end{cases}$$

2-11 (a) $x_1 = -16/10; x_2 = 7/10$;   (c) $x_1 = 2/11; x_2 = 8/11; x_3 = -10/11$

2-13 $x_1 = 2; x_2 = -1$

2-15 $s_1 = 0.622 + i0.730$

$s_2 = -0.308 - i1.249$

2-16 (a) The PIVOT program gives

$x_1 = 1; x_2 = 1$

(c) There is no solution. The third equation is inconsistent with the other two.

(f) The PIVOT program gives

$$\begin{cases} x_2 = 0 \\ x_3 = -3 + 0.333x_1 \\ x_4 = 3 + 0.333x_1 \end{cases}$$

2-20 (a) $x_3 = 5/42; x_2 = -1/6; x_1 = -11/21$

2-23 $x_1 = -0.568; x_2 = 0.242; \quad x_3 = 0.128$

2-24 $I = -1.29$ amperes

Chapter 3

3-1 **(a)**
$$\begin{cases} 31x^3 - 74x^2 + 191/2x - 93/2 = 0 \\ y = -3 + 2x \\ z = 13/2 - 7/2x \end{cases}$$

3-2 **(a)** $x_1 = -.866; x_2 = .866$

3-5 $x_1 = 3.431; x_2 = -0.136$

3-8 **(b)** $x_1 = -3.5558; x_2 = 0; x_3 = 1.254; x_4 = 2.577$

3-9 **(b)** $x_1 = -1.792; x_2 = -1.5033; x_3 = -1.153; x_4 = -0.6666; x_5 = 1;$
$x_6 = 1.486; x_7 = 1.836$

3-11 **(b)** the root is at zero

3-13 **(a)** $x = -2.0136;$ **(c)** $x_1 = 0; x_2 = 2.47$

3-15 **(a)** $x = 2$

3-17 $B = 1024.45$ kg/s

3-19 $x_1 = -0.143; x_2 = 6.140$

3-21 The function has at least one relative minimum with a value more negative than -60.

Chapter 4

4-1 **(a)** The result is an ellipse centered at $(x, y) = 0$ with 2 sqrt(2) unit semimayor axis along the y axis and 4/sqrt(3) unit semiminor axis along the x-axis.

4-4 **(a)** The resulting curve is an ellipse centered at $(x, y) = (2, -1)$ with a 6 unit semimajor axis along the x direction and a 2*sqrt(3) unit semiminor axis along the x-axis.

4-7 $x = 1.5, 1.6, 1.5, \ldots$

4-10 **(a)** $x = -2, 1$

4-11 $f(x, y) = 0 : x = .250, .102, .020, \ldots$
$f(x, y) = 1 : x = .500, .352, .270, \ldots$

4-12 **(b)** $x = -1.633, 1.633, -1.633, 1.633$

4-14 **(b)** $x = 2.695, -2.052$

4-19 The extremum is at $x = 1$, $y = -2$, $z = 1$ and the value of the function at the extremum is $F = 6$.

4-21 **(a)** Relative minimum is at $x = 2$, $y = -3$, and $F = 12$

4-23 **(a)** Maximum $= 2$ at $x = 1$, $y = 2$

4-24 **(b)** Minimum $= -1$ at $x = 1$, $y = 1$

4-28 $n_1 = 9200, n_2 = 4100, F = 88259$

4-29 $n_1 = 5998, n_2 = 3018, F = 57082$

Chapter 5

5-1 The algebraic signs of the coefficients of odd powers of S are reversed. The coefficients of even powers of S are unchanged.

5-3 $p(s') = -2(s'^4 + s'^3 + 1.11s'^2 - 0.889s' - 0.988)$

5-7 **(a)** Resulting value is $6 + i19$

5-9 **(b)** Resulting value is $0 + i0$

5-11 **(a)** $p(s - 2) = s^3 - 9s^2 + 26s - 23$

5-14 **(a)** The remainder polynomial is therefore approximately
$$s^2 + 1.08s + 6.77$$

5-15 **(a)** Roots are $s = -2, -1, 1, 2$

5-17 **(a)** The roots are $s = -2.6316, 0.2658 + i0.57, 0.2658 - i0.573$

5-19 (a) Polynomial has 2 RHP roots, no IA roots, and 1 LHP root.
 (b) Polynomial has 2 RHP, no IA roots, and 2 LHP roots
 (d) Polynomial has 2 RHP, no IA roots, and 2 LHP roots
5-21 (a) Roots are 2.147, -1.088, and $0.97 \pm i1.827$
 (d) Roots are -1, $0.81 \pm i0.587$, and $-0.31 \pm i0.951$

Chapter 6

6-1 $\begin{bmatrix} 3 & 5 & 9 & 17 \\ 4 & 6 & 10 & 18 \\ 5 & 7 & 11 & 19 \end{bmatrix}$

6-3 Only zero (or null) matrices are both symmetric and skew-symmetric.

6-5 Interchanging rows and columns has no effect on a symmetric matrix. It produces the negative of a skew-symmetric matrix.

6-8 $\mathbf{A}^4 = \begin{bmatrix} -125 & 50 \\ -75 & -50 \end{bmatrix}$

6-9 (a) One possibility is

$$\mathbf{A} = \begin{bmatrix} 1 & 2 \\ 0 & 3 \end{bmatrix}; \quad \mathbf{B} = \begin{bmatrix} 1 & 0 \\ -1 & 2 \end{bmatrix}$$

 (c) One possibility is

$$\mathbf{A} = \begin{bmatrix} 1 & 2 \\ -1 & -2 \end{bmatrix}; \quad \mathbf{B} = \begin{bmatrix} 0 & 1 \\ 1 & 1 \end{bmatrix}; \quad \mathbf{C} = \begin{bmatrix} 2 & 1 \\ 0 & 1 \end{bmatrix}$$

 The matrix $\mathbf{A}$ must be singular.

6-11 Two possibilities are

$$\begin{bmatrix} 1 & 0 & \vdots & 3 & -2 \\ -5 & 1 & \vdots & -1 & 4 \\ \hdashline 1 & -2 & \vdots & 3 & 8 \\ -4 & -7 & \vdots & 1 & -1 \end{bmatrix} \begin{bmatrix} -4 & -4 & \vdots & 4 & 1 \\ -5 & -2 & \vdots & 2 & -1 \\ \hdashline 7 & 0 & \vdots & -3 & -5 \\ -1 & 3 & \vdots & 3 & -4 \end{bmatrix}$$

$$= \begin{bmatrix} 19 & -10 & \vdots & -11 & -6 \\ 4 & 30 & \vdots & -3 & -17 \\ \hdashline 19 & 24 & \vdots & 15 & -44 \\ 59 & 27 & \vdots & -36 & 2 \end{bmatrix}$$

and

$$\begin{bmatrix} 1 & 0 & 3 & -2 \\ -5 & 1 & -1 & 4 \\ 1 & -2 & 3 & 8 \\ \hdashline -4 & -7 & 1 & -1 \end{bmatrix} \begin{bmatrix} -4 & \vdots & -4 & 4 & 1 \\ -5 & \vdots & -2 & 2 & -1 \\ 7 & \vdots & 0 & -3 & -5 \\ -1 & \vdots & 3 & 3 & -4 \end{bmatrix}$$

$$= \begin{bmatrix} 19 & \vdots & -10 & -11 & -6 \\ 4 & \vdots & 30 & -3 & -17 \\ 19 & \vdots & 24 & 15 & -44 \\ \hdashline 59 & \vdots & 27 & -36 & 2 \end{bmatrix}$$

6-15 (b) The rank is three.

6-18 (b) $\mathbf{A}^{-1} = 1/74 \begin{bmatrix} 2 & -24 & 4 \\ 14 & 17 & -9 \\ 6 & 2 & 12 \end{bmatrix}$

6-21

$$\mathbf{P} = \begin{bmatrix} 1 & 0 & 0 & 2 \\ 0 & 1 & 0 & 3 \\ 0 & 0 & 1 & 0 \\ 0 & 0 & 0 & 1 \end{bmatrix}$$

6-22

$$\mathbf{P} = \text{adj } (\mathbf{A}) = \begin{bmatrix} 3 & -6 & 3 \\ 12 & -9 & 12 \\ 2 & -4 & 7 \end{bmatrix}$$

6-27 For

$$|\mathbf{A}| = \begin{pmatrix} a_{11} & a_{12} & a_{13} \\ a_{21} & a_{22} & a_{23} \\ a_{31} & a_{32} & a_{33} \end{pmatrix} ; \text{ then}$$

$$|\mathbf{A}| = a_{11} (a_{22}a_{33} - a_{32}a_{23}) - a_{12}(a_{21}a_{33} - a_{31}a_{23}) + a_{13}(a_{21}a_{32} - a_{31}a_{22})$$

6-30 (a) $\|\mathbf{x}\| = \sqrt{1 + 4 + 9} = \sqrt{14}$

(c) $\langle x, y \rangle = 1 + 4 - 3 = 2$

(f) For

$$\mathbf{z} = \begin{bmatrix} z_1 \\ z_2 \\ z_3 \end{bmatrix}$$

the requirement that $\langle x, z \rangle = 0$ is $z_1 + 2z_2 + 3z_3 = 0$; one solution to which is

$$\mathbf{z} = \begin{bmatrix} 1 \\ 1 \\ -1 \end{bmatrix}$$

6-32

$$\begin{cases} k_1 + 2k_4 = 1 \\ k_2 + 2k_3 - k_4 = 0 \\ k_1 + k_2 - k_3 = -2 \\ \quad\quad - k_4 = 3 \end{cases}$$

which has solution

$$k_1 = 7; k_2 = -8; k_3 = 1; k_4 = -3$$

6-34 A vector

$$\mathbf{y} = \begin{bmatrix} y_1 \\ y_2 \\ y_3 \end{bmatrix}$$

that is orthogonal to the given vector satisfies

$$y_1 - y_2 - 2Y_3 = 0$$

one solution to which is:

$$\mathbf{y} = \begin{bmatrix} 2 \\ 0 \\ 1 \end{bmatrix}$$

A vector

$$\mathbf{z} = \begin{bmatrix} z_1 \\ z_2 \\ z_3 \end{bmatrix}$$

that is orthogonal to the given vector and to y satisfies
$$\begin{cases} z_1 = z_2 - 2z_3 = 0 \\ 2z_1 + z_3 = 0 \end{cases}$$
one solution to which is:

Normalizing the three vectors,
$$\begin{bmatrix} 1/\sqrt{6} \\ -1/\sqrt{6} \\ -2/\sqrt{6} \end{bmatrix}, \begin{bmatrix} 2/\sqrt{5} \\ 0 \\ 1/\sqrt{5} \end{bmatrix}, \begin{bmatrix} 1/\sqrt{30} \\ 5/\sqrt{30} \\ -2/\sqrt{30} \end{bmatrix}$$
is an orthonormal basis with one vector in the given direction. There are many other possible solutions of course.

6-36
$$\mathbf{P} \begin{bmatrix} 1 & 0 & 0 \\ 0 & 1 & 0 \\ 0 & 0 & 1 \end{bmatrix} = \begin{bmatrix} -1 & 2 & 1 \\ 0 & 1 & 1 \\ 1 & 0 & 1 \end{bmatrix}$$

$$\mathbf{P} = \begin{bmatrix} -1 & 2 & 1 \\ 0 & 1 & 1 \\ 1 & 0 & 1 \end{bmatrix}$$

6-38 The third column is linearly dependent on the first two columns, so
$$\mathbf{a}_1 = \begin{bmatrix} 1 \\ 2 \\ -1 \end{bmatrix} \text{ and } \mathbf{a}_2 = \begin{bmatrix} 0 \\ 1 \\ -1 \end{bmatrix}$$
are a basis for the range space of the matrix. An orthogonal basis is $\mathbf{a}_1$ and
$$\mathbf{a}_2 - \frac{|\mathbf{a}_1|^2}{(<\mathbf{a}_1, \mathbf{a}_2> \mathbf{a}_1)} = \begin{bmatrix} -1/2 \\ 0 \\ -1/2 \end{bmatrix}$$

Normalizing,
$$\mathbf{b}_1 = \frac{\mathbf{a}_1}{|\mathbf{a}_1|} = \begin{bmatrix} 1/\sqrt{6} \\ 2/\sqrt{6} \\ -1/\sqrt{6} \end{bmatrix} \text{ and } \mathbf{b}_2 = \frac{\mathbf{a}_2}{|\mathbf{a}_2|} = \begin{bmatrix} -1/\sqrt{2} \\ 0 \\ -1/\sqrt{2} \end{bmatrix}$$
are an orthonormal basis. There are many other possibilities, too, of course.

6-40 **(a)** The matrix is orthogonal.
   (b) The matrix is orthogonal.

Chapter 7

7-1 **(a)** $\mathbf{P}^{-1} \mathbf{A} \mathbf{P} = \begin{bmatrix} -1 & 0 \\ 0 & -2 \end{bmatrix} = \Lambda$

7-5 The sum of the eigenvalues is 4. The product of the eigenvalues equals zero.
7-6 Since the constant term of the characteristic polynomial is nonzero, $\mathbf{A}^{-1}$ exists.
7-8 **(a)** The modal matrix is singular.
   (c) $\mathbf{P}^{-1} \mathbf{A} \mathbf{P} = \begin{bmatrix} 2 & 0 \\ 0 & 2 \end{bmatrix} = \Lambda$

7-10 If the matrices are similar, then there is a nonsingular transformation $\mathbf{P}$ such that
   $\mathbf{B} \mathbf{P} = \mathbf{P} \mathbf{A}$
7-13 **(b)** $\exp(\mathbf{A}) = \begin{bmatrix} -0.131 & 0.011 & 0.074 \\ -0.565 & -0.065 & 0.455 \\ -0.066 & -0.444 & 0.749 \end{bmatrix}$

7-15 (a) $\exp(\mathbf{A}) = \begin{bmatrix} 0.0498 & 0.159 \\ 0 & 0.3679 \end{bmatrix}$

7-17 (b) $\mathbf{A}^{-1} = 1/2 \begin{bmatrix} 1 & -4 & 1 \\ 2 & 0 & 0 \\ 0 & 2 & 0 \end{bmatrix}$

7-19 (a) The characteristic equation of $\mathbf{A}$ is
$$\lambda^2 - 5\lambda + 10 = 0$$

7-21 (b) Since the matrix is not singular
$$\begin{bmatrix} -10 & 0 \\ 0 & -10 \end{bmatrix} = |\mathbf{A}|\,\mathbf{I}$$

Chapter 8

8-2 (b) $\mathbf{p}^1 = \begin{bmatrix} 0.0650 \\ -0.624 \\ -0.779 \end{bmatrix}$ $\mathbf{p}^2 = \begin{bmatrix} -0.138 \\ 0.767 \\ -0.626 \end{bmatrix}$ $\mathbf{p}^3 = \begin{bmatrix} 0.988 \\ 0.182 \\ -0.0364 \end{bmatrix}$

The required orthogonal transformation matrix is $\mathbf{P}_0 = [\mathbf{p}^1 \mathbf{p}^2 \mathbf{p}^3]$

8-5 (a) Since all the eigenvalues of $\mathbf{S}$ are positive, $\mathbf{S}$ is positive definite.

(c) The matrix $\mathbf{S}$ thus has two RHP and one LHP eigenvalues and is not sign definite.

8-7 (a) $\Psi_1 = \begin{bmatrix} 1.70 & 0.324 \\ 0.324 & 1.376 \end{bmatrix}$; $\Psi_2 = \begin{bmatrix} 1.05 & 1.375 \\ 1.375 & -0.327 \end{bmatrix}$

$\Psi_3 = \begin{bmatrix} -1.05 & -1.375 \\ -1.375 & 0.327 \end{bmatrix}$; $\Psi_4 = \begin{bmatrix} -1.70 & -0.324 \\ -0.324 & -1.376 \end{bmatrix}$

8-9 $\mathbf{P}_0 = \begin{bmatrix} \cos\theta & -\sin\theta \\ \sin\theta & \cos\theta \end{bmatrix} = \begin{bmatrix} 0.585 & -0.811 \\ 0.811 & 0.585 \end{bmatrix}$

or $\theta = 0.946$ radians or $54.2°$.

8-11 (c) Because $\mathbf{A}$ has more columns than rows,

$$\mathbf{S} = \mathbf{A}\,\mathbf{A}^\dagger = \begin{bmatrix} 3 & 6 & 4 \\ 6 & 27 & 12 \\ 4 & 12 & 8 \end{bmatrix}$$

The singular values of $\mathbf{A}$ are the square roots of the eigenvalues of $\mathbf{S}$. They are approximately 5.87, 1.68, and 0.859.

8-13

$$\frac{d}{dt}(\mathbf{A}^3) = \frac{d\mathbf{A}}{dt}\mathbf{A}^2 + \mathbf{A}\frac{d\mathbf{A}}{dt}\mathbf{A} + \mathbf{A}^2\frac{d\mathbf{A}}{dt}$$

8-15 (b) Least squares estimate vector is -1.92520039 and 2.61990787

8-17 For the given equations and weighting matrix weighted least squares estimate vector is $-1.58494156, 0.656700191, -0.393857026$

8-19 (a) The least squares estimates of a_1 and a_2 are
$$a_1 = -3.6;\ a_2 = 1.7$$
$$y = -3.6 + 1.7\,x$$
(c) $a_1 = -0.784;\ a_2 = 0.183;\ a_3 = -0.366$
$$y = -0.784 + 0.183e^x - 0.366e^{1.5x}$$

8-21 (c) The predicted 1975 GNP is $y = 1.036$. This prediction is too low by 0.194 trillion dollars or nearly 16%. The predicted 1980 GNP is $y = 0.601$ which is low by 0.883 or about 60%.

8-23 The ship's estimated position at $t = 0$

$$\begin{cases} x_0 = 155.08° \\ y_0 = 23.95 \text{ °S} \end{cases}$$

Chapter 9

9-1 (a) $y(t) = -11.15e^{-.382t} + 3.15e^{-2.62t} + 10$

(d) $y(t) = 86/13e^{-3t} - 8/13 \cos 2t + 12/13 \sin 2t$

9-3 $\theta(t) = 10 \cos 1/3t - 6 \sin 1/3t$

9-5 $y(t) = 1/3 - 36.97e^{-3t} + 27.73e^{-4t}, t \geq 0.$

9-7 $x_1(s) = \dfrac{(12s^3 + 32s^2 + 94s + 52)(s^2 + 100) - (2s + 6)(s^3 + 8s^2 + 100s + 740)}{(s^2 + 100)(4s^4 + 10s^3 + 35s^2 + 7s + 22.5)}$

$x_2(s) = \dfrac{(2s + 6)(12s + 8)(s^2 + 100) - (4s^2 + 2s + 9)(s^3 + 8s^2 + 100s + 740)}{(s^2 + 100)(4s^4 + 10s^3 + 35s^2 + 7s + 22.5)}$

9-8 $x(t) = 10.25 - 1/13e^{-3t} - 0.1731 \cos 2t - 0.1154 \sin 2t$

9-11 $h(\tau) = \dfrac{e^{-0.2\tau}[-0.2 \cos 3\tau + 3 \sin 3\tau] + 0.2}{9.04}$

9-14 (a) $x_1(0) = 2, x_2(0) = -5$

9-20 The zeros of $\mathbf{J}_2(t)$ are approximately 0, 5.1358, 8.416771

9-21 $y = \dfrac{3t}{2} + \dfrac{c}{tt}$, where c is a constant

9-26 (a) $y(t) = 5/2 - 137.72e^{-t} + 66.86e^{-2t}, t \geq 0$

INDEX

A

Arithmetic Operations in BASIC, 4
Arithmetic Operations in FORTRAN, 15
Assignment instruction, 5
Axis shifting, 187

B

Back substitution, 41
Backward-difference approximations, 444
Baristow's method, 200
BASIC commands:
  DEL command, 3
  END command, 3
  LIST command, 3
  NEW command, 3
  RUN command, 3
Basis, 290
Bessel's equation, 452
Bisection grid method, 113
Bisection search, 71
Bisection with direct reversal, 78
Bocher's formula, 331, 332
Boundary conditions, 419
Boundary value problem, 466

Broadcast antenna radiation pattern, 78
Built-in FORTRAN function, 24

C

CALL statement, 21
Cayley-Hamilton theorem, 306, 321
Characteristic values, 308
Cholesky's decomposition method, 370
Cofactor, 37
Comment statement, 15
Compiler, 2
Complex conjugate, 28
Complex numbers, 28–35
CONTINUE statement, 21
Control instruction:
  DIM statement, 6
  FOR. . . .NEXT statement, 6
  GOSUB statement, 9
  IF. . . .THEN statement, 6
  INPUT statement, 11
  PRINT statement, 10
  RETURN statement, 9
  SAVE statement, 4
Convergence constant, 97, 424
Convolution, 458
Cramer's rule, 37, 39
Critically damped, 94
Cross product, 287